Fractal Dimensions of Networks

Eric Rosenberg

Fractal Dimensions of Networks

 Springer

Eric Rosenberg
AT&T Labs
Middletown, NJ, USA

ISBN 978-3-030-43171-6   ISBN 978-3-030-43169-3   (eBook)
https://doi.org/10.1007/978-3-030-43169-3

This Springer imprint is published by the registered company Springer Nature Switzerland AG.
The registered company address is: Gewerbestrasse 11, 6330 Cham, Switzerland

To David and Alexa

Acknowledgements

I gratefully acknowledge the comments and suggestions of Lazaros Gallos, Robert Kelley, Robert Murray, Erin Pearse, Curtis Provost, and Brendan Ruskey. Special thanks to Ron Levine for slogging through an early draft of this book and providing enormously helpful feedback. I am of course responsible for any remaining mistakes in this book, and I welcome corrections from readers: send email to FractalDimensionsOfNetworks@gmail.com. Many thanks to the artist Jug Cerovic for providing permission to reproduce their artwork, a beautiful map of the New York City subway system, in the cover of this book. Finally, many thanks to Paul Drougas, Senior Editor at Springer, for his encouragement and support.

About the Author

Eric Rosenberg received a B.A. in Mathematics from Oberlin College and a Ph.D. in Operations Research from Stanford University. He recently retired from AT&T Labs in Middletown, New Jersey, and has joined the faculty of Georgian Court University in New Jersey. Dr. Rosenberg has taught undergraduate and graduate courses in modeling and optimization at Princeton University, New Jersey Institute of Technology, and Rutgers University. He has authored or co-authored 19 patents and has published in the areas of convex analysis and nonlinearly constrained optimization, computer-aided design of integrated circuits and printed wire boards, telecommunications network design and routing, and fractal dimensions of networks. He is the author of *A Primer of Multicast Routing* (Springer, 2012) and *A Survey of Fractal Dimensions of Networks* (Springer, 2018).

Contents

List of Figures

on the segment. Similarly, two numbers are needed to specify the location of a point in $\mathbb{R}^2$.

Another very similar intuitive definition of dimension is "the number of independent degrees of freedom". We learn early in life that the physical world has three dimensions; this world is known as $\mathbb{R}^3$. By our teenage years, we might learn that time is a fourth dimension; that is, spacetime is four-dimensional. With public interest in the origin of the universe sparked by such authors as Stephen Hawking, we might learn that string theory, which treats particles not as points but as patterns of vibration that have length but no height or width, assumes that spacetime has ten dimensions [Hawking 10].

The study of dimension has a long history. About 300 B.C., in *Book 1, On Plane Geometry* of *The Elements*, Euclid wrote [Manin 06]

> *A point is that which has no part.*
> *A line is breadthless length.*
> *The extremities of a line are points.*
> *A surface is that which has length and breadth only.*
> *The extremities of a surface are lines.*

The term *extremity* can be interpreted as *boundary*. In *Book 11, On Spatial Geometry*, Euclid continued with

> *A solid is that which has length, breadth, and depth.*
> *An extremity of a solid is a surface.*

The dimension of a set can also be viewed as the exponent d in the power law that determines how the set's "bulk" varies with its size [Theiler 90]:

$$bulk \sim size^d, \tag{1.1}$$

where the symbol $\sim$ roughly means "behaves like". Here "size" refers to a linear distance, e.g., the diameter of the object, and "bulk" may refer to, for example, mass, volume, or information content. For example, the area (2-dimensional bulk) of a square with side length L is L^2, and the volume (3-dimensional bulk) of the cube with side length L is L^3. The power-law scaling (1.1) will frequently be referred to in this book, since fractal dimensions are typically defined by a power law.

Difficulties in defining dimension have been known for more than a century [Eguiluz 03]. As observed in [Manin 06], Euclid used the slightly refined language of everyday experience to describe these basic geometric notions. However, when applied to a three-dimensional solid ball, we encounter a problem with Euclid's definitions, since the boundary of a ball is a sphere (a two-dimensional surface), but the boundary of a sphere is not a line, since a sphere has no boundary. This example shows that defining dimension may not be as easy as it sounds. As another example of the difficulty of defining dimension, early in the twentieth century Giuseppe Peano constructed a continuous function mapping the unit interval $[0, 1]$ to the unit square $[0, 1] \times [0, 1]$. If a continuous function can change the dimension of an object, clearly a more so-phisticated definition of dimension was required; in particular, an approach to defining the dimension of a geometric fractal was required.

Roughly speaking, a geometric fractal is a set which looks the same at all scales; that is, it looks the same when examined by a microscope zooming in with greater and greater magnification. An intentionally somewhat vague definition [Mandelbrot 86] is:

A fractal is a shape made of parts similar to the whole in some way.

(As we will explore later in this book, this is not a good definition of a fractal net-work, since a fractal network is not necessarily self-similar.) Geometric fractals are theoretical mathematical objects (e.g., the Koch snowflake and Sierpiński triangle [Mandelbrot 83b]), which are approximated by real-life objects such as fern leaves, the network of blood vessels, and broccoli. A picture in [Stanley 92] of the floor of the cathedral in Anagni, Italy shows a fourth-generation Sierpiński triangle (four generations of the Sierpiński triangle are illustrated in Fig. 1.1.[1]) The cathedral and its floor were built in 1104, and Stanley speculated that this is possibly the oldest man-made fractal object.

Figure 1.1: Four generations of the Sierpiński triangle evolution

The study of fractals dates at least as far back as the construction in 1872 by Karl Weierstrass (1815–1897) of a curve so irregular that it is nowhere dif-ferentiable [Falconer 13]. In 1883, Georg Cantor (1845–1918) introduced what is now known as the "middle-third" Cantor set, obtained by starting with the unit interval and repeatedly removing the middle third of each interval.[2] Other well-known fractals are the Koch curve, introduced in 1904 by Helge von Koch (1826–1883), the Sierpiński gasket, introduced in 1915 by Waclaw Sierpiński (1882–1969), the Sierpiński triangle (Fig. 5.3), and the fantastically compli-cated Mandelbrot set, named in honor of Benoit Mandelbrot (1924–2010), who was enormously influential in advancing the theory and application of fractals [Mandelbrot 83b].

The theory of fractal dimensions began with the 1918 work of Felix Hausdorff (1868–1942) [O'Connor 04]. The dimension he defined, now known as the *Hausdorff dimension*, is sometimes called the *Hausdorff–Besicovitch* dimension, due

---

[1] Image from ByWereon-Ownwork,PublicDomain,https://commons.wikimedia.org/w/index.php?curid=1357937.

[2] The "middle-third" Cantor set is also known as the ternary Cantor set, and as the triadic Cantor set.

to the contributions of Abram Samoilovitch Besicovitch (1891–1970).[3] Many
different fractal dimensions of geometric objects have been defined, including
the box counting dimension, the correlation dimension, the information dimension, and generalized dimensions, all of which we will study in this book; we
will also study the meaning and computation of these fractal dimensions for networks. Historically, different scientific disciplines have tended to use different
fractal dimensions to study sets. For example, the biological sciences typically
utilized the box counting dimension, while physicists studying dynamical systems typically utilized the correlation dimension. Fractal dimensions have been
computed for many real-world objects, from cancer tumors to freeze dried carrots to galaxies to paintings. Fractal dimensions have been applied in various
ways, e.g., to study the differences in fractal dimensions between healthy and
diseased cells, or to study the impact of different agricultural policies.

With this qualitative introduction to the different definitions of the dimension of a set, we now turn to networks, for the moment drop the word "fractal",
and ask the simpler question: What are the dimensions of a network? One
immediate answer is the number of nodes and arcs (nodes are also called vertices, and arcs are also called edges or links, but in this book we will use only
the terms "node" and "arc"). However, if we treat the term "dimension" as
being synonymous with "size" or "measure", then there are many other useful
measures of a network. Some of these other measures are now quite well known,
thanks to the explosion of interest in the study of networks. Examples of these
other measures, which will later be formally defined, include:

> *average path length*: find the length of the shortest path between
> each pair of nodes, and compute the average of these lengths,

> *diameter*: find the length of the shortest path between each pair of
> nodes, and pick the longest of these lengths,

> *average node degree*: the average number of arcs connected to a
> node,

> *cluster coefficient*: expressed in social networking terms, the fraction
> of my friends who are friends with each other,

> *node betweenness*: the fraction of all shortest paths that use a given
> node,

> *arc betweenness*: the fraction of all shortest paths that use a given
> arc.

Many additional measures relate to survivability of a network after accidental
failures (e.g., due to a flood or lightning strike) or intentional failures (e.g., due
to a targeted attack) cause a node or arc to fail. Though the time and memory

---

[3] As an interesting tidbit of mathematical genealogy, Besicovitch studied under Andrey
Markov (1856–1922), for whom Markov chains are named.

to compute some of these measures might be significant for a large network, the theory is unambiguous: there is no uncertainty about how to compute the measure.

Unlike the computation of the network measures listed above, the computation and interpretation of fractal dimensions of networks, which roughly dates from the 1988 work of Nowotny and Requardt [Nowotny 88], is not entirely straightforward. This lack of a cookie-cutter or cookbook approach for networks is "inherited" from the lack of a cookie-cutter or cookbook approach for geometric objects: the study of fractal dimensions of a network was inspired by about two decades of intense study of the fractal dimensions of a geometric object, beginning with [Mandelbrot 67], and there is an enormous literature on the challenges in computing and interpreting fractal dimensions of a geometric object. For example, a major issue in computing a fractal dimension of a real-world geometric object is that fractal dimensions are defined in terms of a limit as the number of points tends to infinity, or as the image is viewed with infinitely finer resolution, and real-world objects do not possess an infinite number of points or structure at infinitely small sizes. Another computational issue is determining the range over which the log *bulk* versus log *size* curve, which by (1.1) can be used to estimate d, can be considered approximately linear.[4] Many problems encountered in computing a fractal dimension of a set also arise when computing a fractal dimension of a network. New issues, not encountered with sets, arise with networks, such as the fact that for unweighted networks (networks for which each arc cost is 1) it is meaningless to take a limit as the distance between nodes approaches zero; in contrast, many theoretical results on the fractal dimensions of sets depend on taking a limit as certain distances approach zero.

Starting in about 2003, researchers began thinking about how to extend the concept of the fractal dimensions of a geometric object to a network. As already mentioned, a seminal paper [Nowotny 88] on network fractal dimensions did appear much earlier; we will study these results in Chap. 12. However, this work was unknown to the small group of researchers who, 15 years later, started studying fractal dimensions of networks. Most of the early work on fractal dimensions of networks originated in the physics community, and in particular in the statistical physics community. Much of this initial research focused on the network box counting dimension which, adapting the well-known techniques for computing the box counting dimension of a geometric object, covers a network by the minimal number of "boxes" of a given size. Another important network fractal dimension characterizes how the "mass" (number of nodes) within distance r of a random node varies with r; this fractal dimension is an extension of the correlation dimension of a geometric object. Yet another network fractal dimension is the information dimension, which extends to networks the information dimension of a geometric object.

To present this material on the fractal dimensions of sets and networks, this book weaves together two threads. The first thread is geometric objects, and the

---

[4] By "log *bulk*" we mean the logarithm of *bulk*.

various definitions of dimension of geometric objects. This thread begins with some classical definitions such as the topological dimension. We then study several fractal dimensions of geometric objects, including the Hausdorff dimension, the similarity dimension, and the widely applied box counting dimension. Useful, but not rigorous, definitions of a geometric "fractal" have been proposed in terms of these dimensions. We delve into the information dimension, and the correlation dimension, proposed by physicists in 1983 to study dynamical systems. This first thread also includes geometric multifractals, which are sets whose fractal dimension cannot be characterized by a single number, and instead are characterized by an infinite spectrum of fractal dimensions. As early as 1988 (and this *is* a relatively early date in the history of fractals), [Theiler 88] observed that the literature on computing the dimension of fractal sets is "vast" (and this comment applied only to fractal dimensions of sets, not of networks). There is also a large literature on applications of fractal dimensions of sets, and we review some applications of fractal dimensions in the arts, social sciences, physical sciences, and biological sciences. We also review the literature on the misuse of fractal analysis as applied to sets.

The second thread in this book is networks, and the various definitions of dimension of networks. There has been an explosion of activity in "network science", impacting such areas as ecology and evolution [Proulx 05], the spread of diseases and viruses [Barabási 09], and the science of the brain [Bassett 06]. For example, a 2007 survey of graph methods to neuroscience [Reijneveld 07] noted that

> Evidence has accumulated that functional networks throughout the brain are necessary, particularly for higher cognitive functions such as memory, planning, and abstract reasoning. It is more and more acknowledged that the brain should be conceived as a complex network of dynamical systems, consisting of numerous function interactions between closely related as well as more remote brain areas.

Several books, e.g., [Barabási 02, Watts 03], introduced network science to the non-specialist, and [Costa 11] provided an enormous list of applications of networks to real-world problems and data.

Weaving together these two threads (fractal dimensions of geometric objects, and fractal dimensions of networks), we will provide answers to the question posed at the beginning of this chapter: What are the fractal dimensions of a set or of a network, and how do we compute them? There are abundant reasons for caring about fractal dimensions of a set, as illustrated by the many examples provided in this book. As to why should we care about the fractal dimensions of a network, we provide two motivations. First, network fractal dimensions will increasingly be used to classify and compare networks, both large and small, with applications in, e.g., brain networks, transportation networks, and social networks. Second, the subject of fractal dimensions of networks is fascinating from a theoretical viewpoint, with many avenues still to be explored. It is a wonderful source of research topics for a PhD student or an established researcher, and even undergraduate students can dive into applications of the methodology.

Much of the recent research on networks, especially by physicists, employs the term "complex network". The adjective "complex" is sometimes used as a synonym for "large", sometimes it implies a non-regular topology (e.g., not a regular lattice), and sometimes it refers to the complex dynamics of the network. In [Rosenberg 17a, Rosenberg 17b, Rosenberg 17c, Rosenberg 18b] the term "complex network" denotes an arbitrary network without special structure (e.g., the network need not be a regular lattice), for which all arcs have unit length (so the length of a shortest path, e.g., as computed by Dijkstra's method [Aho 93], between two nodes is the number of arcs in that path), and all arcs are undirected (so the arc between nodes i and j can be traversed in either direction). However, in this book we avoid the adjective "complex" for two reasons. First, one of the objectives of this book is to show that a fractal dimension can be calculated even for "small" networks, and we need not limit attention to "large" networks. Second, even simple networks can exhibit complex behavior, e.g., a five-node "chair" network exhibits many interesting properties, as will be discussed in several later chapters. Hence we eschew the term "complex networks" and instead simply refer to "networks". There are, however, three important classes of networks that have received special attention: *scale-free* networks, *small-world* networks, and *random* networks, so in Chap. 2 we review these three classes, which make frequent appearances throughout this book.

There are two basic approaches that could have been taken in presenting the material in this book. The first approach is to first present the theory, computation, and application of fractal dimensions of geometric objects, and then present the theory, computation, and application of fractal dimensions of networks. The chapters on geometric objects would contain well over 100 pages, and, given the title of this book, it seems terribly unfair to make the reader wait that long before encountering a network. The second approach, which is the path taken, is to discuss a particular geometric fractal dimension and then discuss its network counterpart. Thus, for example, we present the box counting dimension a network right after presenting the box counting dimension of a set, and we present the correlation dimension of a network after presenting the correlation dimension of a set. We make liberal use of citations within the text to published papers and books, rather than including, at the end of each chapter, what would necessarily be a very lengthy "notes and references" section. Since each citation includes the year of publication, the reader immediately learns how recent these results are, and hyperlinks (in e-versions of this book) lead directly to the citation details.

To conclude this introductory section, we quote directly from [Falconer 03]:

> A word of warning is appropriate at this point. It is possible to define the "dimension" of a set in many ways, some satisfactory and others less so. It is important to realize that different definitions may give different values of dimension of the same set, and may also have very different properties. Inconsistent usage has sometimes led to considerable confusion. In particular, warning lights flash in my mind (as in the minds of other mathematicians) whenever the term

"fractal dimension" is seen. Though some authors attach a precise meaning to this, I have known others interpret it inconsistently in a single piece of work. The reader should always be aware of the definition in use in any discussion.

The aim of this book is to present the theory of fractal dimension of sets and networks, algorithms for calculating these dimensions, and applications of these dimensions. If warning lights (as in the above quote) flash in the mind of a student or researcher upon encountering the term "fractal dimension", this book will hopefully allow that person to identify what type of fractal dimension is being discussed, to discern if the term "fractal dimension" is being used appropriately, and to understand how that fractal dimension relates to other fractal dimensions.

1.1   Intended Readership

The intended readership of this book includes

> juniors and seniors at the college or university undergraduate level, Masters and PhD students,
> researchers in the theoretical aspects of fractal dimensions of objects and networks, and
> researchers in applications of fractal dimensions of objects and networks.

The book is intended for a wide range of disciplines, including (but certainly not limited to)

> the physical sciences (e.g., physics, chemistry, geology, and the biological sciences),
> operations research and the management sciences,
> the social sciences (e.g., sociology, where many important network concepts discussed in this book originated),
> the fine arts (e.g., this book discusses the complexity of Jackson Pollock paintings),
> medicine (e.g., in the use of fractals to contrast normal and healthy individuals), and
> mathematics and computer science.

As for mathematical prerequisites, readers should be comfortable with

> differential and integral calculus (at the level of a one-year college course),
> elementary properties of vectors and matrices,
> basic probability theory (e.g., discrete and continuous distributions, mean, and variance), and
> basic network concepts such as Dijkstra's method for finding shortest paths in a network.

It is highly recommended to read this book in the chapter order, since chapters build upon one another. However, it is certainly feasible to jump to particular chapters. For example, the reader wanting to learn about the information dimension should first read Chap. 14, which presents the information dimension of a probability distribution, and then read Chap. 15, which presents the information dimension of a network.

Many examples are provided in this book to illustrate the concepts, and many exercises are provided, some of which are easy and some of which are not.

1.2 Notation and Terminology

Dealing with Quotes: The Queen's English

For enhanced clarity, we adopt the British style of placing punctuation outside of quotation marks. Thus we state a well-known spelling rule as "i" before "e", except after "c", rather than "i" before "e," except after "c."

Following standard practice, when altering words in a direct quotation in order to improve clarity or better integrate the quotation into the surrounding material, square brackets [ ] are placed around the change. Thus, e.g., if the quoted text contains the phrase "proteins A, B, C, D, E, and F were studied", the text might be replaced with "[six proteins] were studied".

Relating to Sets

- Normally we describe a set by writing, e.g., $\{y \mid y^3 \leq 17\}$. Occasionally this notation yields less than pleasing results, as, e.g., the set $\{y \mid |x - y| \leq \epsilon\}$ defining those real numbers y whose distance from the real number x does not exceed ϵ. For such cases we use the notation $\{y : |x - y| \leq \epsilon\}$.

- By $\mathbb{R}^E$ we mean ordinary E-dimensional Euclidean space, where E is a positive integer. By $\mathbb{R}$ we mean $\mathbb{R}^1$.

- If X and Y are sets, by $X \times Y$ we mean the set of ordered pairs (u, v) where $u \in X$ and $v \in Y$.

- By Ω we mean a subset of $\mathbb{R}^E$. If $\Omega \subset \mathbb{R}^E$ and if $y \in \mathbb{R}^E$, then by $\Omega + y$ we mean the set $\{x + y \mid x \in \Omega\}$. If $\Omega \subset \mathbb{R}^E$ and if $\alpha \in \mathbb{R}$, then by $\alpha \Omega$ we mean the set $\{\alpha x \mid x \in \Omega\}$.

- The cardinality of a finite set X (i.e., the number of elements in the set) is denoted by $|X|$. If X and Y are finite sets and $X \subseteq Y$, by $Y - X$ we mean the set of elements of Y that are not in X.

- Let S be a set of real numbers. Then α is an upper bound for S if $x \leq \alpha$ for $x \in S$, and β is a lower bound for S if $x \geq \beta$ for $x \in S$. By $\sup S$ or $\sup_{x \in S} x$ or $\sup\{x \mid x \in S\}$ we mean the least upper bound of S. Similarly, by $\inf S$ or $\inf_{x \in S} x$ or $\inf\{x \mid x \in S\}$ we mean the greatest lower bound of S.

- The *limit superior* [Royden 68] of the sequence $\{x_n\}$ of real numbers, denoted by $\overline{\lim} \, x_n$, is defined by $\overline{x_n} \equiv \inf_n \sup_{k \geq n} x_k$. The real number α is the limit superior of $\{x_n\}$ (written as $\alpha = \overline{\lim} \, x_n$) if (i) given $\epsilon > 0$, there exists an n such that $x_k < \alpha + \epsilon$ for $k \geq n$, and (ii) given $\epsilon > 0$, and given n, there exists a $k \geq n$ such that $x_k > \alpha - \epsilon$. Similarly, the *limit inferior* of the sequence $\{x_n\}$ of real numbers, denoted by $\underline{\lim} \, x_n$, is defined by $\underline{\lim} \, x_n \equiv \sup_n \inf_{k \geq n} x_k$.

Relating to Functions

- It is sometimes useful, e.g., in Sect. 23.5 on the Legendre transform, to distinguish between a function and the value of a function at a point in its domain. A function will be denoted by, e.g., f or $f(\cdot)$, while the value of the function at the point x is $f(x)$. By $\exp x$ we mean e^x.

- By $f : X \to Y$ we mean the domain of the function is X, and the range of the function is Y.

- For $\Omega \subset \mathbb{R}^E$ and $f : \Omega \to \mathbb{R}$, by $f(\Omega)$ we mean the set $\{f(x) \mid x \in \Omega\}$.

- Since the symbol d is reserved for a fractal dimension, the first derivative of f at x is denoted by $f'(x)$ or by $\frac{\mathrm{d}}{\mathrm{d}x} f(x)$, and the second derivative at x is denoted by $f''(x)$ or by $\frac{\mathrm{d}^2}{\mathrm{d}x^2} f(x)$. The differential of the variable x is $\mathrm{d}x$.

- The function $f : \mathbb{R}^E \to \mathbb{R}$ is *convex* if for $x, y \in \mathbb{R}^E$ and $\alpha \in (0,1)$ we have

$$f\big(\alpha x + (1 - \alpha)y\big) \leq \alpha f(x) + (1 - \alpha)f(y).$$

A convex function a *strictly convex* if the inequality holds as a strict inequality. The function $f : \mathbb{R}^E \to \mathbb{R}$ is *concave* if $(-f)$ is convex and f is strictly concave if $(-f)$ is strictly convex. Figure 1.2 (left) illustrates a convex function, and Fig. 1.2 (right) illustrates a concave function. A sufficient condition ensuring convexity is that the second derivative exists and is everywhere nonnegative, and a sufficient condition ensuring strict convexity is that the second derivative exists and is everywhere positive. Examples of strictly convex functions whose domain is $\mathbb{R}$ are $f(x) = e^x + e^{-x}$ and $f(x) = x^4 + x^2$. There is at most one point minimizing a strictly convex function.

- If f and g are functions defined on $\mathbb{R}$, we say that f is $\mathcal{O}(g)$ if there exist positive constants α and β such that $|f(x)| \leq \beta |g(x)|$ whenever $x \geq \alpha$. In particular, f is $\mathcal{O}(1)$ if there exist positive constants α and β such that $|f(x)| \leq \beta$ whenever $x \geq \alpha$, so the run time of an algorithm is $\mathcal{O}(1)$ if the run time is bounded regardless of the size of the problem instance.

- The function f of a single variable is *unimodal* if for some $x^\star$ the values $f(x)$ are monotone increasing for $x \leq x^\star$ and monotone decreasing for $x \geq x^\star$, or if the values $f(x)$ are monotone decreasing for $x \leq x^\star$ and monotone increasing for $x \geq x^\star$. Roughly speaking, a function is unimodal if it has exactly one "hump". An example of a unimodal function is $f(x) = e^{-x^2}$; this function is neither convex nor concave.

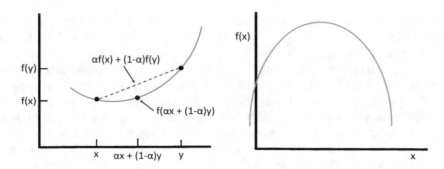

Figure 1.2: Convex and concave functions

Relating to Probability

- Let $\mathcal{X}$ be a sample space (the set of all possible outcomes). An event X is a set of zero or more outcomes. A probability measure $\mu : \mathcal{X} \to \mathbb{R}$ is a function that assigns to an event X a probability $\mu(X) \in [0,1]$ such that (i) $\mu(\emptyset) = 0$, (ii) $\mu(\mathcal{X}) = 1$, and (iii) For every countable collection $\{X_i\}$ of pairwise disjoint sets, we have $\mu(\bigcup_i X_i) = \sum_i \mu(X_i)$.

- The *support* of the probability measure $\mu : \mathbb{R}^E \to \mathbb{R}$ is the set of points $x \in \mathbb{R}^E$ such that for each positive number ϵ we have $\mu\big(B(x,\epsilon)\big) > 0$, where $B(x,\epsilon)$ is the closed ball of radius ϵ centered at x. For example, the support of the uniform probability measure defined on the interval $[a,b]$ is the interval $[a,b]$. The support of the normal (Gaussian) probability density function is $\mathbb{R}$.

- The expected value of a random variable X is denoted by $\langle X \rangle$ or by $\mathcal{E}(X)$.

Relational Symbols

Unfortunately, there is no consistent usage of relational symbols among the material cited in this book. For example, in the Appendix in [Fisher 98], and in [Jizba 04], $x \simeq y$ means x is approximately equal to y, as in $\pi \simeq 3.14$. We choose to follow "Symbols, Units, Nomenclature, and Fundamental Constants in Physics, 1987 Revision (2010) Preprint" [Cohen 10], where $\approx$ means approximately equal to, as in $\pi \approx 3.14$. Following [Cohen 10], we use the following other conventions.

- Proportionality is denoted by $\propto$, so $f(x) \propto g(x)$ means that for some constant k we have $f(x) = kg(x)$. For example, $3x^2 \propto 5x^2$.

- Asymptotic equality (or asymptotic equivalence) is denoted by $\sim$, so by $f(x) \sim g(x)$ as $x \to a$ we mean $\lim_{x \to a} f(x)/g(x) = 1$.[5] The value of a might be a finite number, often zero, or ∞, where by ∞ we mean $+\infty$. Typically in the literature on geometric fractals we see such statements as $f(x) \sim x^{-d}$.

---

[5] In [Cohen 10], asymptotic equality is denoted by $\simeq$, but this might be confused with $\approx$, so we conform with most of the fractal literature in using $\sim$ to denote asymptotic equality.

In such statements, the qualifier "as $x \to a$" is omitted, and it is implied that this refers to the limit as $x \to 0$. Thus $f(x) \sim x^{-d}$ should be strictly be interpreted as $\lim_{x \to 0} f(x)/x^{-d} = 1$. However, many authors use "$\sim$" to mean $\lim_{x \to a} f(x)/g(x) = c$ for some constant c which is not necessarily 1. For example, if $f(x)$ follows the power-law distribution $f(x) = cx^{-\alpha}$ for some c and α (such power laws are ubiquitous in the study of fractals), many authors would express this as $f(x) \sim x^{-\alpha}$. In fact, in the fractal literature, the use of "$\sim$" to mean $\lim_{x \to a} f(x)/g(x) = c$ for some constant c is more common than the use of "$\sim$" to mean $\lim_{x \to a} f(x)/g(x) = 1$. It was observed in [Theiler 88] that the "ambiguous" symbol "$\sim$" denotes the "equally ambiguous" phrase "tends to be approximately proportional to"; note that Theiler's informal definition of "$\sim$" does not include the word "asymptotic". Indeed, the reader delving into the many papers cited in this book will unfortunately see the symbol "$\sim$" used freely, typically without any definition.

- By *vanishes* we mean "has the value zero" or "approaches the value zero", so $x^2 - 4$ vanishes when $x = 2$, and "as x vanishes" means "as $x \to 0$".

- By $x \gg c$ we mean x is much larger than c, where "much" is a deliberatively vague adjective. By $x \ll c$ we usually mean x is much smaller than c; however, by $x \ll 1$ we mean x is very small and positive, as in "for $\epsilon \ll 1$".

- The symbol $\equiv$ denotes a definition, as in $B(x, r) \equiv \{y \mid dist(x, y) \leq r\}$.

Other Symbols

By $\lceil x \rceil$ we mean the smallest integer not less than x. By $\lfloor x \rfloor$ we mean the largest integer not more than x. By $\ln x$ we mean the natural logarithm of x, while $\log x$ means the logarithm of x to any (unspecified) base exceeding 1. The end of a definition, theorem statement, proof, example, etc., is denoted by $\square$. Exercises are indicated using color, as in **Exercise 2.1** in Chap. 2. An exercise which is more like a research problem is indicated by a trailing star, as in **Exercise 7.1*** in Chap. 7.

Reserved Symbols

In Tables 1.1, 1.2, and 1.3 below, the third column provides the section number in which the symbol or notation was first introduced. If two section numbers are provided, the first number tells where the definition was given for a geometric object, and the second number tells where the definition was given for a network.

Reserved Symbols for Dimensions

Table 1.1: Reserved symbols for dimensions

Symbol	Meaning	Section
d_B	Box counting dimension	4.2, 7.2
d_C	Correlation dimension	9.4, 11.1
d_E	Network trans-fractal dimension	12.2
d_H	Hausdorff dimension	5.1, 18.1
d_I	Information dimension	14.2, 15.2
d_K	Minkowski dimension	4.5
d_L	Lyapunov dimension	20.2
d_M	Mass dimension of the sequence $\{\mathbb{G}_t\}$	12
d_P	Packing dimension	5.4
$d(x)$	Pointwise dimension at x	9.2, 11.2
d_S	Similarity dimension	5.2
d_T	Topological dimension	4.1
d_U	Surface dimension	12.3
d_V	Volume dimension	12.3
d_Z	Zeta dimension	22.5
D_q	Generalized dimension of order q	16.2, 17.1
$D_q(L,U)$	Secant estimate of D_q	16.2, 17.2
D_q^H	Generalized Hausdorff dimension of order q	16.9, 18.3
$D_q^{sandbox}$	Sandbox dimension of order q	16.6, 17.5

Reserved Lowercase Letters

Table 1.2: Reserved lowercase letters

Symbol	Meaning	Section
δ_n	Node degree of node n	2.1
μ_L	Lebesgue measure	5.1
$dist(x,y)$	Distance between x and y	4.1, 2.1
$dist_{\mathcal{H}}(S,T)$	Hausdorff distance between sets S and T	3.4
h	Hop count	2.1
$m(\mathbb{G},s,d)$	d-dimensional Hausdorff measure for box size s	18.1
$m(\mathbb{G},s,q,d)$	d-dimensional Hausdorff measure of order q for box size s	18.3
n	Node	2.1
$p_j(s)$	Probability of box $B_j \in \mathcal{B}(s)$	14.3, 15.2
r	Radius	3.2, 7.2
r_g	Radius of gyration	10.4
s	Box size	4.2, 7.1
$\sigma(\mathbb{G},\mathcal{S},d)$	Standard deviation of the Hausdorff measures	18.1
$\sigma(\mathbb{G},\mathcal{S},q,d)$	Standard deviation of the Hausdorff measures of order q	18.3
$skel(\mathbb{G})$	Skeleton of $\mathbb{G}$	2.5
t	Time	3.1
$v(\Omega,s,d)$	d-dimensional volume of Ω	5.1
$x(s)$	Vector summarizing the covering $\mathcal{B}(s)$	17.1

Reserved Uppercase Letters

Table 1.3: Reserved uppercase letters

Symbol	Meaning	Section
Δ	Network diameter	2.1
Λ	Lacunarity	19.1
Ω	Geometric object in $\mathbb{R}^E$	1.2
$\Omega(x, r)$	Neighborhood of Ω centered at x of radius r	9.2
$\mathbb{A}$	The set of arcs of $\mathbb{G}$	2.1
A	Cardinality of $\mathbb{A}$	2.1
$\mathcal{B}(s)$	s-covering of Ω or $\mathbb{G}$	7.1
$B(s)$	Minimal number of boxes of size s needed to cover Ω or $\mathbb{G}$	4.2, 7.1
B_j	Box in $\mathcal{B}(s)$	9.5
$B_D(s)$	Minimal number of diameter-based boxes of size at most s needed to cover $\mathbb{G}$	7.2
$B_P(r)$	Maximum number of disjoint balls with centers in Ω and radii not exceeding r	5.4
$B_R(r)$	Minimal number of radius-based boxes of radius at most r needed to cover $\mathbb{G}$	7.2
$C(r)$	Correlation sum	9.3, 11.1
$CC(n)$	Cluster coefficient of node n	2.1
CC	Cluster coefficient of $\mathbb{G}$	2.1
E	The integer dimension of Euclidean space	1.2
$\mathbb{G}$	A network	2.1
$\mathbb{G}(s)$	Subnetwork of $\mathbb{G}$ of diameter $s - 1$	7.2
$\mathbb{G}(n, r)$	Subnetwork of $\mathbb{G}$ with center n and radius r	7.2
$H(s)$	Entropy of the probability distribution $p_j(s)$	14, 15.1
H_q	Rényi generalized entropy of order q	14.5
$\mathbb{I}^+$	The set of positive integers	22.5
$I(z)$	Heaviside step function with $I(0) = 0$	11.1
$I_0(z)$	Heaviside step function with $I_0(0) = 1$	11.1
L_p	The L_p norm	3.4
$M(n, r)$	The mass of the box centered at x with radius r	2.2
$\mathbb{N}$	The set of nodes of $\mathbb{G}$	2.1
N	Number of nodes in $\mathbb{N}$	2.1
$N_j(s)$	Number of nodes in box $B_j \in \mathcal{B}(s)$	9.5
$\mathbb{R}^E$	E-dimensional Euclidean space	1.2
$\mathcal{S}$	Set of integer values for the box size s	18.1
$T(r)$	Takens estimator	9.4
$\mathcal{X}$	A space, e.g., $\mathbb{R}^E$	3.4
$Z_q(\mathcal{B}(s))$	Partition function value for the covering $\mathcal{B}(s)$	16.2
$Z(x(s), q)$	Partition function value for the summary vector x	17.1

Chapter 2

Networks: Introductory Material

Only within the 20th Century has biological thought been focused on ecology, or the relation of the living creature to its environment. Awareness of ecological relationships is—or should be—the basis of modern conservation programs, for it is useless to attempt to preserve a living species unless the kind of land or water it requires is also preserved. So delicately interwoven are the relationships that when we disturb one thread of the community fabric we alter it all - perhaps almost imperceptibly, perhaps so drastically that destruction follows.
Rachel Carson (1907–1964), pioneering American marine biologist and conservationist, author of the 1962 book "Silent Spring", in "Essay on the Biological Sciences" (1958)

In this chapter we first provide basic definitions related to networks. Then we consider three important classes of networks: small-world networks, random networks, and scale-free networks. We will frequently refer to these three classes in subsequent chapters. The final topic in this chapter is the skeleton of a network.

2.1 Basic Network Definitions

A network $\mathbb{G} = (\mathbb{N}, \mathbb{A})$ is a set $\mathbb{N}$ of nodes connected by a set $\mathbb{A}$ of arcs. For example, a node might represent an airport, and an arc might represent the logical connection taken by a plane flying nonstop from one airport to another. In a social network [Feld 91], a node might represent a person and an arc indicates that two people are friends. In a co-authorship network, a node represents an author, and an arc connecting two authors means that they co-authored (possibly with other authors) at least one paper. In a road network, a node might represent an intersection of two or more roads, and

© Springer Nature Switzerland AG 2020
E. Rosenberg, *Fractal Dimensions of Networks*,
https://doi.org/10.1007/978-3-030-43169-3_2

an arc represents a road connecting two intersections. In manufacturing, a node might represent a station in an assembly line, and an arc might represent the logical flow of a product being assembled as it moves from one station to the subsequent station. Networks have seen widespread use in modelling physical [Albert 02], public health [Baish 00, Luke 07], communications [Comellas 02, Rosenberg 12, Rosenberg 13a], and social [Watts 99b] networks. A large number of network applications are surveyed in [Costa 11].

We use the terms *graph* and *network* interchangeably. Since in popular usage the term "graph" is also commonly used to mean the set of points $(x, f(x))$ for a function f, to avoid confusion we use the term "plot" or "curve", rather than "graph", to refer to the set of points $(x, f(x))$. An *undirected* arc is an arc that can be traversed in either direction. For example, each arc in a co-authorship network is undirected. A *directed* arc is an arc that can be traversed only in one direction, for example, a one-way street in a road network. In a food chain network using directed arcs, an arc from node x to node y might indicate that x eats y (e.g., if x represents "rabbits" and y represents "carrots", then there is a directed arc from x to y but no directed arc from y to x). A manufacturing network typically uses directed arcs, since an item being assembled might move from assembly station 1 (node x) to assembly station 2 (node y), but not from y to x. If all arcs are undirected, the network is called an *undirected* network; if all the arcs are directed, the network is called a *directed* network. We will primarily be concerned with undirected networks in this book, so *henceforth, unless otherwise specified, each network is assumed to be undirected*.

Let N denote the number of nodes in $\mathbb{G} = (\mathbb{N}, \mathbb{A})$, so $N = |\mathbb{N}|$, where $|\mathbb{N}|$ denotes the cardinality of the set $\mathbb{N}$. Let A denote the number of arcs in $\mathbb{G}$, so $A = |\mathbb{A}|$. An arc $a \in \mathbb{A}$ connecting nodes i and j is denoted by (i, j), and i and j are the *endpoints* of a. We assume that $\mathbb{G}$ has no self-loops, where a *self-loop* is an arc connecting a node to itself. We also assume that any two nodes can be connected by at most one arc. Two nodes i and j are *neighbors* or *adjacent* if they are connected by an arc in $\mathbb{A}$. A node is a *leaf node* if it has only one neighbor. A node is *isolated* if it has no neighbors. A *hub* is a node that is connected to a significant fraction of the nodes in the network [Costa 07]; this is a qualitative term rather than a formally defined term. A graph is *complete* if there is an arc between each pair of nodes; a complete graph has $N(N-1)/2$ arcs. Two often-used qualitative terms used to describe networks are *sparse* and *dense*. A network is *sparse* if $A \ll N(N-1)/2$, i.e., if the number of arcs is much smaller than for a complete graph with the same number of nodes. A network is *dense* if the number of arcs is a "significant fraction" of the number of arcs in a complete graph with the same number of nodes. A *module* of $\mathbb{G}$ is a dense subgraph of $\mathbb{G}$ which is sparsely connected to other modules of $\mathbb{G}$.

For $a \in \mathbb{A}$, we associate with arc $a = (i, j)$ a nonnegative length which we denote by c_a or by c_{ij}. If each arc length is 1, the network is said to be *unweighted*; otherwise, the network is said to be *weighted*. In a weighted network, the weights might indicate distances between nodes; in neuroscience applications, weight can be used to differentiate stronger and weaker connections [Bassett 06]. For many networks, e.g., for a social network or for a co-authorship network, there are no natural lengths assigned to the arcs. For such networks, we assign a length of 1 to each arc, thus treating the network as unweighted. A network is *spatially embedded* if its nodes occupy particular fixed positions in space, e.g., if its nodes can be represented by coordinates in $\mathbb{R}^2$ or $\mathbb{R}^3$. Examples of spatially embedded networks are railroad networks, where nodes and arcs have a physical presence, and airline networks, where nodes have a physical presence but arcs do not. Since in general the distance between two adjacent nodes in a spatially embedded network is not 1, spatially embedded networks are, in general, weighted networks.

The *degree* of node n, denoted by δ_n, is the number of arcs incident to n, i.e., the number of arcs having n as one of its endpoints. Thus a node with degree δ_n has δ_n neighbors. In chemistry, the node degree is sometimes called the *coordination number*. The node-node adjacency matrix M of $\mathbb{G}$ is the square $N \times N$ matrix defined by $M_{ij} = 1$ if arc (i,j) exists, and $M_{ij} = 0$ otherwise. Since $\mathbb{G}$ is undirected and has no self-loops, then the diagonal elements of M are zero, and M is symmetric ($M_{ij} = M_{ji}$).

A *path* in $\mathbb{G}$ between two distinct nodes i and j is an ordered sequence of one or more arcs (i, n_1), (n_1, n_2), (n_2, n_3), ..., (n_{k-1}, n_k), (n_k, j) connecting i and j. The *length* of a path is the sum, over all arcs in the path, of the arc lengths. If $\mathbb{G}$ is unweighted, the length of a path is the number of arcs in the path. The network $\mathbb{G}$ is *connected* if there is a path, using arcs in $\mathbb{A}$, between each pair of nodes. A network is *disconnected* if for some distinct nodes i and j there is no path, using arcs in $\mathbb{A}$, from i to j. Clearly, any network with an isolated node is disconnected. A path of length at least 2 connecting a node to itself is a *cycle*; i.e., a cycle is a path of two or more arcs forming a loop. A path between nodes i and j is a *shortest path* if its length is less than or equal to the length of any other path between i and j. We denote by $dist(i,j)$ the length of the shortest path between nodes i and j. We also refer to $dist(i,j)$ as the *distance* between i and j, omitting the adjective "shortest". For an unweighted network, the length of a shortest path between two nodes is often called the *chemical distance* between the nodes, which is an apt description when the nodes are atoms and the arcs are atomic bonds, or when the nodes are molecules and the arcs are molecular bonds. We also use the term *hop count*, ubiquitous in the telecommunications literature, to refer to the length of a shortest path in an unweighted network. The hop count has the properties of a metric: for any nodes x, y, z we have $dist(x,x) = 0$, $dist(x,y) = 0$ implies $x = y$, and $dist(x,y) \leq dist(y,z) + dist(x,z)$; this last inequality is the familiar *triangle inequality*.

Exercise 2.1 Show that, for a weighted network, the number of arcs in a shortest path between two nodes is not a well-defined number.   □

The network $\widetilde{\mathbb{G}}$ is a *subnetwork* of $\mathbb{G}$ if $\widetilde{\mathbb{G}}$ can be obtained from $\mathbb{G}$ by deleting nodes and arcs. We use the terms *subnetwork* and *subgraph* interchangeably. A *tree* in $\mathbb{G}$ is a connected subgraph with no cycles. A *spanning tree* of a connected network $\mathbb{G}$ is a tree $(\mathbb{N}, \widetilde{\mathbb{A}})$ such that $\widetilde{\mathbb{A}} \subseteq \mathbb{A}$, i.e., the tree interconnects all the nodes in $\mathbb{N}$ using a subset of the arcs in $\mathbb{A}$. Define the length of a graph to be the sum of the arc lengths, taken over all arcs in the graph. A *minimal spanning tree* of a connected graph $\mathbb{G}$ is a spanning tree whose length is less than or equal to the length of any other spanning tree of $\mathbb{G}$. A *planar graph* is a graph that can be embedded in two dimensions so that arcs intersect only at nodes. For example, denoting by K_N the complete graph on N nodes, K_4 is planar while K_5 is non-planar (Fig. 2.1).

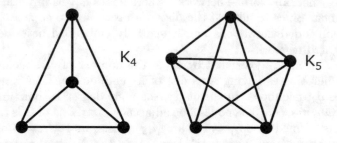

Figure 2.1: Complete graphs K_4 and K_5

This book is about the fractal dimensions of sets and networks. As mentioned in the beginning of Chap. 1, there are many other metrics that can be calculated for a network. We now briefly discuss some of the most important of these other metrics; see also the survey [Costa 07].

Diameter: The *diameter* of $\mathbb{G}$, denoted by $diam(\mathbb{G})$ or simply by Δ, is defined by

$$diam(\mathbb{G}) \equiv \max\{\, dist(x,y) \,|\, x,y \in \mathbb{N} \,\};$$

it is the length of the longest shortest path.

Average Path Length: The *average path length* L is defined by

$$L \equiv \frac{1}{N(N-1)} \sum_{x \in \mathbb{N}} \sum_{\substack{y \in \mathbb{N} \\ y \neq x}} dist(x,y). \qquad (2.1)$$

This definition excludes the zero distance from each node to itself and thus the double sum is divided by $N(N-1)$ rather than by N^2. For an unweighted network, the average path length is the average hop count.

Cluster Coefficient: For a given node n, let $\mathbb{A}^0(n)$ be the set of arcs in $\mathbb{G}$ such that if $a \in \mathbb{A}^0(n)$, then both endpoints of a are neighbors of n. That is, $\mathbb{A}^0(n)$ is the set of arcs directly connecting neighbors of n. Let $A^0(n) = |\mathbb{A}^0(n)|$ be the number of arcs in $\mathbb{A}^0(n)$. Then $A^0(n) \leq \delta_n(\delta_n - 1)/2$ and this upper bound is achieved if each pair of neighbors of n is directly connected. Define

$$CC(n) \equiv \frac{A^0(n)}{\delta_n(\delta_n - 1)/2}. \qquad (2.2)$$

Then $CC(n)$, called the *cluster coefficient* of n, is the fraction of arcs, directly connecting neighbors of n, that actually exist in $\mathbb{G}$. The cluster coefficient of n is typically denoted by $C(n)$, but we reserve $C(\cdot)$ to denote the correlation sum, defined in Chap. 9. In a social network, $CC(n)$ measures the extent to which friends of person n are friends of each other.

It was noted in [Latora 02] that the definition of $CC(n)$ breaks down for a leaf node, since then (2.2) yields the ratio $0/0$. Similarly, (2.2) breaks down for an isolated node. This difficulty is resolved by appealing to the non-mathematical definition of $CC(n)$: the cluster coefficient is the ratio of actual connections between the neighbors of n to the total possible connections between the neighbors of n. If n is isolated, then it has no neighbors, so it a cluster unto itself. Similarly, if n is a leaf node, then the neighbors of n are (trivially) fully interconnected. It is therefore appropriate to define the cluster coefficient $CC(n)$ of an isolated or leaf node to be 1. The *cluster coefficient CC* [Lee 06, Watts 98] of the network $\mathbb{G}$ is the average of $CC(n)$ over all nodes:

$$CC \equiv \frac{1}{N} \sum_{n \in \mathbb{N}} CC(n). \qquad (2.3)$$

Thus CC measures the "cliquishness" of a typical neighborhood. For a complete graph we have $CC = 1$, while for a random graph we have $CC = \delta/N$, where δ is the average node degree [Newman 00].

Betweenness: Let $\mathbb{G} = (\mathbb{N}, \mathbb{A})$ be a weighted network. For $a \in \mathbb{A}$, the *arc betweenness* of a is a measure of the importance of a in calculating shortest paths between all pairs of nodes. For $x, y \in \mathbb{N}$, let $P(x, y)$ be the number of distinct shortest paths between x and y, and let $P_a(x, y)$ be the number of distinct shortest paths between x and y that use arc a. The arc betweenness of a, which we denote by $arc\_bet(a)$, is

$$arc\_bet(a) \equiv \frac{1}{N(N-1)} \sum_{x \in \mathbb{N}} \sum_{\substack{y \in \mathbb{N} \\ y \neq x}} \frac{P_a(x, y)}{P(x, y)}. \qquad (2.4)$$

There are $N(N-1)$ terms in the sum, but at most only $N(N-1)/2$ distinct terms, since $\mathbb{G}$ is undirected. (Sometimes betweenness is defined without normalizing by $N(N-1)$.) We have $0 \leq arc\_bet(a) \leq 1$. If $\mathbb{G}$ consists of the single arc a, then $arc\_bet(a) = 1$. Arc betweenness is illustrated in Fig. 2.2 for two unweighted networks. In (i), there are six ordered pairs of nodes, namely (s, t), (t, s), (t, u), (u, t), (s, u), and (u, s). There is only one shortest path between each pair. The arc (s, t) is used on the shortest path between s and t, and on the shortest path between t and s. Hence $arc\_bet(s, t) = (1 + 1)/[(3)(2)] = 1/3$. In (ii), since there are 6 nodes, there are $(6)(5) = 30$ shortest paths. Six of the shortest paths lie in one of the triangles, and six lie in the other triangle. Hence there are 18 undirected shortest paths that must use arc (u, v), so $arc\_bet(u, v) = 18/30 = 3/5$. We have $arc\_bet(s, t) = 2/30$, since (s, t) is on two shortest paths, namely between s and t and between t and s.

Similarly, for a weighted network and for $n \in \mathbb{N}$, the *node betweenness* of n is measure of the importance of n in calculating shortest paths between all pairs of nodes. For $x, y \in \mathbb{N}$, let $P(x, y)$ again be the number of distinct shortest paths between x and y, and let $P_n(x, y)$ be the number of distinct shortest

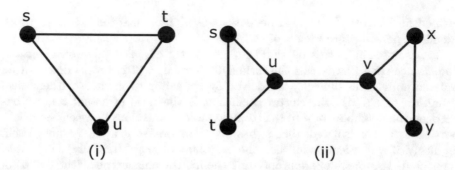

Figure 2.2: Arc betweenness

paths between x and y that touch node n. The node betweenness of n, which we denote by $node\_bet(n)$, is

$$node\_bet(n) \equiv \frac{1}{N(N-1)} \sum_{x\in\mathbb{N}} \sum_{\substack{y\in\mathbb{N}\\ y\neq x}} \frac{P_n(x,y)}{P(x,y)}. \qquad (2.5)$$

The node betweenness values will be used when we study the entropy of a network (Sect. 15.1).

Efficiency: The *efficiency* of the weighted network $\mathbb{G}$ was defined in [Latora 02] as

$$\frac{1}{N(N-1)} \sum_{x\in\mathbb{N}} \sum_{\substack{y\in\mathbb{N}\\ y\neq x}} \frac{1}{dist(x,y)}, \qquad (2.6)$$

so the efficiency of $\mathbb{G}$ is between 0 and 1, the efficiency of a fully connected network is 1, and small-world networks (Sect. 2.2) have high efficiency. Definition (2.6) works even when $\mathbb{G}$ is disconnected, since if no path exists between x and y then $dist(x,y) = \infty$. (In practice, disconnected networks arise through failure, not by design.) The efficiency of the Boston subway network (the "MTA") is 63% of the efficiency of the complete graph with the same number of nodes. Efficiency been used to show that scale-free networks are highly immune to random errors, but quite susceptible to targeted attack on hub nodes [Reijneveld 07].

Motifs: A *motif* is a pattern that occurs with a statistically increased frequency in a network [de Silva 05]. An example of a motif is a triangle connecting three nodes. There are six possible motifs defined by four nodes, as illustrated by Fig. 2.3, taken from [de Silva 05]. To discover the possible motifs of a network, the first step is to count all occurrences of the patterns of interest. Then the network is randomized, so that each arc now connects a randomly selected pair of nodes, but subject to the constraint that, for each node n, the degree of n is the same in the original and in the randomized network. Then the frequencies of the same set of patterns are tabulated for the randomized network. After this is repeated for a large number of randomizations, statistical analysis can

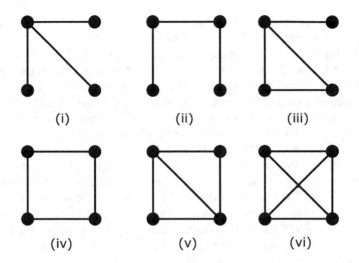

Figure 2.3: Network motifs

determine the patterns that occur at a significantly increased frequency; such patterns are the motifs of the network. For biological networks, the interpretation of motifs and their relevance is somewhat controversial [de Silva 05].

Subgraph Count: Considerably predating the study of motifs in [de Silva 05] is a body of research, described in [Bonchev 05], on subgraph counts. The simplest subgraph of a network is an arc, and

$$\frac{A}{N(N-1)/2}$$

is called the *connectedness* [Bonchev 05]. Since the number of arcs of $\mathbb{G}$ cannot exceed $N(N-1)/2$, connectedness is a measure of the sparsity of $\mathbb{G}$. Next we count connected subgraphs of $\mathbb{G}$ containing two arcs. In chemistry, the importance of two-bond molecular fragments is well established, and the total number of such fragments is known as the *Platt index*, proposed in 1952. The Platt index fails to highlight some complexity structural patterns, and the next step is to count the number of connected subgraphs with three arcs; in chemistry, this is known as the *Gordon-Scantlebury* index, proposed in 1964. A more detailed analysis of three-arc subgraphs separately counts triangles, stars, and linear subgraphs, since these counts are known to correlate with chemical properties.

An even more complete characterization, proposed in the late 1990s, is the *subgraph count SC*, defined as the ordered set $\{SC(1), SC(2), \ldots, SC(A)\}$, where $SC(k)$ is the number of connected subgraphs with k arcs. The complexity of computing $SC(k)$ makes it impractical in general for $k > 3$, although formulas for SC are known for some basic classes of graphs. A further generalization defines an overall graph invariant X by the sum of this invariant over all connected subgraphs, where the contributions of all connected subgraphs having k arcs are combined in a single term $X(k)$. Thus the graph $\mathbb{G}$ is characterized by $X \equiv$

$\{X(1), X(2), \ldots, X(A)\}$. For example, when $X(k)$ is the number of connected subgraphs with k arcs then $X(k) = SC(k)$ [Bonchev 05].

Book Embedding: Any network can be embedded in $\mathbb{R}^3$ by placing the N nodes on any line in $\mathbb{R}^3$ and then drawing the A arcs as curves each of which lies in one of A distinct half-planes having that line as their common boundary. This procedure is called the *book embedding*, where the "spine" of the book is the line containing the nodes, and each "page" of the book is a half-plane containing one arc. Thus one possible, though uninteresting, definition of the dimension of a network is the number of arcs.

Other Metrics: Many other metrics to characterize a graph were defined in [Bonchev 05]. The *total walk count* counts all paths of length h, where $h \leq \Delta$. Another metric is the ratio A/H, where H is the average distance between pairs of nodes; for a given number of nodes, this ratio is maximized for a complete graph and minimized for a chain topology. Since $A/H = 1$ for a complete graph, then for any graph we have $0 \leq A/H \leq 1$. A related measure is $\sum_{n \in \mathbb{N}} \delta_n / H_n$, where δ_n is the degree of node n and $H_n \equiv \sum_{x \in \mathbb{N}} dist(x, n)$. Fractal dimensions of networks were not mentioned in [Bonchev 05].

Routing in a rectilinear grid was studied in [Rosenberg 02], which proposed a metric which reflects how much more capacity is needed to ensure single node survivability than to ensure single link survivability. The metric studied is the ratio Ψ of the additional capacity per tandem node required to make the route immune to a single tandem node failure (which causes all arcs incident to the node to fail), divided by the additional capacity per arc required to make the route immune to a single arc failure. For any route satisfying a mild assumption, $1 \leq \Psi \leq 4/3$. For random routes, the expected value of Ψ is $10/9$ for routes using 2 arcs, $202/189$ for routes using 3 arcs, and has the limit (as the number of arcs increases to infinity) of $7/6$.

The *route factor* of a tree network with a central node $n^\star$ was used in [Buhl 09] to study the foraging trails of various species of ants. Let $\mathcal{D}$ be a set of destination nodes in the tree, where $n^\star \notin \mathcal{D}$. For $n \in \mathcal{D}$, let $dist(n^\star, n)$ be the distance in the tree from $n^\star$ to n, and let $euclid(n^\star, n)$ be the Euclidean distance from $n^\star$ to n. The route factor is $\left(1/|\mathcal{D}|\right) \sum_{n \in \mathcal{D}} dist(n^\star, n)/euclid(n^\star, n)$.

Finally, [Meyer-Ortmanns 04] defined a measure that captures the "diversity" of a network, where "diversity" refers to topologically inequivalent patterns obtained by splitting vertices and rearranging arcs. The above list of network metrics is not an exhaustive list, and other metrics were discussed in [Costa 07].

Exercise 2.2 For a graph of N nodes equally spaced on the perimeter of a circle, with neighboring nodes on the circle are connected by an arc, what is the average distance between two randomly selected nodes?   □

Exercise 2.3 For a two-dimensional square rectilinear grid of N^2 nodes, what is the average distance between two randomly selected nodes? Choosing $n^\star$ to be the node in the center of the grid, and $\mathcal{D}$ to be the set of all other nodes in the grid, calculate the route factor, as defined above.   □

2.2   Small-World Networks

In the next three sections we introduce three important classes of networks: small-world, scale-free, and random. These classes have received enormous attention during the past few decades, for both their theoretical properties and their applicability in modelling real-world systems. Also, and of particular relevance to this book, many of the early papers on fractal dimensions of networks dealt with these classes of networks, especially scale-free networks.

The term *small-world* entered the popular lexicon through the groundbreaking work of Milgram who in 1967 conducted a social psychology experiment to determine how many intermediate acquaintance links are needed to connect two random people [Milgram 67, Travers 69]. The Hungarian writer Frigyes Karinthy was credited in [Barabási 09] with observing the small-world property in 1929, in a short story called "Chains". In what is perhaps the first mention of the small-world property, a character in this story asserts that he could link anyone in the world to himself through at most five acquaintances [Luke 07]. Qualitatively, a network is "small-world" if each pair of nodes is connected by a short path; we will provide a more quantitative definition later in this section.

The actual study of small-world networks appears to have begun in the 1920s when Jacob L. Moreno, a student of the psychiatrist Sigmund Freud, became interested in the sources of the "monsters of the Id" [Uzzi 07]. In a departure from Freud's view that psycho-emotional problems originated from family of origin issues, Moreno believed these problems stemmed from relationships with family, friends, acquaintances, etc. To measure these relationships, Moreno devised a new way, called a *sociogram*, to represent on paper a person's network of connections. A sociogram is a drawing with points representing people, and lines between points representing interpersonal relationships. In 1937 Moreno founded the journal *Sociometry*, which published many of the early studies taking network approaches or developing network methods. A landmark 1957 paper in that journal found that the number and type of social connections of physicians influenced their adoption of a new drug, with close professional ties facilitating the earliest adoptions [Luke 07].

Another early study of the small-world property was a theoretical model [Sola 78] which was parameterized by the number of acquaintances of a random individual. Also, M. Gurevitch determined, by asking a group of people to record all their contacts in a period of 100 days, that the average person has about 500 acquaintances; this yields a better than 50-50 chance that two random people chosen from 200 million (the US population at the time) can be linked with two common acquaintances [Milgram 67]. It was recognized that many of the acquaintances of my contacts are also my acquaintances (e.g., the rich tend to socialize with the rich) and that society is not built on random connections but tends to be fragmented into social classes and cliques; this concept is quantified by the cluster coefficient (Sect. 2.1). In the early 1970s, the sociologist Mark Granovetter, in studying how people find jobs, posited that each person, in addition to having strong ties to neighbors, co-workers, and family, also has weak ties to people such as casual acquaintances and friends of friends. These weak ties create a larger connected network than would be formed by only the strong ties.[1]

Returning to Milgram's famous study, he received a grant of $680 from the Harvard School of Social Relations for his research. In one study, 160 people in Omaha, Nebraska were asked to forward, through the U.S. postal service, a folder to a target person in Sharon, Massachusetts. The rule for each forwarder is to forward the folder to the target person if he/she is personally known to you, otherwise, forward the folder to an acquaintance (whom you know on a first-name basis) who is more likely than you to know the target person. Of the 160 chains started, only 44 were completed (the folder reached the target). For the completed chains, the number of intermediate acquaintances varied from 1 to 11, with the median being five; so the median number of hops is six, hence the famous "six degrees of separation", which entered the popular lexicon from John Guare's 1990 play.[2]

Milgram also observed, in another study, that for 48% of the chains reaching the target, the folder was forwarded to the target by one of three people. Thus not all the acquaintances of a person are equally likely to have outside contacts. Today, such people are known as *hubs*, and considerable research is devoted to their properties (e.g., how a network can be attacked by targeting its hubs). Another observation was that there is a progressive "closing in on the target" as each new person is added to the chain. This is illustrated in Fig. 2.4, adapted from [Milgram 67], which shows the closing in, where each distance is averaged over completed and uncompleted chains to the target.

---

[1] See [Luke 07] for a review of network analysis applications in public health, and also The International Network for Social Network Analysis (INSNA), which sponsors the annual Sunbelt Conference.

[2] The play *Six Degrees of Separation* was the largest commercial success for John Guare (1938). The play is a satire on big city living and the conflict between generations, and on guilt feelings arising in race relations [Guare 04]. This play won the New York Drama Critics Circle Award, the Dramatists Guild Hull-Warriner Award, and an Olivier Best Play Award. Guare also wrote the screenplay for the popular movie *Atlantic City* (1981), starring Burt Lancaster and Susan Sarandon.

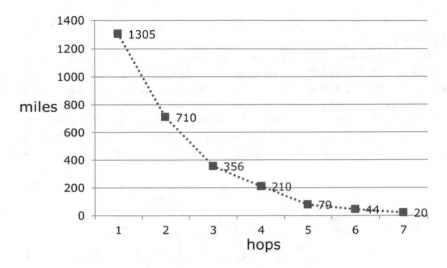

Figure 2.4: Closing in on the target

Although Milgram's experiment contained many possible sources of error [Newman 00], his general conclusion that the network of friends enjoys the small-world property has subsequently been verified, and is now widely accepted.

The pioneering study of Milgram sparked intensive investigation of small-world networks, beginning in the late 1990s and continuing to this day. For example, the "six degrees of Kevin Bacon" game shows that any pair of film actors are connected by a chain of at most eight co-stars [Newman 00]. In discussing both empirical and theoretical reasons why small-world networks present an "attractive model" for brain networks, it was observed in [Bassett 06] that a small-world topology supports both high clustering (compatible with segregated or modular processing by the brain) and short path lengths (compatible with distributed or integrated processing by the brain). The brain architecture has likely evolved to maximize the complexity or adaptivity of function it can support while minimizing "wiring costs". Complete minimization of wiring costs would allow only local connections and not long-range connections, leading to delayed information transfer and metabolic energy depletion, so the brain minimizes energy costs by creating several long-range connections, yielding a small-world network. Small-world networks have been shown to allow higher rates of learning and information processing than random graphs (Sect. 2.3), and cognitive decline due to Alzheimer's disease is associated with increased path length of certain EEG networks [Bassett 06]. It was shown in [Gallos 12a, Gallos 12b] that, similar to Granovetter's model of weak ties binding well-defined social groups, functional networks in the human brain can be viewed as a set of hierarchically organized "large-world" modules interconnected by weak ties. The brain, which exhibits a hierarchical organization at multiple scales, was shown in [Di Ieva 13] to have a fractal structure with modules interconnected in a small-world topology.

The effect of introducing random long-range arcs into a regular lattice was studied in the seminal paper [Watts 98] of Watts and Strogatz. The lattice studied in [Watts 98] is a ring of N nodes, where each node is connected to its k nearest neighbors by an arc, so each node has degree k. Since each of the N nodes has degree k, there are $kN/2$ arcs. For $k < 2N/3$ the graph cluster coefficient is $CC = (3/4)(k-2)/(k-1)$ [Newman 00]. This is illustrated in Fig. 2.5 for $N = 16$ and $k = 4$. There are only 3 arcs directly connecting the 4

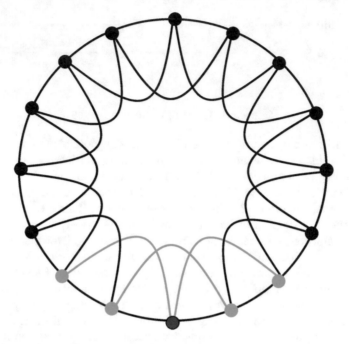

Figure 2.5: Regular lattice with $N = 16$ and $k = 4$

neighbors of each node, as illustrated by considering the red node and the three blue arcs connecting the 4 neighbors of the red node. From (2.2), for each node n we have $CC(n) = \frac{3}{(4)(3)/2} = 1/2$.

Using a fixed probability p, we re-wire this network to introduce a random number of long-range arcs, as follows. Pick a starting node n, and pick the arc that connects n to its nearest clockwise neighbor. With probability p, reconnect the arc so that n is now connected to a new neighbor z chosen uniformly at random over the entire ring, but with the restriction that no parallel arcs (between the same pair of nodes) are allowed. With probability $1 - p$ we leave the arc in place. We repeat this process by moving clockwise around the ring, considering each node and its nearest clockwise neighbor, until we return to the starting node n. Now we consider the arc that connects n to its second-nearest clockwise neighbor, with probability p randomly re-wire that arc, move to the next node clockwise, with probability p randomly re-wire the arc to its second-closest clockwise neighbor, and continue around the ring, considering each node and its second-closest clockwise neighbor, until we return to the starting point. We then continue this process for the third-closest clockwise neighbor, and so on, until finally each arc has been re-wired with probability p. We stop when each arc in the original graph has been considered once. For $p = 0$ this process leaves the original graph unchanged. As p increases, the graph becomes increasingly disordered. When $p = 1$ each edge is re-wired randomly. Define

$$ h_p \equiv \frac{1}{N(N-1)} \sum_{x \in \mathbb{N}} \sum_{\substack{y \in \mathbb{N} \\ y \neq x}} dist(x, y), $$

so h_p is the average number of hops in a shortest path when the graph is re-wired using probability p. Let CC_p be the cluster coefficient, as defined by (2.3) above, of the graph that has been re-wired using probability p. Let CC_r and h_r denote the correlation coefficient and average hop count, respectively, for a random network of N nodes with degree k, i.e., with the Watts and Strogatz model with $p = 1$. The conditions $N \gg k \gg \log(N) \gg 1$ guarantee that a random graph will be connected [Bollobas 85]. Under these conditions, Watts and Strogatz observed that as $p \to 0$ we have $CC_p \sim 3/4$ and $h_p \sim N/(2k) \gg 1$, while as $p \to 1$ we have $CC_p \approx CC_r \sim k/N \ll 1$ and $h_p \approx h_r \sim \log N/\log k$. Thus for small p the cluster coefficient is independent of N and the average distance is *linear* in N, so the re-wired network exhibits high clustering and does not exhibit the small-world property. For p near 1 the cluster coefficient decreases as $1/N$ and the average distance grows as the logarithm of N, so the re-wired network is poorly clustered but does enjoy the small-world property.

The limits as $p \to 0$ and $p \to 1$ suggest that large CC corresponds to large average hop count h, and small CC corresponds to small h. However, there is a broad range of p values for which h_p is almost as small as h_r yet $CC_p \gg CC_r$. In this regime, the creation of a few long-range arcs (known as "shortcuts") establishes the small-world property; these long range arcs can be viewed as connecting neighborhoods (clusters). For small p, the impact of each new shortcut in reducing h_p is highly nonlinear, while the impact on CC_p is at most linear; that is, for small p, each new shortcut leaves CC_p practically unchanged even though h_p decreases rapidly. Thus the transition to a small-world is not accompanied by a significant change in the cluster coefficients.

Based upon their results, Watts and Strogatz called a network *small-world* if its average hop count h satisfies $h \approx h_r$ and its average cluster coefficient CC satisfies $CC \gg CC_r$. This is a different definition than the informal one used by Milgram, since his use of "small-world" considered only the average distance between people (nodes) but not the correlation coefficient. Since for a small-world network we have $h/h_r \approx 1$ and $CC/CC_r \gg 1$, the higher the ratio $(CC/CC_r)/(h/h_r)$, the more the network exhibits the small-world property [Bassett 06]. If $\mathbb{G}$ is small-world, then from $h \approx h_r$ and $h_r \sim \log N/\log k$ we have $\log N \sim h \log k$. Equivalently, $N \sim e^{(h/h_0)}$ for some characteristic length h_0 [Song 05].

The Watts–Strogatz re-wiring model suggests that real-world networks, though sparse, might contain sufficiently many shortcuts to enjoy the small-world property. To test this, Watts and Strogatz computed the average hop count h and cluster coefficient CC for three real-world networks: a collaboration graph for film actors, a power grid graph, and the neural network of the nematode worm *C. elegans*.[3] The computed h and CC were compared with h_r and CC_r, using N and the average k, and all three networks exhibited the small-world property. Among the many studies of small-world networks, [Uzzi 07] reviewed small-world research in the social sciences and management, and [Locci 09] noted that the graph of the interactions between modules in a software system can exhibit the small-world property.[4]

In the Watts–Strogatz model, it may happen that the re-wiring causes the network to become disconnected. A variant of the Watts–Strogatz model which eliminates this possibility was analyzed in [Newman 99b]. In this model, shortcuts were added, but no arcs in the underlying regular lattice were removed, and renormalization (Chap. 21) was used to determine the critical value of the probability p used in the Watts and Strogatz model. Other models have been proposed to explain how the small-world property can arise [Newman 00]. For example, if an additional node is added to the network, the small-world property emerges if the new node is sufficiently highly connected.

There is no formal standardized definition of *small-world*. In [Watts 99a], the definition of [Watts 98] was simplified slightly by saying that $\mathbb{G}$ is a *small-world* network if (i) $diam(\mathbb{G})$ grows as $\log(N)$ and (ii) the cluster coefficients of some nodes are much larger than the cluster coefficient expected for a random network. In [Song 05], the small-world definition requires only (i) but not (ii). In [Garrido 10], a network was defined to be small-world if the average path length L defined by (2.1) satisfies $L \ll N$ for $N \gg 1$. The extensive review paper [Boccaletti 06] mentioned the work of Csanyi and Szendroi who observed that using the relation $L \sim \log N$ is meaningless for networks of a fixed size and proposed that $M(n,r) \sim e^{\alpha r}$ be used to indicate small-world scaling, where α is a positive constant and where $M(n,r)$ is the number of nodes reachable from n in at most r hops. The exponential scaling $M(n,r) \sim e^{\alpha r}$ is quite different than the power-law scaling $M(n,r) \sim r^{d_M}$ which has been found in many geographical

---

[3]Nematodes are the most numerous multicellular animals on earth. A handful of soil will contain thousands of the microscopic worms, many of them parasites of insects, plants, or animals. Free-living species are abundant, including nematodes that feed on bacteria, fungi, and other nematodes, yet the vast majority of species encountered are poorly understood biologically. There are nearly 20,000 described species classified in the phylum *Nemata* [Nematode]. *Caenorhabditis elegans*, or *C. elegans* for short, is the first nervous system to be formally quantified as a small-world network at the microscopic scale of the neuronal network [Bassett 06]. This network has an average path length of 2.65 and $CC = 0.28$.

[4]In a brief retrospective column, [Vespignani 18] recalled the reactions of statistical physicists upon the 1988 publication of [Watts 98], writing "... the model was seen as sort of interesting, but seemed to be merely an exotic departure from the regular, lattice-like network structures we were used to. But the more the paper was assimilated by scientists from different fields, the more it became clear that it had deep implications for our understanding of dynamic behaviour and phase transitions in real-world phenomena ranging from contagion processes to information diffusion. It soon became apparent that the paper had ushered in a new era of research that would lead to the establishment of network science as a multidisciplinary field." Vespignani also noted that, "By the 20th anniversary of the paper, more than 18,000 papers have cited the model, which is now considered to be one of the benchmark network topologies."

networks, including the Western U.S. power grid, the Hungary water network, and the London underground. (The scaling $M(n,r) \sim r^{d_C}$, where d_C is the network correlation dimension, will be studied in Chap. 11, and exponential scaling will be studied in Sect. 12.2.)

It was argued in [Latora 02] that the prevailing definition of "small-world", which is based on comparing the cluster coefficient and average path length of $\mathbb{G}$ to that of a random network, is inadequate. The first argument in [Latora 02] is that the cluster coefficient of a leaf node is not defined. This point is easily addressed by defining the cluster coefficient of a leaf node to be one, as discussed in Sect. 1.2. The second argument is that the small-world definition relies only on hop count and does not utilize the actual length of an arc if $\mathbb{G}$ is a weighted network.

For the remainder of this book, we say that $\mathbb{G}$ is a small-world network if $diam(\mathbb{G}) \ll N$. In Sect. 12.2 we consider a sequence $\{\mathbb{G}_t\}$ of networks such that $\mathbb{G}_{t+1}$ has more nodes and a larger diameter than $\mathbb{G}_t$; for such sequences we will supply a definition of small-world which is a variant of the definition in [Song 05] that states that $\mathbb{G}$ is small-world if $diam(\mathbb{G})$ grows as $\log(N)$.

Constructing Small-World Networks:

Many models for creating small-world networks were mentioned in [Bassett 06]. Here we discuss two deterministic methods supplied in [Comellas 02] for constructing small-world networks. The first method starts with an arbitrary sparse "backbone" network of N nodes such that each node has exactly k neighbors. It was shown in [Bollobas 82] that the expected diameter of such networks is approximately $\log_{k-1} N$. We then replace each backbone node n by K_k (the complete graph of k nodes, which has a diameter of 1), so that each of the k backbone arcs incident (before replacement) to n is now incident to a distinct node in the instance of K_k replacing n. The resulting graph, illustrated in Fig. 2.6 for $N = 6$ and $k = 4$, has kN nodes, each node has k neighbors, and the diameter is approximately $1 + 2\log_{k-1} N$. This is a small-world network, since $1 + 2\log_{k-1} N \ll kN$. This construction has similarities to the method of [Rosenberg 05], which interconnects a set of clusters by a backbone network, subject to diameter and node degree constraints.

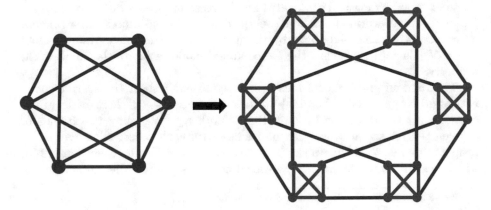

Figure 2.6: Commellas and Sampels first construction for $N = 6$ and $k = 4$

A second method of [Comellas 02] starts with (i) an arbitrary sparse back-bone network of N nodes and diameter Δ, and (ii) a set of N dense networks. Suppose the diameter of dense network j does not exceed some bound Δ_0 which is independent of j. For each j, pick a "border" node n_j of dense network j. We replace each backbone node n by one of the dense networks so that all backbone arcs previously incident to n are now incident to the border node n_j. The resulting graph has diameter $\Delta + 2\Delta_0$, and hence is also small-world. The drawback of this construction is that the failure of border node n_j isolates dense network j.

2.3   Random Networks

Seminal papers on random graphs were published in the 1950s by Gilbert [Gilbert 59], by Rapoport, and by Erdős and Rényi in 1959–1961.[5] In their widely cited 1959 paper, Erdős and Rényi defined a random graph model (known today as the Erdős–Rényi random graph model) that starts with N isolated nodes and a given probability p. With probability p, each pair of nodes is connected by an arc. As $N \to \infty$, the average node degree approaches $p(N-1)$. To keep $p(N-1)$ finite as $N \to \infty$ it suffices to choose $p = \alpha/(N-1)$, where α is a given constant. With this choice, the average node degree approaches α as $N \to \infty$, and the node degree distribution is Poisson [Costa 07]. For $p \approx 0$ the graph is, with high probability, composed of a large number of very small components unreachable from each other. For $p > 1/N$ there is, with high probability, a unique *giant component* with $\mathcal{O}(N)$ nodes (that is, a connected subgraph with most of the nodes in the network), and the remaining components are of size $\mathcal{O}(\log N)$ [Cohen 08].

Random graphs can be used to provide a simple explanation of the small-world property [Newman 00]. Suppose each person has, on average, δ acquaintances. Since each arc has two endpoints, there are $\delta N/2$ arcs in the network. A random person has, on average, δ acquaintances, each of which has δ acquaintances, etc., so a random person is connected by a chain of length h to approximately δ^h other people. The largest h satisfying $\delta^h \leq N$ is the diameter Δ, so $\Delta \approx \log N / \log \delta$. This logarithmic increase of Δ as a function of N explains the small-world effect. The average distance between nodes in a random network, being bounded above by the network diameter, also grows at most as $\log N$ [Garrido 10]. Thus the Erdős–Rényi model exhibits the small-world property.

The random graph model fails in one significant respect to describe social networks [Newman 00]. For real-world social networks, a chain of length h connects a random person to fewer than δ^h other people, since two friends of a given person are also likely to be friends with each other. This clustering property of real-world networks is quantified by the cluster coefficient (2.3), which is significantly higher for real-world networks (e.g., the neural network of

---

[5]See [Costa 07] for the Rapoport and Erdős–Rényi citations.

C. elegans, the movie actor collaboration network, and the Western Power Grid network) than for a random network of the same size.

A useful model for random graphs with a given degree distribution was proposed by Bollobás. This method was used in [Cohen 08] to generate scale-free networks (Sect. 2.4 below) for which the probability p_k that a random node has degree k is given by $p_k = ck^{-\gamma}$. In this *Bollobás configuration model*, all graphs having this degree distribution are equally probable. To construct such graphs, we start with N nodes and use the degree distribution to assign to each node n a node degree δ_n. Node n is then equipped with δ_n arcs, each of which has a "stub" end that leads nowhere, since the other arc endpoint is not yet assigned. Next, random pairs of stubs are connected to each other, forming an arc between the respective nodes. This process continues until no stubs are left. (If $\sum_n \delta_n$ is an odd number, the graph cannot be constructed and we create a new set of node degrees from the distribution.) This construction may yield a graph with *self-loops* (arcs connecting a node to itself), and with *parallel arcs* (more than one arc between the same pair of nodes); if either possibility occurs we discard the graph and start over. The Bollobás configuration model can also generate Erdős–Rényi type random graphs, by using the Poisson distribution so that $p_k = e^{-a}a^k/k!$ is the probability that a random node has degree k. With this choice for p_k, the model does not exactly generate the Erdős–Rényi degree distribution, but the graph generated very closely approximates an Erdős–Rényi random graph.

2.4 Scale-Free Networks

Two 1999 papers launched the field of scale-free networks: a paper in *Nature* [Albert 99] and a paper in *Science* [Barabási 99]. In the *Nature* paper, the authors observed that in some networks, like hyperlinks in the World Wide Web, some nodes had a much higher degree than was expected [Barabási 03]. More than 80% of the web pages had fewer than four links, but a small minority, less than 0.01% of all nodes, had more than 1000 links. A plot of the log of the number of nodes versus the log of the node degree yielded a straight line curve, indicating the power-law relationship

$$p_k \sim k^{-\lambda} \tag{2.7}$$

for some $\lambda > 0$, where p_k is the probability that the node degree is k.

In the *Science* paper [Barabási 99], the authors originated the adjective *scale-free*, applying it to a network if its node degree distribution follows a power law, so $\mathbb{G}$ is scale-free if (2.7) holds for some $\lambda > 0$. Barabási observed that the term "scale-free" is "rooted in the statistical physics literature". More generally, a random variable $\mathcal{X}$ is said to be *scale-free* if its probability distribution follows a power law: for some $\lambda > 0$ and each positive integer k, the probability p_k that $\mathcal{X} = k$ is given by $p_k = ck^{-\lambda}$ for some exponent $\lambda > 0$ and normalization constant $c = 1/\zeta(\lambda)$, where $\zeta(\cdot)$ is the Riemann zeta function defined by $\zeta(\lambda) \equiv \sum_{k=1}^{\infty} \lambda^{-k}$ [de Silva 05]. Three reasons such a distribution is called "scale-free" are: (i) If α and k are positive integers, then

$$\frac{p_{\alpha k}}{p_k} = \frac{(\alpha k)^{-\lambda}}{k^{-\lambda}} = \alpha^{-\lambda} \, ;$$

since this ratio depends only on α but not on k, there is no natural scale to the network. (ii) For $\lambda < 3$ the first moment $\sum_{k=1}^{\infty} k f_k$ of the distribution converges, but the second moment $\sum_{k=1}^{\infty} k^2 f_k$ of the distribution diverges, so there is infinite variation about the mean; hence there is no well-defined scale [Barabási 09]. (iii) Suppose the node degree distribution of $\mathbb{G}$ follows a power law. Then a few nodes have high degree and most nodes have small degree. Hence the average node degree conveys little information. Another perspective on "scale-free" was offered in [Stanley 92]:

> "It is a fact that random fractals abound. Almost any object for which randomness is the basic factor determining the structure will turn out to be fractal over some range of length scales - for much the same reason that the random walk is fractal: there is simply nothing in the microscopic rules that can set a length scale so the resulting macroscopic form is 'scale free' ... scale-free objects obey power laws.

The diameter of scale-free graphs was studied in [Cohen 03], who considered scale-free graphs for which $p_k = ck^{-\lambda}$ over the range $[k_{\min}, k_{\max}]$, where $k_{\max}$ is the "natural" cutoff given by $k_{\max} = k_{\min}N^{1/(\lambda-1)}$ and $c \approx (\lambda-1)k_{\min}^{\lambda-1}$ is a normalization factor. They showed that such networks are *ultrasmall* in that their diameter Δ satisfies

$$\Delta \;\sim\; \begin{cases} \ln\ln N & \text{if } 2 < \lambda < 3 \\ \ln N / \ln\ln N & \text{if } \lambda = 3 \\ \ln N & \lambda > 3 \, . \end{cases}$$

A *preferential attachment* model of dynamic network growth which yields a scale-free network was proposed in [Barabási 99]. In this model, when a new node i is added to the network, the probability that it is connected to an existing node j is proportional to the node degree of j. Specifically, the probability that i is connected to j is $\delta_j / \sum_n \delta_n$, where δ_j is the degree of node j. This dynamic has been described as "the rich get richer", since the more highly connected an existing node is, the more likely that a new node will directly connect to it. Figure 2.7, adapted from [Barabási 09], illustrates this dynamic model, starting with nodes a, b, and c in a triangle, with the degree indicated next to each node.

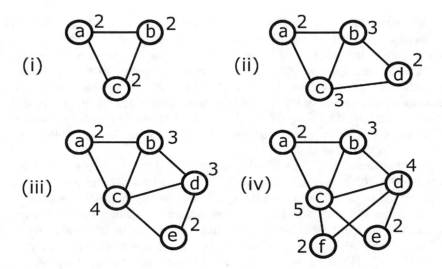

Figure 2.7: Preferential attachment

Suppose each new node connects to two existing nodes using the preferential attachment model. In subfigure (ii), node d attaches to b and c (it is equally likely to connect to any pair of nodes). When e is added in (iii), it favors more highly connected nodes, so it connects to c (with degree 3) and d (with degree 2). Finally, when f is added in (iv), it also favors more highly connected nodes, and connects also to c (with degree 4) and d (with degree 3). The final 6 node network has three nodes with degree 2, and one node with degree 3, one with degree 4, and one with degree 5. This model yields a network with a scale-free node distribution. The fact that preferential attachment yields a scale-free distribution had actually been studied much earlier by Yule [Yule 24] and by a Nobel prize winning economist [Simon 55].

Exercise 2.4 What parameter λ characterizes the scale-free distribution of the node degrees of the final 6 node network in Fig. 2.7?   □

The papers [Albert 99, Barabási 99] mentioned above were quickly followed by six additional papers on scale-free networks, two of which were published in *Nature*. The enormous impact, even in the popular press, of these publications was described in [Keller 05]. The paper "Error and Attack Tolerance of Complex Networks" [Albert 00] made the cover of the July 27, 2000 issue of *Nature* and was immediately picked up by BBC news and CNN. In October of that year, the paper "The Large-Scale Organization of Metabolic Networks", precipitated a flood of applications of network theory to biology. In a 2000 paper in *Nature*, three biologists wrote

> The [human] cell, like the Internet, appears to be a 'scale-free net-
> work': a small subset of proteins are highly connected (linked) and
> control the activity of a large number of other proteins, whereas
> most proteins interact with only a few others. The proteins in this
> network serve as the 'nodes,' and the most highly connected nodes
> are the "hubs".

By the end of 2000, the study of scale-free networks was a nascent "industry" [Keller 05], with ever increasing claims of a new "law of nature" and "universal architecture" applicable to complex networks in the natural, social, and engineering sciences. Even object-oriented software networks, in which the nodes are classes (e.g., C++ or Java classes) and arcs represent class relationships (e.g., inheritance and dependency), can exhibit a power law for the in-degree distribution [Locci 09]. The cover of the April 13, 2002 issue of *New Scientist* asks "How can a single law govern our sex lives, the proteins in our bodies, movie stars, and supercool atoms?".

It was observed in [Keller 05] that since 1999 the term "scale-free" has acquired several different meanings, often leading to considerable confusion. For example, the term has been used to signify self-similarity, loosely meaning that the graph has the same structure when examined at different levels of magnification. The problem with this usage is that a graph can be self-similar without having a power-law degree distribution (an infinite binary tree is a simple example of this). The term "scale-free" has also been used to imply a preferential attachment growth process. The problem with this usage is that the existence of a power-law degree distribution implies nothing about the dynamic process yielding this distribution. For example, as discussed in [Keller 05], Lun Li and her co-authors showed that four different mechanisms for generating a network (including preferential attachment, and a heuristic network design optimization method) all generate a scale-free network. In particular, the growth and preferential attachment mechanism is just one of many mechanisms yielding a power law, and this method is characterized by "a performance so poor as to make it a very unlikely product of evolution". Indeed, concerning social networks, [Watts 03] commented that

> In some respects the behavior of the system is independent of the particulars, but some details still matter. For any complex system, there are many simple models we can invent to understand its behavior. The trick is to pick the right one. And that requires us to think carefully - *to know something* - about the essence of the real thing.

It was also observed in [Keller 07] that, while physicists seek universal laws (e.g., that scale-free networks are a "universal architecture"), biologists have historically been unconcerned about whether their findings constitute a "law" of nature. Because life forms are "contingent" and "specialized", and the product of eons of evolution, any biological law may need to be provisional. For example, exceptions to Mendel's laws of genetics cause no panic or search for better laws, but rather serve as reminders of the complexity of biological systems. Thus, if the focus on scale-free networks stems from the viewpoint that such networks are "universal architectures", then it is important to be aware of differing viewpoints. These other viewpoints do not discount the importance of scale-free networks; they are instead advocating that such networks are not universally applicable, and that they may arise from a dynamic process other than preferential attachment. A retrospective in *Science* [Barabási 09] clearly recognized that the scale-free model is not universal, noting that the randomly bonded atoms in amorphous materials display neither the small-world or the scale-free property, due to the chemical constraints on the bonds.

The desirable and undesirable properties of scale-free networks have received an enormous amount of attention. A desirable property of scale-free networks is that a failure of a random node will inflict minimal harm to the network, since a random node is unlikely to be a hub. An undesirable property is that "even weakly virulent viruses can spread unopposed, a finding that affects all spreading processes, from AIDS to computer viruses" [Barabási 09]. Also, the failure of a hub (e.g., due to a targeted attack) can inflict major damage by either isolating nodes or making their shortest path to other nodes much longer. Following a study [Faloutsos 99] showing that the network of router to router connections in the Internet is scale-free, the susceptibility of the Internet to massive disruption by a hub failure has been termed the "Achilles's heel" of the Internet.[6]

A rebuttal in [Willinger 09] strongly disputed this supposed weakness of the Internet, which, if one believes the many studies citing [Faloutsos 99], apparently went "unnoticed by the engineers and researchers who have designed, deployed, and studied this large-scale, critical complex system". The first issue raised in [Willinger 09] concerns the data used in [Faloutsos 99], the basis of much subsequent research, which characterized the routers in the Internet as having a scale-free distribution (for the router-level topology). The results in [Faloutsos 99] are based upon data collected by Pansiot and Grad [Pansiot 98], whose expressed purpose was "to get some experimental data on the shape of multicast trees one can actually obtain in the [real] Internet" [Pansiot 98].[7] Pansiot and Grad used the `traceroute` tool, which, contrary to popular belief, does *not* generate a list of the Internet Protocol (i.e., IP) routers encountered on the path from a source to a destination. Rather, a router contains hundreds of interfaces (to which cables attach), and `traceroute` generates a list of the input interfaces used by a router to router connection. Thus, without mapping each interface to the router to which it belongs (called *alias resolution*), a naive interpretation of `traceroute` data would convey the impression of many more routers, and a vastly greater router to router connectivity, than actually exists. Referring to Fig. 2.8, adapted from [Willinger 09], the solid circles represent interfaces, each box is a router, the arrows represent interconnections between routers, and interfaces 1 and 2 are aliases. Subfigure (*i*) depicts the actual physical topology, with interfaces 1 and 2 on the same physical router, while (*ii*) shows how not mapping these interfaces to the same router gives the false appearance of additional routers.

---

[6]According to myths and stories composed long after the Iliad, the Greek warrior Achilles was, as a young boy, dipped by his mother, a sea nymph named Thetis, in the river Styx, which flowed through the underworld. The waters of this river would render him invulnerable. She held him by the heel of one foot, so this spot did not receive immunity. In his Iliad, Homer does not explain what happened to Achilles. According to later legends (and bits and pieces of Homer's own Odyssey), Achilles was slain by a well-aimed arrow to his heel delivered by Paris, who had started Trojan War by kidnapping Helen of Troy. http://www.history.com/topics/ancient-history/achilles

[7]Multicast routing [Rosenberg 12] is used to route from one source to many destinations, or from many sources to many destinations.

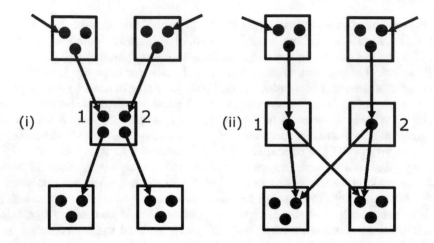

Figure 2.8: IP alias resolution

Ironically, while Pansiot and Grad explained this and other limitations and shortcomings of their measurements, such concerns were ignored in [Faloutsos 99], and subsequent researchers typically cited only [Faloutsos 99] and not [Pansiot 98].

The second issue raised in [Willinger 09] is that `traceroute` only sees IP enabled devices, and has no visibility into Layer 2 (e.g., Frame Relay or Asynchronous Transfer Mode (ATM)) clouds, e.g., as used within an Autonomous System in the Internet. This is illustrated in Fig. 2.9, also adapted from [Willinger 09]. The solid nodes on the boundary of the cloud are IP enabled

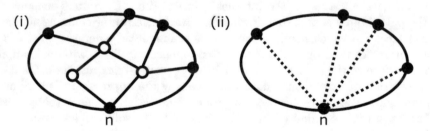

Figure 2.9: ATM cloud

routers, visible to `traceroute`, while the hollow circles inside the cloud are, e.g., ATM switches, and not visible. Thus the output of `traceroute` makes it appear as if the degree of node n is 4 (subfigure (ii)), when in reality it is only 2 (subfigure (i)).

The above two issues are specific to the Internet. A third issue raised in [Willinger 09] has a more general applicability. Even assuming that the router-level view of the Internet does indicate a scale-free distribution (an assumption which is not supported by the Pansiot and Grad measurements), the fact that the preferential attachment model yields a scale-free distribution does not imply that this model is valid. There can be many graph realizations that yield a particular degree distribution, and these realizations can have very different properties. In particular, the Internet topology is not governed by the biased coin tossing fundamental to the preferential attachment model, but rather results from solving the constrained network optimization problem of designing a network that carries the forecasted demand subject to various engineering constraints. More generally, [Bassett 06] observed that many real-world networks have an exponentially truncated power distribution of the form $p_k \sim k^{\alpha-1}e^{k/\beta}$, which implies that the probability of highly connected hubs will be greater than for a random graph but smaller than for a scale-free network. Such networks arise in systems that are physically embedded or constrained, such as transport or infrastructure networks, and in systems in which nodes have a finite life span.

In contrast, [Barabási 09], again citing the Internet as an example of a scale-free network, also referred to traditional network models which "interconnect a fixed number of nodes with cleverly placed links" and cited the Internet as an example of the new paradigm where "to explain a system's topology, we first need to describe how it came into being"—a view which [Willinger 09] strongly refutes for the case of the Internet. Willinger et al. did not say that the preferential attachment model is not valuable, nor did they say that it fails to properly model some real-world networks. They did say that it should not be assumed to apply to the router level (and, by extension, the Autonomous System level) view of the Internet simply because it yields a node distribution supposedly possessed by the Internet. To do so perpetuates a fallacy about the vulnerability of the Internet, and promotes a methodology that fails to validate a proposed model with subject matter experts.

A very recent study [Broido 18] of almost 1000 networks, drawn from social, biological, technological, and informational sources, concluded that, contrary to a central claim in modern network science, scale-free networks are rare. For each of the networks, the methodology employed was to estimate the best-fitting power law, test its statistical plausibility, and then compare it via a likelihood ratio test to alternative non-scale-free distributions. The study found that only 4% of the networks exhibited the strongest possible evidence of scale-free structure, and 52% exhibited the weakest possible evidence. The study also found that social networks are at best weakly scale-free, while a handful of technological and biological networks can be called strongly scale-free. The authors concluded that "These results undermine the universality of scale-free networks and reveal that real-world networks exhibit a rich structural diversity that will likely require new ideas and mechanisms to explain." As described in [Klarreich 18], this study has generated considerable discussion, with one network scientist observing that "In the real world, there is dirt and dust, and this dirt and dust will be on your data. You will never see the perfect power law". However, [Barabási 18] strongly disputed the conclusions of [Broido 18], finding that the study "is oblivious to 18 years of knowledge accumulated in network science", that it employs "a fictional criterion of scale-free networks", and that "their central criterion of scale-freeness fails the most elementary tests".

3.2 Fractals and Power Laws

The scaling (1.1) that *bulk* $\sim$ *size*^{d} shows the connection between fractality and a power-law relationship. However, as emphasized in [Schroeder 91], a power-law relationship does not necessarily imply a fractal structure. As a trivial example, the distance travelled in time t by a mass initially at rest and subjected to the constant acceleration a is $at^2/2$, yet nobody would call the exponent 2 a "fractal dimension". The Stellpflug formula relating weight and radius of pumpkins is *weight* $= k \cdot$ *radius*$^{2.78}$, yet no one would say a pumpkin is a fractal. A pumpkin is a roughly spherical shell enclosing an empty cavity, but a pumpkin clearly is not made up of smaller pumpkins.

To study emergent ecological systems, [Brown 02] examined self-similar ("fractal-like") scaling relationships that follow the power law $y = \beta x^\alpha$, and showed that power laws reflect universal principles governing the structure and dynamics of complex ecological systems, and that power laws offer insight into how the laws governing the natural sciences give rise to emergent features of biodiversity. The goal of [Brown 02] was similar to the goal of [Barabási 09], which showed that processes such as growth with preferential attachment (Sect. 2.4) give rise to scaling laws that have wide application. In [Brown 02] it was also observed that scaling patterns are mathematical descriptions of natural patterns, but not scientific laws, since scaling patterns do not describe the processes that spawn these patterns. For example, spiral patterns of stars in a galaxy are not physical laws but the results of physical laws operating in complex systems.

Other examples, outside of biology, of power laws are the frequency distribution of earthquake sizes, the distribution of words in written languages, and the distribution of the wealth of nations. A few particularly interesting power laws are described below. For some of these examples, e.g., Kleiber's Law, the exponent of the power law may be considered a fractal dimension if it arises from self-similarity or other fractal analysis.

Zipf's Law: One of the most celebrated power laws in the social sciences is Zipf's Law, named after Harvard professor George Kingsley Zipf (1902–1950). The law says that if N items are ranked in order of most common ($n = 1$) to least common ($n = N$), then the frequency of item n is proportional to $1/n^d$, where d is the *Zipf dimension*. This relationship has been used, for example, to study the population of cities inside a region. Let N be the number of cities in the region and for $n = 1, 2, \ldots, N$, let P_n be the population of the n-th largest city, so P_1 is the population of the largest of the N cities and $P_n > P_{n+1}$ for $n = 1, 2, \ldots, N-1$. Then Zipf's law predicts $P_n \propto n^{-d}$. This dimension was computed in [Chen 11] for the top 513 cities in the U.S.A. Observing that the data points on a log–log scale actually follow two trend lines with different slopes, the computed value in [Chen 11] is $d = 0.763$ for the 32 largest of the 513 cites and $d = 1.235$ for the remaining 481 cities.

Kleiber's Law: In the 1930s, the Swiss biologist Max Kleiber empirically determined that for many animals, the basal metabolic rate (the energy expended per unit time by the animal at rest) is proportional to the 3/4 power of the mass of the animal. This relationship, known as Kleiber's law, was found to hold over an enormous range of animal sizes, from microbes to blue whales. Kleiber's law has numerous applications, including calculating the correct human dose of a medicine tested on mice. Since the mass of a spherical object of radius r scales as r^3, but the surface area scales as r^2, larger animals have less surface area per unit volume than smaller animals, and hence larger animals lose heat more slowly. Thus, as animal size increases, the basal metabolic rate must decrease to prevent the animal from overheating. This reasoning would imply that the metabolic rate scales as mass to the 2/3 power [Kleiber 14]. Research on why the observed exponent is 3/4, and not 2/3, has considered the fractal structure of animal circulatory systems [Kleiber 12].

Exponents that are Multiples of 1/3 or 1/4:   Systems characterized by a power law whose exponent α is a multiple of 1/3 or 1/4 were studied in [Brown 02]. Multiples of 1/3 arise in Euclidean scaling: we have $\alpha = 1/3$ when the independent variable x is linear, and $\alpha = 2/3$ for surface area. Multiples of 1/4 arise when, in biological allometries, organism lifespans scale with $\alpha = 1/4$.[6] Organism metabolic rates scale with $\alpha = 3/4$ (Kleiber's Law). Maximum rates of population growth scale with $\alpha = -1/4$. In [Brown 02], the pervasive quarter-power scaling (i.e., multiples of 1/4) in biological systems is connected to the fractal-like designs of resource distribution networks such as rivers. (Fractal dimensions of networks with a tree structure, e.g., rivers, are studied in Sect. 22.1.)

Benford's Law: In its simplest form, Benford's Law [Berger 11, Fewster 09], also known as the "first digit phenomenon", concerns the distribution of the first significant digit in any collection of numbers, e.g., mathematical tables or real-world data. For any base 10 integer or decimal number x, let $first(x)$ be the first significant digit, so $first(531) = 5$, $first(0.0478) = 4$, and $first(\sqrt{2}) = 1$. Letting $p(\cdot)$ denote probability, the law says that for $j = 1, 2, \ldots, 9$ we have

$$p\big(first(x) = j\big) = \log_{10}\left(1 + \frac{1}{j}\right).$$

This yields $p\big(first(x) = 1\big) = 0.301$ and $p\big(first(x) = 2\big) = 0.176$, so the probability is nearly 1/2 that the first significant digit is 1 or 2 [Berger 11]. Benford's Law has many uses, e.g., to detect faked random data [Fewster 09]. We will mention the application of Benford's Law to dynamical systems [Berger 05] after they are introduced in Sect. 9.1.

In the beginning of Chap. 3 we considered a fractal dimension of the coastline of Britain. The fractal dimension of a curve in $\mathbb{R}^2$ was also used to study Alzheimer's Disease [Smits 16], using EEG (electroencephalograph) recordings of the brain. Let the EEG readings at N points in time be denoted by $y(t_1)$, $y(t_2)$, ..., $y(t_N)$. Let $Y(1)$ be the curve obtained by plotting all N points and joining the adjacent points by a straight line; let $Y(2)$ be the curve obtained

---

[6]When the shape or relative sizes of objects (e.g., a puppy's head, body, and paws) changes as a function of size, these objects are said to have an allometric scaling relationship [Hartvigsen 00].

by plotting every other point and joining the adjacent points by a straight line; in general, let $Y(k)$ be the curve obtained by plotting every k-th point and joining the points by a straight line. Let $L(k)$ be the length of the curve $Y(k)$. If $L(k) \sim k^{-d}$, then d is called the *Higuchi* fractal dimension [Smits 16]. This dimension is between 1 and 2, and a higher value indicates higher signal complexity. This study confirmed that this fractal dimension "depends on age in healthy people and is reduced in Alzheimer's Disease". Moreover, this fractal dimension "increases from adolescence to adulthood and decreases from adulthood to old age. This loss of complexity in neural activity is a feature of the aging brain and worsens with Alzheimer's disease".

Exercise 3.2 Referring to the above study of Alzheimer's Disease, is the calculation of the Higuchi dimension an example of the divider method (Sect. 3.1), or is it fundamentally different?   □

3.3   Iterated Function Systems

An *Iterated Function System* (IFS) is a useful way of describing a self-similar geometric fractal $\Omega \subset \mathbb{R}^E$. An IFS operates on some initial set $\Omega_0 \subset \mathbb{R}^E$ by creating multiple copies of Ω_0 such that each copy is a scaled down version of Ω_0. The multiple copies are then translated and possibly rotated in space so that the overlap of any two copies is at most one point. The set of these translated/rotated multiple copies is Ω_1. We then apply the IFS to each scaled copy in Ω_1, and continue this process indefinitely. Each copy is smaller, by the same ratio, than the set used to create it, and the limit of applying the IFS infinitely many times is the fractal Ω. In this section we show how to describe a self-similar set using an IFS. In the next section, we discuss the convergence of the sets generated by an IFS. Iterated function systems were introduced in [Hutchinson 81] and popularized in [Barnsley 12].[7] Besides the enormous importance of iterated function systems for geometric objects, the other reason for presenting (in this section and the following section) IFS theory is that we will encounter an IFS for building an infinite network in Sect. 13.1.

Example 3.1 [Riddle 19] Figure 3.3 illustrates the use of an IFS, using translation and rotation, to generate the Sierpiński triangle in $\mathbb{R}^2$. Let the bottom left corner of equilateral triangle Ω_0 be the origin $(0,0)$, let the bottom right corner be $(1,0)$, and let the top corner be $(1/2, \sqrt{3}/2)$, so each side length is 1. We first scale Ω_0 by $1/2$ and then make two copies, superimposed on each other, as shown in (i). Then one of the copies is rotated $120°$ clockwise, and the other copy is rotated $120°$ counterclockwise, as shown in (ii). Finally, the two rotated copies are translated to their final position, as shown in (iii), yielding Ω_1. Let the point $x \in \mathbb{R}^2$ be represented by a column vector (i.e., a matrix with two rows and one column), so

$$x = \begin{pmatrix} x_1 \\ x_2 \end{pmatrix}.$$

---

[7]The Internet contains many links to free software for playing and experimenting with Iterated Function Systems, e.g., [Bogomolny 98].

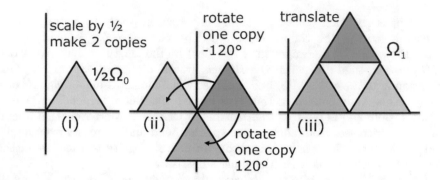

Figure 3.3: IFS for the Sierpiński triangle

This IFS can be represented by

$$f_1(x) = \begin{pmatrix} -1/4 & \sqrt{3}/4 \\ -\sqrt{3}/4 & -1/4 \end{pmatrix} x + \begin{pmatrix} 1/4 \\ \sqrt{3}/4 \end{pmatrix} \tag{3.2}$$

$$f_2(x) = \begin{pmatrix} 1/2 & 0 \\ 0 & 1/2 \end{pmatrix} x + \begin{pmatrix} 1/4 \\ \sqrt{3}/4 \end{pmatrix} \tag{3.3}$$

$$f_3(x) = \begin{pmatrix} -1/4 & -\sqrt{3}/4 \\ \sqrt{3}/4 & -1/4 \end{pmatrix} x + \begin{pmatrix} 1 \\ 0 \end{pmatrix}, \tag{3.4}$$

where the first matrix scales by $1/2$ and rotates $120°$ clockwise, the second matrix scales by $1/2$, and the third matrix scales by $1/2$ and rotates $120°$ counterclockwise. □

Let $f_1(\Omega_0)$ denote the set $\{f_1(x) \mid x \in \Omega_0\}$, and similarly for $f_2(\Omega_0)$ and $f_3(\Omega_0)$. Then Ω_1 can be obtained by scaling three copies of Ω_0 each by a factor of $1/2$, and then translating two of the scaled copies, so

$$\Omega_1 = f_1(\Omega_0) \cup f_2(\Omega_0) \cup f_3(\Omega_0).$$

In general, we apply f_1, f_2, and f_3 to Ω_t, so Ω_{t+1} is given by

$$\Omega_{t+1} = f_1(\Omega_t) \cup f_2(\Omega_t) \cup f_3(\Omega_t). \tag{3.5}$$

The Sierpiński triangle can also be generated by a simpler IFS that eliminates the need for rotation.

Example 3.2 [Riddle 19] Starting with Fig. 3.4(i), the indicated small triangle in the middle of the large triangle Ω_0 is removed, yielding the figure Ω_1 in (ii). Then, as shown in (iii), the middle third of each of the triangles in Ω_1 is removed, yielding the pattern Ω_2 in (iv). This process continues indefinitely. Define

$$M = \begin{pmatrix} 1/2 & 0 \\ 0 & 1/2 \end{pmatrix}.$$

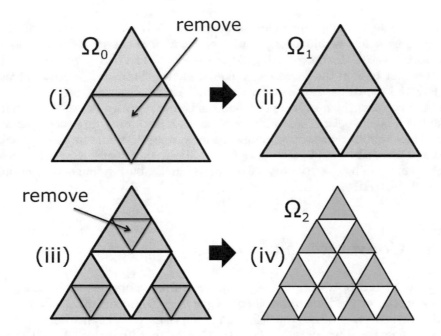

Figure 3.4: Another IFS for the Sierpiński triangle

Let the coordinates of the three corners of Ω_0 be as in Example 3.1. Again writing x as a column vector, the three triangles of Ω_1 are generated by the IFS

$$f_1(x) = Mx, \quad f_2(x) = Mx + \begin{pmatrix} 1/2 \\ 0 \end{pmatrix}, \quad f_3(x) = Mx + \begin{pmatrix} 1/4 \\ \sqrt{3}/4 \end{pmatrix}, \quad (3.6)$$

since the lower corner of the top triangle in Ω_1 has coordinates $(1/4, \sqrt{3}/4)$. The infinite sequence of sets $\{\Omega_t\}_{t=1}^{\infty}$ is generated using the same recurrence (3.5) with f_1, f_2, and f_3 now defined by (3.6). $\quad\square$

We can express the IFS more succinctly. Define $f \equiv f_1 \cup f_2 \cup f_3$ so for any set $C \in \mathbb{R}^2$ we have $f(C) \equiv f_1(C) \cup f_2(C) \cup f_3(C)$. For example, with f_1, f_2, and f_3 defined by either (3.2), (3.3), (3.4) or by (3.6), we have $f(\Omega_0) = \Omega_1$. We can now apply f to $f(\Omega_1)$, obtaining

$$f^2(\Omega_1) = f\big(f(\Omega_1)\big) = f_1\big(f(\Omega_1)\big) \cup f_2\big(f(\Omega_1)\big) \cup f_3\big(f(\Omega_1)\big).$$

Continuing the iteration, we obtain $f^t(\Omega_0) = f\big(f^{t-1}(\Omega_0)\big)$, for $t = 3, 4, \ldots$. In the next section, we will show that f is a *contraction map*, which implies that the sequence of sets $\{f^t(\Omega_0)\}_{t=0}^{\infty}$ converges to some set Ω, called the *attractor* of the IFS f. Moreover, Ω depends only on the function f and not on the initial set Ω_0, that is, the sequence of sets $\{f^t(\Omega_0)\}_{t=0}^{\infty}$ converges to Ω for any non-empty compact set Ω_0.

The use of an IFS to provide a graphic representation of human DNA sequences was discussed in [Berthelsen 92]. DNA is composed of four bases (adenine, thymine, cytosine, and guanine) in strands. A DNA sequence can be represented by points within a square, with each corner of the square representing one of the bases. As described in [Berthelsen 92], "The first point, representing the first base in the sequence, is plotted halfway between the center of the square and the corner representing that base. Each subsequent base is plotted halfway between the previous point and the corner representing the base. The result is a bit-mapped image with sparse areas representing rare subsequences and dense regions representing common subsequences". While no attempt was made in [Berthelsen 92] to compute a fractal dimension for such a graphic representation, there has been considerable interest in computing fractal dimensions of DNA (Sect. 16.6).

3.4   Convergence of an IFS

Since an IFS operates on a set, and we will be concerned with distances between sets, we need some machinery, beginning with the definition of a metric and metric space, and then the definition of the Hausdorff distance between two sets. The material below is based on [Hepting 91] and [Barnsley 12]. Let $\mathcal{X}$ be a space, e.g., $\mathbb{R}^E$.

Definition 3.2 A *metric* is a nonnegative real valued function $dist(\cdot, \cdot)$ defined on $\mathcal{X} \times \mathcal{X}$ such that the following three conditions hold for each x, y, z in $\mathcal{X}$:

1. $dist(x, y) = 0$ if and only if $x = y$,
2. $dist(x, y) = dist(y, x)$,
3. $dist(x, y) \leq dist(y, z) + dist(x, z)$.   □

The second condition imposes symmetry, and the third condition imposes the triangle inequality. Familiar examples of metrics are

1. the "Manhattan" distance $\sum_{i=1}^E |x_i - y_i|$, known as the L_1 metric,
2. the Euclidean distance $\sqrt{\sum_{i=1}^E (x_i - y_i)^2}$, known as the L_2 metric,
3. the "infinity" distance $\max_{i=1}^E |x_i - y_i|$, known as the L_∞ metric.

Example 3.3 These metrics are illustrated in Fig. 3.5, where the L_2 distance between the two points is $\sqrt{7^2 + 4^2} = \sqrt{65}$, the L_1 distance is $7 + 4 = 11$, and the L_∞ distance is $\max\{7, 4\} = 7$.   □

Let $x = (x_1, x_2, \ldots, x_E)$ and $y = (y_1, y_2, \ldots, y_E)$ be points in $\mathbb{R}^E$. The generalization of the $L_1, L_2,$ and L_∞ metrics is the L_p metric, defined by

$$L_p(x, y) = \left(\sum_{i=1}^E |x_i - y_i|^p \right)^{1/p}. \tag{3.7}$$

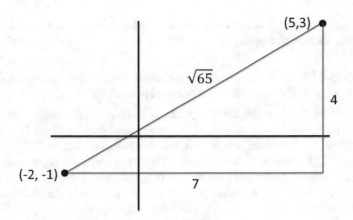

Figure 3.5: The L_1, L_2, and L_∞ metrics

The Euclidean distance is $L_2(x, y)$, the Manhattan distance is $L_1(x, y)$, and the infinity distance is $L_\infty(x, y)$. In Chap. 16 on multifractals, we will encounter expressions similar to (3.7).

Exercise 3.3 Show that $L_\infty = \lim_{p \to \infty} L_p$.   □

Definition 3.3 A *metric space* $\big(\mathcal{X}, dist(\cdot, \cdot)\big)$ is a space $\mathcal{X}$ equipped with a metric $dist(\cdot, \cdot)$.   □

The space $\mathcal{X}$ need not be Euclidean space. For example, $\mathcal{X}$ could be the space of continuous functions on the interval $[a, b]$. Letting f and g be two such functions, we could define $dist(\cdot, \cdot)$ by $dist(f, g) = \max_{x \in [a,b]} |f(x) - g(x)|$.

Definition 3.4 Let $\big(\mathcal{X}, dist(\cdot, \cdot)\big)$ be a metric space. Then $\big(\mathcal{X}, dist(\cdot, \cdot)\big)$ is *complete* if for each sequence $\{x_i\}_{i=1}^{\infty}$ of points in $\mathcal{X}$ that converges to x we have $x \in \mathcal{X}$.   □

In words, the metric space is complete if the limit of each convergent sequence of points in the space also belongs to the space.[8] For example, let $\mathcal{X}$ be the half-open interval $(0, 1]$, with $dist(x, y) = |x - y|$. Then $\big(\mathcal{X}, dist(\cdot, \cdot)\big)$ is not complete, since the sequence $x_i = \frac{1}{i}$ converges to 0, but $0 \notin \mathcal{X}$. However, if now we let $\mathcal{X}$ be the closed interval $[0, 1]$, then $\big(\mathcal{X}, dist(\cdot, \cdot)\big)$ is complete, as is $\big(\mathbb{R}, dist(\cdot, \cdot)\big)$.

Definition 3.5 Let $\big(\mathcal{X}, dist(\cdot, \cdot)\big)$ be a metric space, and let $\Omega \subset \mathcal{X}$. Then Ω is *compact* if every sequence $\{x_i\}$, $i = 1, 2, \ldots$ contains a subsequence having a limit in Ω.   □

---

[8]In [Barnsley 12], a metric space is defined to be complete if each Cauchy sequence converges to a point in the space. A sequence of points $\{x_i\}_{i=1}^{\infty}$ is a *Cauchy* sequence if for each positive number ϵ there exits a number $N(\epsilon)$ such that $dist(x_i, x_j) < \epsilon$ whenever $i > N(\epsilon)$ and $j > N(\epsilon)$. However, since by the Cauchy Convergence Theorem [Cronin 69] a sequence converges if and only if it is Cauchy, for simplicity of presentation we defined a complete metric space without referring to Cauchy sequences.

For example, a subset of $\mathbb{R}^E$ is compact if and only if it is closed and bounded. Now we can define the space which will be used to study iterated function systems.

Definition 3.6 Given the metric space $\left(\mathcal{X}, dist(\cdot, \cdot)\right)$, let $\mathcal{H}(\mathcal{X})$ denote the space whose points are the non-empty compact subsets of $\mathcal{X}$. $\square$

For example, if $\mathcal{X} = \mathbb{R}^E$ then a point $x \in \mathcal{H}(\mathcal{X})$ represents a non-empty, closed, and bounded subset of $\mathbb{R}^E$. The reason we need $\mathcal{H}(\mathcal{X})$ is that we want to study the behavior of an IFS which maps a compact set $\Omega \subset \mathbb{R}^E$ to two or more copies of Ω, each reduced in size. To study the convergence properties of the IFS, we want to be able to define the distance between two subsets of $\mathbb{R}^E$. Thus we need a space whose points are compact subsets of $\mathbb{R}^E$ and that space is precisely $\mathcal{H}(\mathbb{R}^E)$. However, Definition 3.6 applies to any metric space $\left(\mathcal{X}, dist(\cdot, \cdot)\right)$; we are not restricted to $\mathcal{X} = \mathbb{R}^E$.

Now we define the distance between two non-empty compact subsets of $\mathcal{X}$, that is, between two points of $\mathcal{H}(\mathcal{X})$. This is known as the *Hausdorff distance* between two sets. The definition of the Hausdorff distance requires defining the distance from a point to a set.

In Definition 3.7 below, we use $dist(\cdot, \cdot)$ in two different ways: as the distance $dist(x, y)$ between two points in $\mathcal{X}$, and as the distance $dist(x, T)$ between a point $x \in \mathcal{X}$ and an element $T \in \mathcal{H}(\mathcal{X})$ (i.e., a non-empty compact subset of $\mathcal{X}$). Since capital letters are used to denote elements of $\mathcal{H}(\mathcal{X})$, the meaning of $dist(\cdot, \cdot)$ in any context is unambiguous.

Definition 3.7 Let $\left(\mathcal{X}, dist(\cdot, \cdot)\right)$ be a complete metric space. For $x \in \mathcal{X}$ and $Y \in \mathcal{H}(\mathcal{X})$, the distance from x to Y is

$$dist(x, Y) \equiv \min_{y \in Y}\{dist(x, y)\}. \tag{3.8}$$

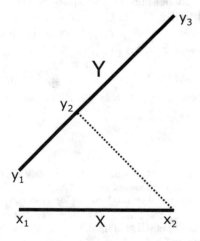

Figure 3.6: Illustration of Hausdorff distance

For $X \in \mathcal{H}(\mathcal{X})$ and $Y \in \mathcal{H}(\mathcal{X})$, the *directed distance* from X to Y is

$$\overrightarrow{dist}(X,Y) \equiv \max_{x \in X}\{dist(x,Y)\} \,. \tag{3.9}$$

The *Hausdorff distance* between X and Y is

$$dist_{\mathcal{H}}(X,Y) \equiv \max\{\overrightarrow{dist}(X,Y), \overrightarrow{dist}(Y,X)\} \,. \quad \square \tag{3.10}$$

Example 3.4 Figure 3.6 shows two line segments, X and Y. For $x \in X$ and $y \in Y$, let $dist(x,y)$ be the Euclidean distance between x and y. Then $dist(x,Y)$ is the shortest Euclidean distance from x to the line segment Y. For example, $dist(x_1,Y) = dist(x_1,y_1)$, and $dist(x_2,Y) = dist(x_2,y_2)$, where the vector y_2-y_1 is perpendicular to the vector y_2-x_2. From Fig. 3.6 we also see that $\overrightarrow{dist}(X,Y) = dist(x_2,y_2)$, that $dist(y_1,X) = dist(y_1,x_1)$, that $dist(y_3,X) = dist(y_3,x_2)$, and that $\overrightarrow{dist}(Y,X) = dist(y_3,x_2)$. The two directed distances are not equal, since $\overrightarrow{dist}(Y,X) = dist(y_3,x_2) > dist(x_2,y_2) = \overrightarrow{dist}(X,Y)$. Finally,

$$dist_{\mathcal{H}}(X,Y) = \max\{\overrightarrow{dist}(X,Y), \overrightarrow{dist}(Y,X)\} = \overrightarrow{dist}(Y,X) = dist(y_3,x_2) \,. \quad \square$$

It is proved in [Barnsley 12] that if $(\mathcal{X}, dist(\cdot,\cdot))$ is a complete metric space, then $(\mathcal{H}(\mathcal{X}), dist_{\mathcal{H}})$ is a complete metric space. The value of this result is that all the results and techniques developed for complete metric spaces, in particular relating to convergent sequences, can be extended to $\mathcal{H}(\mathcal{X})$ with the Hausdorff distance. First, some notation. Let $f : \mathcal{X} \to \mathcal{X}$ and let $x \in \mathcal{X}$. Define $f^1(x) = f(x)$. For $t = 2,3,\ldots$, by $f^t(x)$ we mean $f(f^{t-1}(x))$. Thus $f^2(x) = f(f^1(x)) = f(f(x))$, $f^3(x) = f(f^2(x)) = f\big(f(f(x))\big)$, etc.

Definition 3.8 Let $(\mathcal{X}, dist(\cdot,\cdot))$ be a metric space. The function $f : \mathcal{X} \to \mathcal{X}$ is a *contraction map* if for some positive constant $\alpha < 1$ and all $s,t \in \mathcal{X}$ we have $dist(f(s),f(t)) \leq \alpha\, dist(s,t)$. The constant α is called the *contraction factor*. $\square$

Definition 3.9 A *fixed point* of a function $f : \mathcal{X} \to \mathcal{X}$ is a point $x^\star \in \mathcal{X}$ such that $f(x^\star) = x^\star$. $\square$

A proof of the following classical result can be found in [Barnsley 12].

Theorem 3.1 (Contraction Mapping Theorem) Let $(\mathcal{X}, dist(\cdot,\cdot))$ be a metric space, and let $f : \mathcal{X} \to \mathcal{X}$ be a *contraction map*. Then there is exactly one fixed point $x^\star$ of f. Moreover, for each $x_0 \in \mathcal{X}$ the sequence $f^t(x_0)$ converges, and $\lim_{t \to \infty} f^t(x_0) = x^\star$.

With Theorem 3.1 we can prove, under appropriate assumptions, the convergence of the sets Ω_t generated by an IFS. Recall that $\mathcal{H}(\mathcal{X})$ is the space of non-empty compact subsets of the metric space $\big(\mathcal{X}, dist(\cdot,\cdot)\big)$. For $k = 1, 2, \ldots, K$, let $f_k : \mathcal{H}(\mathcal{X}) \to \mathcal{H}(\mathcal{X})$. Assume that each f_k is a contraction map with contraction factor $\alpha_k < 1$, so

$$\sup_{x \neq y} \frac{dist\big(f_k(x), f_k(y)\big)}{dist(x,y)} = \alpha_k < 1 \, .$$

For $C \in \mathcal{H}(\mathcal{X})$, define $f : \mathcal{H}(\mathcal{X}) \to \mathcal{H}(\mathcal{X})$ by

$$f(C) \equiv \bigcup_{k=1}^{K} f_k(C)$$

so $f(C)$ is the union of the K sets $f_k(C)$. Then f is a contraction map, with contraction factor $\alpha = \max_{k=1}^{K} \alpha_k$, and $\alpha < 1$. By Theorem 3.1, f has a unique fixed point. The fixed point, which we denote by Ω, is an element of $\mathcal{H}(\mathcal{X})$, so Ω is non-empty and compact, and $\lim_{t \to \infty} \Omega_t = \Omega$ for any choice of Ω_0. We call Ω the *attractor* of the IFS. The attractor Ω is the unique non-empty solution of the equation $f(\Omega) = \Omega$, and $\Omega = \bigcup_{k=1}^{K} f_k(\Omega)$.

Example 3.2 (continued) For the iterated function system of Example 3.2, the metric space $\big(\mathcal{X}, dist(\cdot,\cdot)\big)$ is $\mathbb{R}^2$ with the Euclidean distance L_2, $K = 3$, and each f_k has the contraction ratio $1/2$. The attractor is the Sierpiński triangle. □

Sometimes slightly different terminology [Glass 11] is used to describe an IFS. A function f defined on a metric space $\big(\mathcal{X}, dist(\cdot,\cdot)\big)$ is a *similarity* if for some $\alpha > 0$ and each $x, y \in \mathcal{X}$ we have $dist\big(f(x), f(y)\big) = \alpha \, dist(x,y)$. A *contracting ratio list* $\{\alpha_k\}_{k=1}^{K}$ is a list of numbers such that $0 < \alpha_k < 1$ for each k. An *iterated function system* (IFS) realizing a ratio list $\{\alpha_k\}_{k=1}^{K}$ in a metric space $\big(\mathcal{X}, dist(\cdot,\cdot)\big)$ is a list of functions $\{f_k\}_{k=1}^{K}$ defined on $\mathcal{X}$ such that f_k is a similarity with ratio α_k. A non-empty compact set $\Omega \subset X$ is an *invariant* set for an IFS if $\Omega = f_1(\Omega) \cup f_2(\Omega) \cup \cdots \cup f_K(\Omega)$. Theorem 3.1, the Contraction Mapping Theorem now, can now be stated as: If $\{f_k\}_{k=1}^{K}$ is an IFS realizing a contracting ratio list $\{\alpha_k\}_{k=1}^{K}$ then there exists a unique non-empty compact invariant set for the IFS.

The Contraction Mapping Theorem shows that to compute Ω we can start with any set $\Omega_0 \in \mathcal{H}(\mathcal{X})$, and generate the iterates $\Omega_1 = f(\Omega_0)$, $\Omega_2 = f(\Omega_1)$, etc. In practice, it may not be easy to determine $f(\Omega_h)$, since it may not be easy to represent each $f_k(\Omega_t)$ if f_k is a complicated function. Another method of computing Ω is a randomized method, which begins by assigning a positive probability p_k to each of the K functions f_k such that $\sum_{k=1}^{K} p_k = 1$. Let $x \in \mathcal{X}$ be arbitrary, and set $C = \{x\}$. In iteration t of the randomized method, select one of the functions f_k, using the probabilities p_k, select a random point x from the set C, compute $f_k(x)$, and set $C = C \cup \{f_k(x)\}$.

For example, for the IFS (3.2), (3.3), (3.4), suppose we set $p_1 = \frac{3}{6}$, $p_2 = \frac{2}{6}$, $p_3 = \frac{1}{6}$, and choose $x = (\frac{1}{3}, \frac{1}{4})$, so initially $C = \{(\frac{1}{3}, \frac{1}{4})\}$. To compute the next point, we roll a "weighted" 3-sided die with probabilities p_1, p_2, p_3. Suppose we select f_2. We compute

$$f_2(x) = \begin{pmatrix} 1/2 & 0 \\ 0 & 1/2 \end{pmatrix} \begin{pmatrix} 1/3 \\ 1/4 \end{pmatrix} + \begin{pmatrix} 0 \\ \sqrt{3}/4 \end{pmatrix}$$

and add this point to the set $\mathcal{C}$. We start the next iteration by rolling the die again.

Chapter 4

Topological and Box Counting Dimensions

A mind once stretched by a new idea never regains its original dimension.
Oliver Wendell Holmes, Jr. (1841–1935), American, U.S. Supreme Court justice from 1902 to 1932.

A fractal is often defined as a set whose fractal dimension differs from its topological dimension. To make sense of this definition, we start by defining the topological dimension of a set. After that, the remainder of this chapter is devoted to the box counting dimension, which is the first fractal dimension we will study.

4.1 Topological Dimension

The first dimension we will consider is the *topological dimension* of a set. This subject has a long history, with contributions as early as 1913 by the Dutch mathematician L. Brouwer (1881–1966). (The extensive historical notes in [Engelking 78] trace the history back to 1878 work of Georg Cantor.) Roughly speaking, the topological dimension is a nonnegative integer that defines the number of "distinct directions" within a set [Farmer 82]. There are actually several definitions of topological dimension. The usual definition encountered in the fractal literature is the *small inductive dimension* (also called the *Urysohn–Menger* dimension or the *weak inductive dimension*). Other topological dimensions [Edgar 08] are the *large inductive dimension* (also known as the *Čech dimension* or the *strong inductive dimension*), which is closely related to the small inductive dimension, and the *covering dimension* (also known as the *Lebesgue dimension*).[1]

---

[1] In [Engelking 78], the covering dimension is also called the Čech-Lebesgue dimension.

© Springer Nature Switzerland AG 2020
E. Rosenberg, *Fractal Dimensions of Networks*,
https://doi.org/10.1007/978-3-030-43169-3_4

The notion of a topological dimension of a set Ω requires that $\Omega \subset \mathcal{X}$, where $(\mathcal{X}, dist(\cdot, \cdot))$ is a metric space. For example, if $\mathcal{X}$ is $\mathbb{R}^E$ and $x, y \in \mathbb{R}^E$, then $dist(x, y)$ might be the L_1 distance $\sum_{i=1}^E |x_i - y_i|$, the Euclidean distance $\sqrt{\sum_{i=1}^E (x_i - y_i)^2}$, the L_∞ distance $\max_{i=1}^E |x_i - y_i|$, or the L_p distance (3.7). Our first definition of topological dimension is actually a definition of the small inductive dimension.

Definition 4.1 A set Ω has topological dimension d_T if for each $x \in \Omega$ and for each sufficiently small positive number ϵ, the set

$$\{y \in \Omega \mid dist(x, y) = \epsilon\}$$

has topological dimension $d_T - 1$, and d_T is the smallest nonnegative integer for which this holds. The set Ω has topological dimension 0 if for each $x \in \Omega$ and each sufficiently small positive number ϵ we have

$$\{y \in \Omega \mid dist(x, y) = \epsilon\} = \emptyset . \quad \square$$

In words, this recursive definition says that Ω has topological dimension d_T if the boundary of each sufficiently small neighborhood of a point in Ω intersects Ω in a set of dimension $d_T - 1$, and d_T is the smallest integer for which this holds; Ω has topological dimension 0 if the boundary of each sufficiently small neighborhood of a point in Ω is disjoint from Ω.

Example 4.1 Let Ω be the unit ball in $\mathbb{R}^3$ defined by $\Omega = \{x : \|x\| \leq 1\}$. Choose $\tilde{x} \in \Omega$ such that $\|\tilde{x}\| < 1$. Then for all sufficiently small ϵ the set

$$\Omega_1 \equiv \{y \in \Omega \mid dist(\tilde{x}, y) = \epsilon\} \cap \Omega$$

is the surface of a ball centered at $\tilde{x}$ with radius ϵ. Now choose $\tilde{y} \in \Omega_1$. The set

$$\Omega_2 \equiv \{y \in \Omega_1 \mid dist(y, \tilde{y}) = \epsilon\}$$

is the set of points on this surface at a distance ϵ from $\tilde{y}$, so Ω_2 is a circle. Finally, choose $\bar{z} \in \Omega_2$. The set

$$\Omega_3 \equiv \{z \in \Omega_2 \mid dist(z, \bar{z}) = \epsilon\}$$

is the set of the two points $\{z_1, z_2\}$ on the circle at a distance ϵ from $\bar{z}$. The set Ω_3 has topological dimension 0 since for all sufficiently small positive ϵ we have

$$\{z \in \Omega_3 \mid dist(z, z_1) = \epsilon\} = \emptyset \quad \text{and} \quad \{z \in \Omega_3 \mid dist(z, z_2) = \epsilon\} = \emptyset .$$

Thus for Ω we have $d_T = 3$, for Ω_1 we have $d_T = 2$, for Ω_2 we have $d_T = 1$, and for Ω_3 we have $d_T = 0$. $\square$

Exercise 4.1 What is the topological dimension of the set of rational numbers? □

Definition 4.2 Let $\Omega \subset \mathcal{X}$, where $\big(\mathcal{X}, dist(\cdot, \cdot)\big)$ is a metric space, let $x \in \Omega$, and let ϵ be a positive number. The ϵ-*neighborhood* of x is the set

$$\{y \mid y \in \mathcal{X} \text{ and } dist(x, y) < \epsilon\}. \qquad \Box$$

Definition 4.3 Let $\big(\mathcal{X}, dist(\cdot, \cdot)\big)$ be a metric space, and let $\Omega \subset \mathcal{X}$. Then Ω is *open* if for each $x \in \Omega$ there is an $\epsilon > 0$ such that the ϵ-neighborhood of x is contained in Ω. □

For example, $(0, 1)$ is open but $[0, 1]$ is not open. The set $\mathbb{R}^E$ is open. The set $\{x \mid x \in \mathbb{R}^E \text{ and } ||x|| < \epsilon\}$ is open but $\{x \mid x \in \mathbb{R}^E \text{ and } ||x|| \leq \epsilon\}$ is not open. Figure 4.1 illustrates the cases where Ω is open and where Ω is not open. In (*ii*), if x is on the boundary of Ω, then for each positive ϵ the ϵ-neighborhood of x is not a subset of Ω.

The notion of a covering of a set is a central concept in the study of fractal dimensions. A set is *countably infinite* if its members can be put in a 1:1 correspondence with the natural numbers 1, 2, 3, $\cdots$.

Definition 4.4 Let Ω be a subset of a metric space. A *covering* of Ω is a finite or countably infinite collection $\{S_i\}$ of open sets such that $\Omega \subset \bigcup_i S_i$. A *finite covering* is a cover containing a finite number of sets. □

For example, the interval $[a, b]$ is covered by the single open set $(a - \epsilon, b + \epsilon)$ for any $\epsilon > 0$. Let $\mathbb{Q}$ be the set of rational numbers, and for $\epsilon > 0$ and $x \in \mathbb{Q}$ let $B(x, \epsilon)$ be the open ball with center x and radius ϵ. Then $[a, b]$ is also covered by the countably infinite collection $\bigcup_{x \in \mathbb{Q}} B(x, \epsilon)$.[2]

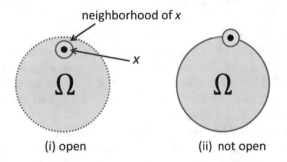

(i) open   (ii) not open

Figure 4.1: A set that is open, and a set that is not open

There is no requirement that the sets in the cover of Ω have identical shape or orientation, as illustrated in Fig. 4.2, which illustrates the covering of a square by six ovals, each of which is an open set.

---

[2]http://www.math.lsa.umich.edu/~kesmith/oct29.pdf.

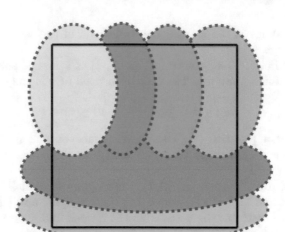

Figure 4.2: Covering of a square by six open sets

Our second definition of topological dimension, taken from [Eckmann 85], is actually a definition of the covering dimension.

Definition 4.5 The topological dimension d_T of a subset Ω of a metric space is the smallest integer K (or ∞) for which the following is true: For every finite covering of Ω by open sets $S_1, S_2, \ldots, S_J$, one can find another covering $S_1', S_2', \ldots, S_J'$ such that $S_j' \subset S_j$ for $j = 1, 2, \ldots, J$ and any $K + 2$ of the S_j' will have an empty intersection:

$$S_{j_1}' \cap S_{j_2}' \cap \cdots \cap S_{j_{K+2}}' = \emptyset. \quad \square$$

Another way of stating this definition, which perhaps is more intuitive, is the following. The topological dimension d_T of a subset Ω of a metric space is the smallest integer K (or ∞) for which the following is true: For every finite covering of Ω by open sets $S_1, S_2, \ldots, S_J$, one can find another covering $S_1', S_2', \ldots, S_J'$ such that $S_j' \subset S_j$ for $j = 1, 2, \ldots, J$ and each point of Ω is contained in at most $K + 1$ of the sets S_j'.

Example 4.2 Figure 4.3 illustrates Definition 4.5 when Ω is a line segment (the bold black line in the figure). In (i), Ω is covered by five open sets: three ovals and two squares. These five sets are the S_j, so $J = 5$. In (ii), we show the five sets S_j', which have been chosen so that $S_j' \subset S_j$ and each point of Ω lies in at most 2 of the sets S_j'. Since this can be done for each finite open cover $S_1, S_2, \ldots, S_J$, then $d_T = 1$ for the line segment. $\square$

Example 4.3 Let Ω be the square with a black border in Fig. 4.4. (for clarity, we have only shown the black border of Ω, but consider Ω to be the entire square, not just the displayed perimeter). Suppose we cover the lower part of Ω using S_1 and S_2, as illustrated in Fig. 4.4 (left). Since these are open sets, if they are to cover the bottom part of Ω, then they must overlap slightly, and hence we can always find a point p, indicated in the figure, such that $p \in S_1 \cap S_2$. We intentionally choose p to be near the top of the intersection $S_1 \cap S_2$, since we

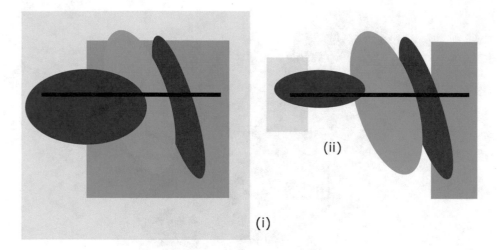

Figure 4.3: Line segment has $d_T = 1$

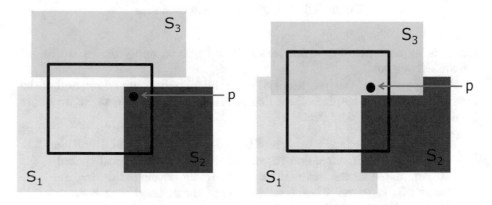

Figure 4.4: Topological dimension of a square

will now consider S_3, which currently does not cover that part of Ω not covered by $S_1 \cup S_2$. Suppose we slide S_3 downwards so that it now covers $\Omega - (S_1 \cup S_2)$, as illustrated in Fig. 4.4 (right). Then S_3 must overlap $S_1 \cup S_2$, and hence if p is sufficiently close to the top of $S_1 \cup S_2$, then $p \in S_1 \cap S_2 \cap S_3$. Moreover, for each covering S_1', S_2', S_3' satisfying $S_j' \subset S_j$ we can always find a point p such that $p \in S_1' \cap S_2' \cap S_3'$. If we cover the square with the 4 sets shown in Fig. 4.5, then each point in the square is contained in at most 3 of the S_j, and $S_1 \cap S_2 \cap S_3 \cap S_4 = \emptyset$. $\square$

As a final example of Definition 4.5, a solid cube in $\mathbb{R}^3$ has $d_T = 3$ because in any covering of the cube by smaller cubes there always exists a point belonging to at most four of the smaller cubes. Clearly, Definition 4.5 of d_T does not lend itself to numerical computation. Indeed, [Farmer 82] gave short shrift to topological dimension, noting that he knows of no general analytic or numerical

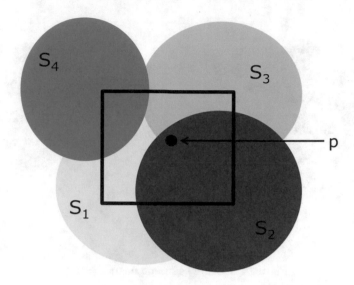

Figure 4.5: Covering a square by 4 sets

methods to compute d_T, which makes it difficult to apply d_T to studies of dynamical systems.

Having presented the definitions of the small inductive dimension, and of the covering dimension, to place these definitions in a historical context we quote from [Engelking 78], who wrote

> Dimension theory is a branch of topology devoted to the definition and study of of dimension in certain classes of topological spaces. It originated in the early [1920s] and rapidly developed during the next fifteen years. The investigations of that period were concentrated almost exclusively on separable metric spaces $\cdots$.[3] After the initial impetus, dimension theory was at a standstill for ten years or more. A fresh start was made at the beginning of the fifties, when it was discovered that many results obtained for separable metric spaces can be extended to larger classes of spaces, provided that the dimension is properly defined. $\cdots$ It is possible to define the dimension of a topological space $\mathcal{X}$ in three different ways, the small inductive dimension $ind\mathcal{X}$, the large inductive dimension $Ind\mathcal{X}$, and the covering dimension $dim\mathcal{X}$. $\cdots$ The covering dimension dim was formally introduced in [a 1933 paper by Čech] and is related to a property of covers of the n-cube I^n discovered by Lebesgue in 1911. The three-dimension functions coincide in the class of separable metric spaces, i.e., $ind\mathcal{X} = Ind\mathcal{X} = dim\mathcal{X}$ for every separable metric space $\mathcal{X}$. In larger classes of spaces, the dimensions ind, Ind, and

---

[3]A metric space $\mathcal{X}$ is *separable* if it has a subset $\mathcal{Y}$ which has a countable number of points such that $\bar{\mathcal{Y}} = \mathcal{X}$, where $\bar{\mathcal{Y}}$ is the closure of $\mathcal{Y}$ [Royden 68].

dim diverge. At first, the small inductive dimension *ind* was chiefly used; this notion has a great intuitive appeal and leads quickly and economically to an elegant theory. [It was later realized that] the dimension *ind* is practically of no importance outside of the class of separable metric spaces and that the dimension *dim* prevails over the dimension *Ind*. $\cdots$ The greatest achievement in dimension theory during the fifties was the discovery that $Ind\mathcal{X} = dim\mathcal{X}$ for every metric space $\mathcal{X}$.

To conclude this section, we note that some authors have used a very different definition of "topological dimension". For example, [Mizrach 92] wrote that Ω has topological dimension d_T if subsets of Ω of size s have a "mass" that is proportional to s^{d_T}. Thus a square of side length s has a "mass" (i.e., area) of s^2 and a cube of side s has a "mass" (i.e., volume) of d^3, so the square has $d_T = 2$ and the cube has $d_T = 3$. We will use a definition of this form when we consider the correlation dimension d_C (Chap. 9) and the network mass dimension d_M (Chap. 12). It is not unusual to see different authors use different names to refer to the same definition of dimension; this is particularly true in the fractal dimension literature.

4.2 Box Counting Dimension

The simplest fractal dimension is the box counting dimension, which we introduced in the beginning of Chap. 3. This dimension is based on covering Ω by equal-sized "boxes". Consider a line segment of length L. If we measure the line segment using a ruler of length s, where $s \ll L$, as illustrated in Fig. 4.6(i), the number of rule lengths $B(s)$ needed is given by $B(s) = \lceil L/s \rceil \approx Ls^{-1}$ for $s \ll L$. We call $B(s)$ the "number of boxes" of size s needed to cover the

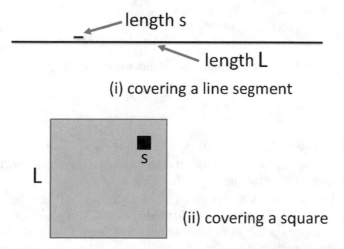

Figure 4.6: Box counting for a line and for a square

segment. In this example, a "box" is a one-dimensional ruler of length s. Since the exponent of s in $B(s) \approx Ls^{-1}$ is -1, we say that a line segment has a *box counting* dimension of 1.

Now consider a two-dimensional square with side length L. We cover the large square by squares of side length s, where $s \ll L$, as illustrated in Fig. 4.6(ii). The number $B(s)$ of small squares needed is given by $B(s) = \lceil L^2/s^2 \rceil \approx L^2s^{-2}$. Since the exponent of s in $B(s) = Ls^{-2}$ is -2, we say that the square has box counting dimension 2. Now consider an E-dimensional hypercube with side length L, where E is a positive integer. For $s \ll L$, the number $B(s)$ of hypercubes of side length s needed to cover the large hypercube is $B(s) = \lceil L^E/s^E \rceil \approx L^Es^{-E}$, so the E-dimensional hypercube has box counting dimension E.

We now provide a general definition of the box counting dimension of a geometric object $\Omega \subset \mathbb{R}^E$. By the term "box" we mean an E-dimensional hypercube. By the "linear size" of a box B we mean the diameter of B (i.e., the maximal distance between any two points of B), or the maximal variation in any coordinate of B, i.e., $\max_{x,y\in B} \max_{1\leq i \leq E} |x_i - y_i|$. By a box of size s we mean a box of linear size s. A set of boxes is a *covering* of Ω (i.e., the set of boxes *covers* Ω) if each point in Ω belongs to at least one box.

Definition 4.6 [Falconer 03] Let $B(s)$ be either (i) the minimal number of sets of diameter at most s needed to cover Ω, or (ii) the smallest number of closed balls of radius s needed to cover Ω. If

$$\lim_{s\to 0} \frac{\log B(s)}{\log(1/s)} \qquad (4.1)$$

exists, then the limit is called the *box counting dimension* of Ω and is denoted by d_B.   □

The subscript B in d_B denotes "box counting". Although the limit (4.1) may not exist, the *lower box counting dimension* $\underline{d_B}$ defined by

$$\underline{d_B} \equiv \liminf_{s\to 0} \frac{\log B(s)}{\log(1/s)}, \qquad (4.2)$$

and the *upper box counting dimension* $\overline{d_B}$ defined by

$$\overline{d_B} \equiv \limsup_{s\to 0} \frac{\log B(s)}{\log(1/s)} \qquad (4.3)$$

always exist. When $\underline{d_B} = \overline{d_B}$ then d_B exists, and $d_B = \underline{d_B} = \overline{d_B}$. In practice, e.g., when computing the box counting dimension of a digitized image, we need not concern ourselves with the possibility that $\underline{d_B} \neq \overline{d_B}$.

The fact that conditions (i) and (ii) of Definition 4.6 can both be used in (4.1) to define d_B is established in [Falconer 03]. In practice, it is often impossible to verify that $B(s)$ satisfies either condition (i) or (ii). The following result provides a computationally extremely useful alternative way to compute d_B.

Theorem 4.1 If d_B exists, then d_B can also be computed using (4.1) where $B(s)$ is the number of boxes that intersect Ω when Ω is covered by a grid of boxes of size s.

Proof [Falconer 03, Pearse] For $\Omega \subset \mathbb{R}^E$, let $\widetilde{B}(s)$ be the number of boxes of size s in some (not necessarily optimal) covering of Ω, and let $B(s)$ be (as usual) the minimal number of boxes of size s required to cover Ω. For $s > 0$, a box of size s in the E-dimensional grid has the form

$$[x_1 s, (x_1 + 1)s] \times [x_2 s, (x_2 + 1)s] \times \cdots \times [x_E s, (x_E + 1)s] \,,$$

where $(x_1, x_2, \ldots, x_E)$ are integers. For example, if $E = 2$ and $(x_1, x_2) = (2, -5)$, then one such box (a square, since $E = 2$) is $[2s, 3s] \times [-5s, -4s]$, which is the set $\{(x_1, x_2) \mid 2s \leq x_1 \leq 3s \text{ and } -5s \leq x_2 \leq -4s\}$. Since the length of each of the E sides is s, the diameter of each box is $(\sum_{i=1}^{E} s^2)^{1/2} = s\sqrt{E}$.

Since $s\sqrt{E} \geq s$ and the minimal number $B(s)$ of boxes of size s needed to cover Ω is a non-increasing function of s, then

$$B(s\sqrt{E}) \leq B(s) \leq \widetilde{B}(s) \,,$$

so $B(s\sqrt{E}) \leq \widetilde{B}(s)$. For all sufficiently small positive s we have $s\sqrt{E} < 1$ and hence

$$\frac{\log B(s\sqrt{E})}{-\log(s\sqrt{E})} = \frac{\log B(s\sqrt{E})}{-\log s - \log \sqrt{E}} \leq \frac{\log \widetilde{B}(s)}{-\log s - \log \sqrt{E}} \,.$$

Letting $s \to 0$ yields

$$d_B \leq \liminf_{s \to 0} \frac{\log \widetilde{B}(s)}{-\log s} \,. \tag{4.4}$$

To prove the reverse inequality, the key observation is that any set of size s in $\mathbb{R}^E$ is contained in at most 3^E boxes of size s. To see this, choose an arbitrary point in the set, let B be the box containing that point, and then select the $3^E - 1$ boxes that are neighbors of B. This is illustrated for $E = 2$ in Fig. 4.7, which shows a set of size s contained in $3^2 = 9$ boxes of size s; the arbitrarily chosen point is indicated by the small black circle. This yields the inequality

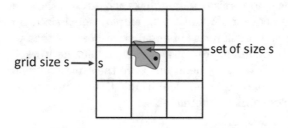

Figure 4.7: 3^E boxes cover the set

$\widetilde{B}(s) \leq 3^E B(s)$, from which we obtain

$$\frac{\log \widetilde{B}(s)}{-\log s} \leq \frac{\log B(s)}{-\log s} + \frac{\log 3^E}{-\log s} .$$

Letting $s \to 0$, we obtain

$$\limsup_{s \to 0} \frac{\log \widetilde{B}(s)}{-\log s} \leq d_B . \qquad (4.5)$$

From (4.4) and (4.5) we have

$$d_B \leq \liminf_{s \to 0} \frac{\log \widetilde{B}(s)}{-\log s} \leq \limsup_{s \to 0} \frac{\log \widetilde{B}(s)}{-\log s} \leq d_B ,$$

so $\lim_{s \to 0} \log \widetilde{B}(s)/(-\log s)$ exists and is equal to d_B. $\square$

Thus we can compute d_B by counting, for each chosen value of s, the number of boxes that intersect Ω when Ω is covered by a grid of boxes of size s. In practice, for a given s the value $B(s)$ is often computed by picking the minimal $B(s)$ over a set of runs, where each run uses a different offset of the grid (i.e., by shifting the entire grid by a small amount). However, this is time consuming, and usually does not reduce the error by more than 0.5% [Jelinek 98].

For a real-life example of box counting using a grid, 19 boxes are needed to cover the image of the starfish in Fig. 4.8.[4] Actually, the image is not quite completely covered by these 19 boxes, since there are two starfish tips (pointed to by two arrows) that are not covered. So we could use two more boxes to cover these two tips. Or perhaps we could use only one additional box, if we translate each box by the same amount (i.e., if we shift the grid coordinates slightly). In general, it is computationally infeasible to find the minimal number of boxes required to cover Ω by a grid of boxes.

In the fractal literature (e.g., [Baker 90, Barnsley 12, Falconer 13, Feder 88, Schroeder 91]) d_B is usually defined assuming that $B(s)$ is either the smallest number of closed balls of radius s needed to cover Ω (i.e., condition (ii) of Definition 4.6), or that $B(s)$ is the number of boxes that intersect Ω when Ω is covered by a grid of boxes of size s (as in Theorem 4.1); both of these choices assume that, for a given s, all boxes have the same size. The definitions of d_B in several well-known books also assume that each box has the same size. All of the examples in this chapter illustrating d_B employ equal-sized boxes. The reason for not defining d_B only in terms of equal-sized boxes is that when we consider, in Chap. 7, the box counting dimension of a complex network $\mathbb{G}$, the boxes that we use to "cover" $\mathbb{G}$ will in general not have the same linear size. Since we will base our definition of d_B of $\mathbb{G}$ on Definition 4.6, we want the definition of d_B for Ω to permit boxes to have different sizes.

---

[4]The image is from "The Most Beautiful Starfish In The World", www.buzzfeed.com/ melismashable/the-most-beautiful-starfish

Figure 4.8: Starfish box counting

In [Falconer 13], a 1928 paper by Georges Bouligand (1889–1979) is credited with introducing box counting; [Falconer 13] also noted that "the idea underlying an equivalent definition had been mentioned rather earlier" by Hermann Minkowski (1984–1909).[5] The "more usual definition" of d_B was given by Pontryagin and Schnirelman in 1932 [Falconer 03]. A different historical view, in [Vassilicos 91], is that d_B was first defined by Kolmogorov in 1958; indeed, [Vassilicos 91] called d_B the *Kolmogorov capacity*, and observed that the Kolmogorov capacity is often referred to in the literature as the box dimension, similarity dimension, Kolmogorov entropy, or ϵ-entropy. The use of the term "Kolmogorov capacity" was justified in [Vassilicos 91] by citing [Farmer 83] and a 1989 book by Ruelle.

Another view, in [Lauwerier 91], is that the definition of fractal dimension given by Mandelbrot is a simplification of Hausdorff's and corresponds exactly with the 1958 definition of Kolmogorov (1903–1987) of the "capacity" of a geometrical figure; in [Lauwerier 91] the term "capacity" was used to refer to the box counting dimension. Further confusion arises since Mandelbrot observed [Mandelbrot 85] that the box counting dimension has been sometimes referred to as the "capacity", and added that since "capacity" has at least two other quite distinct competing meanings, it should not be used to refer to the box counting dimension. Moreover, [Frigg 10] commented that the box counting dimension has also been called the *entropy* dimension, e.g., in [Mandelbrot 83b] (p. 359). Fortunately, this babel of terminology has subsided, and the vast majority of recent publications (since 1990 or so) have used the term "box counting dimension", as we will.

Exercise 4.2 Let $f(x)$ be defined on the positive integers, and let L be the length of the piecewise linear curve obtained by connecting the point $\big(i, f(i)\big)$ to the point $\big(i+1, f(i+1)\big)$ for $i = 1, 2, \ldots, N-1$. Show that d_B of this curve can be approximated by $1 + \big(\log L / \log N\big)$. $\qquad\square$

---

[5] Minkowski is well-known for his work on the theory of relativity. In 1907 he showed that Einstein's 1905 special theory of relativity could be elegantly expressed using 4-dimensional spacetime. Minkowski died suddenly at the age of 44 from a ruptured appendix.

The existence of d_B does not imply that $B(s)$ is proportional to s^{-d_B}. By (4.1), the existence of d_B does imply that for each $\epsilon > 0$ there is an $\tilde{s} < 1$ such that for $0 < s < \tilde{s}$ we have

$$\left| \frac{\log B(s)}{\log(1/s)} - d_B \right| \leq \epsilon$$

$$\Leftrightarrow d_B - \epsilon \leq \frac{\log B(s)}{\log(1/s)} \leq d_B + \epsilon$$

$$\Leftrightarrow (d_B - \epsilon)\log(1/s) \leq \log B(s) \leq (d_B + \epsilon)\log(1/s)$$

$$\Leftrightarrow s^{-d_B + \epsilon} \leq B(s) \leq s^{-d_B - \epsilon}. \tag{4.6}$$

Following [Theiler 90], another way of interpreting d_B is to define the function $\theta(\cdot)$ by $\theta(s) \equiv B(s)/s^{-d_B}$ for $s > 0$. Then $B(s) = \theta(s)s^{-d_B}$ so

$$d_B = \lim_{s \to 0} \frac{\log B(s)}{\log(1/s)} = \lim_{s \to 0} \frac{\log \theta(s) - d_B \log s}{\log(1/s)} = d_B - \lim_{s \to 0} \frac{\log \theta(s)}{\log s}.$$

Thus if d_B exists, then

$$\lim_{s \to 0} \frac{\log \theta(s)}{\log s} = 0.$$

Richardson's empirical result on the length of the coast of Britain, presented in the beginning of Chap. 3, provided the box counting dimension of the coast. Recall that Richardson showed that $L(s) \approx L_0 s^{1-d}$, where $d \approx 1.25$. Thus $1 - d = \lim_{s \to 0} \log L(s)/\log s$. Since the minimal number $B(s)$ of boxes of size s needed to cover the coast is given by $B(s) \approx L(s)/s$, then

$$d_B = \lim_{s \to 0} \frac{\log B(s)}{\log(1/s)} = \lim_{s \to 0} \frac{\log\big(L(s)/s\big)}{\log(1/s)} = \lim_{s \to 0} \frac{\log L(s) - \log s}{-\log s}$$

$$= 1 - \lim_{s \to 0} \frac{\log L(s)}{\log s} = d,$$

so $d_B = d$.

In the above application of box counting to a starfish, a box is counted if it has a non-empty intersection with the starfish. More generally, in applying Theorem 4.1 to compute d_B of Ω, a box is counted if it has a non-empty intersection with Ω. The amount of intersection (e.g., the area of the intersection in the starfish example) is irrelevant. For other fractal dimensions such as the information dimension (Chap. 14) and the generalized dimensions (Chap. 16), the amount of intersection is significant, since "dense" regions of the object are weighted differently than "sparse" regions of the object. The adjective "morphological" is used in [Kinsner 07] to describe a dimension, such as the box counting dimension, which is a purely shape-related concept, concerned only with the geometry of an object, and which utilizes no information about the distribution of a measure over space, or the behavior of a dynamical system over time. Having now discussed both the topological dimension d_T and the box counting dimension d_B, we can present the definition of a fractal given in [Farmer 82]; he attributed this definition to [Mandelbrot 77].

Definition 4.7 A set is a *fractal* if its box counting dimension exceeds its topological dimension. □

Definition 4.7 is illustrated by Fig. 4.9, which shows the first six iterations of the Hilbert curve.[6] This space-filling curve has topological dimension 1

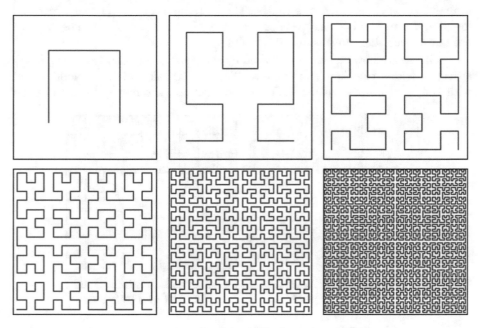

Figure 4.9: Six iterations of the Hilbert curve

($d_T = 1$). Since in generation t this curve consists of $2^{2t} - 1$ segments, each of length 2^{-t}, its box counting dimension is 2 ($d_B = 2$) [Schroeder 91]. However, regarding Definition 4.7, [Falconer 03] wrote "This definition has proved to be unsatisfactory, in that it excluded a number of sets that clearly ought to be regarded as fractals. Various other definitions have been proposed, but they all seem to have this same drawback".

4.3 A Deeper Dive into Box Counting

In this section we provide more insight into the box counting dimension. A "natural and direct" approach to computing d_B for a real-world object is to take the smallest s available, call it s_{min}, and estimate d_B by $\log B(s_{min})/\log(1/s_{min})$. The problem with this approach is the very slow convergence of $\log B(s)/\log(1/s)$

---

[6] This image is https://upload.wikimedia.org/wikipedia/commons/3/3a/Hilbert_curve. png. It was drawn by Zbigniew Fiedorowicz, taken from http://www.math.ohio-state.edu/ ~fiedorow/math655/Peano.html.

to d_B [Theiler 90]. To see this, assume $B(s) = cs^{-d_B}$ holds for some constant c. Then

$$\frac{\log B(s)}{\log(1/s)} = \frac{\log(cs^{-d_B})}{\log(1/s)} = d_B - \frac{\log c}{\log s},$$

which approaches d_B with logarithmic slowness as $s \to 0$, so this approach is not used in practice [Theiler 90]. Instead, to compute d_B, typically $B(s)$ is evaluated over some range of s values. Then a range $[s_{min}, s_{max}]$ over which the $\log B(s)$ versus $\log(1/s)$ curve is approximately linear is identified. The slope of the curve over this range is an estimate of d_B.

To illustrate how d_B can be obtained by the slope of a line in $(\log s, \log B(s))$ space, consider the Sierpiński carpet. The construction of this fractal starts with the filled-in unit square. Referring to Fig. 4.10, the first generation of the

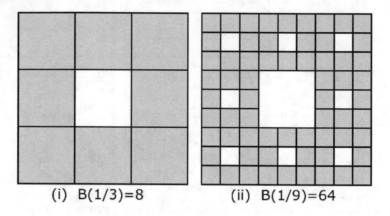

$$\text{(i) } B(1/3)=8 \qquad\qquad \text{(ii) } B(1/9)=64$$

Figure 4.10: Two generations of the Sierpiński carpet

construction removes the center 1/9 of the unit square. What remains is the shaded part of Fig. 4.10(i). This shaded area is covered by 8 of the small squares. Since the side length of the small squares is 1/3, then $s = 1/3$ and $B(1/3) = 8$. In generation 2, we consider each of the 8 small squares produced in generation 1. For each small square, we remove the center 1/9 of the square. The result of the generation 2 construction is illustrated in (ii). The shaded area in (ii) is covered by 8^2 squares of side length 1/9. Since each small square now has side length 1/9, then $s = 1/9$ and $B(1/9) = 8^2$. In each subsequent iteration, we remove the center 1/9 of each square sub-region. The slope of the line through $(\log(1/9), \log 64)$ and $(\log(1/3), \log 8)$ is

$$\frac{\log 64 - \log 8}{\log(1/9) - \log(1/3)} = \frac{\log 8}{\log(1/3)} = \frac{-\log 8}{\log 3},$$

so d_B of the Sierpiński carpet is $d_B = \log 8/\log 3 \approx 1.893$.

Exercise 4.3 The above calculation of d_B just used the two box sizes $s = 1/3$ and $s = 1/9$. Show that $d_B = \log 8/\log 3$ is obtained using any two powers of the box size 1/3.   □

Instead of always removing the center 1/9 of each square region in the
Sierpiński carpet, we can randomly choose, in each region, one of the nine
sub-regions, and delete that randomly selected sub-region [Strogatz 94]. This
randomized construction is illustrated in Fig. 4.11. In Ω_1, a random 1/9 of the

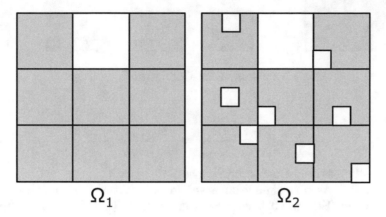

$$\Omega_1 \qquad\qquad \Omega_2$$

Figure 4.11: Randomized Sierpiński carpet

original solid square is deleted. In Ω_2, in each of the 8 remaining smaller squares,
a random 1/9 is deleted. Let Ω_t be the pre-fractal object obtained after t generations of this construction. Each Ω_t is not self-similar, since randomization
was used to select the small squares to delete. Rather, each Ω_t is *statistically
self-similar*; a loosely defined term which expresses the fact that the object is
"almost" self-similar, the "almost" referring to a randomization which does not
change the dimension of the object.[7] Indeed, we can apply box counting to find
the box counting dimension d_B of the randomized Sierpiński carpet, and the calculation is identical to the calculation of d_B for the non-random Sierpiński carpet
of Fig. 4.10; for the randomized Sierpiński carpet we also have $d_B = \log 8/\log 3$.
Example 4.4 For another box counting example, consider the two-dimensional
Cantor dust illustrated in Fig. 4.12. In each generation, each square is replaced
by four squares, each 1/3 the linear size of the previous generation. Each of the
4 second generations squares is 1/3 the linear size of the first generation. Each
of the 16 third generation squares is 1/9 the linear size of the first generation.
The slope of the line through the points $\big(\log(1/9), \log 16\big)$ and $\big(\log(1/3), \log 4\big)$

---

[7] The term *statistical self-similarity* was described in [Di Ieva 13]: "Statistical self-
similarity, also indicated with the term 'self-affinity,' concerns biological objects, including
all anatomic forms. Small pieces that constitute anatomic systems are rarely identical copies
of the whole system. If we consider a portion of tree branches or vascular vessels, they are not
a copy of the whole tree but represent the same self-similarity and structural 'complexity' (i.e.,
roughness and spatial pattern). Various statistically self-similar anatomic structures include
not only the general circulatory system, the bronchial tree, and the biliary tree of the liver
but also the dendritic structure of the neuronal cells, the ductal system of a gland, the cell
membrane, and the fibrous portion in chronic liver disease."

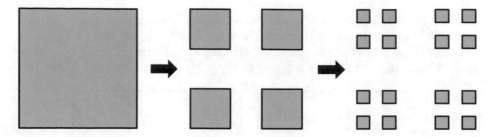

Figure 4.12: Cantor dust

is

$$\frac{(\log 16 - \log 4)}{(\log(1/9) - \log(1/3))} = \frac{\log 4}{\log(1/3)} = -\frac{\log 4}{\log 3},$$

so for two-dimensional Cantor dust we have $d_B = \log 4/\log 3$.   □

Example 4.5 As an example of a set for which the box-counting method behaves badly [Falconer 03], consider the set $\Omega = \{0, 1, \frac{1}{2}, \frac{1}{3}, \ldots\}$. This is a bounded countably infinite set. Although we might expect the box counting dimension of Ω to be 0, in fact it is $1/2$. To see this, pick $s < 1/2$ and let k be the integer such that $\frac{1}{k(k+1)} \leq s < \frac{1}{(k-1)k}$. Then $\log \frac{1}{k(k+1)} \leq \log s$, so $\log k(k+1) \geq -\log s$. Similarly $\log s < \frac{1}{(k-1)k}$ implies $-\log s > \log k(k-1)$.

Suppose we cover Ω by boxes (i.e., line segments) of size s. Define $\Omega_k = \{1, \frac{1}{2}, \frac{1}{3}, \ldots, \frac{1}{k}\}$. For $k \geq 2$, the distance between consecutive terms in Ω_k is at least $\frac{1}{k-1} - \frac{1}{k} = \frac{1}{k(k-1)} > s$. Thus a box of size s can cover at most one point of Ω_k, so k boxes are required to cover Ω_k. This implies

$$\frac{\log B(s)}{-\log s} \geq \frac{\log k}{-\log s} \geq \frac{\log k}{\log k(k+1)} = \frac{\log k}{\log k + \log(k+1)}.$$

Since $k \to \infty$ as $s \to 0$, letting $s \to 0$ yields $\underline{d_B} \geq 1/2$.

To prove the reverse inequality, first observe that since $s \geq \frac{1}{k(k+1)}$ then $k+1$ boxes of size s cover the interval $[0, \frac{1}{k}]$. These $k+1$ boxes do not cover the $k-1$ points $1, \frac{1}{2}, \ldots, \frac{1}{k-1}$, but these $k-1$ points can be covered by another $k-1$ boxes. Hence $B(s) \leq (k+1) + (k-1) = 2k$ and

$$\frac{\log B(s)}{-\log s} \leq \frac{\log 2k}{-\log s} < \frac{\log 2k}{\log k(k-1)}.$$

Letting $s \to 0$ yields $\overline{d_B} \leq 1/2$. Thus for $\Omega = \{0, 1, \frac{1}{2}, \frac{1}{3}, \cdots\}$ we have $1/2 \leq \underline{d_B} \leq \overline{d_B} \leq 1/2$, so the box counting dimension of this countably infinite set is $1/2$.   □

The fact that $d_B = 1/2$ for $\Omega = \{0, 1, \frac{1}{2}, \frac{1}{3}, \cdots\}$ violates one of the properties that should be satisfied by any definition of *dimension*, namely the property that the dimension of any finite or countably infinite set should be zero. This violation does not occur for the Hausdorff dimension (Chap. 5). Such theoretical problems with the box counting dimension have spurred alternatives to the box counting dimension, as discussed in [Falconer 03]. In practice, these theoretical issues have not impeded the widespread application of box counting (Sect. 6.5).

As observed in [Schroeder 91], the definition

$$d_B = \lim_{s \to 0} \frac{\log B(s)}{-\log s}$$

of d_B could alternatively be given by the implicit form

$$\lim_{s \to 0} B(s) s^{d_B} = c \tag{4.7}$$

for some constant c satisfying $-\infty < c < \infty$. In (4.7), d_B is the unique value of the exponent d that keeps the product $B(s)s^d$ finite and nonzero as $s \to 0$. Any other value of d will result in the product $B(s)s^d$ approaching either 0 or ∞ as $s \to 0$. This way of viewing the box counting dimension is fully developed when we explore the Hausdorff dimension in Chap. 5.

4.4 Fuzzy Fractal Dimension

If the box-counting method is applied to a 2-dimensional array of pixels, membership is binary: a pixel is either contained in a given box B, or it is not contained in B. The *fuzzy box counting dimension* is a variant of d_B in which binary membership is replaced by fuzzy membership: a "membership function" assigns to a pixel a value in $[0, 1]$ indicating the degree to which the pixel belongs to B. Similarly, for a 3-dimensional array of voxels, binary membership is replaced by assigning a value in $[0, 1]$ indicating the degree to which the voxel belongs to B.[8]

Fuzzy box counting is useful for medical image processing, e.g., to determine the edges of heart images obtained from an echocardiogram [Zhuang 04]. Suppose that for pixel i in the 2-dimensional array we have the associated grayscale level g_i, which is between 0 and 256. Then we can choose $g_i/256$ as the membership function. Suppose the pixel array is covered by a set $\mathcal{B}(s)$ of boxes of a given size s, and consider a given box B in the covering. Let $\mathcal{I}(B)$ be the set of pixels covered by B. Whereas for "classical" box counting we count B if $\mathcal{I}(B)$ is non-empty, for fuzzy box counting we compute

$$f(B) \equiv \max_{i \in \mathcal{I}(B)} \frac{g_i}{256}$$

$$f(s) \equiv \sum_{B \in \mathcal{B}(s)} f(B). \tag{4.8}$$

---

[8] A volumetric pixel (volume pixel or voxel) is the three-dimensional (3D) equivalent of a pixel and the tiniest distinguishable element of a 3D object. It is a volume element that represents a specific grid value in 3D space. However, like pixels, voxels do not contain information about their position in 3D space. Rather, coordinates are inferred based on their designated positions relative to other surrounding voxels. https://www.techopedia.com/definition/2055/volume-pixel-volume-pixel-or-voxel.

If the plot of $\log f(s)$ versus $\log(1/s)$ is roughly linear over a range of s, the slope of the line is the estimate of the fuzzy box counting dimension.

The local fuzzy box counting dimension [Zhuang 04] restricts the computation of the fuzzy box counting dimension to the neighborhood $\mathbb{N}(j)$ of all pixels within a given radius of a given pixel j, where we might define the distance between two pixels to be the distance between the pixel centers. Instead of (4.8), we calculate

$$f\big(\mathbb{N}(j), B\big) \equiv \max_{i \in \mathcal{I}(B) \cap \mathbb{N}(j)} \frac{g_i}{256}$$

$$f\big(\mathbb{N}(j), s\big) \equiv \sum_{B \in \mathcal{B}(s)} f\big(\mathbb{N}(j), B\big).$$

If the plot of $\log f\big(\mathbb{N}(j), s\big)$ versus $\log(1/s)$ is roughly linear over a range of s, the slope of the line is the estimate of the local fuzzy box counting dimension in the neighborhood $\mathbb{N}(j)$ of pixel j.

4.5   Minkowski Dimension

For $\Omega \subset \mathbb{R}^E$ and $x, y \in \mathbb{R}^E$, let $dist(x, y)$ be the Euclidean distance between x and y. For $r > 0$, the set

$$\Omega_r \equiv \{y \in \mathbb{R}^E \mid dist(x, y) \leq r \text{ for some } x \in \Omega\}$$

is the set of points within distance r of Ω. We call Ω_r the r-neighborhood of Ω or simply a *dilation* of Ω. Let $vol(\Omega_r)$ be the volume (in the usual geometric sense) of Ω_r, and let d_T be the topological dimension of Ω.

Example 4.6 Let $E = 3$ and let Ω be the single point x. Then $d_T = 0$ and Ω_r is the closed ball (in $\mathbb{R}^3$) of radius r centered at x. We have $vol(\Omega_r) = (4/3)\pi r^3 = c_0 r^{E-d_T}$ for some constant c_0, so $vol(\Omega_r)/r^{E-d_T} = c_0$.   □

Example 4.7 Let $E = 3$ and let Ω be a line (straight or curved). Then Ω_r is a "sausage" of diameter $2r$ (except at the ends) traced by the closed balls (in $\mathbb{R}^3$) whose centers follow the line. This construction is a standard method used by mathematicians to "tame" a wildly irregular curve [Mandelbrot 83b]. If Ω is a straight line segment of length L, then $d_T = 1$ and $vol(\Omega_r) \approx \pi r^2 L = c_1 r^{E-d_T}$ for come constant c_1, so $vol(\Omega_r)/r^{E-d_T} \approx c_1$.   □

Example 4.8 Let $E = 3$ and let Ω be a flat 2-dimensional sheet. Then Ω_r is a "thickening" of Ω. If the area of Ω is A, then $d_T = 2$ and $vol(\Omega_r) \approx 2rA = c_2 r^{E-d_T}$ for some constant c_2, so $vol(\Omega_r)/r^{E-d_T} \approx c_2$.   □

Motivated by these examples, let Ω be an arbitrary subset of $\mathbb{R}^E$ and consider $vol(\Omega_r)/r^{E-d}$ as $r \to 0$. Suppose for some nonnegative d the limit

$$\lim_{r \to 0} \frac{vol(\Omega_r)}{r^{E-d}} \tag{4.9}$$

exists. Denote this limit, called the *Minkowski content* of Ω, by c; the Minkowski content is a measure of the length, area, or volume of Ω as appropriate [Falconer 03]. Pick $\epsilon > 0$. For all sufficiently small $r > 0$ we have

$$c - \epsilon \le vol(\Omega_r)r^{d-E} \le c + \epsilon$$
$$\Leftrightarrow \log(c - \epsilon) \le \log vol(\Omega_r) + (d - E)\log r \le \log(c + \epsilon)$$
$$\Leftrightarrow \frac{\log(c - \epsilon)}{\log r} \le \frac{\log vol(\Omega_r)}{\log r} + (d - E) \le \frac{\log(c + \epsilon)}{\log r}.$$

Letting $r \to 0$ yields

$$d = E - \lim_{r \to 0} \frac{\log vol(\Omega_r)}{\log r}, \tag{4.10}$$

and d is called the *Minkowski* dimension, or the *Minkowski-Bouligand* dimension, and is denoted by d_K.[9] Since Ω_r is a dilation of Ω, the technique of using (4.10) to estimate d_K is called the *dilation method*.

The three examples above considered the Minkowski dimension of a point, line segment, and flat region, viewed as subsets of $\mathbb{R}^3$. We can also consider their Minkowski dimensions as subsets of $\mathbb{R}^2$, in which case we replace "volume" with "area".

Example 4.9 Figure 4.13 illustrates this dilation in $\mathbb{R}^2$ for a curve and for a

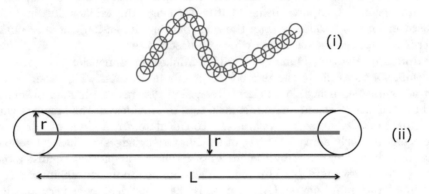

Figure 4.13: Minkowski sausage in two dimensions

straight line of length L. Since now $E = 2$, the Minkowski dimension d_K of a straight line of length L is

$$d_K = 2 - \lim_{r \to 0} \frac{\log area(\Omega_r)}{\log r}. \tag{4.11}$$

We have $area(\Omega_r) = 2rL + 2(\pi r^2/2) = 2rL + \pi r^2$. Using L'Hôpital's rule,

$$d_K = 2 - \lim_{r \to 0} \frac{\log(2rL + \pi r^2)}{\log r} = 2 - \lim_{r \to 0} \frac{(2L + 2\pi r)/(2rL + \pi r^2)}{1/r}$$
$$= 2 - \lim_{r \to 0} \frac{2rL + 2\pi r^2}{2rL + \pi r^2} = 1. \quad \square$$

---

[9] It would be nice to denote the Minkowski dimension by d_M, but we reserve that symbol for the mass dimension (Chap. 12), which has a greater role than the Minkowski dimension in this book.

If the limit (4.9) does not exist, we can still define

$$\underline{d_K} = E - \lim_{r\to 0} \sup \frac{\log vol(\Omega_r)}{\log r} \qquad (4.12)$$

and

$$\overline{d_K} = E - \lim_{r\to 0} \inf \frac{\log vol(\Omega_r)}{\log r} . \qquad (4.13)$$

It is proved in [Falconer 03] that the lower box counting dimension $\underline{d_B}$, defined by (4.2), is equal to $\underline{d_K}$, and that the upper box counting dimension $\overline{d_B}$, defined by defined by (4.3), is equal to $\overline{d_K}$. If $\underline{d_B} = \overline{d_B}$, then the box counting dimension d_B and the Minkowski dimension d_K both exist and $d_B = d_K$. Thus, while defined differently, the box counting dimension and the Minkowski dimensions, if they exist, are the same. For this reason, the box counting dimension is sometimes called the Minkowski dimension or the Minkowski-Bouligand dimension. In this book we will use d_B when the dimension is computed using (4.1), and d_K when the dimension is computed using (4.10). Although the Minkowski dimension is used much less frequently than the box counting dimension, there are many interesting applications, a few of which we describe below.

Estimating People Density: The Minkowski dimension was used in [Marana 99] to estimate the number of people in an area. This task is usually accomplished using closed-circuit television systems and human observers, and there is a need to automate this task, e.g., to provide real-time reports when the crowd density exceeds a safe level. To calculate d_K, a circle of size r was swept continuously along the edges of the image (the edges outline the people). The total area $A(r)$ swept by the circles of radius r was plotted against r on a log–log plot to estimate d_K. This method was applied to nearly 300 images from a railway station in London, UK, to classify the crowd density as very low, low, moderate, high, or very high, and achieved about 75% correct classification. However, the method was not able to distinguish between the high and very high crowd densities.   □

Texture Analysis of Pixelated Images: Fractal analysis was applied in [Florindo 13] to a *color adjacency graph* (CAG), which is a way to represent a pixelation of a colored image. They considered the adjacency matrix of the CAG as a geometric object, and computed its Minkowski dimension.   □

Cell Biology: The dilation method was used in [Jelinek 98] to estimate the length of the border of a 2-dimensional image of a neuron. Each pixel on the border of the image was replaced by a circle of radius r centered on that pixel. Structures whose size is less than r were removed. Letting $area(r)$ be the area of the resulting dilation of the border, $area(r)/(2r)$ is an estimate of the length of the border. A plot of $\log area(r)/(2r)$ versus $\log 2r$ yields an estimate of d_K.   □

Acoustics: As described in [Schroeder 91], the Minkowski dimension has applications in acoustics. The story begins in 1910, when the Dutch physicist

Hendrik Lorentz conjectured that the number of resonant modes of an acoustic resonator depends, up to some large frequency f, only on the volume V of the resonator and not its shape. Although the mathematician David Hilbert predicted that this conjecture, which is important in thermodynamics, would not be proved in his lifetime, it was proved by Herman Weyl, who showed that for large f and for resonators with sufficiently smooth but otherwise arbitrary boundaries, the number of resonant modes is $(4\pi/3)V(f/c)^3$, where c is the velocity of sound. For a two-dimensional resonator such as a drum head, the number of resonant modes is $\pi A(f/c)^2$, where A is the surface area of the resonator. These formulas were later improved by a correction term involving lower powers of f. If we drop Weyl's assumption that the boundary of the resonator is smooth, the correction term turns out to be $(Lf/c)^{d_K}$. The reason the Minkowski dimension governs the number of resonant modes of an acoustic resonator is that normal resonance modes need a certain area or volume associated with the boundary. □

Music Signal Analysis: The Minkowski dimension was used in [Zlatintsi 11] to analyze music signals for such applications as music retrieval, automatic music transcription, and indexing of multimedia databases. (The idea of using fractal dimension to discriminate different genres of music was proposed as early as 2000.) Suppose a window of size r (where the range of r corresponds to a time scale of up to about 50 ms.) is moved along the plot of the music signal, thus creating a dilation of the plot. For a given r, [Zlatintsi 11] computed

$$f(r) = \frac{\log[(\text{area of graph dilated by disks of radius } r)/r^2]}{\log(1/r)}$$
$$= 2 - \frac{\log(\text{area of graph dilated by disks of radius } r)}{\log r}.$$

The Minkowski dimension $d_K = \lim_{r\to 0} f(r)$ was estimated by fitting a line to the $f(r)$ values. □

As described in [Eins 95], Flook in 1978 noted that when computing d_K for a figure containing "open ends" (i.e., end points of lines), using the dilation method can overestimate area; by (4.10) an overestimate of area causes an underestimate of d_K. For example, an astrocyte, illustrated in Fig. 4.14, contains many open ends.[10] The underestimate of d_K arises since in [Eins 95] the length $L(r)$ corresponding to a dilation was estimated from the dilated area $A(r)$ using $L(r) = A(r)/(2r)$. Therefore, using the Richardson relation $L(r) \approx r^{1-d}$ (see (3.1)), which we rewrite as $d = 1 - \frac{\log L(r)}{\log r}$, an overestimate of $L(r)$ yields an underestimate of d. The degree of underestimation depends on the number of end points and the image complexity. To eliminate the underestimation of d_K, specially designed "anisotropic" shapes, to be used at the open ends of such figures, were proposed in [Eins 95].

---

[10] An *astrocyte* is a star-shaped glial cell of the central nervous system. The image Fig. 4.14 is from https://wiki.brown.edu/confluence/display/BN0193S04/Astrocytes.

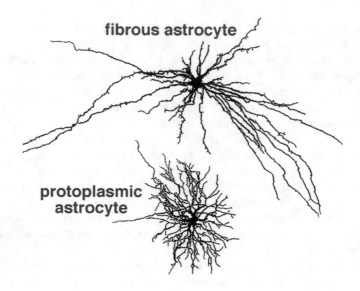

Figure 4.14: astrocytes

Statistical issues in estimating d_K were considered in [Hall 93] for the curve $X(t)$ over the interval $0 < t < 1$, where $X(t)$ is a stochastic process observed at N regularly spaced points. Group these N points into J contiguous blocks of width r. Let $\mathbb{N}_j$ be the points in block j and define $L_j \equiv \min_{j \in \mathbb{N}_j} X(j)$ and $U_j \equiv \max_{j \in \mathbb{N}_j} X(j)$. Then $A = \sum_{1 \le j \le J} r\,(U_j - L_j)$ is an approximation to the area of the dilation Ω_r, and $2 - (\log A)/(\log r)$ estimates d_K. It was shown in [Hall 93] that this estimator suffers from excessive bias, and a regression-based estimator based on least squares was proposed.[11]

---

[11] The bias of an estimator is the difference between the expected value of the estimator and the true value of the parameter being estimated.

Chapter 5

Hausdorff, Similarity, and Packing Dimensions

There is a fifth dimension, beyond that which is known to man. It is a dimension as vast as space and as timeless as infinity. It is the middle ground between light and shadow, between science and superstition.
Rod Sterling (1924–1975), American screenwriter, playwright, television producer, and narrator, best known for his television series "The Twilight Zone", which ran from 1959 to 1964

In this chapter we consider three fractal dimensions of a geometric object $\Omega \subset \mathbb{R}^E$:

the *Hausdorff dimension*, a generalization of the box counting dimension, which has enormous theoretical importance, even though it has been rarely used for computational purposes,

the *similarity dimension*, which provides a very useful way to calculate the box counting dimension without actually performing box counting, and

the *packing dimension*, which provides a "dual" way of looking at dimension.

This chapter lays the foundation for our study of the similarity dimension of a network (Chap. 13) and our study of the Hausdorff dimension of a network (Chap. 18).

5.1   Hausdorff Dimension

In this section we study the Hausdorff dimension of a bounded set $\Omega \subset \mathbb{R}^E$. Recall first that Definition 4.6 of d_B (Sect. 4.2) allows us to assume that each box in the covering of Ω has the same size s, where $s \to 0$. Therefore, d_B, which assigns the same "weight" to each non-empty box in the covering of Ω, "may be thought of as indicating the efficiency with which a set may be covered by small sets of equal size" [Falconer 03]. In contrast, the *Hausdorff dimension* assumes a covering of Ω by sets of size at most s, where $s \to 0$, and assigns a weight to each set that depends on the size of the set. The Hausdorff dimension was introduced in 1919 by Felix Hausdorff [Hausdorff 19], and historically preceded the box counting dimension.[1] Our presentation is based on [Farmer 83, Schleicher 07, Theiler 90]. Let $(\mathcal{X}, dist(\cdot,\cdot))$ be a metric space and let Ω be a bounded subset of $\mathcal{X}$.

Definition 5.1 For $s > 0$, an s-covering of Ω is a finite collection of J sets $\{X_1, X_2, \cdots, X_J\}$ that cover Ω (i.e., $\Omega \subseteq \cup_{j=1}^J X_j$) such that for each j we have $diam(X_j) \leq s$.   □

In Definition 5.1, we follow [Falconer 03] and require $diam(X_j) \leq s$; we could alternatively have followed [Schleicher 07] and required $diam(X_j) < s$, so strict inequality holds. A definition using strict inequality will be useful in our presentation in Chap. 7 of the network box counting dimension. Let $\mathcal{C}(s)$ be the set of all s-coverings of Ω. For $s > 0$ and for $d \geq 0$, define

$$v(\Omega, s, d) \equiv \inf_{\mathcal{C}(s)} \sum_{j=1}^J \big(diam(X_j)\big)^d, \qquad (5.1)$$

where the infimum is over all s-coverings $\mathcal{C}(s)$ of Ω. We take the infimum since the goal is to cover Ω with small sets X_j as efficiently as possible.

We can think of $v(\Omega, s, d)$ as the d-dimensional volume of Ω. For all values of d other than one critical value, the limit $\lim_{s \to 0} v(\Omega, s, d)$ is either 0 or ∞. For example, suppose we cover the unit square $[0,1] \times [0,1]$ by small squares of side length s. We need $1/s^2$ small squares, the diameter of each square is $\sqrt{2}s$, and $v(\Omega, s, d) = (1/s^2)(\sqrt{2}s)^d = \sqrt{2}^d s^{d-2}$. We have

$$\lim_{s \to 0} s^{d-2} = \begin{cases} \infty & \text{if } d < 2 \\ 1 & \text{if } d = 2 \\ 0 & \text{if } d > 2. \end{cases}$$

Thus, for example, if $d = 3$, then the unit square $[0,1] \times [0,1]$ has zero volume; if $d = 1$, then the unit square has infinite length.

---

[1]The idea of defining measures using covers of sets was introduced in 1914 by Constantin Carathéodory [Falconer 03]. Due to the contributions of Abram Samoilovitch Besicovitch, the Hausdorff dimension is sometimes called the *Hausdorff–Besicovitch* dimension. "For a long time, Besicovitch was the author or co-author of nearly every paper on [fractal dimensions]. While Hausdorff is the father of nonstandard dimension, Besicovitch made himself its mother." [Mandelbrot 83b].

For a given d, as s decreases, the set of available s-coverings of Ω shrinks, so $v(\Omega, s, d)$ increases as s decreases. Thus

$$v^\star(\Omega, d) \equiv \lim_{s \to 0} v(\Omega, s, d) \qquad (5.2)$$

always exists in $[0, \infty) \cup \{\infty\}$; that is, $v^\star(\Omega, d)$ might be ∞. The value $v^\star(\Omega, d)$ is called the d-dimensional Hausdorff measure of Ω. For the case where $\mathcal{X} = \mathbb{R}^E$ and the distance metric $dist(\cdot, \cdot)$ is the Euclidean distance and d is a positive integer, $v^\star(\Omega, d)$ is within a constant scaling factor of the Lebesgue measure of Ω.[2] If Ω is countable, then $v^\star(\Omega, d) = 0$.

Since for each fixed $s < 1$ the function $v(\Omega, s, d)$ is non-increasing with d, then $v^\star(\Omega, d)$ is also non-increasing with d [Falconer 03]. For $d \geq 0$ and $d' \geq 0$, definition (5.2) implies [Schleicher 07]

If $v^\star(\Omega, d) < \infty$ and $d' > d$, then $v^\star(\Omega, d') = 0$.

If $v^\star(\Omega, d) > 0$ and $d' < d$, then $v^\star(\Omega, d') = \infty$.

These two assertions imply the existence of a unique value of d, called the *Hausdorff dimension* of Ω and denoted by $d_H(\Omega)$, such that $v^\star(\Omega, d) = \infty$ for $d < d_H(\Omega)$ and $v^\star(\Omega, d) = 0$ for $d > d_H(\Omega)$. If there is no ambiguity about the set Ω, for brevity we write d_H rather than $d_H(\Omega)$. The following alternative succinct definition of d_H was provided in [Falconer 03].

Definition 5.2

$$d_H \equiv \inf\{d \geq 0 \,|\, v^\star(\Omega, d) = 0\}. \qquad \square \qquad (5.3)$$

In [Mandelbrot 83b] (p. 15), the term "fractal dimension" is used to refer to the Hausdorff dimension; we will use the term "fractal dimension" only in a generic manner, to refer to any fractal dimension. The Hausdorff dimension d_H might be zero, positive, or ∞. A set $\Omega \subset \mathbb{R}^E$ with $d_H < 1$ is totally disconnected [Falconer 03]. (A set Ω is totally disconnected if for $x \in \Omega$, the largest connected component of Ω containing x is x itself.) If $\Omega \subset \mathbb{R}^E$ is an open set, then $d_H = E$. If Ω is countable, then $d_H = 0$. A function $f : \Omega \to \mathbb{R}^E$ is *bi-Lipschitz* if for some positive numbers α and β and for each $x, y \in \Omega$, we have

$$\alpha \|x - y\| \leq |f(x) - f(y)| \leq \beta \|x - y\|.$$

If $f : \Omega \to \Omega$ is bi-Lipschitz, then $d_H\big(f(\Omega)\big) = d_H(\Omega)$; that is, the Hausdorff dimension of Ω is unchanged by a bi-Lipschitz transformation. In particular, the Hausdorff dimension of a set is unchanged by rotation or translation.

If $\widetilde{\Omega}$ is a measurable subset of Ω, then we can similarly form $v^\star(\widetilde{\Omega}, d)$, and in particular, choosing $d = d_H(\Omega)$ we can form $v^\star(\widetilde{\Omega}, d_H(\Omega))$. The function

$$v^\star\big(\,\cdot\,, d_H(\Omega)\big),$$

---

[2]The Lebesgue measure μ_L of a bounded subset $B \subset \mathbb{R}^E$ corresponds to the usual geometric notion of volume, so the Lebesgue measure of the set $\{x \in \mathbb{R}^E \,|\, L_i \leq x_i \leq U_i\}$ is $\prod_{i=1}^E (U_i - L_i)$ [Schleicher 07].

which assigns a value to measurable subsets of Ω, is a measure concentrated on Ω; this measure is called the *Hausdorff measure* [Grassberger 85].

For ordinary geometric objects, the Hausdorff and box counting dimensions are equal: $d_H = d_B$. However, in general we only have $d_H \leq d_B$, and there are many examples where this inequality is strict [Falconer 03]. To prove this inequality, we will use Definition 4.6 of d_B with $B(s)$ defined as the minimal number of boxes of size s needed to cover Ω. Recall that the lower box counting dimension $\underline{d_B}$ is defined by (4.2) and if d_B exists then $\underline{d_B} = d_B$.

Theorem 5.1 $d_H \leq \underline{d_B}$.

Proof [Falconer 03] Suppose Ω can be covered by $B(s)$ boxes of diameter s. By (5.1), for $d \geq 0$ we have $v(\Omega, s, d) \leq B(s)s^d$. Since $v^\star(\Omega, d) = \infty$ for $d < d_H$, then for $d < d_H$ we have $v^\star(\Omega, d) = \lim_{s\to 0} v(\Omega, s, d) > 1$. Hence for $d < d_H$ and s sufficiently small, $B(s)s^d \geq v(\Omega, s, d) > 1$, which implies $\log B(s) + d\log s > 0$. Rewriting this as $d < \log B(s)/\log(1/s)$, it follows that $d \leq \liminf_{s\to 0} \log B(s)/\log(1/s) = \underline{d_B}$. Since this holds for $d < d_H$, then $d_H \leq \underline{d_B}$. $\square$

Example 5.1 Let Ω be the unit square in $\mathbb{R}^2$. Cover Ω with J^2 squares, each with side length $s = 1/J$. Using the L_1 metric (i.e., $\|x\| = \max_{i=1,2} |x_i|$), the diameter of each box is $1/J$. We have $v^\star(\Omega, d) \leq \sum_{j=1}^{J^2}(1/J)^d = J^2(1/J)^d = J^{2-d}$, and $\lim_{J\to\infty} J^{2-d} = \infty$ for $d < 2$ and $\lim_{J\to\infty} J^{2-d} = 0$ for $d > 2$. Since $v^\star(\Omega, d) = 0$ for $d > 2$, then by Definition 5.2 we have $d_H \leq 2$. $\square$

Example 5.2 [Schleicher 07] If Ω is an E-dimensional hypercube, then the number $B(s)$ of boxes of size s needed to cover Ω satisfies $B(s) \leq c(1/s)^E$ for some constant c, so

$$v(\Omega, s, d) \leq B(s)s^d \leq c(1/s)^E s^d = cs^{d-E}.$$

We have $\lim_{s\to 0} cs^{d-E} = 0$ if $d > E$, so $v^\star(\Omega, d) = 0$ if $d > E$. Thus $d_H \leq E$. $\square$

Self-similar sets provide an upper bound on d_H [Schleicher 07]. For suppose Ω is the union of K pairwise disjoint subsets, each of which is obtained from Ω by scaling with the factor $\alpha < 1$, followed by translation and/or rotation. In turn, suppose each instance of $\alpha\Omega$ is decomposed into K objects, where each sub-object is a translation and possible rotation of the set $\alpha^2\Omega$, and this construction continues infinitely many times. Let $\Delta = diam(\Omega)$. Then Ω can be covered by one box of size Δ, so $v^\star(\Omega, d) \leq \Delta^d$. Or Ω can be covered by K boxes of size $\alpha\Delta$, so $v^\star(\Omega, d) \leq K(\alpha\Delta)^d$. Or Ω can be covered by K^2 boxes of size $\alpha^2\Delta$, so $v^\star(\Omega, d) \leq K^2(\alpha^2\Delta)^d$. In general, Ω can be covered by K^t boxes of size $\alpha^t\Delta$, so $v^\star(\Omega, d) \leq K^t(\alpha^t\Delta)^d$. For any d we have

$$v^\star(\Omega, d) \leq \lim_{t\to\infty} K^t(\alpha^t\Delta)^d = \lim_{t\to\infty}(K\alpha^d)^t\Delta^d$$

and $\lim_{t\to\infty}(K\alpha^d)^t\Delta^d = 0$ if $K\alpha^d < 1$. We rewrite this last inequality as $d > \log K/\log(1/\alpha)$. Thus $d_H \leq \log K/\log(1/\alpha)$. However, this argument provides no information about a lower bound for d_H. For example, if the self-similar set Ω is countable, then its Hausdorff dimension is 0. The above examples illustrate that it is easier to obtain an upper bound on d_H than a lower bound. Obtaining a lower bound on d_H requires estimating all possible coverings of Ω. Interestingly, there is a way to compute a lower bound on the box counting dimension of a network (Sect. 8.8).

In [Manin 06] it was observed that "the general existence of d_H is a remarkable mathematical phenomenon, akin to those that appear in the description of phase transitions and critical exponents in physics." (We will encounter phase transitions and critical exponents in Chap. 21.) Manin also provided a slightly different definition of the Hausdorff dimension of a set $\Omega \subset \mathbb{R}^E$. Given that the volume $V_E(r)$ of a ball with radius r in E-dimensional Euclidean space is[3]

$$V_E(r) = \frac{\left(\Gamma(1/2)\right)^E}{\Gamma\left(1 + (E/2)\right)} r^E,$$

where Γ is the Gamma function (Sect. 23.2), we *declare* that, for any real number d, the volume $V_d(r)$ of the d-dimensional ball with radius r is

$$V_d(r) = \frac{\left(\Gamma(1/2)\right)^d}{\Gamma\left(1 + (d/2)\right)} r^d.$$

We cover Ω by a finite collection $\{X_1, X_2, \cdots, X_J\}$ of J balls, where the radius of ball X_j is r_j. Define

$$V(d) = \lim_{r \to 0} \inf_{r_j < r} \sum_{j=1}^{J} V_d(r_j).$$

Then d_H is the unique value of d for which $V(d) = 0$ for $d > d_H$ and $V(d) = \infty$ for $d < d_H$. This definition of d_H is equivalent to Definition 5.2.

In [Bez 11] it was observed that d_H is a local, not global, property of Ω, since it is defined in terms of a limit as $s \to 0$. That is, even though d_H is defined in terms of a cover of Ω by sets of diameter at most s, since $s \to 0$ then d_H is still a local property of Ω. The term "roughness" was used in [Bez 11] to denote that characteristic of Ω which is quantified by d_H. Also, [Bez 11] cautioned that the fact that, while fractal sets are popular mainly for their self-similar properties, computing a fractal dimension of Ω does *not* imply that Ω is self-similar. In applications of fractal analysis to ecology, there are many systems which exhibit roughness but not self-similarity, but the systems of strong interest exhibit either self-similarity or they exhibit self-similarity and roughness. However, assessing self-similarity is a challenging if not impossible task without an a priori reason to believe that the data comes from a model exhibiting self-similarity (e.g. fractionary Brownian processes). The notion that the ability to compute a fractal dimension does *not* automatically imply self-similarity holds not only for a geometric object, but also for a network (Chaps. 18 and 21).

---

[3]We use r rather than s here since r is a radius rather than a box size. The distinction between a ball of radius r and a box of size s will be very important when we consider box counting for networks (Chap. 7).

The Hausdorff dimension is the basis for some differing definitions of a fractal. For example, [Manin 06] mentioned a definition of Mandelbrot that a set is a fractal if its Hausdorff dimension is not an integer. Another definition proposed by Mandelbrot is that a set is a fractal if its Hausdorff dimension exceeds its topological dimension; this is the definition preferred in [Schleicher 07]. Interestingly, Mandelbrot observed that this definition, while rigorous, is tentative. He wrote [Mandelbrot 85] "the Hausdorff-Besicovitch dimension is a very nonintuitive notion; it can be of no use in empirical work, and is unduly complicated in theoretical work, except in the case of self-similar fractals." In his landmark paper [Mandelbrot 67] on the length of the coast of Britain, Mandelbrot warned that

> The concept of "dimension" is elusive and very complex, and is far from exhausted by [these simple considerations]. Different definitions frequently yield different results, and the field abounds in paradoxes. However, the Hausdorff-Besicovitch dimension and the capacity dimension, when computed for random self-similar figures, have so far yielded the same value as the similarity dimension.[4]

In [Jelinek 06] it was observed that although the Hausdorff dimension is not usable in practice, it is important to be aware of the shortcomings of methods, such as box counting, for estimating the Hausdorff dimension. In studying quantum gravity, [Ambjorn 92] wrote that the Hausdorff dimension is a very rough measure of geometrical properties, which may not provide much information about the geometry.[5] The distinction between d_H and d_B is somewhat academic, since very few experimenters are concerned with the Hausdorff dimension. Similarly, [Badii 85] observed that, even though there are examples where $d_H \neq d_B$, "it is still not clear whether there is any relevant difference in physical systems."[6] In practice the Hausdorff dimension of a geometric object has been rarely used, while the box counting dimension has been widely used, since it is rather easy to compute.

So far, we have studied two fractal dimensions: the box counting dimension and the Hausdorff dimension. Although we have yet to introduce other important fractal dimensions (e.g., the similarity, correlation, and information dimensions), it is useful to now examine the list in [Falconer 03] of properties

---

[4]The *capacity dimension* is Mandelbrot's name for the box counting dimension.

[5]As described for the lay person in [Rovelli 16], two pillars of physics, namely general relativity and quantum mechanics, work remarkably well. Yet they contradict each other, since general relativity describes the world as a curved space where everything is continuous, while quantum mechanics describes a flat space where quanta of energy leap. Quantum gravity refers to the attempt to find a theory to resolve this paradox. In particular, the central result of "loop quantum gravity" is that space is not continuous, but rather is composed of extremely minute grains, or "atoms of space", each a billion billion times smaller than the smallest atomic nuclei. The theory describes these grains, and their evolution. The term "loop" is used since these grains "are linked to one another, forming a network of relations that weaves the texture of space, like the rings of a finely woven immense chain mail," e.g., as worn by medieval knights in the days of King Arthur.

[6]Such an example is the set of all rational numbers within the closed interval $[0, 1]$ for which we have $d_H = 0$ and $d_B = 1$.

that should be satisfied by any definition of *dimension*. Consider the function $dim(\cdot)$ which assigns a nonnegative number to each $\Omega \subseteq \mathbb{R}^E$. To be called a *dimension* function, $dim(\cdot)$ should satisfy the following for any subsets $\mathcal{X}$ and $\mathcal{Y}$ of Ω:

> **monotonicity:** If $\mathcal{X} \subset \mathcal{Y}$ then $dim(\mathcal{X}) \leq dim(\mathcal{Y})$.
> **stability:** If $\mathcal{X}$ and $\mathcal{Y}$ are subsets then $dim(\mathcal{X} \cup \mathcal{Y}) = \max\big(dim(\mathcal{X}),$ $dim(\mathcal{Y})\big)$.
> **countable stability:** If $\{\mathcal{X}_i\}_{i=1}^{\infty}$ is an infinite collection of subsets then $dim(\bigcup_{i=1}^{\infty} \mathcal{X}_i) = \sup_i dim(\mathcal{X}_i)$.
> **countable sets:** If $\mathcal{X}$ is finite or countably infinite then $dim(\mathcal{X}) = 0$.
> **open sets:** If $\mathcal{X}$ is an open subset of $\mathbb{R}^E$ then $dim(\mathcal{X}) = E$.
> **smooth manifolds:** If $\mathcal{X}$ is a smooth (i.e., continuously differentiable) E-dimensional manifold (curve, surface, etc.) then $dim(\mathcal{X}) = E$.

Falconer observed that all definitions of dimension are monotonic, most are stable, but some common definitions do not exhibit countable stability and may have countable sets of positive dimension. The *open sets* and *smooth manifolds* conditions ensure that the classical definition of dimension is preserved. The *monotonicity* condition was used in [Rosenberg 16b] to study the correlation dimension of a rectilinear grid network (Sect. 11.4). All the above properties are satisfied by the Hausdorff dimension d_H. As to whether these properties are satisfied by the box counting dimension, we have [Falconer 03] (*i*) the lower and upper box counting dimensions are monotonic; (*ii*) the upper box counting dimension is finitely stable, i.e., $\overline{d_B}(\Omega_1 \cup \Omega_2) = \max\{\overline{d_B}(\Omega_1), \overline{d_B}(\Omega_2)\}$; (*iii*) the box counting dimension of a smooth E-dimensional manifold in $\mathbb{R}^E$ is E; also, the lower and upper box counting dimensions are bi-Lipschitz invariant.

5.2 Similarity Dimension

The next definition of dimension we study is the *similarity* dimension [Mandelbrot 83b], applicable to an object which looks the same when examined at any scale. We have already encountered, in the beginning of Chap. 3, the idea of self-similarity when we considered the Cantor set. For self-similar fractals, the similarity dimension is a way to compute a fractal dimension without computing the slope of the $\log B(s)$ versus $\log s$ curve.

In [Mandelbrot 67], Mandelbrot studied geometric figures that can be decomposed into K parts such that each part is similar to the whole by the ratio α. For example, consider Fig. 5.1, taken from [Mandelbrot 67]. Start with the segment $[0, 1]$ and from it create a figure with $K = 8$ line segments, each of which is one-fourth the length of the original segment, so $\alpha = 1/4$. In iteration t, use the same process to replace each segment of length $1/4^t$ with $K = 8$ segments, each of length $1/4^{t+1}$. In the limit, the curve has dimension $d = -\log K / \log \alpha = \log 8 / \log 4$.

Generalizing this notion from a line to any geometric object, suppose Ω can be decomposed into K pairwise disjoint, or "just touching", identical sub-

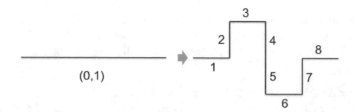

Figure 5.1: Self-similarity with $K = 8$ and $\alpha = 1/4$

objects, each of which is a reduced version of Ω. By "just touching" sub-objects we mean that any two sub-objects intersect only at a set of measure zero (e.g., they intersect only at a point or at a line). By a reduced version of Ω we mean a translation and possibly rotation of the set $\alpha\Omega$, for some α satisfying $0 < \alpha < 1$. In turn each instance of $\alpha\Omega$ is decomposed into K objects, where each sub-object is a translation and possible rotation of the set $\alpha^2\Omega$. This generic construction is illustrated in Fig. 5.2. This construction corresponds to an iterated function

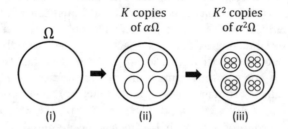

Figure 5.2: Three generations of a self-similar fractal

system (Sect. 3.3) for which each of the K functions $f_k : \mathcal{H}(\mathcal{X}) \to \mathcal{H}(\mathcal{X})$ uses the same contraction factor α, and the functions f_k differ only in the amount of rotation or translation they apply to a point in $\mathcal{H}(\mathcal{X})$ (recall that a point in $\mathcal{H}(\mathcal{X})$ is a non-empty compact subset of a metric space $\mathcal{X}$). The IFS theory guarantees the existence of an attractor Ω of the IFS. The heuristic argument (not a formal proof) leading to the similarity dimension is as follows. Let $B(s)$ be the minimal number of boxes of size s needed to cover Ω.

Assumption 5.1 For some positive number d_B, some c, and all sufficiently small positive s, we have

$$B(s) = cs^{-d_B}. \quad \square \tag{5.4}$$

Let $B_\alpha(s)$ be the minimal number of boxes of size s needed to cover $\alpha\Omega$. Since Ω can be decomposed into K instances of $\alpha\Omega$, then for small s we have

$$B(s) = KB_\alpha(s). \tag{5.5}$$

Since $\alpha\Omega$ is a scaled down copy of Ω, the minimal number of boxes of size s needed to cover $\alpha\Omega$ is equal to the the minimal number of boxes of size s/α needed to cover Ω, that is,

$$B_\alpha(s) = B(s/\alpha). \tag{5.6}$$

To illustrate (5.6), let Ω be a square with side length L. If we cover $\alpha\Omega$ with boxes of size s, the minimal number of boxes required is $\left((\alpha L)^2\right)/s^2 = \alpha^2(L/s)^2$. If we cover Ω with boxes of size s/α, the minimal number of boxes required is $L^2/\left((s/\alpha)^2\right) = \alpha^2(L/s)^2$.

Since (5.4) is assumed to hold as an equality, from (5.4), (5.5), and (5.6), we have

$$B(s) = cs^{-d_B} = KB_\alpha(s) = KB(s/\alpha) = Kc(s/\alpha)^{-d_B}$$

so $cs^{-d_B} = Kc(s/\alpha)^{-d_B}$, and this equality can be written as

$$d_B = \frac{\log K}{\log(1/\alpha)}. \tag{5.7}$$

The ratio (5.7) uses only K, the number of copies of Ω, and α, the scaling factor; it has no explicit dependence on s or $B(s)$. For this reason, the ratio deserves its own name and symbol.

Definition 5.3 For a self-similar fractal Ω constructed by taking K pairwise disjoint or "just touching" instances of Ω, each scaled by the factor $\alpha < 1$, the *similarity dimension* d_S is defined by

$$d_S \equiv \frac{\log K}{\log(1/\alpha)}. \qquad \square \tag{5.8}$$

Example 5.3 Consider the Sierpiński triangle of Fig. 3.3. In each iteration we shrink the object by the factor $\alpha = 1/2$, and take $K = 3$ copies of the reduced figure. From (5.8) we have $d_S = \log 3/\log 2$. $\qquad \square$

Example 5.4 Consider the Sierpiński carpet of Fig. 4.10. In each iteration we shrink the object by the factor $\alpha = 1/3$, and take $K = 8$ copies of the reduced figure. Thus $d_S = \log 8/\log 3$. $\qquad \square$

Example 5.5 Consider the Cantor set of Fig. 3.2. In each iteration we shrink the object by the factor $\alpha = 1/3$, and take $K = 2$ copies of the reduced figure. Thus $d_S = \log 2/\log 3$. $\qquad \square$

Example 5.6 We can generalize the Cantor set construction, as described in [Schleicher 07]. We start with the unit interval, and rather than remove the middle third, we remove a segment of length $1 - 2x$ from the middle, where $x \in (0, 1/2)$. This leaves two segments, each of length x. In subsequent iterations, we remove from each segment an interval in the middle with the same fraction of length. Let C_t be the pre-fractal set obtained after t iterations of this construction. From (5.8) we have $d_S = \log 2/\log(1/x)$. Since $d_S \to 1$ as $x \to 1/2$ and $d_S \to 0$ as $x \to 0$, the similarity dimension can assume any value in the open interval $(0, 1)$. $\qquad \square$

Example 5.7 Consider the Cantor dust of Fig. 4.12. This figure can be described as the Cartesian product $\mathcal{C} \times \mathcal{C}$ of the Cantor set $\mathcal{C}$ with itself. We construct Cantor dust by taking, in generation t, the Cartesian product $\mathcal{C}_t \times \mathcal{C}_t$, where $\mathcal{C}_t$ is defined in Example 5.6 above. The similarity dimension of the generalized Cantor dust is $d_S = \log(2^2)/\log(1/x)$, where $x \in (0,1)$. Since $d_S \to 2$ as $x \to 1/2$ and $d_S \to 0$ as $x \to 0$, the similarity dimension can assume any value in the open interval $(0,2)$. $\square$

Example 5.8 Two-dimensional Cantor dust can be generalized further [Schleicher 07]. We can take the E-fold Cartesian product

$$\underbrace{\mathcal{C}_t \times \mathcal{C}_t \times \ldots \times \mathcal{C}_t}_{E \text{ times}}$$

of C_h with itself. For this E-dimensional generalized Cantor dust we have $d_S = \log(2^E)/\log(1/x)$, where $x \in (0,1)$. Since $d_S \to E$ as $x \to 1/2$ and $d_S \to 0$ as $x \to 0$, the similarity dimension can assume any value in the open interval $(0,E)$. This example demonstrates why defining a fractal to be a set with a non-integer dimension may not be a particularly good definition. It seems unreasonable to consider E-dimensional generalized Cantor dust to be a fractal only if $\log(2^E)/\log(1/x)$ is not an integer. $\square$

Example 5.9 Wool fiber (from sheep) has remarkable properties, and it is difficult to synthesize a material like wool fiber that has perspiration and moisture absorption and warmth retention. The structure of wool fiber is rather complex, with several hierarchical levels. At the lowest hierarchy level (call it level 0), three α-helix protein macromolecules are compactly twisted together to form a level-1 *protofibril*. Then eleven level-1 protofibrils form a level-2 *microfibril*. Then multiple level-2 microfibrils form a level-3 macrofibril, multiple level-3 macrofibris form a level-4 cell, and finally multiple level-4 cells form a wool fiber. Using (5.8), the similarity dimension d_S was calculated in [Fan 08] for the three lowest levels:

$$d_S = \frac{\log\left(\text{number of lowest level units within the higher level unit}\right)}{\log\left(\text{ratio of linear size of higher level unit to the lowest level unit}\right)}.$$

Considering levels 0 and 1, the ratio of the protofibril diameter to the α-helix diameter is 2.155, and there are 3 α-helix macromolecules per protofibril. Thus the first similarity dimension is $d^{0\leftrightarrow1} = \log 3/\log 2.155 = 1.431$. Considering levels 0 and 2, the ratio of the microfibril diameter to the α-helix diameter is 8.619, and there are 33 α-helix macromolecules per microfibril. Thus the next similarity dimension is $d^{0\leftrightarrow2} = \log 33/\log 8.619 = 1.623$. Finally, considering levels 0 and 3, the ratio of the macrofibril diameter to the α-helix diameter is 17.842, and there are 99 α-helix macromolecules per microfibril. Thus that similarity dimension is $d^{0\leftrightarrow3} = \log 99/\log 17.842 = 1.595$. The authors noted that $d^{0\leftrightarrow2}$ and $d^{0\leftrightarrow3}$ are close to the "golden mean" value $(1+\sqrt{5})/2 \approx 1.618$, and conjectured that this is another example of the significant role of the golden mean in natural phenomena. $\square$

The above discussion of self-similar fractals assumed that each of the K instances of Ω is reduced by the same scaling factor α. We can generalize this to the case where Ω is decomposed into K non-identical parts, and the scaling factor α_k is applied to the k-th part, where $0 < \alpha_k < 1$ for $1 \le k \le K$. Some or all of the α_k values may be equal; the self-similar case is the special case where all the α_k values are equal. The construction for the general case corresponds to an IFS for which each of the K functions $f_k : \mathcal{H}(\mathcal{X}) \to \mathcal{H}(\mathcal{X})$ does not necessarily use the same contraction factor, but aside from the possibly different contraction factors, the functions f_k differ only in the amount of rotation or translation they apply to a point in $\mathcal{H}(\mathcal{X})$. That is, up to a constant scaling factor, the f_k differ only in the amount of rotation or translation.

To compute the similarity dimension d_S of Ω in this case, assume that any two of the K parts are disjoint, or possibly just touching (as defined above). It was proved in [Hutchinson 81] that d_S is the unique value satisfying

$$\sum_{k=1}^{K} \alpha_k^{d_S} = 1. \tag{5.9}$$

This equation is known as the *Moran equation* and dates from 1946. If all the α_k values are equal to the common value α, (5.9) simplifies to $K\alpha^{d_S} = 1$, yielding $d_S = -\log K / \log \alpha$, which is precisely (5.8).

An informal proof of (5.9), provided in [Barnsley 12, Tél 88], yields considerable insight. Suppose Ω can be decomposed into the K parts Ω_k, $1 \leq k \leq K$, where part k is obtained by scaling Ω by α_k. Assume any two parts are disjoint or just touching. Let $B_k(s)$ be the minimal number of boxes of size s needed to cover Ω_k. For $s \ll 1$ we have

$$B(s) = \sum_{k=1}^{K} B_k(s). \tag{5.10}$$

From the similarity property (5.6) we have

$$B_k(s) = B(s/\alpha_k) = c \left(\frac{s}{\alpha_k} \right)^{-d_B}, \tag{5.11}$$

where the second equality holds from (5.4) applied to Ω_k. From (5.10) and (5.11) we have

$$B(s) = c\, s^{-d_B} = \sum_{k=1}^{K} B(s/\alpha_k) = \sum_{k=1}^{K} c \left(\frac{s}{\alpha_k} \right)^{-d_B}.$$

This yields

$$c\, s^{-d_B} = \sum_{k=1}^{K} c \left(\frac{s}{\alpha_k} \right)^{-d_B}$$

which simplifies to

$$\sum_{k=1}^{K} \alpha_k^{d_B} = 1. \tag{5.12}$$

Since the exponent in (5.12) is derived from a similarity argument, we will refer to this exponent as the similarity dimension d_S. There is a unique solution d_S to the Moran equation (5.9). To see this, define

$$f(d) \equiv \sum_{k=1}^{K} \alpha_k^d.$$

Since $0 < \alpha_k < 1$, then $f'(d) < 0$. Hence f is strictly decreasing for $d > 0$. Since $f(0) = K$ and $\lim_{d \to \infty} f(d) = 0$, there is a unique d satisfying $f(d) = 1$.

Example 5.10 Suppose the Cantor set construction of Fig. 3.2 is modified so that we take the first $1/4$ and the last $1/3$ of each interval to generate the next generation intervals. Starting in generation 0 with $[0, 1]$, the generation-1 intervals are $[0, \frac{1}{4}]$ and $[\frac{2}{3}, 1]$. The four generation-2 intervals are (i) $[0, (\frac{1}{4})(\frac{1}{4})]$, (ii) $[\frac{1}{4} - (\frac{1}{3})(\frac{1}{4}), \frac{1}{4}]$, (iii) $[\frac{2}{3}, \frac{2}{3} + (\frac{1}{4})(\frac{1}{3})]$, and (iv) $[1 - (\frac{1}{3})(\frac{1}{3}), 1]$. From (5.12), with $K = 2$, $\alpha_1 = 1/4$, and $\alpha_2 = 1/3$, we obtain $(1/4)^{d_B} + (1/3)^{d_B} = 1$, which yields $d_S \approx 0.560$; this is smaller than the value $d_S = \log 2 / \log 3 \approx 0.631$ obtained for $K = 2$ and $\alpha_1 = \alpha_2 = 1/3$. $\square$

In general, the similarity dimension of Ω is not equal to the Hausdorff dimension of Ω. However, if the following *Open Set Condition* holds, then the two dimensions are equal [Glass 11].

Definition 5.4 [Glass 11] Let $\big(\mathcal{X}, dist(\cdot, \cdot)\big)$ be a metric space, and let Ω be a non-empty compact subset of $\mathcal{X}$. Suppose for the IFS $\{f_k\}_{k=1}^{K}$ with contraction factors $\{\alpha_k\}_{k=1}^{K}$ we have $\Omega = \cup_{k=1}^{K} f_k(\Omega)$. Then the IFS $\{f_k\}_{k=1}^{K}$ satisfies the *Open Set Condition* if there exists a non-empty open set $U \subseteq \mathbb{R}^E$ such that

$$\bigcup_{k=1}^{K} f_k(U) \subseteq U \text{ and } f_i(U) \cap f_j(U) = \emptyset \text{ for } i \neq j. \quad \square \qquad (5.13)$$

In practice, the Open Set Condition can be easy to verify.

Example 5.11 The Cantor Set can be generated by an IFS using two similarities defined on $\mathbb{R}$. Let $f_1(x) = \frac{x}{3}$ and $f_2(x) = \frac{x}{3} + \frac{2}{3}$. The associated ratio list is $\{\frac{1}{3}, \frac{1}{3}\}$. The invariant set of this IFS is the Cantor set. The similarity dimension of the Cantor set satisfies $\left(\frac{1}{3}\right)^{d_S} + \left(\frac{1}{3}\right)^{d_S} = 1$, so $d_S = \log 2 / \log 3$. To verify the Open Set Condition, choose $U = (0, 1)$. Then $f_1(U) = (0, \frac{1}{3})$ and $f_2(U) = (\frac{2}{3}, 1)$, so $f_1(U) \cap f_2(U) = \emptyset$ and $f_1(U) \cup f_2(U) \subset U$. Thus the Open Set Condition holds. $\square$

Example 5.12 Figure 5.3, from [Zhao 16], illustrates the Sierpiński triangle, which is obtained by scaling each triangle by $\alpha = 1/2$ and taking $K = 3$ copies. By (5.8) its similarity dimension is $\log 3 / \log 2$. Treating $x \in \mathbb{R}^2$ as a column vector (i.e., a 2×1 matrix), the IFS generating this fractal uses the three similarities

$$f_1(x) = Mx, \quad f_2(x) = Mx + \begin{pmatrix} \frac{1}{2} \\ 0 \end{pmatrix}, \quad f_3(x) = Mx + \begin{pmatrix} \frac{1}{4} \\ \frac{\sqrt{3}}{4} \end{pmatrix},$$

Figure 5.3: Sierpiński triangle

where

$$M = \begin{pmatrix} \frac{1}{2} & 0 \\ 0 & \frac{1}{2} \end{pmatrix}.$$

To show that this IFS satisfies the Open Set Condition, let U be the open equilateral triangle with corner points $(0,0)$, $\left(\frac{1}{2}, \frac{\sqrt{3}}{2}\right)$, and $(1,0)$. As illustrated in Fig. 5.4, $f_1(U)$ is the open triangle with corner points $(0,0)$, $\left(\frac{1}{4}, \frac{\sqrt{3}}{4}\right)$, and $\left(\frac{1}{2},0\right)$; $f_2(U)$ is the open triangle with corner points $\left(\frac{1}{2},0\right)$, $\left(\frac{3}{4}, \frac{\sqrt{3}}{4}\right)$, and $(1,0)$; and $f_3(U)$ is the open triangle with corner points $\left(\frac{1}{4}, \frac{\sqrt{3}}{4}\right)$, $\left(\frac{1}{2}, \frac{\sqrt{3}}{2}\right)$, and $\left(\frac{3}{4}, \frac{\sqrt{3}}{4}\right)$. Thus $f_i(U) \cap f_j(U) = \emptyset$ for $i \neq j$ and $f_1(U) \cup f_2(U) \cup f_3(U) \subset U$, so the condition holds. $\square$

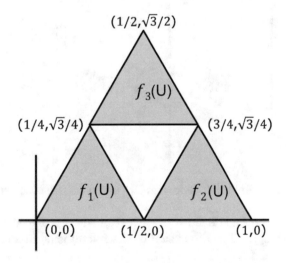

Figure 5.4: The Open Set Condition for the Sierpiński gasket

Example 5.13 An example of an IFS and a set U for which (5.13) does not hold was given by McClure.[7] Pick $r \in (\frac{1}{2}, 1)$. Define $f_1(x) = rx$ and $f_2(x) = rx + (1-r)$. Choosing $U = (0,1)$, we have $f_1(U) = (0,r)$ and $f_2(U) = (1-r,1)$. Since $r \in (\frac{1}{2}, 1)$, the union of $(0,r)$ and $(1-r,1)$ is $(0,1)$ and the intersection of $(0,r)$ and $(1-r,1)$ is the non-empty set $(1-r,r)$, so (5.13) does not hold for this U. Note that blindly calculating the similarity dimension with $K = 2$, we obtain $d_S = \log 2 / \log(1/r) > 1$, and the box counting dimension of $[0,1]$ is 1. □

Exercise 5.1 Construct an IFS and a set U in $\mathbb{R}^E$ for which (5.13) does not hold. □

5.3 Fractal Trees and Tilings from an IFS

Mandelbrot [Mandelbrot 83b] first hypothesized that rivers are fractal, and introduced fractal objects similar to river networks. With the goal of unifying different approaches to the computation of the fractal dimension of a river, [Claps 96] described a tree generation method using an IFS (Sect. 3.3), in which an *initiator* (usually a unit-length line segment) is replaced by a *generator* (which is a tree-like structure of M arcs, each of length L). After a replacement, each segment of the generator becomes an initiator and is replaced by the generator; this process continues indefinitely.

In the construction of [Claps 96], each generator is the composition of a *sinuosity* generator, which turns a straight segment into a jagged segment which emulates the meandering of rivers, and a *topology* generator which emulates the branching of rivers. An example of a sinuosity generator is the well-known Koch curve, created by Helge von Koch in 1904, and illustrated in Fig. 5.5, where the initiator is the straight segment on the left.

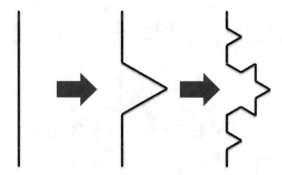

Figure 5.5: Sinuosity generator $\mathcal{I}_U$ based on the Koch fractal

In each iteration, the middle third of each segment is replaced by two sides of an equilateral triangle. The dimension of the fractal generated from this sinu-

---

[7]Example 5.13 expands upon https://math.stackexchange.com/questions/647896.

osity generator is $\log 4/\log 3$. Figure 5.6, from [Claps 96], illustrates a topology generator. In this figure, each of the three branches has length one-fourth the

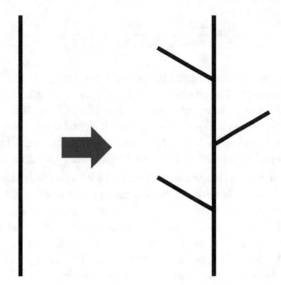

Figure 5.6: Topology generator $\mathcal{I}_T$

length of the initiator on the left. Since the tree on the right has seven equal length segments, then the dimension of this fractal generated from this topology generator is $\log 7/\log 4$. Let $\mathcal{I}_U$ denote the above sinuosity generator, and let $\mathcal{I}_T$ denote the above topology generator. Now define $\mathcal{I} \equiv \mathcal{I}_U\,\mathcal{I}_T$, which means we first apply $\mathcal{I}_T$ and then apply $\mathcal{I}_U$. As applied to the straight line initiator, the result of applying $\mathcal{I}$ is illustrated by Fig. 5.7.

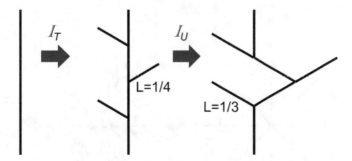

Figure 5.7: $\mathcal{I} \equiv \mathcal{I}_U\,\mathcal{I}_T$ applied to a straight line initiator

The effect of $\mathcal{I}_S$ is to make the length of each of the seven line segments created by $\mathcal{I}_T$ equal to $1/3$. Since in this figure there are seven segments, each of which is one-third the length of the initiator, then this fractal has similarity dimension $\log 7/\log 3$. The similarity dimension of $\mathcal{I}$ is equal to the product of the similarity dimensions of $\mathcal{I}_U$ and $\mathcal{I}_T$; that is,

$$d_S = \frac{\log 7}{\log 3} = \left(\frac{\log 4}{\log 3}\right)\left(\frac{\log 7}{\log 4}\right).$$

This product relationship does not always hold; it does holds if M_U, the number of equal length segments in the sinuosity iterator, is equal to $1/L_T$, where L_T is the length of each segment in the topology iterator [Claps 96]. In the example above, we have $M_U = 4$ and $L_T = 1/4$. A "clean application" of the composition $\mathcal{I}_U\,\mathcal{I}_T$ requires that the initial tree have equal arc lengths, and all angles between arcs be a multiple of $\pi/2$ radians (i.e., all angles must be right angles).

Although the above analysis was presented in [Claps 96] without reference to nodes and arcs, it could have been presented in a network context, by defining an arc to be a segment and a node to be the point where two segments intersect. Suppose $M_U = 1/L_T$, so the similarity dimension of $\mathcal{I}$ is equal to the product of the similarity dimensions of $\mathcal{I}_U$ and $\mathcal{I}_T$. We express this relationship as $d_S(\mathcal{I}) = d_S(\mathcal{I}_U)d_S(\mathcal{I}_T)$. The scaling factor of the IFS is $\alpha = 1/L_T$, and $d_S(\mathcal{I})$ is the fractal dimension of the fractal tree generated by this IFS. For $t > 1$, the tree network $\mathbb{G}_t$ constructed in generation t by this IFS is a weighted network, since each arc length is less than 1.

Example 5.14 Another example of generating a tree from an IFS is the model of [Gao 09] for the structure of goose down. The top of Fig. 5.8 shows the basic motif, in which a line (i) is replaced by the pattern (ii) of 10 lines, each $1/4$ the length of the original line. Then, starting with (iii), we apply the motif,

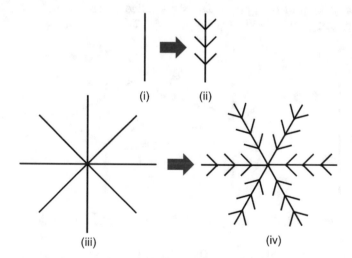

Figure 5.8: IFS for goose down

yielding (iv). The next iteration applies the transformation $(i) \to (ii)$ to each line segment in (iv). If this construction is repeated indefinitely, the resulting fractal tree has similarly dimension $d_S = \log 10 / \log 4 \approx 1.66$.   □

Exercise 5.2 Describe the above goose down construction by an IFS, explicitly specifying the similarities and contraction factors.   □

A Penrose tiling [Austin 05] of $\mathbb{R}^2$ can be viewed as network that spans all of $\mathbb{R}^2$. These visually stunning tilings can be described using an IFS, as described in [Ramachandrarao 00], which considered a particular tiling based on a rhombus, and observed that quasi crystals have been shown to pack in this manner.[8] An interesting feature of this tiling is that the number of node and arcs can be represented using the Fibonacci numbers. Letting $N(s)$ be the number of arcs of length s in the tiling, they defined the fractal dimension of the tiling to be $\lim_{s \to 0} \log N(s) / \log(1/s)$ and called this dimension the "Hausdorff dimension" of the tiling. Since the distance between two atoms in a quasi crystal cannot be zero, when considering the Penrose packing from a crystallographic point of view, a limit has to be placed on the number of iterations of the IFS; for this tiling, the atomic spacing limit was reached in 39 iterations. In [Ramachandrarao 00], a fractal dimension of 1.974 was calculated for the tiling obtained after 39 iterations, and it was observed that the fractal dimension approaches 2 as the number of iterations tends to infinity, implying a "non-fractal space filling structure." Thus [Ramachandrarao 00] implicitly used the definition that a fractal is an object with a non-integer dimension.

Exercise 5.3* Pick a Penrose tiling, and explicitly specify the IFS needed to generate the tiling.   □

5.4   Packing Dimension

The Hausdorff dimension (Sect. 5.1) is based upon covering a geometric object Ω by small sets X_j in the most efficient manner, where "more efficient" means a smaller value of $\sum_{j=1}^{J} \left(diam(X_j) \right)^d$. A "dual" approach to defining the dimension of Ω is to pack Ω with the *largest* number of pairwise disjoint balls whose centers lie in Ω, and where the radius of each ball does not exceed r. Our presentation of the packing dimension is based on [Edgar 08]. By a *ball* we mean a set

$$X(\widetilde{x}, r) \equiv \{x \mid dist(\widetilde{x}, x) \leq r\}$$

with center $\widetilde{x}$ and radius r.

Definition 5.5 Let $\{X(x_j, r_j)\}_{j=1}^{B_P(r)}$ be a finite collection of $B_P(r)$ pairwise disjoint balls such that for each ball we have $x_j \in \Omega$ and $r_j \leq r$.[9] We do not require that $X(x_j, r_j) \subset \Omega$. We call such a collection an *r-packing* of Ω. Let $\mathcal{P}(r)$ denote the set of all *r-packings* of Ω.   □

---

[8]There are many web sites with beautiful picture of Penrose tilings and a discussion of how to construct them. See, e.g., http://mathworld.wolfram.com/PenroseTiles.html.

[9]The subscript "P" in $B_P(s)$ distinguishes $B(s)$, used in defining d_B, from $B_P(s)$.

Example 5.15 Figure 5.9, where Ω is the diamond shaped region, illustrates an r-packing for which $B_P(r) = 4$. Each ball center lies in Ω, but the entire ball

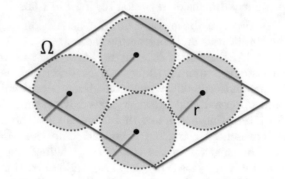

Figure 5.9: Packing the diamond region with four balls

is not contained in Ω.   □

For $d > 0$ and $r > 0$, our goal, roughly speaking, is to maximize

$$\sum_{j=1}^{B_P(r)} \Big[diam \big(X(x_j, r_j) \big) \Big]^d. \tag{5.14}$$

The reason for packing with balls, rather than with arbitrary sets, is illustrated by Fig. 5.10, where Ω is a square in $\mathbb{R}^2$. By choosing long skinny ovals, for $d > 0$

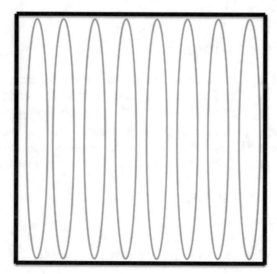

Figure 5.10: Packing a square with long skinny rectangles

we can make the sum (5.14) arbitrarily large. While packing with hypercubes works well for a set $\Omega \subset \mathbb{R}^E$, to have a definition that works in a general metric space, we pack with balls. Since a subset of a metric space may contain no balls at all, we drop the requirement that each ball in the packing be contained in Ω. However, to ensure that the packing provides a measure of Ω, we require the center of each ball in the packing to lie in Ω. The balls in the packing are closed balls, but open balls could also be used.

Maximizing, over all r-packings, the sum (5.14) works well in $\mathbb{R}^E$, since the diameter of a ball of radius r is $2r$. However, there are metric spaces for which this is not true. Thus, rather than maximizing (5.14), we maximize, over all r-packings, the sum

$$\sum_{j=1}^{B_P(r)} (2r_j)^d.$$

That is, we are interested in

$$\sup_{\mathcal{P}(r)} \sum_{j=1}^{B_P(r)} (2r_j)^d, \tag{5.15}$$

where $\sup_{\mathcal{P}(r)}$ denotes the supremum over all r-packings of Ω. As r decreases to 0, the set $\mathcal{P}(r)$ shrinks, so the supremum in (5.15) decreases. Define

$$f(\Omega, d) = \lim_{r \to 0} \sup_{\mathcal{P}(r)} \sum_{j=1}^{B_P(r)} (2r_j)^d. \tag{5.16}$$

There is a critical value $d_0 \in [0, \infty]$ such that

$$f(\Omega, d) = \begin{cases} \infty & \text{if } d < d_0 \\ 0 & \text{if } d > d_0. \end{cases}$$

We do not want to call d_0 the packing dimension of Ω since, as the following example shows, it is possible to have the undesirable outcome that $d_0 > 0$ for a countable set Ω.

Example 5.16 Let $\Omega = \{0, 1, 1/2, 1/3, 1/4, \cdots\}$. Let k be odd, choose $r = 2^{-k}$, and let $J = 2^{(k-1)/2}$. Since

$$\frac{1}{J-1} - \frac{1}{J} > \frac{1}{J^2} = 2r,$$

the J balls with centers 1, 1/2, 1/3, $\cdots$, $1/J$ and radius r are disjoint. Also, $J(2r)^{1/2} = 1$. Choosing $d = 1/2$, we have

$$\sup_{\mathcal{P}(r)} \sum_{j=1}^{B_p(r)} (2r_j)^d > J(2r)^{1/2} = 1.$$

It follows from (5.16) that $f(\Omega, d) \geq 1$. $\square$

The details of the resolution of this problem were provided in [Edgar 08]; here we provide only a brief sketch of the resolution. Define

$$f^\star(\Omega, d) \equiv \inf_{\mathcal{C}} \sum_{C \in \mathcal{C}} f(C, d),$$

where the infimum is over all countable covers $\mathcal{C}$ of Ω by closed sets, and where $f(C, d)$ is defined by (5.16). If we restrict Ω to be a measurable set, then $f^\star(\Omega, d)$ is a measure, called the d-dimensional packing measure. Finally, if Ω is a *Borel* set,[10] there is a critical value $d_P \in [0, \infty]$ such that

$$f^\star(\Omega, d) = \begin{cases} \infty & \text{if } d < d_P \\ 0 & \text{if } d > d_P. \end{cases}$$

The value d_P is called the *packing dimension* of Ω. The packing dimension has many of the same properties as the Hausdorff dimension. For example, if A and B are Borel sets, then $A \subseteq B$ implies $d_P(A) \le d_P(B)$, and $d_P(A \cup B) = \max\{d_P(A), d_P(B)\}$. Also, if $\Omega \subset \mathbb{R}^1$ then the one-dimensional packing measure coincides with the Lebesgue measure [Edgar 08].

Since d_H considers the smallest number of sets that can cover Ω, and d_P considers the largest number of sets that can be packed in Ω, one might expect a duality theory to hold, and it does. It can be shown [Edgar 08] that if Ω is a Borel subset of a metric space, then $d_H \le d_P$. This leads to another suggested definition [Edgar 08] of the term "fractal": the set $\Omega \subset \mathbb{R}^E$ is a fractal if $d_H = d_P$; it was mentioned in [Cawley 92] that this definition is due to S.J. Taylor. Under this definition, the real line $\mathbb{R}$ is a fractal, so this definition is not particularly enlightening.

The packing dimension of a fractal Ω generated by an IFS was considered in [Lalley 88].[11] A finite subset $\widetilde{\Omega}$ of Ω is said to be *r-separated* if $dist(x, y) \ge r$ for all distinct points $x, y \in \widetilde{\Omega}$. In Fig. 5.9, the centers of the four circles are $(2r)$-separated. Let $B_P(r)$ be the maximal cardinality of an r-separated subset of Ω. Then d_P satisfies

$$d_P = -\lim_{r \to 0} \frac{\log B_P(r)}{\log r},$$

provided this limit exists. Recalling that d_S is the similarity dimension defined by (5.8), Lalley proved that if a certain *Strong Open Set Condition* holds then $d_S = d_P = d_B$. The *Strong Open Set Condition* is stronger than the *Open Set Condition* given by Definition 5.4. The Strong Open Set Condition holds for

---

[10]The class of Borel sets is the smallest collection of subsets of $\mathbb{R}^E$ such that (*i*) every open set and every closed set is a Borel set, and (*ii*) the union of every finite or countable collection of Borel sets is a Borel set, and the intersection of every finite or countable collection of Borel sets is a Borel set [Falconer 03]. The class of Borel sets includes all commonly encountered sets; indeed, referring to his book [Falconer 03], Falconer wrote that "Throughout this book, virtually all of the subsets of $\mathbb{R}^E$ that will be of any interest to us will be Borel sets."

[11]As further evidence of the lack of standardization of names of fractal dimensions, in this 1988 paper d_B was called the "covering dimension" and denoted by d_C; however, in this book d_C will always refer to the correlation dimension introduced in Chap. 9. Lalley also mentioned that d_B (using our notation) is usually called the "metric entropy" and that d_P is often called the "capacity."

the Cantor set, the Koch snowflake, and the Sierpiński gasket fractals generated by an IFS.

5.5 When Is an Object Fractal?

We have already discussed some suggested definitions of "fractal." In this section we present, in chronological order, additional commentary on the definition of a "fractal."

By 1983 [Mandelbrot 83a], Mandelbrot had abandoned the definition of a fractal as a set for which $d_H > d_T$, writing

> But how should a fractal set be defined? In 1977, various pressures had made me advance the "tentative definition", that a fractal is "a set whose Hausdorff-Besicovitch dimension strictly exceeds its topological dimension". But I like this definition less and less, and take it less and less seriously. One reason resides in the increasing importance of the "borderline fractals", for example of sets which have the topologically "normal" value for the Hausdorff dimension, but have anomalous values for some other metric dimension. I feel - the feeling is not new, as it had already led me to abstain from defining "fractal" in my first book of 1975 - that the notion of fractal is more basic than any particular notion of dimension. A more basic reason for not defining fractals resides in the broadly-held feeling that the key factor to a set's being fractal is invariance under some class of transforms, but no one has yet pinned this invariance satisfactorily. Anyhow, I feel that leaving the fractal geometry of nature without dogmatic definition cannot conceivably hinder its further development.

An interesting historical perspective on this was provided in [Bez 11], who wrote

> Mandelbrot [Mandelbrot 75] himself, the father of the concept, did not provide the scientific community with a clear and unique definition of fractal in his founding literature which was written in French. This was indeed deliberate as stated by Mandelbrot [Mandelbrot 77] himself in the augmented English version published 2 years later: "the French version deliberately avoided advancing a definition" (p. 294). The mathematical definition finally coined in 1977 stated that "an object is said to be fractal if its Hausdorff-Besicovitch dimension d_H is greater than its topological dimension d_T".

A convenient working definition of fractals was provided in [Theiler 90]:

> Fractals are crinkly objects that defy conventional measures, such as length and area, and are most often characterized by their fractal dimension.

Theiler divided fractals into either solid objects or strange attractors; [Theiler 90] predated the application of fractal methods to networks, with the significant exception of [Nowotny 88], studied in Chap. 12. Examples of solid object fractals are coastlines, electrochemical deposits, porous rocks, spleenwort ferns, and mammalian brain folds.[12] Strange attractors, on the other hand, are conceptual fractal objects that exist in the state space of chaotic dynamical systems (Chap. 9).

Mandelbrot's 1977 definition excluded some self-similar geometric objects which might be deserving of being called fractals. Such an example is the Lévy curve, named after the French mathematician Paul Lévy (1886–1971). The steps for constructing this curve are illustrated in Fig. 5.11. The basic step in the con-

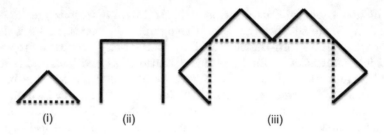

(i)   (ii)   (iii)

Figure 5.11: Constructing the Lévy curve

struction is described by (i) each line segment (the dotted line, with length 1) is replaced by an angled pair of lines (the solid lines, each of length $1/\sqrt{2}$). To generate the Lévy curve using the basic step, start with (ii) and replace each of the three lines by an angled pair of lines, yielding (iii). After 12 iterations, we obtain Fig. 5.12.[13] The similarity dimension is $d_S = \log(2)/\log(1/\sqrt{2}) = 2$, so this curve is "space-filling" in $\mathbb{R}^E$, and $d_B = d_S$. It follows from a theorem in [Falconer 89] that $d_H = d_B$. In [Lauwerier 91], it was deemed "a pity" that, under Mandelbrot's 1977 definition, the Lévy curve does not get to be called a fractal. Lauwerier continued with "Let it be said again: the essential property of a fractal is indefinitely continuing self-similarity. Fractal dimension is just a by-product." In a later edition of *The Fractal Geometry of Nature*, Mandelbrot regretted having proposed a strict definition of fractals [Briggs 92]. In 1989, [Mandelbrot 89] observed that fractal geometry "... is not a branch of mathe-

---

[12]Fractal applications in neuroscience were discussed in [Di Ieva 13]. They quoted [Mandelbrot 67], who observed "Zoologists argue that the proportion of white matter (formed by the neuron axons) to gray matter (where neurons terminate) is approximately the same for all mammals, and that in order to maintain this ratio a large brain's cortex must necessarily become folded. Knowing that the extent of folding is of purely geometric origin relieves Man of feeling threatened by Dolphin or Whale: they are bigger than us but need not be more highly evolved. ··· A quantitative study of such folding is beyond standard geometry but fits beautifully in fractal geometry."

[13]Maurits Escher, the artist known for his recursive geometries, said "Since a long time I am interested in patterns with 'motives' getting smaller and smaller until they reach the limit of infinite smallness" [Briggs 92]. Pictures of fractals occurring in nature, fractals generated by artists, and fractals generated by computer programs can be found in [Briggs 92].

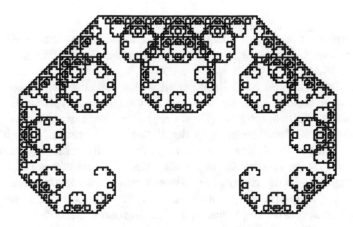

Figure 5.12: The Lévy curve after 12 iterations

matics like, for example, the theory of measure and integration. It fails to have a clean definition and unified tools, and it fails to be more or less self contained."

Making no reference to the concept of dimension, fractals were defined in [Shenker 94] "... as objects having infinitely many details within a finite volume of the embedding space. Consequently, fractals have details on arbitrarily small scales, revealed as they are magnified or approached." Shenker stated that this definition is equivalent to Mandelbrot's definition of a fractal as an object having a fractal dimension greater than its topological dimension. As we will discuss in Sect. 20.4, Shenker argued that fractal geometry does *not* model the natural world; this opinion is most definitely not a common view.

A perspective similar to that of Lauwerier was offered in [Jelinek 98], who noted that not all self-similar objects are fractal. For example, the Pythagoras tree (the tree after 15 generations is shown in Fig. 5.13) is self-similar, has Hausdorff dimension 2, and thus cannot be called a fractal if the defining

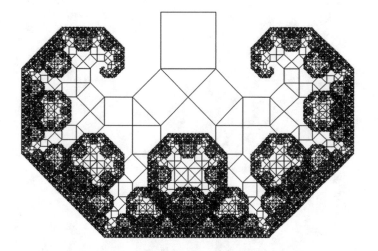

Figure 5.13: Pythagoras tree after 15 iterations

characteristic is a non-integer dimension. The conclusion in [Jelinek 98] was that self-similarity or the fractal dimension are not "absolute indicators" of whether an object is fractal.

It was also observed in [Jelinek 98] that the ability to calculate a fractal dimension of a biological image does not necessarily imply that the image is fractal. Natural objects are not *self-similar*, but rather are *scale invariant* over a range, due to slight differences in detail between iteration (i.e., hierarchy) levels. The limited resolution of any computer screen necessarily implies that any biological image studied is pre-fractal. Also, the screen resolution impacts the border roughness of an object and thus causes the computed fractal dimension to deviate from the true dimension. Thus the computed dimension is not very precise or accurate, and is not significant beyond the second decimal place. However, the computed dimension *is* useful for object classification.

We are cautioned in [Halley 04] that, because fractals are fashionable, there is a natural tendency to see fractals everywhere. Moreover, it was observed in [Halley 04] that the question of deciding whether or not an object is fractal has received little attention. For example, any pattern, whether fractal or not, when analyzed by box counting and linear regression may yield an excellent linear fit, and logarithmic axes tend to obscure deviations from a power-law relationship that might be quite evident using linear scales. In [Halley 04], some sampling and Monte Carlo approaches were proposed to address this question, and it was noted that higher-order Rényi dimensions (Chap. 16) and maximum likelihood schemes can be applied to estimate dimension, but that these methods often utilize knowledge of the process generating the pattern, and hence cannot be used in an "off the shelf" manner as box counting methods can. Indeed, [Mandelbrot 97] observed that fractal geometry "cannot be automated" like analysis of variance or regression, and instead one must always "look at the picture."

Chapter 6

Computing the Box Counting Dimension

My momma always said, "Life was like a box of chocolates. You never know what you're gonna get." from the 1994 movie *Forest Gump*.

From roughly the late 1980s to the mid-1990s, a very large number of papers studied computational issues in determining fractal dimensions of geometric objects, or provided variants of algorithms for calculating the fractal dimensions, or applied these techniques to real-world problems. There are many surveys of results from those years, e.g., [Rigaut 90]. In this chapter we study the computation of the box counting dimension d_B for geometric objects. Thus equipped, we are prepared to study, in Chap. 7, the computation of d_B of a network.

6.1 Lower and Upper Box Sizes

For real-world fractals, the scaling $B(s) \sim s^{-d_B}$ (see (5.4)) holds only over some range of s. To compute d_B, typically $B(s)$ is evaluated over some range of s values, and from a plot of $\log B(s)$ versus $\log s$, linear regression is used to estimate d_B. In this section we consider the range of box sizes that can be used to compute d_B.

It was suggested in [Mandelbrot 83b] that strict scale invariance (i.e., self-similarity) is likely to apply only over a range of sizes for which a particular process determines structure. Similarly, [Berntson 97] observed that unlike ideal, abstract fractals, real-world fractals exhibit self-similarity only over a finite range of scales. For example, a plant root and rhizome[1] system grows by iterative growth of modules, specifically, a branching event and the elongation of

---

[1]A botanical term, referring to an underground plant stem capable of producing the shoot and root systems of a new plant. A rhizome grows perpendicularly to the force of gravity. Examples include ginger, asparagus, and irises.

© Springer Nature Switzerland AG 2020

E. Rosenberg, *Fractal Dimensions of Networks*,

https://doi.org/10.1007/978-3-030-43169-3_6

a meristem.[2] Thus the module size is the minimal scale over which the root system structure exhibits self-similarity, and the maximal scale is the overall size of the entire root system. The same considerations apply to other biological examples such as nerve ganglia and blood vessels.

It is common in the box-counting method to increase the box size s by a factor of 2 in each iteration, so the box sizes are $s_i = s_0 2^i$ for some range of positive integers i, where s_0 is the small initial size [Halley 04, Jelinek 98]. Alternatively, we can start with a large initial size and decrease the box size by a factor of 2 in each iteration. With either choice, the values of $\log s_i$ will be evenly spaced on the horizontal axis.

Suppose the range of box sizes used to compute d_B is $[s_{min}, s_{max}]$. Often this range is quantified by the corresponding number of *decades*, defined as $\log_{10}(s_{max}/s_{min})$ [Panico 95]. Thus one decade corresponds to $s_{max}/s_{min} = 10$, or equivalently, $\ln(s_{max}/s_{min}) = 2.3$; two decades correspond to $s_{max}/s_{min} = 100$, or equivalently, $\ln(s_{max}/s_{min}) = 4.6$, etc. A standard guideline discussed in [Halley 04] is that a linear fit in a log–log plot must be seen over at least two decades to assert fractal behavior. However, actual observations of fractal behavior spanning a scale of more than three decades are rare. When we consider the box counting dimension of networks (Chap. 7), we are often lucky to get a range of box sizes of one decade, much less two or three decades.

As an example of what can go wrong when the scale is less than three decades, [Halley 04] discussed the analysis of Hamburger, who randomly distributed 100 discs of radius 10^{-4} on the unit square. Since the discs are themselves two-dimensional, and since the fractal dimension of a random point pattern on $\mathbb{R}^2$ is also 2, then the pattern of discs is not fractal. However, for box sizes varying from $s = 0.0004$ to $s = 0.026$, a range of 1.8 decades, the plot of $\log B(s)$ versus $\log s$ is almost perfectly linear (suggesting a fractal object) with $d_B < 2$. This "fallacious" detection occurred since the range of box sizes is too small, and [Halley 04] cautioned that a scaling relationship sustained over two orders of magnitude or less is not strong evidence of fractality.

An example of a real-world study with a small range of box sizes is the analysis in [Lenormand 85] of the displacement of one fluid by another in a porous media. In their plot in log–log space, the scale on the horizontal axis ranges from 1.6 to 2.4. Assuming that $\log = \log_{10}$, this is a scale range of $10^{1.6} = 40$ to $10^{2.4} = 251$, which is not even one decade.

A 1998 debate between Mandelbrot and other researchers [Mandelbrot 98] concerned the number of decades needed to establish fractality. Biham et al. argued that the Richardson coastline of Britain study cited by Mandelbrot demonstrates power-law behavior only over about 1.3 decades. Moreover, a well-known study of cloud fractality, claiming fractality over six decades, actually supports fractality over three decades, since the six decades is for a perimeter to area relationship, not for a perimeter to box side length relationship. Mandelbrot's response included the statement that "Many claims that are questioned by Avnir

---

[2] A plant tissue made up of cells that are not specialized for a particular purpose, are capable of dividing any number of times, and can produce cells that specialize to form the fully developed plant tissues and organs (from http://www.wordcentral.com).

et al. are best understood as unfortunate side effects of enthusiasm, imperfectly controlled by refereeing, for a new tool that was (incorrectly) perceived as simple." Mandelbrot's point, which we will explore later in more detail, is that the rush to apply and extend fractal methodology has sometimes led to publications claiming, without sufficient justification, that a fractal relationship holds.

When using box counting to determine d_B, it is necessary to exclude box sizes for which the $(\log s, \log B(s))$ values deviate from a linear relationship. Such methods have routinely been applied in exploratory data analysis to determine break points in linear regression or deviations from linearity. This warning was echoed in [Brown 02], who noted that some statistical distributions, such as the log normal and exponential distribution, can appear linear on a log–log plot, especially for a limited range of data.

Since the box counting dimension is defined by a limit as $r \to 0$, [Bez 11] recommended that in practice the behavior of the $\log B(s)$ versus $\log s$ curve should be considered only for small s, even though there might be a larger region over which the plot is linear.[3] However, we cannot let the box size get too small, since for all sufficiently small s each non-empty box contains only one point and at this scale $B(s) = N$, where N is the number of observations. In this range of s, the slope of the $\log B(s)$ versus $\log s$ curve is zero. A related issue is that when digitized images of real-world structures are used, the resolution of the image needs to be considered. For example, studies of plants and algae have shown that the line thickness in digitized images can significantly impact the estimated fractal dimension. This digitization issue leads to discarding box counts for the smallest box sizes.

We also cannot let the box sizes get too large, since above a certain threshold value of s all observations are contained in a single box (e.g., for $s \geq \Delta$ we have $B(s) = 1$). Let $B_g(s)$ be the number of boxes of size s needed to cover the entire grid in which the object lies, and consider the ratio $B(s)/B_g(s)$. As s increases, there tends to be an increase in this ratio, and as s increases to some size s_{max}, we observe $B(s_{max}) = B_g(s_{max})$. For example, consider Fig. 6.1 which shows the Sierpiński carpet illustrated in Fig. 4.10. When covering this figure with boxes of size s_{max} we have $B_g(s_{max}) = 4$, and $B(s_{max}) = 4$. No larger boxes should be used. Even before s reaches s_{max}, the rise in the ratio $B(s)/B_g(s)$ results in a gradual rise in the computed d_B value. It might also be that at some size s_{max} the entire object Ω is covered by one box; i.e., $B(s_{max}) = 1$. This is illustrated in Fig. 6.2. When this occurs, no box size larger than s_{max} should be used. Also, when superimposing a grid of boxes on the object, the precise location of the grid can impact the results, especially for large boxes, since then there are fewer boxes. Thus it is recommended [Halley 04] to repeat the box counting analysis by translating the grid (with the same s), and taking the average of the results over the different translations of the grid; this technique was mentioned above in Sect. 4.3.

---

[3]The extensive reference lists in [Bez 11, Kenkel 13] are excellent guides to the use of fractals in biology and ecology.

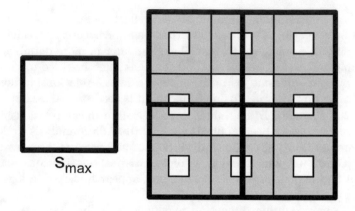

Figure 6.1: Maximal box size for the Sierpiński carpet

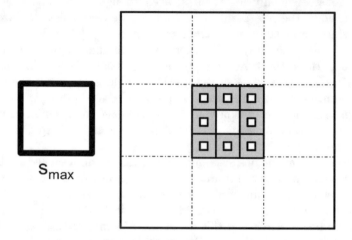

Figure 6.2: s_{max} is reached when $B(s_{max}) = 1$

Linear regression is typically used to estimate the slope of the $\log B(s)$ versus $\log s$ curve. If the slope is estimated to be d, then $-d$ is the estimate of d_B. Unfortunately, even if regression yields a good estimate of d_B, the associated standard regression statistics such as p-values and confidence intervals cannot be applied, since the points are not independent as the *same* object is being analyzed at multiple scales. Also, the deviations do not typically exhibit constant variance, nor do they typically possess a Gaussian distribution [Halley 04].

The residuals arising from a least squares fit to the $\log B(s)$ versus $\log s$ curve were studied in [Berntson 97]. A second-order polynomial was fitted to the residuals over the range of box sizes and tested for curvilinearity using a p-value of 0.1; any residual plot that displayed curvilinearity at a p value of less than 0.1 was taken as evidence that the image did not exhibit fractal behavior over the range. If this occurred, then either the largest box sizes or the smallest box sizes, or both, were removed until either too few points remained to perform a regression or until the deviation from linearity was sufficiently small. To avoid these issues relating to s_{max} for an object Ω consisting of a discrete set of points, the usual recommendation is that the maximal box size s_{max} should be much less than $\Delta/2$, and the minimal box size s_{min} should be much greater than the minimal inter-point separation [Halley 04].

Computational aspects of computing d_B of a 2-dimensional object from a digitized image were discussed in [Gonzato 00]. Through examples of a digitized straight line, and a digitized Koch curve, [Gonzato 00] showed that the the length, orientation, and placement of an image are all potential sources of error that can lead to a significant bias in the estimate of d_B. Two of the recommendations in [Gonzato 00] were: (*i*) an adequate number (15–20, say) of box sizes are required for "credible application" of linear regression and (*ii*) the final estimate of d_B should be the average of d_B computed for at least 20 random image placements. Even when all their recommendations are followed, since the estimate of d_B can be expected to be 10–20% higher than the true value, the final estimate should be multiplied by 0.8–0.9 [Gonzato 00].

A *local scale variation estimator* (LSV) for computing d_B was proposed in [Da Silva 06], who used LSV to compare d_B for artificial plants (generated from three-dimensional iterated function systems) and real-life peach trees. Assume that for each s the box counting scaling $B(s) \propto s^{-d_B}$ holds. Consider two box sizes s and $s + \delta$, where δ is not necessarily small. Define $\tilde{s} \equiv \delta/s$ and $\widetilde{B} \equiv \big(B(s + \delta) - B(s)\big)/B(s)$, so $\tilde{s}$ and $\widetilde{B}$ are the relative increases in the box size and the number of boxes, respectively, when the box size increases from s to $s + \delta$. From $B(s) \propto s^{-d_B}$ we obtain

$$\widetilde{B} \propto \frac{(s + \delta)^{-d_B} - s^{-d_B}}{s^{-d_B}} = (1 + \tilde{s})^{-d_B} - 1.$$

This yields

$$\log(1 + \widetilde{B}) \propto -d_B \log(1 + \tilde{s}),$$

so d_B can be estimated by linear regression of $\log(1+\widetilde{B})$ versus $\log(1+\tilde{s})$. There may be statistical advantages to computing d_B in this manner.

6.2   Computing the Minimal Box Size

As described above, when computing d_B from N points in Ω, we cannot make the box size s arbitrarily small since then $B(s)$ approaches N, and the slope of the $\big(\log s, \log B(s)\big)$ curve approaches 0. In this section we present the method of [Kenkel 13] for choosing the minimal box size and the generalization of Kenkel's method in [Rosenberg 16a].

Kenkel's analysis begins with the 1-dimensional case. If we randomly select N points on $[0, 1]$, then the probability that a given box (i.e., interval) of size s (where $s < 1$) does not contain any of these points is $(1 - s)^N$. The probability that a given box of size s contains at least one of N randomly selected points is $1 - (1 - s)^N$. There are s^{-1} intervals of size s in the unit interval. Hence the expected number $B(s)$ of non-empty boxes of size s is given by $B(s) = [1 - (1 - s)^N]/s$.

Considering next the 2-dimensional case, if we randomly select N points on the unit square $[0, 1] \times [0, 1]$, then the probability that a given box of size s (where $s < 1$) does not contain any of these points is $(1 - s^2)^N$. The probability that a given box of size s contains at least one of N randomly selected points is $1 - (1 - s^2)^N$. There are s^{-2} boxes of size s in the unit square. Hence the expected number $B(s)$ of non-empty boxes of size s is given by $B(s) = [1 - (1 - s^2)^N]/s^2$.

Finally, consider the case where the N points are sampled from a self-similar fractal set with box counting dimension d_B. The exponent of s is now d_B, and the expected number $B(s)$ of non-empty boxes of size s is given by

$$B(s) = \frac{1 - \left(1 - s^{d_B}\right)^N}{s^{d_B}}. \tag{6.1}$$

From (6.1), Kenkel numerically computed the "local slope" of the $\log B(s)$ versus $\log s$ curve, where the "local slope" $m(s)$ is defined by

$$m(s) \equiv \left(\log B(s + \delta) - \log B(s)\right)/\left(\log(s + \delta) - \log s\right)$$

for some small increment δ. Then s_{min} is the value for which $m(s) + d_B = 0.001$. Using 28 simulations in which s was decreased until the local slope is within 0.001 of d_B, Kenkel obtained the following approximation for the minimal usable box size:

$$s_{min} \approx \left(\frac{N}{10}\right)^{-1/d_B}. \tag{6.2}$$

From (6.2) we obtain $(s_{min})^{d_B} = 10/N$. Substituting this in (6.1), we have

$$B(s_{min}) = \frac{1 - \left(1 - (10/N)\right)^N}{10/N},$$

which yields, for large N, the estimate $B(s_{min}) = N/10$. This means that to estimate d_B, no box size s should be used for which $B(s) > N/10$. Since (6.2) was obtained numerically, and only for the accuracy 0.001, this result provides no guidance on how s_{min} varies as a function of the accuracy. Indeed, in many applications, e.g., in the analysis of neurons [Karperien 13], d_B is only estimated to two decimal places.

Kenkel's model was generalized in [Rosenberg 16a] by determining the minimal box size when the constant 0.001 is replaced by a positive parameter ϵ. Rather than work with a local slope of the $(\log s, \log B(s))$ curve, [Rosenberg 16a] worked with a derivative, which can be obtained in closed form. Using the chain rule for derivatives,

$$\frac{\mathrm{d}\log B(s)}{\mathrm{d}\log s} = \left(\frac{\mathrm{d}\log B(s)}{\mathrm{d}s}\right)\left(\frac{\mathrm{d}s}{\mathrm{d}\log s}\right) = \left(\frac{\mathrm{d}\log B(s)}{\mathrm{d}s}\right)\left(\frac{\mathrm{d}\log s}{\mathrm{d}s}\right)^{-1}$$

$$= \left(\frac{\mathrm{d}\log B(s)}{\mathrm{d}s}\right)s. \tag{6.3}$$

From (6.1), taking the derivative of $\log B(s)$ we have

$$\frac{\mathrm{d}\log B(s)}{\mathrm{d}\,s} = \frac{d_B N s^{-1}\left(1 - s^{d_B}\right)^{N-1} - d_B s^{-d_B-1}\left(1 - \left(1 - s^{d_B}\right)^N\right)}{\left(1 - \left(1 - s^{d_B}\right)^N\right)s^{-d_B}}. \tag{6.4}$$

Combining (6.3) and (6.4) yields

$$\frac{\mathrm{d}\log B(s)}{\mathrm{d}\log s} = \frac{d_B N\left(1 - s^{d_B}\right)^{N-1} - d_B s^{-d_B}\left(1 - \left(1 - s^{d_B}\right)^N\right)}{\left(1 - \left(1 - s^{d_B}\right)^N\right)s^{-d_B}}. \tag{6.5}$$

Denoting the right-hand side of (6.5) by $F(s)$, the task is to find the value $s^\star$ such that $F(s^\star) = -d_B + \epsilon$. For $s < s^\star$ we have $F(s) > -d_B + \epsilon$, so such box sizes should not be used in the estimation of d_B. Thus $s^\star$ is the minimal usable box size. As a first step towards calculating $s^\star$, define

$$\alpha \equiv \left(1 - s^{d_B}\right)^{N-1}. \tag{6.6}$$

While α is actually a function of s, for notational simplicity we write α rather than $\alpha(s)$. The equation $F(s) = -d_B + \epsilon$ can now be rewritten as

$$\frac{d_B N\alpha - d_B s^{-d_B}\left(1 - \alpha(1 - s^{d_B})\right)}{\left(1 - \alpha(1 - s^{d_B})\right)s^{-d_B}} + d_B = \epsilon. \tag{6.7}$$

After some algebra, from (6.6) and (6.7) we obtain an implicit equation for $s^\star$:

$$\left(\left(1 - s^{d_B}\right)^{1-N} - 1\right)s^{-d_B} = \epsilon^{-1}N d_B - 1. \tag{6.8}$$

Equation (6.8) is exact; no approximations were made. The parameter ϵ appears only in the right-hand side, which is independent of s.

Exercise 6.1 Derive Eq. (6.8).   □

Binary search can be used to compute a solution $s^\star$ of (6.8). A Python implementation is provided in Fig. 6.3. The code performs binary search over the interval $[0, 1]$ to compute $s^\star$ to an accuracy of 10^{-7} for $\epsilon = 0.001$ and for d_B from 1 to 2, in increments of $1/200$. The variable `Estimate` is the estimate given by (6.2). The binary search is halted when the width of the interval containing $s^\star$ is less than 1.0×10^{-7}, which took 24 iterations. For $N = 1000$ and $\epsilon = 0.001$, the percent error $100(s_{min} - s^\star)/s^\star$, where s_{min} is the estimate defined by (6.2), is plotted in Fig. 6.4 as a function of d_B for $1 \le d_B \le 2$. The error in the approximation (6.2) is highest (about 10%) when $d_B = 1$. As noted above, in many applications d_B is estimated to only two decimal places, and Fig. 6.5 compares, for $N = 1000$, the value $s^\star$ for $\epsilon = 0.1$, $\epsilon = 0.01$, and $\epsilon = 0.001$, all for the same range $1 \le d_B \le 2$.

We now derive a closed-form approximation to $s^\star$, and compare our approximation to (6.2). Defining $x = s^{d_B}$, rewrite (6.8) as

$$x^{-1}\left((1 - x)^{1-N} - 1\right) = \epsilon^{-1}N d_B - 1. \tag{6.9}$$

Binary Search

```
1       import math
2       N = 1000.
3       epsilon = .001
4       for i in range(201):
5               d = 1.0 + float(i)/200.
6               RHS = -1.0 + N*d/epsilon
7               Estimate = (0.1*N)**(-1.0/d)
8               low = 10**(-8)
9               high = 1.0
10              while (low + 10**(-7) <= high):
11                      s = (low + high)/2.0
12                      x = s**d
13                      LHS = ((1.0-x)**(1.0-N) -1.0)/x
14                      if LHS < RHS:
15                              low = s
16                      else:
17                              high = s
18              error = 100*(Estimate-s)/s
19              print " d ", d, " s ", s, " error ", error
```

Figure 6.3: Python code for binary search

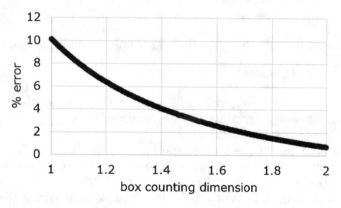

Figure 6.4: % error in s_{min} for $N = 1000$ and $\epsilon = 0.001$, for $1 \leq d_B \leq 2$

We have $x < 1$ since $s < 1$ and $d_B > 0$. Ignoring terms of degree four and higher, the Taylor series expansion of $(1 - x)^{1-N}$ yields

$$(1 - x)^{1-N} \approx 1 - (1 - N)x + (1 - N)(-N)x^2/2 - (1 - N)(-N)(-N - 1)x^3/6.$$

For large N we have $(1-x)^{1-N} \approx 1 + Nx + (N^2/2)x^2 + (N^3/6)x^3$. Substituting this in (6.9) yields

$$x^{-1}\left((1-x)^{1-N} - 1\right) \approx x^{-1}\left(1 + Nx + (N^2/2)x^2 + (N^3/6)x^3 - 1\right)$$
$$= N + (N^2/2)x + (N^3/6)x^2 = \epsilon^{-1}Nd_B - 1 \approx \epsilon^{-1}Nd_B.$$

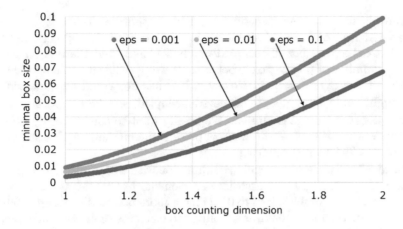

Figure 6.5: s_{min} for $N = 1000$ and for three values of ϵ, for $1 \leq d_B \leq 2$

Dividing by N, we obtain $(N^2/6)x^2 + (N/2)x + (1 - \epsilon^{-1}d_B) \approx 0$. Using $1 - \epsilon^{-1}d_B \approx -\epsilon^{-1}d_B$ we obtain

$$x = \frac{-\frac{N}{2} \pm \sqrt{\frac{N^2}{4} - 4\frac{N^2}{6}(-\epsilon^{-1}d_B)}}{(2/6)N^2} \approx \frac{N\sqrt{\frac{2}{3}\epsilon^{-1}d_B}}{(1/3)N^2} = \frac{1}{N}\sqrt{\frac{6d_B}{\epsilon}}.$$

By definition, $x = s^{d_B}$, so $s^{d_B} = (1/N)\sqrt{6d_B/\epsilon}$. Solving for s yields an approximation $\tilde{s}$ for the minimal usable box size:

$$\tilde{s} = \left(N\sqrt{\frac{\epsilon}{6\,d_B}}\right)^{-1/d_B}. \qquad (6.10)$$

The estimate $\tilde{s}$, which is a decreasing function of ϵ, looks identical to (6.2), except that the empirically derived constant 0.1 in (6.2) is replaced by $\sqrt{\epsilon/(6d_B)}$ in (6.10). However, $\tilde{s}$ is not a good approximation to $s^\star$. For example, with $N = 1000$, $d_B = 2$, and $\epsilon = 0.001$, we have $s^\star \approx 0.099$, while $\tilde{s} \approx 0.331$. The error arises from the third-order Taylor series approximation, since for $x = (s^\star)^2$ we have $(1-x)^{1-N} \approx 19{,}700$, while $1 + Nx + (N^2/2)x^2 + (N^3/6)x^3 \approx 219$. Thus, while (6.10) yields the functional form obtained by Kenkel, in practice $s^\star$ should be computed using (6.8) and binary search.

6.3   Coarse and Fine Grids

It is trivial to write a fast box counting algorithm: first compute a histogram giving the number of points or pixels in each box, and then scan the histogram [Grassberger 93]. However, this crude approach can result in very inefficient memory usage. Consider a set $\Omega \subset \mathbb{R}^E$, and without loss of generality, assume

Ω is contained in the unit hypercube in $\mathbb{R}^E$. Suppose we cover Ω with an E-dimensional grid of boxes of size $s \ll 1$. To compute d_B, we need only one bit per box (i.e., the box is occupied or unoccupied), and there is no need to store the number of points/pixels per box. If we pre-allocate memory, we need an array of size $(1/s)^E = s^{-E}$. However, if Ω has box counting dimension d_B, then roughly only $B(s) \propto s^{-d_B}$ boxes are occupied, so when $d_B < E$, much of the pre-allocated storage is unused. We ideally want to consume only $\mathcal{O}(B(s))$ memory. With this motivation, we now present the method of [Grassberger 93] for memory-efficient box counting.

The idea is to use a coarse grid in conjunction with a fine grid. Corresponding to box j of a coarse grid, we associate an element j of the array `coarse`. The element `coarse[j]` is initialized to 0. If one or more points of the fractal fall into box j, then `coarse` now contains a pointer to elements of another array, called `fine`, which provides further information about that part of the fine grid which lies inside box j of the coarse grid. This is illustrated in Fig. 6.6, a picture of red clover. Suppose the original coarse-grained image (i) lies in the unit square.

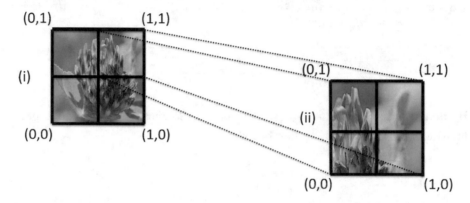

Figure 6.6: Coarse-grained and fine-grained images of clover

Dividing image (i) into $2^E = 4$ pieces, we focus on the upper right-hand piece; zooming in on this piece yields the fine-grained image (ii). In this example, all four quadrants of (i) contain points, and each of the four elements of `coarse` contains a pointer to a (distinct) `fine` array, providing information on that quadrant of (i).

Let s_c be the side length of a box in the coarse grid, and let s_f be the side length of a box in the fine grid. Define $\beta \equiv s_c/s_f$, so $\beta > 1$. Recalling that Ω lies in the unit hypercube, define $L \equiv 1/s_f$, so the number of fine grid boxes needed to cover the unit hypercube is L^E. Then $L/\beta = 1/s_c$, and the number of coarse grid boxes needed to cover the unit hypercube is $(L/\beta)^E$.

Assume that $B(s_c)$, the number of non-empty boxes in the coarse grid, satisfies $B(s_c) = k s_c^{-d_B}$ for some positive constant k. (This is a strong version of the assumption that d_B exists.) Then $B(s_c) = k s_c^{-d_B} = k(L/\beta)^{d_B}$. For each occupied box in the coarse grid we have β^E fine grid boxes, i.e., boxes of size s_f. Thus the memory required for all the fine grid boxes is $k(L/\beta)^{d_B}\beta^E$, and the total memory required is

$$M(\beta) = (L/\beta)^E + k(L/\beta)^{d_B}\beta^E, \qquad (6.11)$$

where the first term is the number of coarse boxes and the second term is the total number of fine boxes corresponding to all the occupied coarse boxes. We write $M(\beta)$ since our goal is to determine the value of β that minimizes the total memory.

Since $M(\beta)'' > 0$ for $\beta > 0$, then $M(\beta)$ is strictly convex. Thus if $M'(\beta^\star) = 0$ for some $\beta^\star > 0$, then $\beta^\star$ is the unique point at which $M(\beta)$ achieves its global minimum. A first derivative and some algebra yields

$$\beta^\star = aL^\phi, \qquad (6.12)$$

where

$$a = \left(\frac{E}{k(E-d_B)}\right)^{\frac{1}{2E-d_B}} \quad \text{and} \quad \phi = \frac{E-d_B}{2E-d_B}. \qquad (6.13)$$

For efficient programming, β should be chosen as the power of 2 closest to (6.12), but for this analysis we work with the minimal value (6.12). Using (6.11), (6.12), and (6.13), the minimal memory $M(\beta^\star)$ is given by

$$
\begin{aligned}
M(\beta^\star) &= (L/\beta^\star)^E + k(L/\beta^\star)^{d_B}(\beta^\star)^E \\
&= [L/(aL^\phi)]^E + k[L/(aL^\phi)]^{d_B}(aL^\phi)^E \\
&= (aL^{\phi-1})^{-E} + k(aL^{\phi-1})^{-d_B}(aL^\phi)^E.
\end{aligned}
\qquad (6.14)
$$

Considering the exponent of L in the two terms in (6.14), from (6.13) we have

$$
\begin{aligned}
(\phi-1)(-E) &= \frac{E^2}{2E-d_B} \\
(\phi-1)(-d_B) + \phi E &= \frac{E^2}{2E-d_B}.
\end{aligned}
$$

Since the exponent of L in both terms in (6.14) is the same, we have the very pretty result

$$M(\beta^\star) \propto L^{\left(\frac{E^2}{2E-d_B}\right)}. \qquad (6.15)$$

Exercise 6.2 Derive Eqs. (6.12) and (6.13).   □

Using this technique, [Grassberger 93] reported a 94% reduction in memory consumption. Grassberger also observed that ideally the minimal memory would scale as the number of non-empty coarse grid boxes, i.e., the minimal memory would be $\mathcal{O}\left(L^{d_B}\right)$. Using (6.15) we can calculate the deviation from the optimal scaling, by determining the exponent ω such that

$$L^{\frac{E^2}{2E-d_B}} = L^{d_B}L^{\omega}.$$

We obtain

$$\omega = \frac{(E - d_B)^2}{2E - d_B}.$$

For example, with $E = 2$ and $d_B = 3/2$ we have $\omega = 1/10$, and with $E = 3$ and $d_B = 5/2$ we obtain $\omega = 1/14$, so this method is close to optimal.

So far we have only considered memory. We also want to examine the impact of Grassberger's method on the compute time. Suppose there are N known points of the fractal. The compute time is the sum of the following two components. The first component is the time to scan the N points and update the array `coarse`; this time is $\mathcal{O}(N)$. The second component is the time to process each non-empty box in the coarse grid and update the array `fine` corresponding to each non-empty box in the coarse grid. The number of non-empty boxes in the coarse grid is $\mathcal{O}\left((L/\beta)^{d_B}\right)$. Each box in the coarse grid contains β^E fine boxes, and each of these fine boxes must be examined to see if it is occupied. Hence the total time complexity of updating the array `fine` for all the non-empty coarse-grid boxes is $\mathcal{O}\left((L/\beta)^{d_B}\beta^E\right) = \mathcal{O}\left(\beta^{E-d_B}\right)$. Grassberger reported that this scheme reduced compute time by a factor of about 28. For computing d_B of high-dimensional objects, such as strange attractors of dynamical systems (Chap. 9), it is memory and not compute time that is critical, which provides the justification for choosing β to minimize the memory $M(\beta)$.

We can further reduce the memory consumption and approach the desired goal of $\mathcal{O}\left(L^{d_B}\right)$ memory, by using a hierarchy of nested grids, so that nonzero elements of the array `coarse0`, corresponding to the coarsest grid, contain pointers to the array `coarse1`, corresponding to a finer grid. The nonzero elements of `coarse1` contain pointers to `coarse2`, corresponding to a yet finer grid, etc., and the last pointer points to the array for the finest grid. Since for each of the N points in the fractal we follow a chain of $\log L/\log \beta$ pointers, the time complexity is essentially proportional to $N \log L/\log \beta$ [Grassberger 93]. Assuming the coarsest grid contains $\mathcal{O}(1)$ boxes, the required memory $M(\beta)$ is given by

$$
\begin{aligned}
M(\beta) &= k\beta^E(L/\beta)^{d_B}\left(1 + \beta^{-d_B} + \beta^{-2d_B} + \beta^{-3d_B} + \cdots\right) \\
&= k\beta^E(L/\beta)^{d_B}\frac{1}{1 - \beta^{-d_B}} = k\beta^E(L/\beta)^{d_B}\frac{\beta^{d_B}}{\beta^{d_B} - 1} \\
&= \frac{k\beta^E L^{d_B}}{\beta^{d_B} - 1} \tag{6.16}
\end{aligned}
$$

so we have reached the goal of a box-counting method whose memory requirement and compute time are both $\mathcal{O}\left(L^{d_B}\right)$. With this hierarchy of pointers, β can be much less than the value given by (6.12), and good results are obtained even with $\beta = 2$ [Grassberger 93].

Exercise 6.3 What value of β minimizes (6.16)? □

Other researchers have proposed box counting methods that follow a similar strategy of zooming into each non-empty box. The recursive method of [Feeny 00] partitions the 2-dimensional square into four smaller squares, using two nested loops. In E dimensions, the method requires E nested loops, and the overall method has time complexity $\mathcal{O}(2^E)$. Feeny observed that this complexity has been observed in numerical examples, and so the method may not be efficient if E is large. Many other papers devoted to algorithmic improvements in box counting were written in the early 1990s, as cited in [Feeny 00].

6.4 Box Counting Dimension of a Spiral

We now present the analysis of [Vassilicos 91], who analytically determined the box counting dimension of the intersection of a spiral and a line intersecting it. We present this analysis since, other than when Ω is generated from an IFS (Sect. 3.3), it is unusual to be able to analytically determine d_B of a geometric object. The spiral was studied in [Vassilicos 91] because of its relevance to the study of turbulence in fluid flows, e.g., describing the spread of a blob of dye in turbulent fluid.

A spiral centered at the origin is a curve defined by $z(\theta) = c\theta^{-\alpha}$, where c and α are positive constants. Here (z, θ) are polar coordinates so $x = z\cos\theta$ and $y = z\sin\theta$ in Euclidean rectilinear coordinates.[4] When $\alpha > 0$, the spiral converges to the origin as $\theta \to \infty$. Figure 6.7 (left) illustrates four turns (from $\theta = 8\pi$ to $\theta = 16\pi$) of the spiral $z(\theta) = (1/200)\theta^{-0.2}$. Draw a line through the

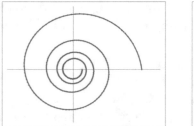 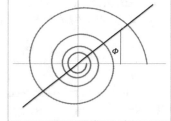

Figure 6.7: Spiral (left) and spiral and line (right)

origin that makes an angle Φ with the x-axis, as illustrated in Fig. 6.7 (right). Our goal is to compute the box counting dimension of the intersection of the spiral with the line.

For $j = 1, 2, \cdots$, let (z_j, θ_j) be the polar coordinates of the points in Ω. Then $\theta_j = \Phi + 2\pi j$ and $z_j = c\theta_j^{-\alpha} = c(\Phi + 2\pi j)^{-\alpha}$. For j sufficiently large $(j \gg \Phi/2\pi)$ we have $z_j \sim j^{-\alpha}$. Since $j^{-\alpha}$ is independent of Φ for $\Phi \in [0, 2\pi]$,

---

[4]With polar coordinates, r is typically used to denote the radius, but we use z since we reserve the symbol r to refer to the radius of a ball.

then d_B must be independent of Φ. Hence we can choose $\Phi = 0$, in which case the line intersecting the spiral is the x axis. Considering only the points where the spiral intersects $\{x \in \mathbb{R} \mid x \geq 0\}$, define

$$x_j \equiv c\,(2\pi j)^{-\alpha} \tag{6.17}$$

for $j \geq 1$, so each $x_j > 0$. Define $\Omega = \{x_j\}_{j=1}^{\infty}$. Let the box size s be a fixed small positive number. To compute d_B we must cover Ω with boxes of size s.

Let J be the smallest integer for which $x_j - x_{j+1} \leq s$ for $j \geq J$. We use one box to cover each x_j for $1 \leq j \leq J$, while the points x_j for $j > J$ will be covered by x_J/s adjacent boxes.[5] This is illustrated in Fig. 6.8 for the case $J = 4$. By

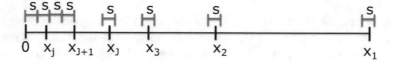

Figure 6.8: Spiral radius for $J = 4$

definition, J satisfies

$$c[2\pi J]^{-\alpha} - c[2\pi(J+1)]^{-\alpha} \leq s. \tag{6.18}$$

Since $J \to \infty$ as $s \to 0$, then J scales according to $J^{-\alpha} - (J+1)^{-\alpha} \sim s$, or

$$1 - [(J+1)/J]^{-\alpha} \sim s/J^{-\alpha}. \tag{6.19}$$

Using the approximation $[(J+1)/J]^{-\alpha} \approx 1 - \alpha/J$, we obtain $1 - [1 + (1/J)]^{-\alpha} \approx \alpha/J$. Thus (6.19) yields $\alpha/J \sim s/J^{-\alpha}$, which implies $s \sim \alpha J^{-(1+\alpha)}$, which we rewrite as

$$J \sim s^{-1/(1+\alpha)}. \tag{6.20}$$

Recall that one box is needed to cover each x_j for $1 \leq j \leq J$ (for a total of J boxes), while the remaining points x_j for $j > J$ can be covered by x_J/s adjacent boxes. Hence the number $B(s)$ of boxes needed to cover Ω is approximately

$$B(s) \approx J + x_J/s. \tag{6.21}$$

From (6.17) and (6.20) we have $x_J/s = c\,(2\pi J)^{-\alpha}/s \sim J^{-\alpha}/s \sim [s^{-1/(1+\alpha)}]^{-\alpha}/s = s^{-1/(1+\alpha)}$. Since also $J \sim s^{-1/(1+\alpha)}$, then from (6.21) we have $B(s) \sim s^{-1/(1+\alpha)}$. Assuming that $B(s)$ is indeed the minimal number of boxes of size s needed to cover x_j for $j \geq 1$, the box counting dimension d_B of Ω is given by $d_B = 1/(1+\alpha)$.

The box counting dimension of the entire spiral $z(\theta) = c\,\theta^{-\alpha}$ is also analytically determined in [Vassilicos 91]. Let s be a given box size. Let θ_0 be the starting angle of the spiral. Since $z(\theta) \to 0$ as $\theta \to \infty$, we can consider two

---

[5]This is the same approach as used in Example 4.11.

regions of the spiral. The first region contains the outer coils of the spiral for which $\theta \geq \theta_0$ and $z(\theta) - z(\theta + 2\pi) > s$. The second region contains the inner coils for which $z(\theta) - z(\theta + 2\pi) \leq s$. The critical angle θ_s dividing these two regions satisfies $z(\theta_s) - z(\theta_s + 2\pi) = s$, and $\theta_s \to \infty$ as $s \to 0$. Since $z(s) = c\theta^{-\alpha}$, we must solve

$$\theta^{-\alpha} - (\theta + 2\pi)^{-\alpha} = s/c,$$

which we rewrite as

$$1 - \left(\frac{\theta + 2\pi}{\theta}\right)^{-\alpha} = 1 - \left(1 + \frac{2\pi}{\theta}\right)^{-\alpha} = \frac{s}{c\theta^{-\alpha}}.$$

Using the approximation $(1 + x)^{-\alpha} \approx 1 - \alpha x$ for small x, if $\theta \gg 2\pi$ (which, as noted above, is satisfied for all sufficiently small s) we obtain

$$\frac{2\pi\alpha}{\theta} \approx \frac{s}{c\theta^{-\alpha}}.$$

The critical value θ_s is thus given by

$$\theta_s \approx \left(\frac{s}{2\pi\alpha c}\right)^{-1/(\alpha+1)}. \tag{6.22}$$

As for the length of the outer coils of the spiral, we have $dz(\theta)/d\theta = (-\alpha)c\theta^{-\alpha-1} = z(\theta)(-\alpha/\theta)$. Using the formula (23.3) derived in Sect. 23.1, the length $L(\theta_s)$ of the outer coils of the spiral is given by

$$
\begin{aligned}
L(\theta_s) &= \int_{\theta_0}^{\theta_s} \sqrt{z(\theta)^2 + (dz(\theta)/d\theta)^2}\, d\theta = \int_{\theta_0}^{\theta_s} \sqrt{z(\theta)^2 + [z(\theta)(-\alpha/\theta)]^2}\, d\theta \\
&= \int_{\theta_0}^{\theta_s} z(\theta)\sqrt{1 + (\alpha/\theta)^2}\, d\theta = c\int_{\theta_0}^{\theta_s} \theta^{-\alpha}\sqrt{1 + (\alpha/\theta)^2}\, d\theta.
\end{aligned}
$$

The number of boxes needed to cover the outer coils scales as the length of the outer coils divided by the box size, which is $L(\theta_s)/s$. The number of boxes needed to cover the inner coils scales as the area of the inner coils divided by the box area, which is $\pi[z(\theta_s)]^2/s^2$. The total number of boxes $B(s)$ needed to cover the spiral thus scales as $B(s) \sim L(\theta_s)/s + \pi[z(\theta_s)]^2/s^2$.

To complete the analysis, we consider two cases. If $\theta_s \gg \theta_0$, then $L(\theta_s) \approx c\theta_s^{1-\alpha}/(1 - \alpha)$ if $\alpha < 1$ and $L(\theta_s) \approx c\theta_0^{1-\alpha}/(\alpha - 1)$ if $\alpha > 1$, from which it can be shown [Vassilicos 91] that $d_B = \max\{1, 2/(\alpha + 1)\}$. For the other case $\theta_s - \theta_0 \ll \theta_0$, we have $L(\theta_s) \approx c(\theta_s - \theta_0)\theta_0^{-\alpha}$, from which it can be shown that $d_B = 1 + (1/(\alpha + 1))$.

For an entirely different example [Vassilicos 91], let Ω be the intersection of the logarithmic spiral $z(\theta) = ce^{-\theta}$ and a line through the origin. The number of boxes of size s needed to cover Ω scales as $B(s) \sim \ln(1/s)$ as $s \to 0$. Since $B(s)$ does not exhibit a power-law scaling, then Ω does not have a box counting dimension. This occurs since the logarithmic spiral converges onto its center too quickly.

Exercise 6.4 Prove that $B(s) \sim \ln(1/s)$ for the intersection of the logarithmic spiral and a line through the origin.   □

Another curve whose box counting dimension can be analytically calculated is the *clothoid* [6] double spiral, illustrated in Fig. 6.9. This curve is defined

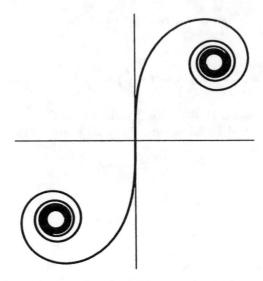

Figure 6.9: Clothoid double spiral

parametrically, for $t \in \mathbb{R}$, by

$$x(t) = \int_0^t \cos(\theta^2)\, \mathrm{d}\theta, \quad y(t) = \int_0^t \sin(\theta^2)\, \mathrm{d}\theta.$$

It was proved in [Korkut 09] that the box counting dimension of the clothoid is 4/3.

6.5   Some Applications of the Box Counting Dimension

There are many alternatives to the box counting dimension, and later chapters of this book will study several of these alternatives. Nonetheless, box counting remains very popular [Gui 16]. It provides a way to quantify shape, which leads to a variety of applications, across a wide variety of disciplines. In cancer

---

[6]Clotho was one of the three Fates who spun the thread of human life, by winding it around a spindle. From this poetic reference the Italian mathematician Ernesto Cesàro, at the beginning of the twentieth century, gave the name "Clothoid" to a curve with this shape, which reminded him of the thread wound around the spindle and the distaff. This curve was studied earlier, by Euler in 1700, in connection with a problem posed by Jakob Bernoulli. Clothoid curves are the ideal shapes for roller coasters (http://www.phy6.org/stargaze/Sclothoid.htm).

research, the difference in the fractal dimension between normal and cancerous blood vessels revealed important aspects of the underlying mechanisms of tumor growth [Baish 00]. In crop science, the fractal dimension of plants was used in determining the effectiveness of different crop management strategies [Jaradat 07]. In geology, it was used to study the shear patterns of mineral deposits [Hassani 09]. In biology, shape is used to describe differences between species, and shape strongly influences function. For example, the shape of a male moth's antennae strongly influences his ability to detect female moth pheromones. The fact that organisms are fractal over some scale range may imply the presence of important developmental constraints and evolutionary processes governing the shape of the organisms. Moreover, the current limited understanding of the relationship between structure and function should not impede efforts to estimate the shape of biological structures [Hartvigsen 00].

There are certainly hundreds, and probably thousands, of published applications of the box counting dimension. Below we describe a few particularly interesting applications. Besides the intrinsic interest of these applications, an awareness of the range of applications of the box counting dimension of a geometric object will, we expect, inspire a wide range of applications for the many fractals dimensions of networks that will be presented in the following chapters. In many of the applications described in this section, the range of box sizes is not even close to the two orders of magnitude recommended in [Halley 04]. At the end of this section we present the results of a study [Hadzieva 15] comparing the results of ten open source software packages for estimating d_B. The following applications are presented in no particular order.

Road Networks: The box counting dimension of the road network of 93 Texas cities with a population exceeding 2500 was calculated in [Lu 04]. The goal of this study was to determine if the complexity of the road network of a city is correlated to the population and number of houses in the city. Seven box sizes (successive box sizes differed by a factor of 2) were used to compute d_B. The computed d_B values ranged from less than 1 to 1.745, with Fort Worth having the highest d_B and Dallas having the third highest. There are three cities for which d_B is less than 1 (the three anomalous d_B values are between 0.78 and 0.88) and [Lu 04] suggested that for these cities, each with population less than 4100, the chosen box counting grid may be inappropriate. After discarding these three results, for the remaining cities they determined that for some constants a and b we have $\log(P_i) \approx a + bd_i$, where P_i is the population of city i and d_i is the box counting dimension of the road network of city i. Moreover, d_i and the number of houses in city i are also approximately linearly related. This 2004 study, which treats a road network as a geometric object, appeared just before the widespread introduction of network-based methods to compute d_B of a network.

The d_B of the road network of Dalian, a city in China, was calculated in [Sun 12]. A grid of size 2022×1538 was used to cover the study area, which contains 3452 road links (sections of roads between intersections). Using seven different values of the box size, they determined $d_B = 1.24$. They noted that this value, which considers the city as a whole, is useful when comparing Dalian against other cities, but is not useful when investigating the "heterogeneity" of different locations within a single urban road network. This study, though computing d_B of a road network, also makes no use of a network structure and instead treats the network as a geometric object.

Botany: The value $d_B \approx 1.7$ was calculated in [Pita 02] for the self-similar branches of the African plant *asparagus plumosus*, by applying box counting to a 640×640 pixel image, using box sizes from 1 to 200 pixels.

Food Science: Box counting was used in [Kerdpiboon 06] to quantify structural changes in the micro-structural features of dried food products, to better understand and predict the property changes of foods undergoing drying. Digital images of carrot sections, with a 520×520 pixel resolution, were covered with square boxes of size 4, 5, 10, 13, 26, 65, 130, and 260 pixels. They determined that d_B increased with drying time, corresponding to less orderly cell structures. The fractal dimension was then used in a neural network model to predict shrinkage and rehydration ratios of carrots undergoing different drying techniques.

The d_B of food products containing solid fat was calculated in [Tang 06]. The fat crystals interact with each other to form a 3-dimensional crystal network, which percolates throughout the food. Computer simulations were used to generate the crystal network, in order to better assess how individual structural factors impact d_B. They analyzed 512×512 pixel images, using 18 box sizes, starting with 128 pixels, and reducing the size by the factor 1.3 to get the next box size. Boxes containing fewer than m particles, for a specified parameter m, were discarded.

Electrical Engineering: In [Huang 03], d_B was computed to determine if the waveform produced by a power system had experienced a degradation in voltage, current, or frequency. They computed d_B every one-quarter cycle of the sinusoidal signal and compared it to the d_B expected each one-quarter cycle for a trouble-free signal.

Character Recognition: The box counting dimension was used in [Alsaif 09] for Arabic character recognition. A set of reference Arabic character was scanned, and d_B of the bitmap image was computed for each character. To identify an arbitrary character, d_B of its bitmap image was computed and then compared with the stored d_B of each reference character. Since the gray-scale value of a pixel becomes the third dimension of a two-dimensional image, the boxes are three-dimensional. (Many existing box-counting methods for two-dimensional images assume a square array of pixels, and box sizes independent of the range of gray-scale values; more general methods were presented in, e.g., [Long 13].)

Mechanical Engineering: The d_B of the complicated speckling pattern produced by the reflected waves when a laser beam illuminates a rough surface was computed in [Ayala-Landeros 12]. The speckle patterns were recorded on an X-ray plate, digitized to 2048×1536 pixels, stripped of noise, and converted to black and white. A commercial box counting package was used, with a starting box size of $s = 195$ which was reduced by a factor of 1.3 in each iteration. The box counting dimension can be used to determine the forms of vibration damping coefficients of materials, e.g., as required in the aerospace industry to isolate the strong vibrations during takeoff.

Biology: Box counting was used in [Martin-Garin 07] to quantify and classify coral. The ability to recognize patches of coral, algae, or other organisms by their fractal dimension facilitates detecting bottom communities, quantifying the coral cover, and evaluating surfaces damaged by hurricanes. The boxes sizes used vary from 2 to 100 pixels. For each sample subjected to box counting, the set of $(\log s, \log B(s))$ values was divided into two ranges; for $s \geq s_0$ (where s_0 depends on the sample), the values were used to compute a structural box counting dimension d_B^S, while for $s < s_0$ the values were used to compute a textural box counting dimension d_B^T which describes the fine-grained texture. A plot of d_B^S versus d_B^T was then used to differentiate between species of coral. The authors also provided references to many biological uses of fractals, including distinguishing between malignant and benign tumor cells and characterizing leaf shapes. They noted that fractals are used in biology both as mathematical descriptors of growth processes (e.g., growth of corals, sponges, and seaweeds) and as the basis of simulations.

Electric Power Networks: Various electric networks (e.g., the 500 kV main power grid of the China Southern Power Grid and the 500 kV main power grid in Guangdong province in China) were treated in [Hou 15] as geometric objects (rather than as networks). For the nine networks they studied, box counting and regression were used to compute d_B. The d_B values ranged from 1.15 to 1.61, and a higher d_B may indicate a higher failure rate.

Architecture: Box counting was used in [Ostwald 09] to compare the architectural styles of major twentieth century architects and to compare buildings of a single architect. The works of Peter Eisenman[7] were studied and compared with designs by Frank Lloyd Wright, Le Corbusier, Eileen Gray, and Kazuyo Sejima. Box counting was applied to four elevations (views from the North, South, East, West) to five Eisenman houses. The digital images are 1200×871 pixels, with 13 box sizes used. The average d_B of Eisenman's houses (built between 1968 and 1976) is 1.425. It is rare in modern architecture to see a box counting dimension exceeding 1.6; typically such values are seen only in the highly decorative designs of the Arts and Crafts movement of the late nineteenth and early twentieth centuries. The study showed that, compared with Eisenman's works, Frank Lloyd Wright's Prairie-style architecture is more complex ($d_B = 1.543$), and Le Corbusier's Arts and Crafts style designs are also slightly more complex ($d_B = 1.481$ to 1.495). Kazuyo Sejima's Minimalist works are less complex ($d_B = 1.310$). The variation of dimension within the work of each architect was low for Frank Lloyd Wright and Le Corbusier, but higher for Eileen Gray and Kazuyo Sejima, which supports a visual reading of their designs as less consistent in their visual complexity.

Urban Studies: In [Shen 02], d_B was computed for the urbanized areas of 20 of the most populous cities in the USA. The range of box sizes was 1–500 pixels, and for each urban area, multiple values of $B(s)$ were obtained

---

[7]Eisenman was known for abstract formal designs that refused to acknowledge site conditions or respond directly to the needs of human inhabitants. He was in the vanguard of the Deconstructivist movement, which is regarded as having high visual complexity.

by using different origins for the grid. Linear least squares regression of the $\left(\log s, \log B(s)\right)$ pairs was used to compute d_B. New York City had the highest d_B (1.7014) and Omaha had the smallest d_B (1.2778); since New York City's area is much larger than Omaha's, a higher d_B indicates a more complex or dispersed urban area. Since the inland urban area of Dallas-Fort Worth and the waterfront city of Chicago had nearly the same d_B (1.6439 versus 1.6437), d_B gives little information about the specific orientation and configuration of an urban area. Rather, d_B is useful as an aggregate measure of overall urban form. Moreover, cities with nearly identical d_B and urban areas may have quite different population sizes, as observed by comparing Chicago and Dallas and by comparing Pittsburgh and Nashville. Thus d_B is a fair indicator of the urban area but not a good measure of urban population density. By computing d_B for Baltimore for 12 time periods over the years 1792–1992, it was determined that while d_B always increases with the logarithm of the city area, d_B is not always positively correlated to the urban population.

The variation over time of the box counting dimension of the spatial form of various cities was studied in [Chen 18]. For cities in developed countries, the shape of the $d_B(t)$ versus t curve (where t is in years) takes the shape of a sigmoid ("S" shaped) curve, often chosen to be the logistic function. For cities in developing countries, the shape of the $d_B(t)$ versus t curve takes the shape of a "J" shaped curve, and a generalized logistic function was proposed in [Chen 18] to model such curves.

Medicine: Box counting was used in [Esteban 07] to help detect the presence of white matter abnormalities arising from multiple sclerosis (MS) in the human brain. Magnetic resonance images of 77 people were analyzed. For the different box counting analyses made, the box size ranges were 3–12 pixels, 3–15 pixels, and 3–20 pixels; box sizes yielding points deviating sufficiently from a linear fit on a log–log scale were discarded. The computed dimensions of skeletonized images exhibited small differences: $d_B = 1.056$ for clinically isolated syndrome MS, $d_B = 1.059$ for relapsing–remitting MS, $d_B = 1.048$ for secondary progressive MS, and $d_B = 1.060$ for primary progressive MS, compared with $d_B = 1.073$ for healthy control people. Despite these small differences, the study showed that patients with MS exhibited a significant decrease in d_B for the brain white matter, as compared with healthy controls, and d_B might become a useful marker of diffuse damage in the central nervous system in MS patients.

Box counting in $\mathbb{R}^3$ was used in [Helmberger 14] to compute d_B of the human lung, in order to determine if d_B is correlated with pulmonary hypertension (PH), a life-threatening condition characterized by high blood pressure in the arteries leading to the lungs and heart. Using the lungs of 24 deceased individuals (18 with PH and 6 without PH), for each person the linear portion of the $\log B(s)$ versus $\log s$ plot was identified; this was accomplished by iteratively discarding $\left(\log s, \log B(s)\right)$ pairs until a good fit was obtained. They computed $d_B \approx 2.37$ for the 6 people with no PH and $d_B \approx 2.34$ for the 18 with PH. Using appropriate statistical tests, they concluded that d_B may not be suitable for detection or explanation of PH in adult patients.

Dentistry: Box counting was used in [Yu 09] to determine if significant changes occur in the bony structures of reactive bone following root canal treatment (RCT). Two radiographs for each of 19 subjects were taken a week before RCT and 6 months after. The radiographs were scanned, converted to images 120×120 pixels, and the skeleton pattern extracted. Box sizes of 2, 4, 8, and 15 were used since they are all factors of 120 and almost all are powers of two (yielding equal intervals between points on a log–log scale). Of the 19 patients, 17 showed decreased fractal dimension of the bone in the first group of images, and 13 in the second group; the calculated dimensions were not provided.

Marine Science: Marine snow particles form aggregates, and a better understanding of the aggregate properties would aid in coagulation and vertical flux models and calculations. The particles themselves are known to be fractals. The particle fractal dimensions were shown in [Kilps 94] to be a function of the type of particle forming the aggregate, and fractal geometrical relationships were used to determine how properties of marine snow, such as density and settling velocity, change with aggregate size.

Music: Box counting was applied in [Das 05] to three categories of Indian music: classical, semi-classical, and light. Waveforms were extracted from musical excerpts and converted to ASCII data. Since the data for each file contains about 120,000 points, each file was divided into six chunks of 20,000 points, and d_B was calculated for each of the six chunks. For each song, the maximal and minimal values of d_B (over all six chunks for the song) were calculated. They concluded that classical songs have the highest values (between 3.14 and 4.09) of maximal d_B, which reflects that in this genre the singer stresses, producing a wide range of frequencies or repetition of frequencies over a short period. For light songs, the maximal d_B ranged from 0.81 to 2.45, and for semi-classical songs the range was 2.21–3.14.

Agriculture: Soybean yield was predicted in [Jaradat 06] using d_B. In a 2 year study comparing five crop management strategies (such as conventional versus organic), grain yield per square meter was positively and closely associated with d_B (and also with a small set of other factors). For example, plants grown under a conventional system with moldboard tillage have $d_B = 1.477$ and a yield of 11.2 g per plant, compared with $d_B = 1.358$ and a yield of 2.32 g per plant for plants grown under an organic system with strip tillage. These results are particularly important for the upper Midwestern United States, where the short growing season limits crop growth, so the crop needs to establish and maximize canopy coverage rapidly to exploit the available light.

"Optimal" Fractal Dimension of Art and Architecture: It was suggested in [Salingaros 12] that people respond best to fractals whose box counting dimension is between 1.3 and 1.5 (compare this with [Draves 08], discussed in Sect. 10.4 on applications of the correlation dimension). For example, a dense forest scene ($d_B = 1.6$) lowered stress somewhat, and a savanna landscape of isolated trees ($d_B = 1.4$) lowered stress considerably.

Biochemistry: Silicatein proteins, which are formed by a diffusion limited aggregation (DLA) process (Sect. 10.5), govern the synthesis of silica in marine sponges. Silicateins may facilitate the assembly of nanofibers, which in turn could facilitate the fabrication of inorganic materials for use in semiconductor and optoelectronic devices. The value $d_B \approx 1.7$ was calculated in [Murr 05] using linear regression and box sizes of $s = 2^i$, for $1 \leq i \leq 9$. This value is in close agreement with the theoretical value for structures formed during a DLA process.

Astrophysics: The vast majority of box-counting occurs in two dimensions. Box counting in three dimensions was used by [Murdzek 07] to compute the large scale structure of the distribution of galaxies. The d_B of two sets was computed, one with 8616 galaxies and one with 11,511 galaxies. The coordinates of a galaxy are given in "redshift" space using three numbers (right ascension, declination, and redshift). The redshift was mapped to a distance, using the Einstein–de Sitter model of the universe.[8] The spherical coordinates were then transformed to Euclidean coordinates. Twenty different box sizes were used, where $s_{max}/s_{min} \approx 100$. The computed dimension was 2.024 for the first set and 2.036 for the second set.

The idea of computing the dimension of a fractal set in $\mathbb{R}^3$ from the perspective of a single point was used in [Eckmann 03] to study the fractal dimension of the visible universe. They imagined shining a spotlight from a given point and showed that the part of a fractal visible from the point can in general have d_B not exceeding 2, since galaxies hide behind each other.

Other: The use of fractals in archeology was explored in [Brown 05], which mentioned many other applications. For example, a box counting dimension of 1.67 was calculated for a Mayan funerary vase from Guatemala, indicating that at least some Mayan art is fractal.[9] The d_B of drip paintings of the artist Jackson Pollock changed over time, reflecting the evolution of his style [Brown 05]; this analysis can also be used to detect forgeries. Fractal analysis has been used in remote sensing and searching to differentiate low dimensional Euclidean objects from the higher dimensional fractal natural background. Some information processing applications are mentioned in [Wong 05].

Conclusions:

The most important conclusion to be extracted from the above discussion of selected applications is that the value of the fractal dimension is, by itself, of relatively little interest. What is valuable, for a given application, is (*i*) how the dimension of the object compares to the dimension of other objects being studied (e.g., how the dimension of a part of the brain varies between healthy and diseased individuals) and (*ii*) how the dimension of the object changes over time. These conclusions will be equally applicable when we calculate the dimension of a network. That is, the numerical value of the dimension of a network will be of relatively little value by itself; what is valuable is how the dimension compares to the dimension of similar networks, or how the network dimension changes over time.

We close this chapter with the results in [Hadzieva 15] of a comparison of ten open source software packages. They observed that there is significant in-

---

[8]Proposed in 1932, this simplest possible model of the universe assumes that the kinetic energy of expansion due to the big bang is exactly countered by the gravitational attraction of the matter in the universe.

[9]See the Montgomery rollout drawing at http://www.famsi.org/.

consistency in terminology among the packages; e.g., sometimes d_B is called the Hausdorff or the similarity dimension. They compared the software on the Sierpiński triangle, the Koch curve, the Minkowski sausage (these three have known d_B), and one medical image of a skin disease lesion. They ran a variety of tests, e.g., using white images on a black background and black images on a while background. One package provides the option of using triangles rather than boxes. Some of the packages also can compute the correlation dimension (Chap. 9). For 9 of the 10 packages, the average error for the three fractals with known d_B ranged from 1.03% to 7.81%, and the outlier software had 53% error. For 4 of the 10 packages, the average error for these fractals was below 2%. For the analysis of the skin lesion, the estimates of d_B ranged from 1.39 to 1.73. Given this wide range, they suggested that medical researchers carefully choose the images, do cautious preprocessing, and define relevant parameters. Considering other criteria such as ease of use, image preprocessing, and consistency, they concluded that only 4 of the 10 packages are useful for research.

Chapter 7

Network Box Counting Dimension

Little boxes on the hillside, little boxes made of ticky tacky,
Little boxes on the hillside, little boxes all the same.
There's a green one and a pink one, and a blue one and a yellow one,
And they're all made out of ticky tacky, and they all look just the same.
"Little Boxes" by Malvina Reynolds (1900–1978), American folk/blues singer-songwriter and political activist.

In this chapter we begin our detailed study of fractal dimensions of a network $\mathbb{G}$. There are two approaches to calculating a fractal dimension of $\mathbb{G}$. One approach, applicable if $\mathbb{G}$ is a spatially embedded network, is to treat $\mathbb{G}$ as a geometric object and apply techniques, such as box counting or modelling $\mathbb{G}$ by an IFS, applicable to geometric objects. For example, in Sect. 5.3 we showed how to calculate the similarity dimension d_S of a tree described by an IFS. Also, several studies have computed d_B for networks embedded in $\mathbb{R}^2$ by applying box counting to the geometric figure. This was the approach taken to calculate d_B in [Benguigui 92] for four railroad networks, in [Arlt 03] for the microvascular network of chicks, in [Lu 04] for road networks in cities in Texas, and in [Hou 15] for electric power networks; later in this book we will consider those studies in more detail. These approaches to computing d_B make no use of network properties. The focus of this chapter is how to compute a fractal dimension for $\mathbb{G}$ using methods that treat $\mathbb{G}$ as a network and not simply as a geometric object.

Just as there is no one definition of the "fractal dimension" of a geometric object, there is no one definition of "the fractal dimension" of $\mathbb{G}$. The network fractal dimensions we will consider include:

> the *box counting dimension* (this chapter), based on the box-counting method for geometric fractals, describes how the number of boxes of size s needed to cover $\mathbb{G}$ varies with s,

© Springer Nature Switzerland AG 2020
E. Rosenberg, *Fractal Dimensions of Networks*,
https://doi.org/10.1007/978-3-030-43169-3_7

the *correlation dimension* (Chapter 11), which describes how the number of nodes within r hops of a random node varies with r, extends to networks the correlation dimension of geometric objects (Chapter 9),

the *information dimension* (Chapter 15) which extends to networks the information dimension of geometric objects (Chapter 14),

the *generalized dimensions* (Chapter 17) which extends to networks the generalized dimensions of geometric objects (Chapter 16).

As discussed in the beginning of Chap. 3 and in Sect. 5.5, several definitions of "fractal" have been proposed for a geometric object Ω. None of them has proved entirely satisfactory. As we saw in Sect. 6.5, applications of d_B of Ω compare d_B to the dimension of similar objects (e.g., healthy versus unhealthy lungs), or examine how d_B changes over time. Although focusing on the definition of "fractality" for Ω does not concern us, in later chapters we will need to focus on the definition of "fractality" for $\mathbb{G}$. The reason this is necessary for $\mathbb{G}$ is that, while the terms "fractality" and "self-similarity" are often used interchangeably for Ω, they have been given very different meanings for $\mathbb{G}$. With this caveat on terminology, in this chapter and in the next chapter we study the box counting dimension of $\mathbb{G}$.

Analogous to computing d_B of a geometric fractal, computing d_B of $\mathbb{G}$ requires covering $\mathbb{G}$ by a set of "boxes", where a box is a subnetwork of $\mathbb{G}$. The methods for computing a covering of $\mathbb{G}$ are also called *box counting* methods, and several methods have been proposed. Most methods are for unweighted networks for which "distance" means chemical distance (i.e., hop count); a few methods are applicable to unweighted networks, e.g., spatially embedded networks for which "distance" means Euclidean distance. Most of this chapter will focus on unweighted networks, and we begin there. Recall that throughout this book, except where otherwise indicated, we assume that $\mathbb{G} = (\mathbb{N}, \mathbb{A})$ is a connected, unweighted, and undirected network.

7.1   Node Coverings and Arc Coverings

In this section we consider what it means to cover a network $\mathbb{G} = (\mathbb{N}, \mathbb{A})$ of N nodes and A arcs.

Definition 7.1 The network B is a *subnetwork* of $\mathbb{G}$ if B can be obtained from $\mathbb{G}$ by deleting nodes and arcs. A *box* is a subnetwork of $\mathbb{G}$. A box is *disconnected* if some nodes in the box cannot be connected by arcs in the box.   □

Let $\{B_j\}_{j=1}^J \equiv \{B_1, B_2, \dots, B_J\}$ be a collection of boxes. Two types of coverings of $\mathbb{G}$ have been proposed: *node coverings* and *arc coverings*. Let s be a positive integer.

Definition 7.2 (i) The set $\{B_j\}_{j=1}^J$ is a *node s-covering* of the network $\mathbb{G}$ if for each j we have $diam(B_j) < s$ and if each node in $\mathbb{N}$ is contained in exactly one B_j. (*ii*) The set $\{B_j\}_{j=1}^J$ is an *arc s-covering* of $\mathbb{G}$ if for each j we have $diam(B_j) < s$ and if each arc in $\mathbb{A}$ is contained in exactly one B_j. □

If B_j is a box in a node or arc s-covering of $\mathbb{G}$, then the requirement $diam(B_j) < s$ in Definition 7.2 implies that B_j is connected. However, as discussed in Sect. 8.6, this requirement, which is a standard assumption in defining the box counting dimension of $\mathbb{G}$ [Gallos 07b, Kim 07a, Kim 07b, Rosenberg 17b, Song 07], may for good reasons frequently be violated in some methods for determining the fractal dimensions of $\mathbb{G}$.

It is possible to define a node covering of $\mathbb{G}$ to allow a node to be contained in more than one box; coverings with overlapping boxes are used in [Furuya 11, Sun 14]. The great advantage of non-overlapping boxes is that they immediately yield a probability distribution, as discussed in Chap. 15. The probability distribution obtained from a non-overlapping node covering of $\mathbb{G}$ is the basis for computing the information dimension d_I and the generalized dimensions D_q of $\mathbb{G}$ (Chap. 17). Therefore, in this book each covering of $\mathbb{G}$ is assumed to use non-overlapping boxes, as specified in Definition 7.2.

Since a network of diameter 0 contains only a single node, a node 1-covering contains N boxes. Since a node 1-covering provides no useful information other than N itself, we consider node s-coverings only for $s \geq 2$. Figure 7.1(*i*) illustrates a node 3-covering with $J = 2$; both boxes in this covering have diameter 2. Figure 7.1(*ii*) illustrates an arc 3-covering of the same network using 3 boxes.

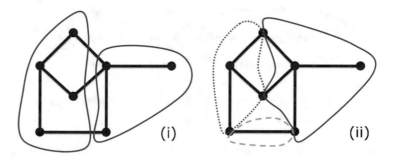

Figure 7.1: A node 3-covering (*i*) and an arc 3-covering (*ii*)

The box indicated by the solid line contains 4 arcs and has diameter 2, the box indicated by the dotted line contains 3 arcs and has diameter 2, and the box indicated by the dashed line contains 1 arc and has diameter 1. Figure 7.2 illustrates arc s-coverings for $s = 2$, $s = 3$, and $s = 4$, using 6, 3, and 2 boxes, respectively.

A node covering is not necessarily an arc covering, as illustrated by Fig. 7.3. The nodes are covered by B_1 and B_2, but the arc in the middle belongs to neither B_1 nor B_2. Also, an arc covering is not necessarily a node covering. For example, as shown in Fig. 7.4, an arc 2-covering of the 3 node chain network consists of the boxes B_1 and B_2; node n is contained in both B_1 and B_2, which violates Definition 7.2.

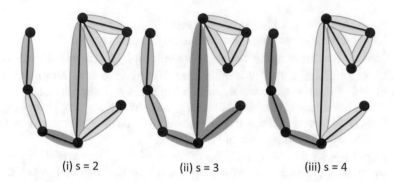

(i) s = 2   (ii) s = 3   (iii) s = 4

Figure 7.2: Arc s-coverings for $s = 2, 3, 4$

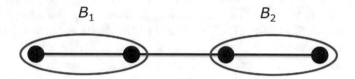

Figure 7.3: A node covering which is not an arc covering

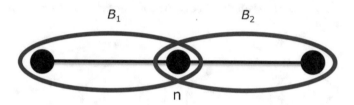

Figure 7.4: An arc covering which is not a node covering

Definition 7.3 (i) An arc s-covering $\{B_j\}_{j=1}^{J}$ is *minimal* if for any other arc s-covering $\{B'_j\}_{j=1}^{J'}$, we have $J \leq J'$. Let $B_A(s)$ be the number of boxes in a minimal arc s-covering of $\mathbb{G}$. (ii) A node s-covering $\{B_j\}_{j=1}^{J}$ is *minimal* if for any other node s-covering $\{B'_j\}_{j=1}^{J'}$, we have $J \leq J'$. Let $B_N(s)$ be the number of boxes in a minimal node s-covering of $\mathbb{G}$. $\square$

That is, a covering is minimal if it uses the fewest possible number of boxes. For $s > \Delta$, the minimal node or arc s-covering consists of a single box, which is $\mathbb{G}$ itself. Thus $B_A(s) = B_N(s) = 1$ for $s > \Delta$.

How large can the ratio $B_A(s)/B_N(s)$ be? For $s = 2$, the ratio can be arbitrarily large. For consider $\mathcal{K}_N$, the complete graph on N nodes, which has $N(N-1)/2$ arcs. For $s = 2$, we have $B_N(2) = \lceil N/2 \rceil$, since each box in the node 2-covering contains one arc, which covers two nodes. We have $B_A(2) = N(N-1)/2$, since each box in the arc 2-covering contains one arc. Thus $B_A(2)/B_N(2)$ is $\mathcal{O}(N)$.

For $s = 3$, the ratio $B_A(s)/B_N(s)$ can also be arbitrarily large, as illustrated by Fig. 7.5. This network has $2N + 2$ nodes and $3N$ arcs. We have $B_N(3) =$

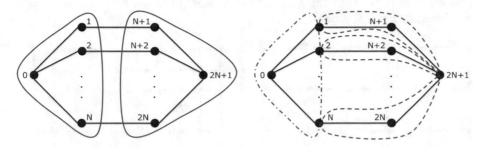

Figure 7.5: Comparing $B_N(2)$ and $B_A(2)$

2; the two boxes in the minimal node cover are shown on the left. We have $B_A(3) = N + 1$, as shown on the right. Thus $B_A(3)/B_N(3)$ is $\mathcal{O}(N)$.

Exercise 7.1* For each integer $s \geq 2$, is there a graph with N nodes such that $B_A(s)/B_N(s)$ is $\mathcal{O}(N)$? □

Virtually all research on network coverings has considered node coverings; only a few studies, e.g., [Jalan 17, Zhou 07], use arc coverings. The arc covering approach of [Jalan 17] uses the $N \times N$ adjacency matrix M (Sect. 2.1). For a given integer s between 2 and $N/2$, the method partitions M into non-overlapping square $s \times s$ boxes, and then counts the number $m(s)$ of boxes containing at least one arc. The network has fractal dimension d if $m(s) \sim s^{-d}$. Since $m(s)$ depends in general on the order in which the nodes are labelled, $m(s)$ is averaged over many shufflings of the node order.

The reason arc coverings are rarely used is that, in practice, computing a fractal dimension of a geometric object typically starts with a given set of points in $\mathbb{R}^E$ (the points are then covered by boxes, or the distance between each pair of points is computed, as discussed in Chap. 9), and nodes in a network are analogous to points in $\mathbb{R}^E$. Having contrasted arc coverings and node coverings for a network, we now abandon arc coverings; henceforth, all coverings of $\mathbb{G}$ are node coverings, and by covering $\mathbb{G}$ we mean covering the nodes of $\mathbb{G}$. Also, henceforth by an s-covering we mean a node s-covering, and by a covering of size s we mean an s-covering.

7.2 Diameter-Based and Radius-Based Boxes

There are two main approaches used to define boxes for use in covering $\mathbb{G}$: diameter-based boxes and radius-based boxes.

Definition 7.4 (*i*) A *radius-based box* $\mathbb{G}(n, r)$ with center node $n \in \mathbb{N}$ and radius r is the subnetwork of $\mathbb{G}$ containing all nodes whose distance to n does not exceed r. Let $B_R(r)$ be the minimal number of radius-based boxes of radius at most r needed to cover $\mathbb{G}$. (*ii*) A *diameter-based box* $\mathbb{G}(s)$ of size s is a subnetwork of $\mathbb{G}$ of diameter $s - 1$. Let $B_D(s)$ denote the minimal number of diameter-based boxes of size at most s needed to cover $\mathbb{G}$. □

Example 7.1 For a rectilinear grid, the diameter-based box containing the largest number of nodes for $s = 3$ is shown in Fig. 7.6 (*left*); this box contains 5 nodes. The boxes containing the largest number of nodes for $s = 5$ and $s = 7$ are shown in Fig. 7.6 (*right*).   □

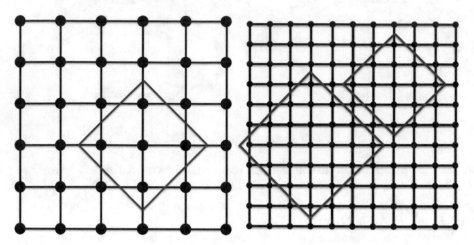

Figure 7.6: Optimal $B_D(3)$ box (left), and optimal $B_D(5)$ and $B_D(7)$ boxes (right)

The subscript "R" in $B_R(n, r)$ denotes "radius". Thus the node set of $\mathbb{G}(n, r)$ is $\{x \in \mathbb{N} \,|\, dist(n, x) \leq r\}$. Radius-based boxes are used in the *Maximum Excluded Mass Burning* and *Random Sequential Node Burning* methods which we will study in Chap. 8. Interestingly, the above definition of a radius-based box may frequently be violated in the *Maximum Excluded Mass Burning* and *Random Sequential Node Burning* methods. In particular, some radius-based boxes created by those methods may be disconnected, or some boxes may contain only some of the nodes within distance r of the center node n.

For a given s there is in general not a unique $\mathbb{G}(s)$; e.g., for a long chain network and s small, there are many diameter-based boxes of size s. A diameter-based box $\mathbb{G}(s)$ is not defined in terms of a center node; instead, for nodes $x, y \in \mathbb{G}(s)$ we require $dist(x, y) < s$.[1] Diameter-based boxes are used in the *Greedy Node Coloring*, *Box Burning*, and *Compact Box Burning* heuristics described in Chap. 8. The above definition of a diameter-based box also may frequently be violated in the *Box Burning* and *Compact Box Burning* methods. Also, since each node in $\mathbb{G}$ must belong to exactly one B_j in an s-covering $\{B_j\}_{j=1}^{J}$ using diameter-based boxes, then in general we will not have $diam(B_j) = s - 1$ for all j. To see this, consider a chain of 3 nodes (call them x, y, and z), and let $s = 2$. The minimal 2-covering using diameter-based boxes requires two boxes, B_1 and B_2. If B_1 covers x and y, then B_2 covers only z, so the diameter of B_2 is 0.

---

[1] There is a slight abuse of notation here. We should properly write "for $x, y \in \mathbb{N}(s)$", where $\mathbb{N}(s)$ is the node set of $\mathbb{G}(s)$. To avoid introducing extra notation, we simply write "for nodes $x, y \in \mathbb{G}(s)$".

The subscript "D" in $B_D(s)$ denotes "diameter". The minimal number of diameter-based boxes of size at most $2r+1$ needed to cover $\mathbb{G}$ is, by definition, $B_D(2r+1)$. We have $B_D(2r+1) \leq B_R(r)$ [Kim 07a]. To see this, let $\mathbb{G}(n_j, r_j)$, $j = 1, 2, \ldots, B_R(r)$ be the boxes in a minimal covering of $\mathbb{G}$ using radius-based boxes of radius at most r. Then $r_j \leq r$ for all j. Pick any j, and consider box $\mathbb{G}(n_j, r_j)$. For any nodes x and y in $\mathbb{G}(n_j, r_j)$ we have

$$dist(x,y) \leq dist(x, n_j) + dist(n_j, y) \leq 2r_j \leq 2r \,,$$

so $\mathbb{G}(n_j, r_j)$ has diameter at most $2r$. Thus these $B_R(r)$ boxes also serve as a covering of size $2r+1$ using diameter-based boxes. Therefore, the minimal number of diameter-based boxes of size at most $2r+1$ needed to cover $\mathbb{G}$ cannot exceed $B_R(r)$; that is, $B_D(2r+1) \leq B_R(r)$.

The reverse inequality does not in general hold, since a diameter-based box of size $2r+1$ can contain more nodes than a radius-based box of radius r. For example, consider the network $\mathbb{G}$ of Fig. 7.7. The only nodes adjacent to n are

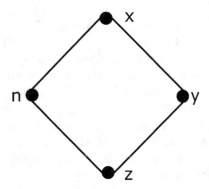

Figure 7.7: Diameter-based versus radius-based boxes

x and z, so $\mathbb{G}(n,1) = \{n, x, z\}$ and $B_R(1) = 2$. Yet the diameter of $\mathbb{G}$ is 2, so it can be covered by a single diameter-based box of size 3, namely $\mathbb{G}$ itself, so $B_D(3) = 1$. Thus $B_R(r)$ and $B_D(2r+1)$ are not in general equal. Nonetheless, for the *C. elegans* and Internet backbone networks studied in [Song 07], the calculated fractal dimension was the same whether radius-based or diameter-based boxes were used. Similarly, both radius-based and diameter-based boxes yielded a fractal dimension of approximately 4.1 for the WWW (the World Wide Web) [Kim 07a].

As applied to a network $\mathbb{G}$, the term *box counting* refers to computing a minimal s-covering of $\mathbb{G}$ for a range of values of s, using either radius-based boxes or diameter-based boxes. Conceivably, other types of boxes might be used to cover $\mathbb{G}$. Once we have computed $B_D(s)$ for various values of s (or $B_R(r)$ for various values of r) we can try to compute d_B.

Computing any fractal dimension for a finite network is necessarily open to the same criticism that can be directed to computing a fractal dimension of a real-world geometric object, namely that a fractal dimension is defined only for a fractal (a theoretical construct with structure at infinitely many levels), and is not defined for a pre-fractal. This criticism has not impeded the usefulness of d_B of real-world geometric objects. Applying the same reasoning to networks, we can try to compute d_B for a finite network, and even for a small finite network.

In the literature on fractal dimensions of networks, d_B is often (e.g., [Song 07]) informally defined by the scaling $B_D(s) \sim s^{-d_B}$. The drawback of this definition is that the scaling relation "$\sim$" is not well defined, e.g., often in the statistical physics literature "$\sim$" indicates a scaling as $s \to 0$ (Sect. 1.2), which is not meaningful for a network. In the context of network fractal dimensions, the symbol "$\sim$", is usually interpreted to mean "approximately behaves like". Definition 7.5 below provides a more computationally useful definition of d_B. Recall that Δ is the diameter of $\mathbb{G}$.

Definition 7.5 [Rosenberg 17a] $\mathbb{G}$ has box counting dimension d_B if over some range of s and for some constant c we have

$$\log B_D(s) \approx -d_B \log(s/\Delta) + c. \qquad (7.1)$$

If the box counting dimension of $\mathbb{G}$ exists, then $\mathbb{G}$ enjoys the *fractal scaling property*, or, more simply, $\mathbb{G}$ is *fractal*.   □

Alternatively and equivalently, sometimes (7.1) is written as $\log B_D(s) \approx -d_B \log s + c$. If $\mathbb{G}$ has box counting dimension d_B, then over some range of s we have $B_D(s) \approx \alpha s^{-d_B}$ for some constant α. Since each box in the covering need not have the same size, it could be argued that this fractal dimension of $\mathbb{G}$ should instead be called the Hausdorff dimension of $\mathbb{G}$; a different definition of the Hausdorff dimension of $\mathbb{G}$ [Rosenberg 18b] is studied in Chap. 18. Calling $\mathbb{G}$ "fractal" if d_B exists is the terminology used in [Gallos 07b].

Example 7.2 The *dolphins* network [Lusseau 03] is a social network describing frequent associations between 62 dolphins in a community living off Doubtful Sound, New Zealand. The network, which has 62 nodes, 159 arcs, and $\Delta = 8$, is illustrated in Fig. 7.8. Figure 7.9 provides the node degree histogram; clearly

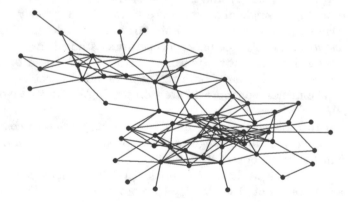

Figure 7.8: Frequent associations among 62 dolphins

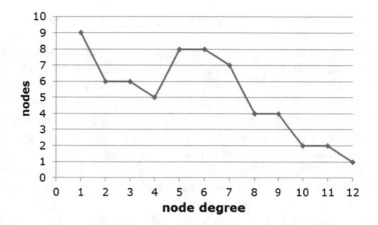

Figure 7.9: Node degrees for the *dolphins* network

this is not a scale-free network. The results from the *Greedy Node Coloring* box counting method (to be described in Sect. 8.1), which uses diameter-based boxes, are shown in Fig. 7.10 This figure suggests that d_B should be estimated

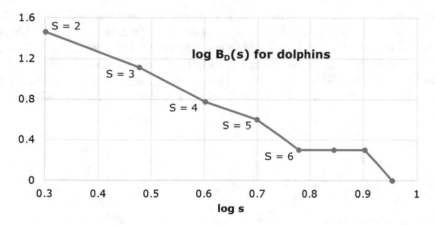

Figure 7.10: Diameter-based box counting results for the *dolphins* network

over the range $2 \leq s \leq 6$. Applying regression to the $\left(\log s, \log B_D(s) \right)$ values over this range of s yields the estimate $d_B = 2.38$. □

Exercise 7.2* In Sect. 5.4 we defined the packing dimension of a geometric object. Can we define a packing dimension of a network? □

7.3   A Closer Look at Network Coverings

In describing their box counting method (which we present in Chap. 8), [Kim 07a] wrote "The particular definition of box size has proved to be inessential for fractal scaling." On the contrary, it can make a great deal of difference which box definition is being used when computing d_B. To see this, and recalling Definition 7.4, suppose we create two new definitions: (i) a *wide box* $\widetilde{\mathbb{G}}(s)$ of size s is a subnetwork of $\mathbb{G}$ of diameter s, and (ii) let $\widetilde{B}(s)$ denote the minimal number of wide boxes of size at most s needed to cover $\mathbb{G}$. With these two new definitions, the difference between a wide box $\widetilde{\mathbb{G}}(s)$ of size s and a diameter-based box $\mathbb{G}(s)$ of size s is that the diameter of $\widetilde{\mathbb{G}}(s)$ is s and the diameter of $\mathbb{G}(s)$ is $s-1$.

Consider a network of N nodes, arranged in a chain, as illustrated by Fig. 7.11 for $N = 5$. For $N \gg 1$ we would expect such a 1-dimensional net-

Figure 7.11: Chain network

work to have $d_B = 1$, or at least $d_B \approx 1$. Suppose we cover the chain of N nodes with wide boxes of size s. Since a wide box of size s contains $s+1$ nodes, then $\widetilde{B}(s) = \lceil N/(s+1) \rceil$. Let $N = K!$ for some integer K. Then for $1 \le s \le K-1$, if we cover $\mathbb{G}$ by wide boxes of size s, each box contains exactly $s+1$ nodes, so there are no partially filled boxes in the covering. Figure 7.12 plots $\log \widetilde{B}(s)$ versus $\log s$ for $K = 15$ and this range of s. The curve is concave, not linear,

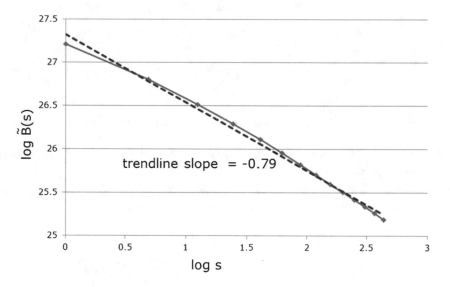

Figure 7.12: $\log \widetilde{B}(s)$ versus $\log s$ using wide boxes

and a trend line for these points has a slope of -0.79, rather than a straight line with a slope of -1.

This example numerically illustrates that, for a chain network of $N = K!$ nodes, when using wide boxes the relation $\widetilde{B}(s) = \beta s^{-d_B}$ cannot be satisfied, for some constant β, with $d_B = 1$ for some range of s. To show analytically that for this chain network $\widetilde{B}(s) = \beta s^{-d_B}$ cannot be satisfied with $d_B = 1$, suppose $s\widetilde{B}(s) = \beta$ for some range of s. For $s \leq K - 1$ each wide box in the covering of the chain contains exactly $s + 1$ nodes, so $\widetilde{B}(s) = K!/(s+1)$ and

$$s\widetilde{B}(s) = K! \left(\frac{s}{s+1} \right)$$

which is not independent of s. So $\widetilde{B}(s) = \beta s^{-d_B}$ with $d_B = 1$ cannot hold, which provides the counterexample to the claim in [Kim 07a] that the particular definition of box size is inessential. However, if we use diameter-based boxes $\mathbb{G}(s)$ as defined by Definition 7.4, then for a chain network of $K!$ nodes the problem disappears, since then each box of size s contains s nodes, not $s + 1$. This leads to the following rule, applicable to any $\mathbb{G}$, not just to a chain network.

Rule 7.1 When using diameter-based boxes of size s, the box counting dimension d_B of $\mathbb{G}$ should be calculated using the ordered pairs $\big(\log s, \log B_D(s)\big)$. $\square$

Now consider covering a chain network with radius-based boxes. By Definition 7.4, for a chain network a radius-based box $\mathbb{G}(n, r)$ contains $2r+1$ nodes, assuming we are not close to either end of the chain. This is illustrated is Fig. 7.13, where for $r = 2$ the box covers 5 nodes. Assume again that the chain network

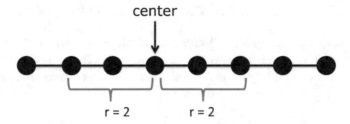

Figure 7.13: Radius-based box

has $N = K!$ nodes, for some K. Suppose we cover the nodes of this chain with radius-based boxes of radius r, where $r \leq (K - 1)/2$, so $2r + 1 \leq K$. By definition, $B_R(r)$ is the number of radius-based boxes of radius at most r needed to cover the chain. Since each box in the covering of the chain contains exactly $2r + 1$ nodes, then for $r \leq (K-1)/2$ we have $B_R(r) = N/(2r+1) = K!/(2r+1)$, which is an integer. Setting $d_B = 1$ in $B_R(r) = \beta r^{-d_B}$, where β is a constant, yields $B_R(r) = \beta r^{-1}$, which we rewrite as $rB_R(r) = \beta$. Thus for $r \leq (K-1)/2$,

$$rB(r) = K! \left(\frac{r}{2r+1} \right) = N \left(\frac{r}{2r+1} \right)$$

which is not independent of r. The problem disappears when the scaling relation is written as $B_R(r) = \beta(2r+1)^{-1}$, since then for $r \le (K-1)/2$ we obtain the desired identity

$$\beta = (2r+1)B_R(r) = (2r+1)\left(\frac{K!}{2r+1}\right) = N\,.$$

This leads to the following rule, applicable to any $\mathbb{G}$, not just to a chain network.
Rule 7.2 When using radius-based boxes with radius r, the box counting dimension d_B of $\mathbb{G}$ should be calculated using the ordered pairs $\big(\log(2r+1), \log B_R(r)\big)$. $\square$

Exercise 7.3 Since $B_D(s) = 1$ when $s > \Delta$, does it follow that $B_R(r) = 1$ when $2r + 1 > \Delta$? $\square$

Exercise 7.4 For a positive integer K, let $G(K)$ denote a square $K \times K$ rectilinear grid. The diameter of $G(K)$ is $2(K-1)$. If $K = K_1K_2$, then we can cover $G(K)$ using K_1^2 copies of $G(K_2)$, or using K_2^2 copies of $G(K_1)$. This is illustrated in Fig. 7.14 for $K = 6$, $K_1 = 2$, and $K_2 = 3$. We can cover $G(6)$

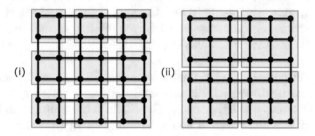

Figure 7.14: Covering a 6×6 grid

using 9 copies of $G(2)$, or using 4 copies of $G(3)$. Since $diam\big(G(3)\big) = 4$ and $diam\big(G(2)\big) = 2$, then the 4 copies of $G(3)$ constitute a diameter-based covering of size 5, and the 9 copies of $G(2)$ constitute a diameter-based covering of size 3. Estimating d_B from these two coverings,

$$d_B = \frac{\log B_D(3) - B_D(5)}{\log 5 - \log 3} = \frac{\log(9/4)}{\log(5/3)}\,. \tag{7.2}$$

There is an error in the above analysis. What is the error? Why did we not obtain $d_B = 2$? $\square$

We will return to rectilinear networks in Sect. 11.4, where we study the correlation dimension of a rectilinear grid.

Exercise 7.5 Consider an infinite rectilinear grid in $\mathbb{R}^2$. Show that, when s is odd, a box with the maximal number of nodes has a center node, and the maximal number of nodes in a box of size s is $(s^2 + 1)/2$. $\square$

Exercise 7.6 Let $d_B(\mathbb{G})$ be the box counting dimension of network $\mathbb{G}$. Suppose $\mathbb{G}_1$ is a subnetwork of $\mathbb{G}_2$. Does the strict inequality $d_B(\mathbb{G}_1) < d_B(\mathbb{G}_2)$ hold? $\square$

Exercise 7.7 As discussed in Sect. 5.5, a geometric fractal is sometimes defined as an object whose dimension is not an integer. Is this a useful definition of a "fractal network"? □

Exercise 7.8 Consider an infinite rectilinear lattice in $\mathbb{R}^2$. What does a diameter-based box of size s look like when s is even? □

Exercise 7.9 Let $\mathbb{G}$ be a 6×6 square rectilinear lattice in $\mathbb{R}^2$. Compute $B_D(s)$ for $s = 2, 3, 4, 5$. Compute $B_R(r)$ for $r = 1, 2, 3$. □

7.4 The Origin of Fractal Networks

Recall from Definition 7.5 that $\mathbb{G}$ is fractal if for some d_B and some c we have $\log B_D(s) \approx -d_B \log(s/\Delta) + c$ over some range of s. The main feature apparently displayed by fractal networks is a repulsion between hubs, where a hub is a node with a significantly higher node degree than a non-hub node. That is, the highly connected nodes tend to be not directly connected [Song 06, Zhang 16]. This tendency can be quantified using the joint node degree distribution $p(\delta_1, \delta_2)$ that a node with degree δ_1 and a node with degree δ_2 are neighbors. In contrast, for a non-fractal network $\mathbb{G}$, hubs are mostly connected to other hubs, which implies that $\mathbb{G}$ enjoys the small-world property [Gallos 07b]. (Recall from Sect. 2.2 that $\mathbb{G}$ is a *small-world* network if $diam(\mathbb{G})$ grows as $\log N$.) Also, the concepts of *modularity* and *fractality* for a network are closely related. Interconnections within a module (e.g., a biological subsystem) are more prevalent than interconnections between modules. Similarly, in a fractal network, interconnections between a hub and non-hub nodes are more prevalent than interconnections between hubs. Non-fractal networks are typically characterized by a sharp decay of $B_D(s)$ with s, which is better described by an exponential law $B_D(s) \sim e^{-\beta s}$, where $\beta > 0$, rather than by a power law $B_D(s) \sim s^{-\beta}$, with a similar statement holding if radius-based boxes are used [Gallos 07b]. These two cases are illustrated in Fig. 7.15, taken from [Gallos 07b], where the solid circles are mea-

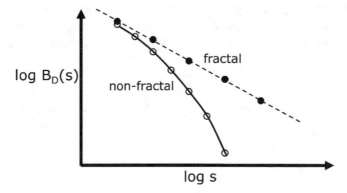

Figure 7.15: Fractal versus non-fractal scaling

surements from a fractal network, and the hollow circles are from a non-fractal network.

Various researchers have proposed models to explain why networks exhibit fractal scaling. One proposed cause [Song 06] is the disassortative correlation between the degrees of neighboring nodes. A *disassortative* network is a network in which nodes of low degree are more likely to connect with nodes of high degree, and so there is a strong disassortativity (i.e., repulsion) between hubs. This can be modelled by Coulomb's Law [Zhang 16].

A different explanation is based on the *skeleton* of $\mathbb{G}$ (Sect. 2.5). Recall that the skeleton $skel(\mathbb{G})$ of $\mathbb{G}$ is often defined to be the betweenness centrality or load-based spanning tree of $\mathbb{G}$. The skeleton of a scale-free network has been found [Kim 07b] to have the same topological properties as a random branching tree. Associated with a random branching tree is the parameter $\mu(r)$, the average number of children of a node whose distance from the root node is r. As $r \to \infty$, we have $\mu(r) \to \mu$ for some μ. The critical case for this branching process corresponds to $\mu = 1$, and for this value the random branching tree is known to be fractal. Specifically, suppose $\mathbb{G}$ is a scale-free random branching tree for which the probability p_k that each branching event produces k offspring scales as $p_k \sim k^{-\gamma}$ and $\langle k \rangle \equiv \sum_{k=0}^{\infty} k\, p_k = 1$. Then the box counting dimension of $\mathbb{G}$ is given by

$$d_B = \begin{cases} (\gamma - 1)/(\gamma - 2) & \text{for } 2 < \gamma < 3 \\ 2 & \text{for } \gamma > 3 \,. \end{cases}$$

Similar results have been found for *supercritical* branching trees, where supercritical means $\langle k \rangle > 1$ [Kim 07b]. The significance of these results is that the original scale-free network can be modelled by first starting with a skeleton network with the properties of a critical or supercritical random branching tree, and then adding shortcuts, but not so many that fractality is destroyed. Since the number of boxes needed to cover $\mathbb{G}$ and $skel(\mathbb{G})$ have been found to be nearly equal [Kim 07b], then the fractal scaling enjoyed by $skel(\mathbb{G})$ is also enjoyed by $\mathbb{G}$. For example, for the World Wide Web (WWW), the number of boxes needed to cover the WWW and its skeleton are nearly equal, and the same fractal dimension of 4.10 is obtained [Kim 07b]. The mean branching value $\mu(r)$ hits a plateau value $\mu \approx 3.5$ at $r \approx 20$. In contrast, for non-fractal networks, the mean branching number of the skeleton decays to zero without forming a plateau.

Chapter 8

Network Box Counting Heuristics

It's just a box of rain,
I don't know who put it there,
Believe it if you need it,
Or leave it if you dare
from the song "Box of Rain", lyrics by Phil Lesh and Robert Hunter,
recorded on the 1970 album "American Beauty"

In this chapter we examine several methods for computing a minimal set of
diameter-based boxes, or a minimal set of radius-based boxes, to cover $\mathbb{G}$. We
will see that some of these methods, which have been shown to be quite effective,
nonetheless require us to bend some of the definitions presented in Chap. 7. In
particular, some of these methods may generate boxes that are not connected
subnetworks of $\mathbb{G}$.

8.1 Node Coloring Formulation

The problem of determining the minimal number $B_D(s)$ of diameter-based
boxes of size at most s needed to cover $\mathbb{G}$ is an example of the NP-hard *graph
coloring problem* [Song 07], for which many good heuristics are available. To
transform the covering problem into the graph coloring problem for a given
$s \geq 2$, first create the auxiliary graph $\widetilde{\mathbb{G}}_s = (\mathbb{N}, \widetilde{\mathbb{A}}_s)$ as follows. The node set
of $\widetilde{\mathbb{G}}_s$ is $\mathbb{N}$; it is independent of s. The arc set of $\widetilde{\mathbb{G}}_s$ depends on s: there is an
undirected arc (u, v) in $\widetilde{\mathbb{A}}_s$ if and only if $dist(u, v) \geq s$, where the distance is in
the original graph $\mathbb{G}$.

Having constructed $\widetilde{\mathbb{G}}_s$, the task is to color the nodes of $\widetilde{\mathbb{G}}_s$, using the minimal number of colors, such that no arc in $\widetilde{\mathbb{A}}_s$ connects nodes assigned the same
color. That is, if $(u, v) \in \widetilde{\mathbb{A}}_s$, then u and v must be assigned different colors.
The minimal number of colors required is called the *chromatic number* of $\widetilde{\mathbb{G}}_s$,
traditionally denoted by $\chi(\widetilde{\mathbb{G}}_s)$.

© Springer Nature Switzerland AG 2020
E. Rosenberg, *Fractal Dimensions of Networks*,
https://doi.org/10.1007/978-3-030-43169-3_8

Theorem 8.1 $\chi(\widetilde{\mathbb{G}}_s) = B_D(s)$.

Proof [Song 07] Suppose that nodes u and v are assigned the same color. Then u and v cannot be the endpoints of an arc in $\widetilde{\mathbb{G}}_s$, because if they were, they would be assigned different colors. Hence $dist(u,v) < s$, so u and v can be placed in a single diameter-based box of size s. It follows that $B_D(s) \leq \chi(\widetilde{\mathbb{G}}_s)$. To prove the reverse inequality, consider any minimal s-covering using $B_D(s)$ boxes, and let B be any box in this covering. For any nodes x and y in this box we have $dist(x,y) < s$, so x and y are not connected by an arc in $\widetilde{\mathbb{G}}_s$. Thus x and y can be assigned the same color, which implies $\chi(\widetilde{\mathbb{G}}_s) \leq B_D(s)$. $\square$

We illustrate the node coloring formulation using the network of Fig. 8.1. For $s = 3$, the auxiliary graph $\widetilde{\mathbb{G}}_3$ is given by Fig. 8.2(i). Arc (x,y) exists in $\widetilde{\mathbb{G}}_3$

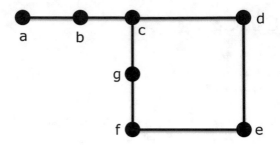

Figure 8.1: Example network for node coloring

if and only if in $\mathbb{G}$ we have $dist(x,y) \geq 3$. Thus node c is isolated in $\widetilde{\mathbb{G}}_3$ since its distance in $\mathbb{G}$ to all other nodes does not exceed 2. Also, the distance in $\mathbb{G}$ from g to all nodes except a does not exceed 2, so arc (g,a) exists in $\widetilde{\mathbb{G}}_3$.

The chromatic number $\chi(\widetilde{\mathbb{G}}_3)$ of the simple network of Fig. 8.2(i) can be exactly computed using the *Greedy Coloring* method [Song 07], which assigns colors based on a random ordering of the nodes. For large networks, *Greedy Coloring* would typically be run many times, using different random orderings of the nodes; using 10,000 random orderings, *Greedy Coloring* has been shown to provide significant accuracy. The method is very efficient, since, for a given random ordering of the nodes, a single pass through all the nodes suffices to compute an s-covering of $\mathbb{G}$ for all box sizes s [Song 07].

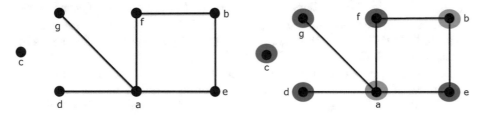

Figure 8.2: (i) Auxiliary graph with s=3, and (ii) its coloring

We illustrate *Greedy Coloring* using Fig. 8.2(ii). Suppose we randomly pick a as the first node, and assign the color green to node a. Then d, e, f, and g cannot be colored green, so we color them blue. We can color b green since it is connected only to nodes already colored blue. Since c is isolated we are free to assign it any color, so we color it blue. We are done; nodes a and b are in the green box and nodes c, d, e, f, and g are in the blue box. This is an optimal coloring, since at least two colors are needed to color any graph with at least one arc. Figure 8.3 illustrates, in the original network, the two boxes in this minimal covering for $s = 3$.

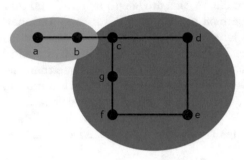

Figure 8.3: Minimal covering for $s = 3$

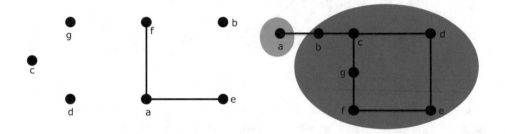

Figure 8.4: (i) Auxiliary graph with s=4, and (ii) minimal covering for s=4

For $s = 4$, the auxiliary graph $\widetilde{\mathbb{G}}_4$ is shown in Fig. 8.4(i).
There is an arc (x, y) in $\widetilde{\mathbb{G}}_4$ if and only if in $\mathbb{G}$ we have $dist(x, y) \geq 4$. We again apply the *Greedy Coloring* heuristic to compute the chromatic number $\chi(\widetilde{\mathbb{G}}_4)$. Suppose we randomly pick a as the first node, and assign the color green to node a. Then e and f cannot be colored green, so we color them blue. The remaining nodes are isolated so we arbitrarily color them blue. We are done; a is in the green box and the remaining nodes are in the blue box. This is also an optimal coloring. Figure 8.4(ii) illustrates, in the original network, the two boxes in the minimal covering for $s = 4$.

8.2   Node Coloring for Weighted Networks

For an unweighted network, the distance between two nodes (also known as the chemical distance, or the hop count) ranges from 1 to the diameter Δ of the network. However, when applying box counting to a weighted network, choosing box sizes between 2 and Δ will not in general be useful.[1] For example, if the network diameter is less than 1, then the entire network is contained in a box of size 1. One simple approach to dealing with box size selection for weighted networks is to multiply each arc length by a sufficiently large constant k. For example, if we approximate each arc length by a rational number, then choosing k to be the least common denominator of all these rational numbers will yield a set of integer arc lengths. However, even if all the arc lengths are integer, a set of box sizes must still be selected. The box sizes could be selected using a default method, such as starting with a box size less than the network diameter, and decreasing the box size by a factor of 2 in each iteration. Alternatively, the box sizes could be determined by an analysis of the set of arc lengths. This is the approach taken in [Wei 13], and we describe their method using the network of Fig. 8.5. The nodes are a, b, c, d, e, f, and each arc length is shown. We

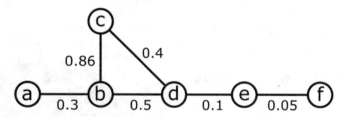

Figure 8.5: A weighted network

first pre-process the data by computing the shortest distance d_{ij} between each pair (i, j) of nodes. The largest d_{ij} is the diameter Δ. For this example we have $\Delta = 1.16$, which is the distance between nodes a and c. The second pre-processing step is to sort all the d_{ij} values in increasing order. For this example, the five smallest d_{ij} values are $0.05, 0.1, 0.3, 0.4, 0.5$. Next we compute the successive sums of the ordered d_{ij} values, stopping when the sum first exceeds Δ. The first sum is $\sigma(1) = 0.05$, the second is $\sigma(2) = \sigma(1) + 0.1 = 0.15$, the third is $\sigma(3) = \sigma(2) + .3 = .45$, the fourth is $\sigma(4) = \sigma(3) + 0.4 = 0.85$, and finally $\sigma(5) = \sigma(4) + 0.5 = 1.35 > \Delta$. We set $s_1 = \sigma(4) = 0.85$ since this is the largest sum not exceeding Δ.

Next we create an auxiliary graph $\widetilde{\mathbb{G}}$ such that an arc in $\widetilde{\mathbb{G}}$ exists between nodes i and j if and only if in $\mathbb{G}$ we have $d_{ij} \geq s_1$. There are four pairs of nodes for which $d_{ij} \geq s_1$, namely (a, c), (a, e), (a, f), and (c, b), so $\widetilde{\mathbb{G}}$ has four arcs. Node d does not appear in $\widetilde{\mathbb{G}}$. The length of arc (i, j) in $\widetilde{\mathbb{G}}$ is d_{ij}, e.g.,

---

[1] This section is one of the few sections in this book where we consider weighted networks.

$d_{af} = 0.95$, which is the length of the shortest path in $\mathbb{G}$ from a to f. Next we assign a weight to each node in $\widetilde{\mathbb{G}}$. For node i in $\widetilde{\mathbb{G}}$, the weight $w(i)$ is

$$w(i) \equiv \sum_{(i,j)\in\widetilde{\mathbb{G}}} d_{ij}.$$

The $w(i)$ values are the underlined values in Fig. 8.6 next to each node. Thus $w(a) = 0.9 + 1.16 + 0.95 = 3.01$ for the three arcs in $\widetilde{\mathbb{G}}$ incident to a, and $w(f) = 0.95$ for the one arc in $\widetilde{\mathbb{G}}$ incident to f. As in Sect. 8.1, we color the

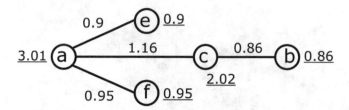

Figure 8.6: Auxiliary graph for box size $s_1 = 0.85$

nodes of $\widetilde{\mathbb{G}}$ so that the endpoints of each arc in $\widetilde{\mathbb{G}}$ are assigned different colors. Each color will correspond to a distinct box, so using the minimal number of colors means using the fewest boxes. We start with the node with the highest weight. This is node a, whose weight is 3.01. Suppose we assign to a the color `yellow`, as indicated in Fig. 8.7. Then nodes c, e, and f cannot be colored `yellow`, so we color them `blue`. The remaining node in $\widetilde{\mathbb{G}}$ to be colored is node b, and it can be colored `yellow`.

The final step in this iteration is to color to each node not in $\widetilde{\mathbb{G}}$. If node i does not appear in $\widetilde{\mathbb{G}}$, then by construction of $\widetilde{\mathbb{G}}$, we have $d_{ij} < s_1$ for $j \in \mathbb{N}$. Thus i can be assigned to any non-empty box, so i can be assigned any previously used color. In our example, only d is not in $\widetilde{\mathbb{G}}$, and we arbitrarily color it `yellow`. Thus for the initial box size s_1 we require only two boxes, a `yellow`

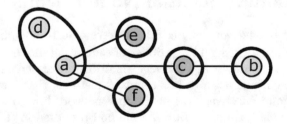

Figure 8.7: Two boxes are required for box size s_1

box containing a, b, and d, and a `blue` box containing c, e, and f. This concludes the iteration for the initial box size s_1.

For the second iteration, we first need the new box size s_2. The method of [Wei 13] simply takes the next smallest of the sums. Since $s_1 = \sigma(4)$, then

$s_2 = \sigma(3) = 0.45$. With this new box size, we continue as before, creating a new auxiliary graph $\widetilde{\mathbb{G}}$ containing each arc (i,j) such that $d_{ij} \geq s_2$, determining the weight of each node in $\widetilde{\mathbb{G}}$, coloring the nodes of $\widetilde{\mathbb{G}}$, and then coloring the nodes not in $\widetilde{\mathbb{G}}$. Then we select $s_3 = \sigma(2)$ and continue in this manner.

This method was applied in [Wei 13] to the *C. elegans* network of 306 nodes and 2148 arcs, and to the *USAir97* network of 332 nodes and 2126 arcs. The original *USAir97* network, illustrated in Fig. 8.8 below, is a directed network, but here the network is assumed to be undirected. The most computationally

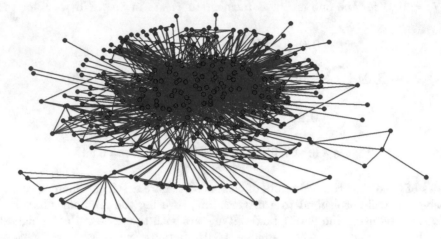

Figure 8.8: *USAir97* network

intensive step is the initial computation of the shortest path between each pair of nodes; with a Fibonacci heap implementation [Aho 93] this computation has $\mathcal{O}(AN + N^2 \log N)$ running time, where A is the number of arcs in $\mathbb{G}$.

We now return to the study of unweighted networks, and examine other methods for box counting.

8.3   Random Sequential Node Burning

Although for the network of Fig. 8.1 we were able to compute $\chi(\widetilde{\mathbb{G}}_3)$ and $\chi(\widetilde{\mathbb{G}}_4)$ by inspection, in general the problem of computing the chromatic number of $\mathbb{G}$ is NP-hard. Thus, except for very small instances such as Fig. 8.1, we must resort to heuristic network box-counting methods that have been found to be both fast and effective. One of the first network box-counting methods is the *Random Sequential Node Burning* method proposed in [Kim 07a]. This method, which utilizes radius-based boxes, was developed for computing the fractal dimension of a scale-free network, but has general applicability.

For a given radius r, the procedure is as follows. Initially all nodes are uncovered (or "unburned", in the terminology of [Kim 07a], i.e., not yet assigned to a box), and the box count $B_R(r)$ is initialized to 0. In each iteration, we first pick a random node n which may be covered or uncovered (i.e., n may be burned or unburned), but which has not previously been selected as the center node of a box. We create the new radius-based box $\mathbb{G}(n, r)$. Next we add to $\mathbb{G}(n, r)$ each uncovered node whose distance from n does not exceed r. If $\mathbb{G}(n, r)$ contains no uncovered nodes, then $\mathbb{G}(n, r)$ is discarded; otherwise, $B_R(r)$ is incremented by 1, and each uncovered node added to $\mathbb{G}(n, r)$ is marked as covered. This concludes an iteration of the method.

If there are still uncovered nodes, the next iteration begins by picking another random node n' which may be covered or uncovered, but which has not previously been selected as the center node of a box. We continue in this manner until all nodes are covered. Pseudocode of this method is given in Fig. 8.9. The set $\mathcal{C}$ is the set of covered nodes, which is initially empty. The node selected in Step 3 should not previously have been selected as the center node of a box. The binary variable δ indicates whether any unburned nodes were added to the box. If the center node selected in Step 3 is unburned, then in Step 5 we set $\delta = 1$, so in Step 7 this box is kept even if no additional unburned nodes were added to it. The method terminates when $\mathcal{C} = \mathbb{N}$.

```
procedure Random Sequential Node Burning(r)
1      initialize: set  B_R(s) = 0 and C = ∅;
2      while (C ≠ N) {
3           select a new random center n ∈ N; create the new box G(n,r) = {n} and set δ = 0;
4           for (x ∈ N − C) {
5                if (dist(n, x) ≤ r) {set δ = 1 and C = C ∪ {x} and G(n,r) = G(n,r) ∪ {x};}
6           }
7           if (δ = 0) then discard G(n,r); else set B_R(s) = B_R(s) + 1;
8      }
9      return B_R(s);
```

Figure 8.9: *Random Sequential Node Burning*

Figure 8.10, adapted from [Kim 07a], illustrates *Random Sequential Node Burning* for $r = 1$. Initially, all eight nodes are uncovered. Suppose node a is randomly selected as the center node, and $\mathbb{G}(a, 1)$ is created. Since a, b, and c are uncovered and are within one hop of a, they are placed in $\mathbb{G}(a, 1)$; nodes a, b, and c are now covered. There are still uncovered nodes. Suppose b is randomly selected as the next center node, and $\mathbb{G}(b, 1)$ is created. There are no uncovered nodes within 1 hop of b, and b is already covered, so we discard $\mathbb{G}(b, 1)$. Suppose c is randomly selected as the next center node, and $\mathbb{G}(c, 1)$ is created. The uncovered nodes d and e are 1 hop away from c, so they are placed in $\mathbb{G}(c, 1)$; nodes d and e are now covered. There are still uncovered nodes, so suppose d is randomly selected as the next center node, and $\mathbb{G}(d, 1)$ is created. The only uncovered node 1 hop from d is f, so we place f in $\mathbb{G}(d, 1)$; node f is now covered. There are still uncovered nodes, so suppose e is randomly selected as the next center node, and $\mathbb{G}(e, 1)$ is created. Node g and h are uncovered and 1 hop from e so we place g and h in $\mathbb{G}(e, 1)$; nodes g and h are now covered. At this point there are no uncovered nodes, and *Random Sequential Node Burning* halts for this radius r; four boxes were created. □

As mentioned earlier, this method can generate disconnected boxes. For example, in Fig. 8.10 there is no path in $\mathbb{G}(c, 1)$ connecting the two nodes in that

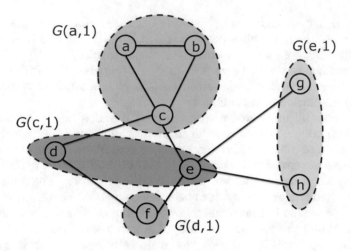

Figure 8.10: Covering using *Random Sequential Node Burning*

box, and there is no path in $\mathbb{G}(e,1)$ connecting the two nodes in that box. Also, a box centered at a node may not even contain the center node, e.g., $\mathbb{G}(e,1)$ does not contain the center node e. Thus a radius-based box $\mathbb{G}(n,r)$ generated by *Random Sequential Node Burning* may fail to satisfy Definition 7.4, which would appear to cast doubt on the validity of d_B calculated by this method. However, computational results [Kim 07a, Song 07] showed that *Random Sequential Node Burning* yields the same value of d_B as obtained using other box-counting methods.

It is perhaps surprising that *Random Sequential Node Burning* does not exclude a node that has been covered from being selected as a center. When *Random Sequential Node Burning* was modified to exclude a covered node from being a box center, a power-law scaling relation was not observed for the WWW [Kim 07a]. However, for another fractal network studied in [Kim 07a], a power law continued to hold with this modification, although the modification led to somewhat different values of $B_R(r)$.

The modular structure of networks can be studied using k-core analysis [Kim 07b], where the k-core of $\mathbb{G}$ is the maximal subgraph of $\mathbb{G}$ containing only nodes of degree at least k. Thus the 1-core is the entire graph $\mathbb{G}$, and the 2-core is the graph with all leaf nodes removed. For $k \geq 3$, the k-core of $\mathbb{G}$ can be disconnected, e.g., if $\mathbb{G}$ is constructed by two instances of the complete graph K_4 connected by a chain of two arcs, then the 3-core of $\mathbb{G}$ is disconnected. In [Kim 07b], *Random Sequential Node Burning* was applied to the k-core of the WWW (a scale-free network), for $1 \leq k \leq 10$. They found that for $k = 1$ and 2, the box-counting dimension of the k-core is 4.1, unchanged from the value for the original network. For $3 \leq k \leq 5$, the behavior changed significantly, yielding $d_B \approx 5$. For $k \geq 10$, the k-core is a highly connected graph with diameter 3, and is no longer a fractal. One of the conclusions of this study was that low degree nodes which connect different modules play an important role in the observed fractal scaling.

Exercise 8.1 Consider applying *Random Sequential Node Burning* to a hub and spoke network. If there are K spokes, then for $r = 1$ in the worst case the method can generate K boxes. As illustrated in Fig. 8.11, this worst case behavior occurs if (i) a leaf node n_1 is chosen as the center of the first box, in which case n_1 and the hub are assigned to the first box, (ii) leaf node n_2 is chosen as the center of the second box, and this box contains only n_2, and so on, until finally (iii) either the hub or the last uncolored leaf node (call it n_K) is chosen of the center of the K-th box, and this box contains only n_K. What is

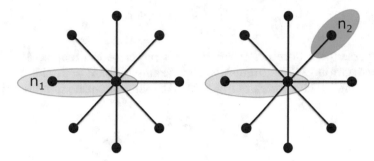

Figure 8.11: *Random Sequential Node Burning* applied to a hub and spoke network

the expected number of boxes *Random Sequential Node Burning* will generate for a hub and spoke network with K spokes?   □

Exercise 8.2 For $r = 1$, what is the expected number of boxes that *Random Sequential Node Burning* will create for the "double ring" network of Fig. 8.12, assuming there are K spokes?   □

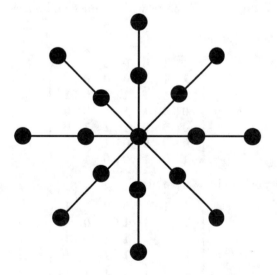

Figure 8.12: Double ring hub and spoke network

8.4   Set Covering Formulation and a Greedy Method

The problem of computing the minimal number $B_R(r)$ of radius-based boxes of radius at most r needed to cover $\mathbb{G}$ can be formulated as a *set covering problem*, a classical combinatorial optimization problem. For simplicity we will refer to node j rather than node n_j, so $\mathbb{N} = \{1, 2, \ldots, N\}$. For a given positive integer r, let M^r be the $N \times N$ matrix defined by

$$M_{ij}^r = \begin{cases} 1 & \text{if } dist(i,j) \leq r, \\ 0 & \text{otherwise}. \end{cases} \tag{8.1}$$

(The superscript r does *not* mean the r-th power of the matrix M.) For an undirected graph, M^r is symmetric. For example, for $r = 1$, the matrix M^1 corresponding to the network of Fig. 8.13 is the same as the node-node adjacency

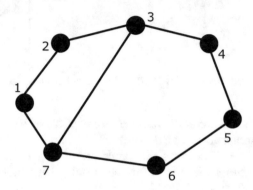

Figure 8.13: Network with 7 nodes and 8 arcs

matrix of the network, namely

$$M^1 = \begin{pmatrix} 1 & 1 & 0 & 0 & 0 & 0 & 1 \\ 1 & 1 & 1 & 0 & 0 & 0 & 0 \\ 0 & 1 & 1 & 1 & 0 & 0 & 1 \\ 0 & 0 & 1 & 1 & 1 & 0 & 0 \\ 0 & 0 & 0 & 1 & 1 & 1 & 0 \\ 0 & 0 & 0 & 0 & 1 & 1 & 1 \\ 1 & 0 & 1 & 0 & 0 & 1 & 1 \end{pmatrix}.$$

For the same network and $r = 2$ we have

$$M^2 = \begin{pmatrix} 1 & 1 & 1 & 0 & 0 & 1 & 1 \\ 1 & 1 & 1 & 1 & 0 & 0 & 1 \\ 1 & 1 & 1 & 1 & 1 & 1 & 1 \\ 0 & 1 & 1 & 1 & 1 & 1 & 1 \\ 0 & 0 & 1 & 1 & 1 & 1 & 1 \\ 1 & 0 & 1 & 1 & 1 & 1 & 1 \\ 1 & 1 & 1 & 1 & 1 & 1 & 1 \end{pmatrix}.$$

For $n \in \mathbb{N}$, let x_n be the binary variable defined by

$$x_n = \begin{cases} 1 & \text{if the box centered at } n \text{ with radius } r \text{ is used in the covering of } \mathbb{G}, \\ 0 & \text{otherwise.} \end{cases}$$

The minimal number $B_R(r)$ of boxes needed to cover $\mathbb{G}$ is the optimal objective function value of the following binary integer program (an *integer program* is a linear optimization problem whose variables are restricted to be integer valued):

$$\text{minimize} \quad \sum_{n=1}^{N} x_n \tag{8.2}$$

$$\text{subject to} \quad \sum_{n=1}^{N} M_{jn}^r x_n \geq 1 \text{ for } j \in \mathbb{N} \tag{8.3}$$

$$x_n = 0 \text{ or } 1 \text{ for } n \in \mathbb{N}. \tag{8.4}$$

Here M_{jn}^r is the element in row j and column n of the matrix M^r.

The objective function (8.2) is the number of center nodes used in the covering, i.e., the number of boxes used in the covering. The left-hand side of constraint (8.3) is the number of boxes covering node j, so this constraint requires that each node be within distance r of at least one center node used in the covering.

To express this formulation more compactly, let $x = (x_1, x_2, \ldots, x_N)$ be the column vector of size N (that is, a matrix with N rows and 1 column) and let x^T be the transpose of x (so x^T is a matrix with 1 row and N columns). Let $\mathbf{1} = (1, 1, \ldots, 1)$ be the column vector of size N each of whose components is 1. Then the above set covering formulation can be written as

$$\text{minimize} \quad x^T \mathbf{1}$$
$$\text{subject to} \quad M^r x \geq \mathbf{1}.$$
$$x_n = 0 \text{ or } 1, \ j \in \mathbb{N}.$$

Let $\widetilde{g}(n)$ be the sum of the entries in column n of the matrix M^r, i.e., $\widetilde{g}(n) = \sum_{j=1}^{N} M_{jn}^r$. Then $\widetilde{g}(n)$ is the number of nodes whose distance from node n does not exceed r. Intuitively, a node n for which $\widetilde{g}(n)$ is high (for example, a hub node) has more value in a covering of $\mathbb{G}$ than a node for which $\widetilde{g}(n)$ is low. However, once some boxes have been selected to be in the covering, to determine which additional boxes to add to the covering, the computation of $\widetilde{g}(n)$ should consider only nodes not yet covered by any box. Therefore, given the binary vector $x \in \mathbb{R}^N$, define

$$U(x) \equiv \left\{ j \in \mathbb{N} \ \middle| \ \sum_{n=1}^{N} M_{jn}^r x_n = 0 \right\},$$

so if $j \in U(x)$, then node j is currently uncovered. For $n \in \mathbb{N}$, define

$$g(n) \equiv \sum_{j \in U(x)} M_{jn}^r,$$

so $g(n)$ is the number of currently uncovered nodes that would be covered if the radius-based box $\mathbb{G}(n, r)$ centered at n were added to the covering. The $g(n)$ values are used in a greedy heuristic to solve the integer optimization problem defined by (8.2)–(8.4). Greedy methods for set covering have been known for decades, and [Chvatal 79] provided a bound on the deviation from optimality of a greedy method for set covering.

For a given r, the greedy *Maximum Excluded Mass Burning* method [Song 07] begins by initializing each component of the vector $x \in \mathbb{R}^N$ to zero, indicating that no node has yet been selected as the center of a radius-based box. The set Z of uncovered nodes is initialized to $\mathbb{N}$. In each iteration of *Maximum Excluded Mass Burning*, the node selected to be the center of a radius-based box is a node j for which $x_j = 0$ and $g(j) = \max_i\{g(i) \,|\, x_i = 0\}$. That is, the next center node j is a node which has not previously been selected as a center, and which, if used as the center of a box of radius r, covers the maximal number of uncovered nodes. There is no requirement that j be uncovered. In the event that more than one node yields $\max_i\{g(i) \,|\, x_i = 0\}$, a node can be randomly chosen, or a deterministic tie-breaking rule can be utilized. We set $x_j = 1$ indicating that j has now been used as a box center. Each uncovered node i within distance r of j is removed from Z since i is now covered. If Z is now empty, we are done. Otherwise, since the newly added box will cover at least one previously uncovered node, we update $g(n)$ for each n such that $x_n = 0$ (i.e., for each node not serving as a box center). Upon termination, the estimate of $B_R(r)$ is $\sum_{n \in \mathbb{N}} x_n$. As with *Random Sequential Node Burning*, a box generated by *Maximum Excluded Mass Burning* can be disconnected [Song 07]. Pseudocode for *Maximum Excluded Mass Burning* is provided by Fig. 8.14.

procedure *Maximum Excluded Mass Burning*
1   **initialize:** $x = 0$ and $U(x) = \mathbb{N}$;
2   **while** $(U(x)$ is non-empty) {
3   pick any j such that $x_j = 0$ and $g(j) = \max_i\{g(i) \,|\, x_i = 0\}$;
4   **if** $(n \in U(x)$ and $dist(j, n) \leq r)$ **then** set $U(x) = U(x) - \{n\}$;
5   set $x_j = 1$ and $B_R(r) = B_R(r) + 1$;
6   **for** $(n \in \mathbb{N})$ {**if** $(x_n = 0)$ **then** update $g(n)$ };
7   }
8   **return** $B_R(r)$;

Figure 8.14: *Maximum Excluded Mass Burning*

Once the method has terminated, each non-center (a node n for which $x_n = 0$) is within distance r of a center (a node n for which $x_n = 1$). We may now want to assign each non-center to a center. One way to make this assignment is to arbitrarily assign each non-center n to any center c for which $dist(c, n) \leq r$. Another way to make this assignment is to assign each non-center n to the closest center, breaking ties randomly. Yet another way was suggested in [Song 07]. Although the number of nodes in each box is not needed to calculate d_B, the number of nodes in each box is required to calculate other network fractal dimensions, as will be discussed in Chaps. 14 and 17.

8.5 Box Burning

The *Box Burning* method [Song 07] is a heuristic for computing the minimal number $B_D(s)$ of diameter-based boxes of size at most s needed to cover $\mathbb{G}$. Let $\mathcal{C}$ be the set of covered nodes, so initially $\mathcal{C} = \emptyset$. Initialize $B_D(s) = 0$. In each iteration, a random uncovered node x is selected to be the initial occupant (i.e., the "seed") of a new box $\mathbb{G}(s)$. Since we have created a new box, we increment $B_D(s)$ by 1. We add x to $\mathcal{C}$. Now we examine each node n not in $\mathcal{C}$; if n is within distance $s - 1$ of each node in $\mathbb{G}(s)$ (which initially is just the seed node x), then n is added to $\mathbb{G}(s)$ and added to $\mathcal{C}$. We again examine each node n not in $\mathcal{C}$; if n is within distance $s - 1$ of each node in $\mathbb{G}(s)$, then n is added to $\mathbb{G}(s)$ and added to $\mathcal{C}$. This continues until no additional nodes can be added to $\mathbb{G}(s)$. At this point a new box is needed, so a random uncovered node x is selected to be the seed of a new box and we continue in this manner, stopping when $\mathcal{C} = \mathbb{N}$. Pseudocode for *Box Burning* is provided in Fig. 8.15.

procedure *Box Burning*(s)
1   **initialize:** $B_D(s) = 0$ and $\mathcal{C} = \emptyset$;
2   **while** $(\mathcal{C} \neq \mathbb{N})$ {
3   randomly select $x \in \mathbb{N} - \mathcal{C}$ and create the new box $\mathbb{G}(s) = \{x\}$;
4              $B_D(s) = B_D(s) + 1$;
5   **for** $(n \in \mathbb{N} - \mathcal{C})$ {
6   **if** $(dist(n, w) < s$ for each $w \in \mathbb{G}(s))$ {
7   add w to $\mathcal{C}$ and to $\mathbb{G}(s)$;
8   }
9   }
10   }
11   **return** $B_D(s)$;

Figure 8.15: *Box Burning*

We illustrate *Box Burning* for $s = 2$ using the network of Fig. 8.16.
Iteration 1 Choose f as the first seed node, create a new box containing only this node, and increment $B_D(2)$. Add node e to the box; this is allowed since the distance from e to each node in the box (which currently contains only f) does not exceed 1. No additional uncovered nodes can be added to box with nodes $\{e, f\}$; e.g., g cannot be added since its distance to e is 2.
Iteration 2 Choose b as the second seed node and create a new box containing only this node. Add node c to the box; this is allowed since the distance from c to each node in the box (which currently contains only b) is 1. No additional uncovered nodes can be added to the box with nodes $\{b, c\}$; e.g., a cannot be added since its distance to c is 2.

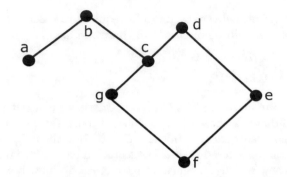

Figure 8.16: Network for illustrating box burning

Iteration 3 Choose a as the third seed node, and create a new box containing only this node. No additional uncovered nodes can be added to this box, since the only uncovered nodes are d and g, and their distance from a exceeds 1.

Iteration 4 Choose d as the fourth seed node; d can be the only occupant of this box.

Iteration 5 Finally, choose g as the fifth seed node; g can be the only occupant of this box.

We created five boxes to cover the network, which is not optimal since we can cover it using the four boxes with node sets $\{a\}$, $\{b,c\}$, $\{d,e\}$, and $\{f,g\}$. The covering created by *Box Burning* and this optimal covering are illustrated in Fig. 8.17.   □

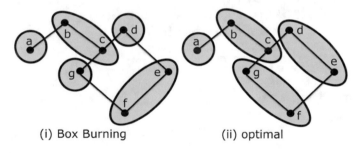

(i) Box Burning   (ii) optimal

Figure 8.17: Box burning 2-covering and a minimal 2-covering

Just as the *Random Sequential Node Burning* method described above in Sect. 8.3 can generate disconnected boxes, the boxes generated by *Box Burning* can be disconnected. (So both radius-based box heuristics for covering $\mathbb{G}$, and diameter-based box heuristics for covering $\mathbb{G}$, can generate disconnected boxes.) This is illustrated by the network of Fig. 8.18 for $s = 3$. All five nodes cannot be in the same box of diameter 2. If w is the first random seed, then the first box contains u, v, and w. If x is selected as the second random seed, the second box,

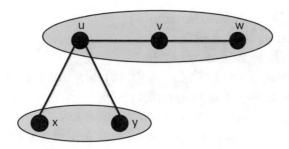

Figure 8.18: Disconnected boxes in a minimal covering

containing x and y, is disconnected. The only path connecting x and y goes through a node in the first box. These two boxes form a minimal 3-covering of $\mathbb{G}$. However, for this network and $s = 3$, there is a minimal 3-covering using two connected boxes: place x, y, and u in the first box, and v and w in the second box.

While the *Box Burning* method is easy to implement, its running time is excessive. A faster *Compact Box Burning* heuristic [Song 07] also uses diameter-based boxes. To begin, initialize $B_D(s) = 0$. Each iteration of *Compact Box Burning* processes the set U of uncovered nodes using the following steps. (*i*) Initialize $Z = \emptyset$, where Z is the set of nodes in the next box created in the covering of $\mathbb{G}$. Initialize $U = \mathbb{N}$. Increment $B_D(s)$ by 1. (*ii*) Select a random node $x \in U$; add x to Z and remove x from U, since x is now covered by the box with node set Z. (*iii*) For each node $y \in U$, if $dist(x,y) \geq s$, then remove y from U, since x and y cannot both belong to Z. (*iv*) Repeat steps (*ii*) and (*iii*) until U is empty.

When U is empty, no more nodes can be added to Z. If $\mathbb{N}$ is empty, we are done. If $\mathbb{N}$ is not empty, another iteration of *Compact Box Burning* is required. To start this new iteration, first remove each node in Z from $\mathbb{N}$, since these nodes are now covered. Now continue the iteration, starting with step (*i*).

Note that in the first iteration of *Compact Box Burning*, when we initialize $U = \mathbb{N}$ in step (*i*), the set $\mathbb{N}$ is the original set of all the nodes in $\mathbb{G}$. In subsequent iterations, $\mathbb{N}$ is no longer the original set of all the nodes in $\mathbb{G}$, since each subsequent iteration begins with removing each node in Z from $\mathbb{N}$. Pseudocode for *Compact Box Burning* is provided in Fig. 8.19.

```
procedure CompactBoxBurning(s)
1      initialize: B_D(s) = 0;
2      while (ℕ is non-empty) {
3            set  U = ℕ, B_D(s) = B_D(s) + 1, and Z = ∅;
4            while (U is non-empty) {
5                  select a random node x ∈ U;
6                  remove x from U and add x to Z;
7                  for (y ∈ U) {
8                        if (dist(x, y) ≥ s) { remove y from U; }
9                  }
10           }
11           set ℕ = ℕ − Z;
12     }
13     return B_D(s);
```

Figure 8.19: *Compact Box Burning*

We illustrate *Compact Box Burning* for $s = 2$, again using the network of Fig. 8.16.

Iteration 1 Initialize $B_D(2) = 0$. To start the first iteration, set $U = ℕ$, increment $B_D(s)$ (since U is non-empty then at least one more box is needed), and create the empty set Z. Choose f in step (ii). Add f to Z and remove f from U. In (iii), remove nodes a, b, c, d from U, since their distance from f exceeds 1. Since U is not empty, return to (ii) and choose e. Add e node to Z, remove it from U, and remove node g from U, since its distance from e exceeds 1. Now U is empty.

Iteration 2 Since $Z = \{e, f\}$, removing these nodes from ℕ yields $ℕ = \{a, b, c, d, g\}$. Since ℕ is not empty, a new iteration is required. To start the second iteration, initialize $U = ℕ$, increment $B_D(s)$, and create the empty set Z. Choose b in step (ii). Add b to Z, remove b from U, and remove nodes d and g from U, since their distance from b exceeds 1. Since U is not empty, return to step (ii) and choose c. Add c to Z, remove c from U, and remove node a, since its distance from c exceeds 1. Now U is empty.

Iteration 3 Since $Z = \{b, c\}$, removing these nodes from ℕ yields $ℕ = \{a, d, g\}$. Since ℕ is not empty, a new iteration is required. To start the third iteration, initialize $U = ℕ$, increment $B_D(s)$, and create the empty set Z. Choose a in step (ii). Add a to Z, remove a from U, and remove nodes d and g since their distance from a exceeds 1. Now U is empty.

Iteration 4 Since $Z = \{a\}$, removing a from ℕ yields $ℕ = \{d, g\}$. Since ℕ is not empty, a new iteration is required. To start the fourth iteration, initialize $U = ℕ$, increment $B_D(s)$, and create the empty set Z. Choose d in step (ii). Add d to Z, remove d from U, and remove node g since its distance from d exceeds 1. Now U is empty.

Iteration 5 Since $Z = \{d\}$, removing d from $\mathbb{N}$ yields $\mathbb{N} = \{g\}$. Since $\mathbb{N}$ is not empty, a new iteration is required. To start the fifth iteration, initialize $U = \mathbb{N}$, increment $B_D(s)$, and create the empty set Z. Choose g in step (ii); there is no choice here, since $\mathbb{N} = \{g\}$. Add g to Z, and remove it from U. Now U is empty.

Termination Since $\mathbb{N} = \{g\}$, removing g from $\mathbb{N}$ makes $\mathbb{N}$ empty, so we are done. For this particular network and random node selections, *Box Burning* and *Compact Box Burning* yielded the same covering using five boxes. □

Compact Box Burning, like the other heuristics we have presented, is not guaranteed to yield the minimal value $B_D(s)$. Since a random node is chosen in step (ii), the method is not deterministic, and executing the method multiple times, for the same network, will in general yield different values for $B_D(s)$. *Compact Box Burning* was applied in [Song 07] to two networks, the cellular *E. coli* network and the *mbone* Internet multicast backbone network.[2] For both networks, the average value of $B_D(s)$ obtained by 10,000 executions of *Compact Box Burning* was up to 2% higher than the average for a greedy coloring method similar to the *Greedy Coloring* method described in Sect. 8.1. The conclusion in [Song 07] was that *Compact Box Burning* provides results comparable with *Greedy Coloring*, but *Compact Box Burning* may be a bit simpler to implement.

Additionally, the four methods *Greedy Coloring*, *Random Sequential Node Burning*, *Maximum Excluded Mass Burning*, and *Compact Box Burning* were compared in [Song 07]. Using 10,000 executions of each method, they found that all four methods yielded the same box-counting dimension d_B. However, the results for *Random Sequential Node Burning* showed a much higher variance than for the other three methods, so for *Random Sequential Node Burning* it is not clear how many randomized executions are necessary, for a given r, for the average values $B_R(r)$ to stabilize.

8.6 Box Counting for Scale-Free Networks

As discussed in Sect. 8.3, *Random Sequential Node Burning* can create disconnected boxes. For example, in the network of Fig. 8.10, when c is selected as the third center node, the resulting box $\mathbb{G}(c, 1)$, which contains d and e but not c, is disconnected. Suppose we modify *Random Sequential Node Burning* so that the nodes inside each box are required to be connected within that box. With this modification, we still obtain $d_B \approx 2$ for a rectilinear lattice (i.e., a primitive square Bravais lattice) in $\mathbb{R}^2$. However, with this modification the power-law scaling $B_R(r) \sim (2r + 1)^{-d_B}$ was not observed for the WWW [Kim 07a].

The reason we must allow disconnected boxes to obtain a power-law scaling for the WWW is that the WWW is a scale-free network. Recall from Sect. 2.4 that a network is scale-free if p_k, the probability that the node degree is k, follows

---

[2] A *multicast* network [Rosenberg 12] facilitates one-to-many transmissions (e.g., a basketball game is streamed to a set of geographically diverse viewers), or many-to-many transmissions (e.g., a set of geographically diverse committee members join a teleconference with both audio and video capabilities).

the power-law distribution $p_k \sim k^{-\lambda}$ for some $\lambda > 0$. For a scale-free network, a few nodes will have high degree. Nodes with high degree (i.e., hubs) will tend to be quickly assigned to a box, since a hub is assigned to a box whenever one of its neighbors is selected as a center node. Once a hub is covered by a box, adjacent nodes will be disconnected if they are connected only via the hub. This is illustrated in Fig. 8.20. Suppose we cover this network using radius-

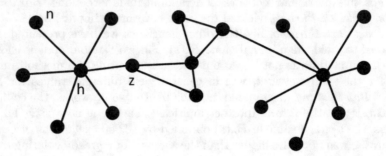

Figure 8.20: Adjacent nodes only connected through a hub

based boxes of radius 1, and spoke n is randomly selected as the first center node. Then hub h is added to box $\mathbb{G}(n, 1)$. If h is now selected as a center node, all five uncovered nodes adjacent to h (e.g., z) will be added to $\mathbb{G}(h, 1)$, even though these five nodes interconnect only through hub h. Counting each spoke connected to h as a separate box creates too many boxes, so *Random Sequential Node Burning* corrects for this phenomenon by allowing nodes in a box to be disconnected within that box. This yields a power-law scaling for the WWW.

It might be argued that allowing the nodes in a box to be disconnected within that box violates the spirit of the Hausdorff dimension (Sect. 5.1), which covers a geometric object $\Omega \subset \mathbb{R}^E$ by connected subsets of $\mathbb{R}^E$. However, with a change in perspective, allowing nodes in a box to be disconnected within that box can be viewed as using connected boxes, but allowing a node to belong to more than one box. To see this, we revisit Fig. 8.10, but now allow overlapping boxes, as shown in Fig. 8.21.

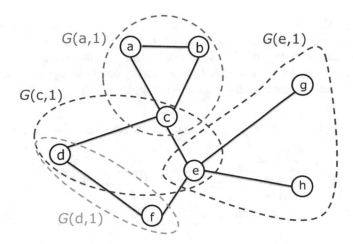

Figure 8.21: Overlapping boxes

Assume the randomly chosen centers are in the same order as before, namely a, b, c, d, e, and again choose $r = 1$. Box $\mathbb{G}(a, 1)$ is unchanged, and $\mathbb{G}(b, 1)$ is again discarded. However, now $\mathbb{G}(c, 1)$ also includes c, box $\mathbb{G}(d, 1)$ also contains d, and $\mathbb{G}(e, 1)$ also contains e. So c, d, and e belong to two boxes. However, as mentioned above, the great advantage of non-overlapping boxes is that they immediately yield a probability distribution (Chap. 14).

8.7   Other Box Counting Methods

In this short section we describe two other box-counting methods.

Merge: The `merge` method of [Locci 09] combines two randomly selected boxes $\mathbb{G}_1$ and $\mathbb{G}_2$ if the distance between them is less than s, where $dist(\mathbb{G}_1, \mathbb{G}_2) \equiv \max\{dist(x, y) \mid x \in \mathbb{G}_1 \text{ and } y \in \mathbb{G}_2\}$ if some arc connects a node in $\mathbb{G}_1$ to a node in $\mathbb{G}_2$, and $dist(\mathbb{G}_1, \mathbb{G}_2) \equiv \infty$ otherwise. Thus `merge` uses diameter-based boxes. Let $\mathcal{X}$ be the current set of boxes. Initially each node is its own box, so there are N boxes. If boxes $\mathbb{G}_1$ and $\mathbb{G}_2$ are combined into a box $\mathbb{G}_3$, then $\mathbb{G}_1$ and $\mathbb{G}_2$ are removed from $\mathcal{X}$, and box $\mathbb{G}_3$ is added to $\mathcal{X}$. The method starts with $s = 2$. When no further merges for randomly chosen $\mathbb{G}_1 \in \mathcal{X}$ and $\mathbb{G}_2 \in \mathcal{X}$ and $s = 2$ are possible, then s is incremented by 1, and we begin a new iteration. The method terminates when s reaches a specified value.

Another method (call it `simulated annealing`) analyzed by Locci et al. [Locci 09] is to first run `merge` and then run a simulated annealing method to improve the covering. A simulated annealing method always accepts a change to a configuration if it improves the "cost" of the configuration; otherwise, the change is accepted with probability $e^{-\delta/T}$, where δ is the increase in cost if we make the change, and T is the "temperature". Three types of changes are considered: moving a node from $\mathbb{G}_1$ to $\mathbb{G}_2$, removing a node from a box to form a new box, and applying `merge` to the current set of boxes.

Three methods for computing d_B were compared in [Locci 09]: a greedy coloring method, merge, and simulated annealing. The results for a software development network with 8499 nodes and 42,048 arcs are given in Table 8.1. The times, on a 2009 vintage computer, nonetheless provide an indication of the relative running times on any computer.

Table 8.1: Comparison of three box-counting methods

Method	Time(s)	d_B
Greedy coloring	410	3.96
Merge	289	4.24
Simulated annealing	8807	4.06

They also applied the three methods 1000 times, for each s value in the range 3–9, to the *E. coli* protein interaction network of 2859 nodes and 6890 arcs. Since the standard deviation of the $B_D(s)$ values for merge is much higher than for the other two methods, they deemed merge unsuitable for the computation of d_B of the software development network. Due to the excessive running time of simulated annealing, they recommended greedy coloring, since its running time is comparable to merge and its accuracy is comparable to simulated annealing.

Cluster Growing: The d_B of nine large object-oriented software systems was computed by Concas et al. [Concas 06] using a cluster growing method. The basic building block of object-oriented systems is the *class*, which implements one or more data structures and methods to access and process the data in the data structures. Classes may be related by various binary relations such as *inheritance* (a derived class inherits all the structures and methods of the parent class). These relationships are directed, e.g., "class A is derived from B" is different than "class B is derived from A". All the relationships between classes yield a directed "class graph" in which the nodes are classes and the arcs are the directed relationships. However, since network box-counting methods typically work only for undirected graphs, they analyzed the undirected version of the class graph.

The cluster growing method of [Concas 06] is the following. Initialize $s = 2$ and let $\delta_{\min}$ be the minimal node degree, taken over all nodes. Considering only those nodes with degree $\delta_{\min}$, select a random node n, and create a box containing n. Add nodes to this box subject to the constraint that $dist(x, y) < s$ for each pair (x, y) of nodes in the box. All nodes in this box are now "burned". Next, select a new random node n from among those unburned nodes with minimal node degree, create a box containing n, and add to this box all unburned nodes provided that $dist(x, y) < s$ for each pair (x, y) of nodes in the box. Continue until all nodes are burned; these boxes form a diameter-based 2-covering of $\mathbb{G}$.

The motivation of [Concas 06] for computing d_B was that object-oriented software architectures exhibit modularity, and that [Song 06] showed that fractal networks are characterized by a modular hierarchical structure. Therefore a well-engineered object-oriented system, "being modular, should show self-similarity." The plots of $\log B_D(s)$ versus $\log s$ in [Concas 06] for three software networks clearly showed that d_B exists for the underlying undirected graph. However, one year after the publication of [Concas 06], [Gallos 07b] explained how for a network $\mathbb{G}$ the definitions of "fractality" and "self-similarity" are quite different; this difference will be explored in Chap. 21.

8.8 Lower Bounds on Box Counting

In general, when using a heuristic box-counting method, whether it uses radius-based boxes or diameter-based boxes, there is no guarantee that the computed covering of $\mathbb{G}$ is optimal or even near optimal. By "optimal" we mean that if using diameter-based boxes then $B_D(s)$ is minimal, or that if using radius-based boxes then $B_R(r)$ is minimal. However, if a lower bound on $B_D(s)$ or $B_R(r)$ is available, we can immediately determine the deviation from optimality for the computed covering.

A method that provides a lower bound $B_R^L(r)$ on $B_R(r)$ was presented in [Rosenberg 13]. The lower bound is computed by formulating box counting as an *uncapacitated facility location problem* (**UFLP**), a classical combinatorial optimization problem. This formulation provides, via the dual of the linear programming relaxation of **UFLP**, a lower bound on $B_R(r)$. The method also yields an estimate of $B_R(r)$; this estimate is an upper bound on $B_R(r)$. Under the assumption that $B_R(r) = a(2r+1)^{-d_B}$ holds for some positive constant a and some range of r, a linear program [Chvatal 83], formulated using the upper and lower bounds on $B_R(r)$, provides an upper and lower bound on d_B. In the event that the linear program is infeasible, a quadratic program [Gill 81] can be used to estimate d_B.[3]

The literature on computable bounds on d_B is rather scant. Confidence intervals were derived in [Taylor 91] for the estimate of d_B of a curve defined by a stochastic process (a randomized version of the classical Weierstrass function which is nowhere differentiable) and for a two-dimensional model in which pixels are "hit" with a given probability. For a network, [Hu 07] and [Gao 08] bounded the relative error in the estimated d_B under the assumption that $B_R(r) = cr^{-d_B}$ holds as a strict equality over a range of r; this strong assumption was not required in [Rosenberg 13] to calculate a lower bound on $B_R(r)$. The remainder of this section presents the method of [Rosenberg 13].

---

[3] An optimization problem is *feasible* if there exists a set of values for the variables such that each constraint is satisfied; the optimization problem is *infeasible* if it is not feasible. An example of an infeasible optimization problem is: minimize x subject to $x \leq 1$ and $x \geq 2$.

Mathematical Formulation

Let the box radius r be fixed. For simplicity we will refer to node j rather than node n_j. Define $\mathbb{N} \equiv \{1, 2, \ldots, N\}$. Let C^r be the symmetric N by N matrix defined by

$$C_{ij}^r = \begin{cases} 0 & \text{if } dist(i, j) \leq r, \\ \infty & \text{otherwise.} \end{cases}$$

(As with the matrix M_{ij}^r defined by (8.1), the superscript r in C_{ij}^r does *not* mean the r-th power of the matrix C.) For example, setting $r = 1$, the matrix C^1 corresponding to the network of Fig. 8.22 is

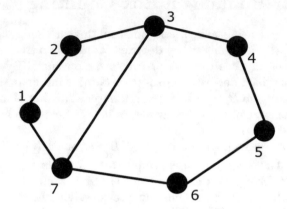

Figure 8.22: Example network with 7 nodes and 8 arcs

$$C^1 = \begin{pmatrix} 0 & 0 & - & - & - & - & 0 \\ 0 & 0 & 0 & - & - & - & - \\ - & 0 & 0 & 0 & - & - & 0 \\ - & - & 0 & 0 & 0 & - & - \\ - & - & - & 0 & 0 & 0 & - \\ - & - & - & - & 0 & 0 & 0 \\ 0 & - & 0 & - & - & 0 & 0 \end{pmatrix},$$

where a dash "–" is used to indicate the value ∞.

For $j \in \mathbb{N}$, let

$$y_j = \begin{cases} 1 & \text{if the box centered at } j \text{ is used to cover } \mathbb{G}, \\ 0 & \text{otherwise.} \end{cases}$$

A given node i will, in general, be within distance r of more than one center node j used in the covering of $\mathbb{G}$. However, we will assign each node i to exactly one node j, and the variables x_{ij} specify this assignment. For $i, j \in \mathbb{N}$, let

$$x_{ij} = \begin{cases} 1 & \text{if } i \text{ is assigned to the box centered at } j, \\ 0 & \text{otherwise.} \end{cases}$$

With the understanding that r is fixed, for simplicity we write c_{ij} to denote element (i, j) of the matrix C^r. The minimal network covering problem is

$$\text{minimize} \quad \sum_{j=1}^{N} y_j + \sum_{i=1}^{N} \sum_{j=1}^{N} c_{ij} x_{ij} \tag{8.5}$$

$$\text{subject to} \quad \sum_{j=1}^{N} x_{ij} = 1 \text{ for } i \in \mathbb{N} \tag{8.6}$$

$$x_{ij} \leq y_j \text{ for } i, j \in \mathbb{N} \tag{8.7}$$

$$x_{ij} \geq 0 \text{ for } i, j \in \mathbb{N} \tag{8.8}$$

$$y_j = 0 \text{ or } 1 \text{ for } j \in \mathbb{N}. \tag{8.9}$$

Let **UFLP** denote the optimization problem defined by (8.5)–(8.9). Constraint (8.6) says that each node must be assigned to the box centered at some j. Constraint (8.7) says that node i can be assigned to the box centered at j only if that box is used in the covering, i.e., only if $y_j = 1$. The objective function is the sum of the number of boxes in the covering and the total cost of assigning each node to a box. Problem **UFLP** is feasible since we can always set $y_i = 1$ and $x_{ii} = 1$ for $i \in \mathbb{N}$; i.e., let each node be the center of a box in the covering. Given a set of binary values of y_j for $j \in \mathbb{N}$, since each c_{ij} is either 0 or ∞, if there is a feasible assignment of nodes to boxes then the objective function value is the number of boxes in the covering; if there is no feasible assignment for the given y_j values, then the objective function value is ∞. Note that **UFLP** requires only $x_{ij} \geq 0$; it is not necessary to require x_{ij} to be binary. This relaxation is allowed since if (x, y) solves **UFLP** then the objective function value is not increased, and feasibility is maintained, if we assign each i to exactly one k (where k depends on i) such that $y_k = 1$ and $c_{ik} = 0$.

The *primal linear programming relaxation* **PLP** of **UFLP** is obtained by replacing the restriction that each y_j is binary with the constraint $y_j \geq 0$. We associate the dual variable u_i with the constraint $\sum_{j=1}^{N} x_{ij} = 1$, and the dual variable w_{ij} with the constraint $x_{ij} \geq 0$. The *dual linear program* [Gill 81] **DLP** corresponding to **PLP** is

$$\text{maximize} \quad \sum_{i=1}^{N} u_i$$

$$\text{subject to} \quad \sum_{i=1}^{N} w_{ij} \leq 1 \text{ for } j \in \mathbb{N}$$

$$u_i - w_{ij} \leq c_{ij} \text{ for } i, j \in \mathbb{N}$$

$$w_{ij} \geq 0 \text{ for } i, j \in \mathbb{N}.$$

Following [Erlenkotter 78], we set $w_{ij} = \max\{0, u_i - c_{ij}\}$ and express **DLP** using only the u_i variables:

$$\text{maximize} \quad \sum_{i=1}^{N} u_i \tag{8.10}$$

$$\text{subject to} \quad \sum_{i=1}^{N} \max\{0, u_i - c_{ij}\} \leq 1 \text{ for } j \in \mathbb{N}. \tag{8.11}$$

Let $v(UFLP)$ be the optimal objective function value of **UFLP**. Then $B_R(r) = v(UFLP)$. Let $v(PLP)$ be the optimal objective function value of the linear programming relaxation **PLP**. Then $v(UFLP) \geq v(PLP)$. Let $v(DLP)$ be the optimal objective function value of the dual linear program **DLP**. By linear programming duality theory, $v(PLP) = v(DLP)$. Define $u \equiv (u_1, u_2, \ldots, u_N)$. If u is feasible for **DLP** as defined by (8.10) and (8.11), then the dual objective function $\sum_{i=1}^{N} u_i$ satisfies $\sum_{i=1}^{N} u_i \leq v(DLP)$. Combining these relations, we have

$$B_R(r) = v(UFLP) \geq v(PLP) = v(DLP) \geq \sum_{i=1}^{N} u_i .$$

Thus $\sum_{i=1}^{N} u_i$ is a lower bound on $B_R(r)$. As described in [Rosenberg 13], to maximize this lower bound subject to (8.11), we use the *Dual Ascent* and *Dual Adjustment* methods of [Erlenkotter 78]; see also [Rosenberg 01].

Dual Ascent and Dual Adjustment

Call the N variables $u_1, u_2, \ldots, u_N$ the *dual variables*. The *Dual Ascent* method initializes $u = 0$ and increases the dual variables, one at a time, until constraints (8.11) prevent any further increase in any dual variable. For $i \in \mathbb{N}$, let $\mathbb{N}_i = \{j \in \mathbb{N} \mid c_{ij} = 0\}$. By definition of c_{ij}, we have $\mathbb{N}_i = \{j \mid dist(i, j) \leq r\}$. Note that $i \in \mathbb{N}_i$. From (8.11), we can increase some dual variable u_i from 0 to 1 only if $\sum_{i=1}^{N} \max\{0, u_i - c_{ij}\} = 0$ for $j \in \mathbb{N}_i$. Once we have increased u_i then we cannot increase u_k for any k such that $c_{kj} = 0$ for some $j \in \mathbb{N}_i$. This is illustrated, for $r = 1$, in Fig. 8.23, where $c_{ij_1} = c_{ij_2} = c_{ij_3} = 0$ and $c_{j_1 k_1} = c_{j_2 k_2} = c_{j_2 k_3} = 0$. Once we set $u_i = 1$, we cannot increase the dual variable associated with k_1 or k_2 or k_3. Recalling that δ_j is the node degree of node j, if $c_{ij} = 0$ then the number of dual variables prevented by node j from increasing when we increase u_i is at least $\delta_j - 1$, where we subtract 1 since u_i is being increased from 0. In general, increasing u_i prevents approximately at least $\sum_{j \in \mathbb{N}_i}(\delta_j - 1)$ dual variables from being increased. This is approximate, since there may be arcs connecting the nodes in $\mathbb{N}_i$, e.g., there may be an arc between j_1 and j_2 in Fig. 8.23. However, we can ignore such considerations since we use $\sum_{j \in \mathbb{N}_i}(\delta_j - 1)$ only as a heuristic metric: we pre-process the data by ordering the dual variables in order of increasing $\sum_{j \in \mathbb{N}_i}(\delta_j - 1)$. We have

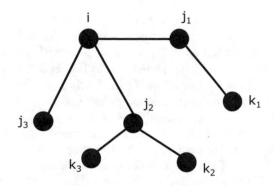

Figure 8.23: Increasing u_i to 1 blocks other dual variable increases

$\sum_{j \in \mathbb{N}_i}(\delta_j - 1) = 0$ only if $\delta_j = 1$ for $j \in \mathbb{N}_i$, i.e., only if each node in $\mathbb{N}_i$ is a leaf node. This can occur only for the trivial case that $\mathbb{N}_i$ consists of two nodes (one of which is i itself) connected by an arc. For any other topology we have $\sum_{j \in \mathbb{N}_i}(\delta_j - 1) \geq 1$. For $j \in \mathbb{N}$, define $s(j)$ to be the slack in constraint (8.11) for node j, so $s(j) = 1$ if $\sum_{i=1}^{N} \max\{0, u_i - c_{ij}\} = 0$ and $s(j) = 0$ otherwise.

Having pre-processed the data, we run the following *Dual Ascent* procedure. This procedure is initialized by setting $u = 0$ and $s(j) = 1$ for $j \in \mathbb{N}$. We then examine each u_i in the sorted order and compute $\gamma \equiv \min\{s(j) \,|\, j \in \mathbb{N}_i\}$. If $\gamma = 0$, then u_i cannot be increased. If $\gamma = 1$, then we increase u_i from 0 to 1 and set $s(j) = 0$ for $j \in \mathbb{N}_i$, since there is no longer slack in those constraints.

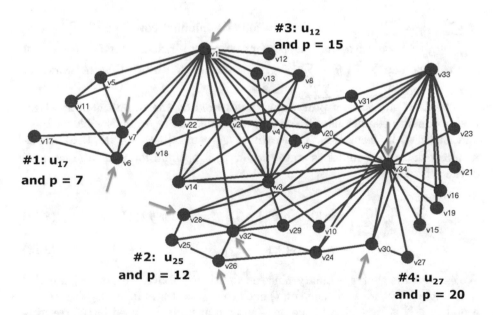

Figure 8.24: Results of applying *Dual Ascent* to *Zachary's Karate Club* network

Figure 8.24 shows the result of applying *Dual Ascent*, with $r = 1$, to *Zachary's Karate Club* network [Newman], which has 34 nodes and 77 arcs. In this figure, node 1 is labelled as "v1", etc. The node with the smallest penalty $\sum_{j \in \mathbb{N}_i} (\delta_j - 1)$ is node 17, and the penalty (p in the figure) is 7. Upon setting $u_{17} = 1$ we have $s(17) = s(6) = s(7) = 0$; in the figure, arrows point to nodes 6 and 7. The node with the next smallest penalty is node 25, and the penalty is 12. Upon setting $u_{25} = 1$ we have $s(25) = s(26) = s(28) = s(32) = 0$; in the figure, arrows point to nodes 26, 28, and 32. The node with the next smallest penalty is node 26, and the penalty is 13. However, u_{26} cannot be increased, since $s(25) = s(32) = 0$. The node with the next smallest penalty is node 12, and the penalty is 15. Upon setting $u_{12} = 1$ we have $s(12) = s(1) = 0$; in the figure, an arrow points to node 1. The node with the next smallest penalty is node 27, and the penalty is 20. Upon setting $u_{27} = 1$ we have $s(27) = s(30) = s(34) = 0$; in the figure, arrows point to nodes 30 and 34. No other dual variable can be increased, and *Dual Ascent* halts, yielding a dual objective function value of 4, which is the lower bound $B_R^L(1)$ on $B_R(1)$.

We can now calculate the upper bound $B_R^U(1)$. For $j \in \mathbb{N}$, set $y_j = 1$ if $s(j) = 0$ and $y_j = 0$ otherwise. Setting $y_j = 1$ means that the box of radius r centered at node j will be used in the covering of $\mathbb{G}$. For *Zachary's Karate Club* network, at the conclusion of *Dual Ascent* with $r = 1$ there are 12 values of j such that $s(j) = 0$; for each of these values we set $y_j = 1$.

We have shown that if u satisfies (8.11), then

$$\sum_{i=1}^{N} u_i = B_R^L(r) \le B_R(r) \le B_R^U(r) = \sum_{j=1}^{N} y_j \,.$$

If $\sum_{i=1}^{N} u_i = \sum_{j=1}^{N} y_j$, then we have found a minimal covering. If $\sum_{i=1}^{N} u_i < \sum_{j=1}^{N} y_j$, then we use a *Dual Adjustment* procedure [Erlenkotter 78] to attempt to close the gap $\sum_{j=1}^{N} y_j - \sum_{i=1}^{N} u_i$. For *Zachary's Karate Club* network, for $r = 1$ we have $\sum_{j=1}^{N} y_j - \sum_{i=1}^{N} u_i = 8$.

The *Dual Adjustment* procedure is motivated by the complementary slackness optimality conditions of linear programming. Let (x, y) be feasible for **PLP** and let (u, w) be feasible for **DLP**, where $w_{ij} = \max\{0, u_i - c_{ij}\}$. The complementary slackness conditions state that (x, y) is optimal for **PLP** and (u, w) is optimal for **DLP** if

$$y_j \left(\sum_{i=1}^{N} \max\{0, u_i - c_{ij}\} - 1 \right) = 0 \text{ for } j \in \mathbb{N} \qquad (8.12)$$

$$(y_j - x_{ij}) \max\{0, u_i - c_{ij}\} = 0 \text{ for } i, j \in \mathbb{N}. \qquad (8.13)$$

We can assume that x is binary, since as mentioned above, we can assign each i to a single k (where k depends on i) such that $y_k = 1$ and $c_{ik} = 0$. We say that a node $j \in \mathbb{N}$ is "open" (i.e., the box centered at node j is used in the covering of $\mathbb{G}$) if $y_j = 1$; otherwise, j is "closed". When (x, y) and u are feasible for **PLP** and **DLP**, respectively, and x is binary, constraints (8.13) have a simple interpretation: if for some i we have $u_i = 1$, then there can be at most one open node j such that $dist(i, j) \leq r$. For suppose to the contrary that $u_i = 1$ and there are two open nodes j_1 and j_2 such that $dist(i, j_1) \leq r$ and $dist(i, j_2) \leq r$. Then $c_{ij_1} = c_{ij_2} = 0$. Since x is binary, by (8.6), either $x_{ij_1} = 1$ or $x_{ij_2} = 1$. Suppose without loss of generality that $x_{ij_1} = 1$ and $x_{ij_2} = 0$. Then

$$(y_{j_1} - x_{ij_1}) \max\{0, u_i - c_{ij_1}\} = (y_{j_1} - x_{ij_1})u_i = 0$$

but

$$(y_{j_2} - x_{ij_2}) \max\{0, u_i - c_{ij_2}\} = y_{j_2} u_i = 1 \,,$$

so complementary slackness fails to hold. This argument is easily extended to the case where there are more than two open nodes such that $dist(i, j) \leq r$. The conditions (8.13) can also be visualized using Fig. 8.23 above, where $c_{ij_1} = c_{ij_2} = c_{ij_3} = 0$. If $u_i = 1$, then at most one node in the set $\{i, j_1, j_2, j_3\}$ can be open.

If $B_R^U(r) > B_R^L(r)$, we run the following *Dual Adjustment* procedure to close some nodes, and construct x, to attempt to satisfy constraints (8.13). Define

$$Y = \{j \in \mathbb{N} \,|\, y_j = 1\} \,,$$

so Y is the set of open nodes. The *Dual Adjustment* procedure, which follows *Dual Ascent*, has two steps.

Step 1. For $i \in \mathbb{N}$, let $\alpha(i)$ be the "smallest" node in Y such that $c_{i,\alpha(i)} = 0$. By "smallest" node we mean the node with the smallest node index, or the alphabetically lowest node name; any similar tie-breaking rule can be used. If for some $j \in Y$ we have $j \neq \alpha(i)$ for $i \in \mathbb{N}$, then j can be closed, so we set $Y = Y - \{j\}$. In words, if the chosen method of assigning each node to a box in the covering results in the box centered at j never being used, then j can be closed.

Applying Step 1 to *Zachary's Karate Club* network with $r = 1$, using the tie-breaking rule of the smallest node index, we have, for example, $\alpha(25) = 25$, $\alpha(26) = 25$, $\alpha(27) = 27$, and $\alpha(30) = 27$. After computing each $\alpha(i)$, we can close nodes 7, 12, 17, and 28, as indicated by the bold **X** next to these nodes in Fig. 8.25. After this step, we have $Y = \{1, 6, 25, 26, 27, 30, 32, 34\}$. This step lowered the primal objective function from 12 (since originally $|Y| = 12$) to 8.

Step 2. Suppose we consider closing j, where $j \in Y$. We consider the impact of closing j on i, for $i \in \mathbb{N}$. If $j \neq \alpha(i)$, then closing j has no impact on i, since i is not assigned to the box centered at j. If $j = \alpha(i)$, then closing j is possible only if there is another open node $\beta(i) \in Y$ such that $\beta(i) \neq \alpha(i)$ and $c_{i,\beta(i)} = 0$

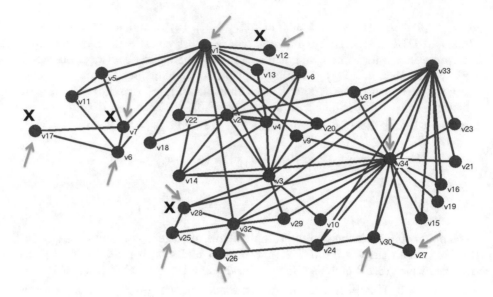

Figure 8.25: Closing nodes in *Zachary's Karate Club* network

(i.e., if there is another open node, distinct from $\alpha(i)$, whose distance from i does not exceed r). Thus we have the rule: close j if for $i \in \mathbb{N}$ either

$$j \neq \alpha(i)$$

or

$$j = \alpha(i) \text{ and } \beta(i) \text{ exists.}$$

Once we close j and set $Y = Y - \{j\}$ we must recalculate $\alpha(i)$ and $\beta(i)$ (if it exists) for $i \in \mathbb{N}$.

Applying Step 2 to *Zachary's Karate Club* network with $r = 1$, we find that, for example, we cannot close node 1, since $1 = \alpha(5)$ and $\beta(5)$ does not exist. Similarly, we cannot close node 6, since $6 = \alpha(17)$ and $\beta(17)$ does not exist. We can close node 25, since $25 = \alpha(25)$ but $\beta(25) = 26$ (i.e., we can reassign node 25 from the box centered at 25 to the box centered at 26), $25 = \alpha(26)$ but $\beta(26) = 26$, $25 = \alpha(28)$ but $\beta(28) = 34$, and $25 = \alpha(32)$ but $\beta(32) = 26$. After recomputing $\alpha(i)$ and $\beta(i)$ for $i \in \mathbb{N}$, we determine that node 26 can be closed. Continuing in this manner, we determine that nodes 27 and 30 can be closed, yielding $Y = \{1, 6, 32, 34\}$. Since now the primal objective function value and the dual objective function value are both 4, we have computed a minimal covering. When we execute *Dual Ascent* and *Dual Adjustment* for *Zachary's Karate Club* network with $r = 2$ we obtain primal and dual objective function values of 2, so again a minimal covering has been found.

Bounding the Fractal Dimension

Assume that for some positive constant a we have

$$B_R(r) = a(2r + 1)^{-d_B} . \tag{8.14}$$

Suppose we have computed $B_R^L(r)$ and $B_R^U(r)$ for $r = 1, 2, \ldots, K$. From

$$B_R^L(r) \le B_R(r) \le B_R^U(r)$$

we obtain, for $r = 1, 2, \ldots, K$,

$$\log B_R^L(r) \le \log a - d_B \log(2r + 1) \le \log B_R^U(r) . \tag{8.15}$$

The system (8.15) of $2K$ inequalities may be infeasible, i.e., it may have no solution a and d_B. If the system (8.15) is feasible, we can formulate a linear program to determine the maximal and minimal values of d_B [Rosenberg 13]. For simplicity of notation, let the K values $\log(2r + 1)$ for $r = 1, 2, \ldots, K$ be denoted by x_k for $k = 1, 2, \ldots, K$, so $x_1 = \log(3)$, $x_2 = \log(5)$, $x_3 = \log(7)$, etc. For $k = 1, 2, \ldots, K$, let the K values of $\log B_R^L(r)$ and $\log B_R^U(r)$ be denoted by y_k^L and y_k^U, respectively. Let $b = \log a$. The inequalities (8.15) can now be expressed as

$$y_k^L \le b - d_B\, x_k \le y_k^U .$$

The minimal value of d_B is the optimal objective function value of **BCLP** (Box Counting Linear Program):

$$
\begin{aligned}
\text{minimize} \quad & d_B \\
\text{subject to} \quad & b - d_B\, x_k \ge y_k^L \text{ for } 1 \le k \le K \\
& b - d_B\, x_k \le y_k^U \text{ for } 1 \le k \le K.
\end{aligned}
$$

This linear program has only 2 variables, b and d_B. Let $d_B^{\min}$ and $b^{\min}$ be the optimal values of d_B and b, respectively. Now we change the objective function of **BCLP** from *minimize* to *maximize*, and let $d_B^{\max}$ and $b^{\max}$ be the optimal values of d_B and b, respectively, for the *maximize* linear program. The box counting dimension d_B, assumed to exist by (8.14), satisfies

$$d_B^{\min} \le d_B \le d_B^{\max} .$$

For the much-studied *jazz* network [Gleiser 03], a collaboration network of jazz musicians with 198 nodes, 2742 arcs, and $\Delta = 6$, the linear program **BCLP** is feasible, and solving the *minimize* and *maximize* linear programs yields $2.11 \le d_B \le 2.59$. When applied to *Lada Adamic's social network*,[4] **BCLP** is feasible, and we obtain the narrow range $3.19 \le d_B \le 3.31$; the results are plotted in Fig. 8.26, where by $B(r)$ we mean $B_R(r)$ [Rosenberg 13].

Feasibility of **BCLP** does not imply that the box counting relationship (8.14) holds, since the upper and lower bounds might be so far apart that alternative

---

[4] supplied by her, for her 2013 Coursera course on "Social Network Analysis"

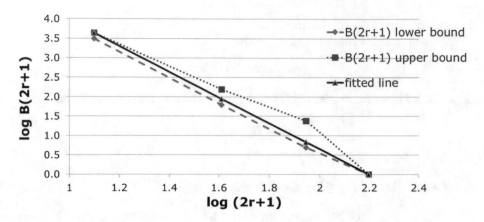

Figure 8.26: Linear program results for *Lada Adamic's social network*

relationships could be posited. If the linear program is infeasible, we can assert that the network does *not* satisfy the box counting relationship (8.14). Yet even if **BCLP** is infeasible, it might be so "close" to feasible that we nonetheless want to calculate d_B. For example, for the *C. elegans* network, **BCLP** is infeasible, but close to feasible, as shown by Fig. 8.27. When **BCLP** is infeasi-

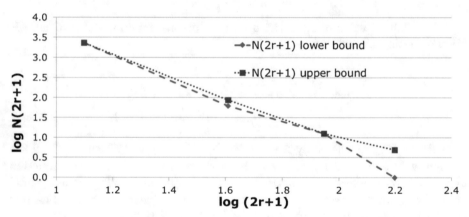

Figure 8.27: BCLP is infeasible for *C. elegans*

ble, we can compute d_B using the solution of **BCQP** (box counting quadratic program), which minimizes the sum of the squared distances to the $2K$ bounds [Rosenberg 13]:

$$\text{minimize} \quad \sum_{k=1}^{K}(u_k^2 + v_k^2)$$

$$\text{subject to} \quad u_k = (b - d_B\, x_k) - y_k^L \text{ for } 1 \le k \le K$$

$$v_k = y_k^U - (b - d_B\, x_k) \text{ for } 1 \le k \le K.$$

Exercise 8.3* The following method for computing d_B of a geometric object Ω was proposed in [Prigarin 13]. If the box counting dimension of Ω is d_B, then for $s \ll 1$ the volume V of Ω is approximated using $V \approx s^{d_B} N(s)$. Replace the approximation by an equality by writing $V = f(s) s^{d_B} N(s)$, where $\lim_{s \to 0} f(s) = 0$. It is reasonable to set $f(s) = e^{-\alpha s}$ where $\alpha > 0$, since the d_B-dimensional volume of Ω should be less than the d_B-dimensional volume of the set of boxes covering Ω. This leads to the equation

$$d_B \log s - \log V - \alpha s = -\log N(s).$$

If we compute $N(s)$ for at least three values of s, then we can use constrained least squares to estimate the three unknowns d_B, $\log V$, and α subject to the constraints that $d_B > 0$, $V > 0$, and $\alpha > 0$. Computational experience in [Prigarin 13] showed that this method outperformed ordinary least squares estimation of d_B of a randomized Sierpiński carpet with known d_B. Can this approach be extended to compute d_B of a network?   □

Exercise 8.4* Recalling from Sect. 4.5 that the Minkowski dimension d_K and the box counting dimension d_B are equal, can definition (4.10) of d_K be extended to spatially embedded networks?   □

Chapter 9

Correlation Dimension

Distance doesn't exist, in fact, and neither does time. Vibrations from love or music can be felt everywhere, at all times.
Yoko Ono (1933-) Japanese multimedia artist, singer, and peace activist, and widow of John Lennon of The Beatles.

Covering a hypercube in $\mathbb{R}^E$, with side length L, by boxes of side length s requires $(L/s)^E$ boxes. This complexity, exponential in E, makes d_B expensive to compute for sets in a high-dimensional space. The *correlation dimension* was introduced in 1983 by Grassberger and Procaccia [Grassberger 83a, Grassberger 83b] and by Hentschel and Procaccia [Hentschel 83] as a computationally attractive alternative to d_B. The correlation dimension has been enormously important to physicists [Baker 90], and even by 1999 the literature on the computation of the correlation dimension was huge [Hegger 99, Lai 98]. The Grassberger and Procaccia papers introducing the correlation dimension were motivated by the study of strange attractors, a topic in dynamical systems. Although this book is not focused on chaos and dynamical systems, so much of the literature on the correlation dimension is motivated by these topics that a very short review, based on [Ruelle 80, Ruelle 90] will provide a foundation for our discussion of the correlation dimension of a geometric object (in this chapter and in the next chapter) and of a network (in Chap. 11).

9.1 Dynamical Systems

A dynamical system (e.g., a physical, chemical, or biological dynamical system) is a system that changes over time. A deterministic dynamical system models the evolution of $x(t)$, where t is time and $x(t)$ is a vector with finite or infinite dimension. A continuous time dynamical system has the form $dx(t)/dt = F(x(t))$, for some function F, while a discrete time dynamical system has the form $x(t+1) = F(x(t))$ for some function F. A physical system can be described using its *phase space*, so that each point in phase space represents an instantaneous description of the system at a given point in time. We will be concerned only with finite dimensional systems for which the phase space is $\mathbb{R}^E$,

© Springer Nature Switzerland AG 2020
E. Rosenberg, *Fractal Dimensions of Networks*,
https://doi.org/10.1007/978-3-030-43169-3_9

and the state of the system is specified by $x = (x_1, x_2, \ldots, x_E)$. For example, for a physical system the state might be the coordinates on a grid, and for a chemical system the state might be temperature, pressure, and concentration of the reactants. The E coordinates of x are assumed to vary with time, and their values at time t are $x(t) = (x_1(t), x_2(t), \ldots, x_E(t))$. We assume [Ruelle 80] the state at time $t + 1$ depends only on the state at time t:

$$
\begin{aligned}
x_1(t+1) &= F_1\big(x(t)\big) \\
x_2(t+1) &= F_2\big(x(t)\big) \\
&\cdots \\
x_E(t+1) &= F_E\big(x(t)\big).
\end{aligned}
$$

Define $F = (F_1, F_2, \ldots, F_E)$, so $F : \mathbb{R}^E \to \mathbb{R}^E$. Given $x(0)$, we can determine $x(t)$ for each time t.

If an infinitesimal change $\delta x(0)$ is made to the initial conditions (at time $t = 0$), there will be a corresponding change $\delta x(t)$ at time t. If for some $\lambda > 0$ we have $|\delta x(t)| \propto |\delta x(0)| e^{\lambda t}$ (i.e., if $\delta x(t)$ grows exponentially with t), then the system is said to exhibit *sensitive dependence on initial conditions*. This means that small changes in the initial state become magnified, eventually becoming distinct trajectories [Mizrach 92]. The exponent λ is called a *Lyapunov exponent*.

A dynamical system is said to be *chaotic* if for all (or almost all) initial conditions $x(0)$ and almost all $\delta x(0)$ the system exhibits sensitive dependence on initial conditions. The evolution of chaotic systems is complex and essentially random by standard statistical tests. Only short-term prediction of chaotic systems is possible, even if the model is known with certainty [Barkoulas 98]. Chaos was first observed by J. Hadamard in 1898, and H. Poincaré in 1908 realized the implications for weather predictability, a subject addressed by Edward Lorenz [Chang 08] in a famous 1963 paper that revitalized interest in chaos.[1] Lorenz and others have shown that theoretically it is not possible to make reasonable weather forecasts more than one or 2 weeks ahead. An example of a continuous time chaotic dynamical system is the system studied by Lorenz in 1963:

$$
\begin{aligned}
\mathrm{d}x_1/\mathrm{d}t &= 10(x_2 - x_1) \\
\mathrm{d}x_2/\mathrm{d}t &= x_1(28 - x_3) - x_2 \\
\mathrm{d}x_3/\mathrm{d}t &= x_1 x_2 - (8/3)x_3.
\end{aligned}
$$

These equations approximate the behavior of a horizontal fluid layer heated from below. As the fluid warms, it tends to rise, creating convection currents. If the heating is sufficiently intense, the convection currents become irregular and turbulent. Such behavior happens in the earth's atmosphere, and since

---

[1] An obituary [Chang 08] of Lorenz mentioned that he is best known for the notion of the "butterfly effect", the idea that a small disturbance like the flapping of a butterfly's wings can induce enormous consequences. J. Doyne Farmer, a professor at the Santa Fe Institute in New Mexico, was quoted in [Chang 08] as saying "The paper he wrote in 1963 is a masterpiece of clarity of exposition about why weather is unpredictable".

a sensitive dependence on initial conditions holds, this model provides some theoretical justification for the inability of meteorologists to make accurate long-range weather forecasts.[2]

There are two types of dynamical physical systems. When the system loses energy due to friction, as happens in most natural physical systems, the system is *dissipative*. If there is no loss of energy due to friction, the system is *conservative*. A dissipative system always approaches, as $t \to \infty$, an asymptotic state [Chatterjee 92]. Consider a dissipative dynamical system $F : \mathbb{R}^E \to \mathbb{R}^E$, and suppose that at time 0, all possible states of the system lie in a set Ω_0 having positive volume (i.e., positive Lebesgue measure, as defined in Sect. 5.1). Since the system is dissipative, Ω_0 must be compressed due to loss of energy, and Ω_0 converges to a compact set Ω as $t \to \infty$. The set Ω lives in a lower dimensional subspace of the phase space, and thus has Lebesgue measure zero in the phase space. The set Ω satisfies $F(\Omega) = \Omega$, and Ω can be extremely complicated, often possessing a fractal structure [Hegger 99]. Interestingly, Benford's Law (Sect. 3.2) has been shown to apply to one-dimensional dynamical systems [Fewster 09].

The term *strange attractor* was introduced by Ruelle and Takens in 1971. Informally, an attractor of a dynamical system is the subset of phase space to which the system evolves, and a strange attractor is an attractor that is a fractal [Theiler 90]. More formally, a bounded set $\Omega \subset \mathbb{R}^E$, with infinitely many points, is a strange attractor for the map $F : \mathbb{R}^E \to \mathbb{R}^E$ if there is a set U with the following four properties [Ruelle 80]:

1. U is an E-dimensional neighborhood of Ω, i.e., for each point $x \in \Omega$, there is a ball centered at x and entirely contained in U. In particular, $\Omega \subset U$.

2. For each initial point $x(0) \in U$, the point $x(t) \in U$ for all $t > 0$, and $x(t)$ is arbitrarily close to Ω for all sufficiently large t. (This means that Ω is *attracting*.)

3. There is a sensitive dependence on initial condition whenever $x(0) \in U$. (This makes Ω is *strange* attractor.)

4. One can choose a point $x(0)$ in Ω such that, for each other point y in Ω, there is a point $x(t)$ that is arbitrarily close to y for some positive t. (This *indecomposability* condition says that Ω cannot be split into two different attractors.)

To see how a strange attractor arises in dissipative systems, imagine we begin with a small sphere. After a small time increment, the sphere may evolve into an ellipsoid, such that the ellipsoid volume is less than the sphere volume. However, the length of longest principle axis of the ellipsoid can exceed the

---

[2] Although Lorenz has been generally regarded as the discoverer of chaos, the mathematician Mary Cartwright described the same phenomena 20 years before Lorenz, in a 1943 lecture at Cambridge. She discovered chaos in the solutions of the van der Pol equation, which describes the oscillations of nonlinear amplifiers, which were important in WWII since they fed power to early radar systems [Dyson 09].

sphere diameter. Thus, over time, the sphere may compress in some directions and expand in other directions. For the system to remain bounded, expansion in a given direction cannot continue indefinitely, so folding occurs. If the pattern of compressing, expansion, and folding continues, trajectories that are initially close can diverge rapidly over time, thus creating the sensitive dependence on initial conditions. The amounts of compression or stretching in the various dimensions are quantified by the Lyapunov exponents [Chatterjee 92].

A well-known example of a discrete time dynamical system was introduced by Hénon. For this two-dimensional system, we have $x_1(t+1) = x_2(t) + 1 - ax_1(t)^2$ and $x_2(t+1) = bx_2(t)$. With $a = 1.4$ and $b = 0.3$, the first 10^4 points distribute themselves on a complex system of lines that are a strange attractor (see [Ruelle 80] for pictures); for $a = 1.3$ and $b = 0.3$ the lines disappear and instead we obtain 7 points of a *periodic attractor*.

Chaos in human physiology was studied in many papers by Goldberger, e.g., [Goldberger 90]. In the early 1980s, when investigators began to apply chaos theory to psychological systems, they expected that chaos would be most apparent in diseased or aging systems (e.g., [Ruelle 80] stated that the normal cardiac regime is periodic). Later research found instead that the phase-space representation of the normal heartbeat contains a strange attractor, suggesting that the dynamics of the normal heartbeat may be chaotic. There are advantages to such dynamics [Goldberger 90]: "Chaotic systems operate under a wide range of conditions and are therefore adaptable and flexible. This plasticity allows systems to cope with the exigencies of an unpredictable and changing environment."

Several other examples of strange attractors were provided in [Ruelle 80]. Strange attractors arise in studies of the earth's magnetic field, which reverses itself at irregular intervals, with at least 16 reversals in the last 4 million years. Geophysicists have developed *dynamo equations*, which possess chaotic solutions, to describe this behavior. The Belousov–Zhabotinski chemical reaction turns ferroin from blue to purple to red; simultaneously the cerium ion changes from pale yellow to colorless, so all kinds of hues are produced (this reaction caused astonishment and some disbelief when it was discovered). For a biological example, discrete dynamical models of the population of multiple species yield strange attractors. Even for a single species, the equation $x(t+1) = Rx(t)\big(1 - x(t)\big)$, where $R > 0$, gives rise to non-periodic behavior. Ruelle observed that the non-periodic fluctuations of a dynamical system do not necessarily indicate an experiment that was spoiled by mysterious random forces, but rather may indicate the presence of a strange attractor. Chaos in ecology was studied in [Hastings 93].

Physicists studying dynamical systems want to distinguish behavior that is irregular, but low-dimensional, from behavior that is irregular because it is essentially stochastic, with many effective degrees of freedom [Theiler 90]. Calculating the fractal dimension of the attractor helps make this distinction, since the fractal dimension roughly estimates the number of independent variables involved in the process [Badii 85]. That is, the fractal dimension can be interpreted as the number of degrees of freedom of the system [Ruelle 90]. A finite non-integer fractal dimension indicates the presence of complex deterministic dynamics [Krakovska 95]. The connection between strange attractors and fractals was succinctly described in [Grassberger 83b]:

> In a system with F degrees of freedom, an attractor is a subset of F-dimensional phase space toward which almost all sufficiently close trajectories get "attracted" asymptotically. Since volume is contracted in dissipative flows, the volume of an attractor is always zero, but this leaves still room for extremely complex structures.
>
> Typically, a strange attractor arises when the flow does not contract a volume element in all directions, but stretches it in some. In order to remain confined to a bounded domain, the volume element gets folded at the same time, so that it has after some time a multisheeted structure. A closer study shows that it finally becomes (locally) Cantor-set like in some directions, and is accordingly a fractal in the sense of Mandelbrot [Mandelbrot 77].

We could determine the fractal dimension of an attractor by applying box counting. However, a major limitation of box counting is that it counts only the number of occupied boxes, and is insensitive to the number of points in an occupied box. Thus d_B is more of a geometrical measure, and provides very limited information about the clumpiness of a distribution. Moreover, if the box size s is small enough to investigate the small-scale structure of a strange attractor, the relation $B_D(s) \sim s^{-d_B}$ (assuming diameter-based boxes) implies that the number $B_D(s)$ of occupied boxes becomes very large. The most interesting cases of dynamical systems occur when the dimension of the attractor exceeds 3. As observed in [Farmer 82], when $d_B = 3$ and $s = 10^{-2}$ from $B_D(s) = s^{-d_B}$ we obtain one million boxes, which in 1982 taxed the memory of a computer. While such a requirement would today be no burden, the point remains: box-counting methods in their basic form have time and space complexity exponential in d_B. Moreover, for a dynamical system, some regions of the attractor have a very low probability, and it may be necessary to wait a very long time for a trajectory to visit all of the very low probability regions. These concerns were echoed in [Eckmann 85], who noted that the population of boxes is usually very uneven, so that it may take a long time for boxes that should be occupied to actually become occupied. For these reasons, box counting was not (as of 1985) used to study dynamical systems [Eckmann 85]. Nine years later, [Strogatz 94] also noted that d_B is rarely used for studying high-dimensional dynamical processes.

An early (1981) alternative to box counting was proposed in [Froehling 81]. Given a set of points in $\mathbb{R}^E$, use regression to find the best hyperplane of dimension m that fits the data, where m is an integer in the range $1 \le m \le E-1$. The goodness of fit of a given hyperplane is measured by χ^2, the sum of deviations from the hyperplane divided by the number of degrees of freedom. If m is too small, the fit will typically be poor, and χ^2 will be large. As m is increased, χ^2 will drop sharply. The m for which χ^2 is lowest is the approximate fractal dimension.

A better approach, proposed two years later in 1983, is the correlation dimension. Our discussion of the correlation dimension begins with the natural invariant measure and the pointwise mass dimension.

9.2   Pointwise Mass Dimension

The *pointwise mass* dimension is based upon the *measure* of a subset of $\mathbb{R}^E$. The study of measures of sizes of sets dates at least as far back as Borel (1885), and was continued by Lebesgue (1904), Carathéodory (1914), and Hausdorff (1919) [Chatterjee 92]. Recall that the Lebesgue measure μ_L was defined in Chap. 5. Formalizing the qualitative definition of "dissipative" provided above, we say that a dynamical physical system g_t, $t = 1, 2, \ldots$, where $g_t : \mathbb{R}^E \to \mathbb{R}^E$, is *dissipative* if the Lebesgue measure μ_L of a bounded subset tends to 0 as $t \to \infty$, i.e., if for each bounded $B \subset \mathbb{R}^E$ we have

$$\lim_{t\to\infty} \mu_L[g_t(B)] = 0 \, . \tag{9.1}$$

It can be shown that only dissipative systems have strange attractors [Theiler 90]. Suppose the dissipative system has the strange attractor Ω. Since by (9.1) we have $\mu_L(\Omega) = 0$, then to study Ω we require a new measure defined on subsets of Ω; the measure of a subset should reflect how much of the subset is contained in Ω. The new measure μ of a subset $B \subset \Omega$ should count how often and for how long a typical trajectory visits B, and an appropriate definition is

$$\mu(B) \equiv \lim_{t\to\infty} \frac{1}{t} \int_0^t I_B[g_t(x_0)]\mathrm{d}t \, , \tag{9.2}$$

where $x_0 \in \mathbb{R}^E$ is the starting point of the system g_t, and $I_B[x]$ is the indicator function whose value is 1 if $x \in B$ and 0 otherwise [Theiler 90]. The measure defined by (9.2) is the *natural measure* for almost all x_0. For a discrete dynamical system, the natural measure of B is [Peitgen 92]

$$\mu(B) = \lim_{t\to\infty} \frac{1}{t+1} \sum_{i=0}^{t} I_B[x(t)] \, .$$

The natural measure is an invariant measure, which means [Peitgen 92] that for the system $x_{t+1} = f(x_t)$ if $f^{-1}(B)$ is the set of points which after one iteration land in B, then $\mu\big(f^{-1}(B)\big) = \mu(B)$. For $x \in \Omega$ and $r > 0$, define

$$\Omega(x, r) \equiv \{y \in \Omega : \, dist(x, y) \leq r\} \, ,$$

where the choice of norm determines the shape of $\Omega(x, r)$ (e.g., a hypersphere for the Euclidean norm, and a hypercube for the L_∞ norm). Then $\mu\big(\Omega(x, r)\big)$ is the natural measure of the ball $\Omega(x, r)$.

Definition 9.1 The *pointwise dimension* of μ at x, denoted by $d(x)$, is given by

$$d(x) \equiv \lim_{r\to 0} \frac{\log \mu\big(\Omega(x, r)\big)}{\log r} \, . \qquad \square \tag{9.3}$$

Thus $\mu\big(\Omega(x,r)\big) \sim r^{d(x)}$ as $r \to 0$. The pointwise dimension at x is also called the *local Hölder exponent* at x. In contrast to the Hausdorff and box counting dimensions, which are global quantities whose computation requires covering all of Ω, the pointwise dimension is a local quantity, defined at a single point. If $d(x)$ is independent of x, then the probability (i.e., "mass") of $\Omega(x,r)$ depends only on r but not on x; this holds, e.g., when points in Ω are uniformly distributed. If $d(x)$ is a non-constant function of x, then the mass of $\Omega(x,r)$ depends on x and r; this holds, e.g., for the Gaussian distribution. (See [Simpelaere 98] for additional theory of the pointwise dimension.) The *average pointwise dimension*, denoted by $\langle d \rangle$, is given by

$$\langle d \rangle \equiv \int_\Omega d(x)\mathrm{d}\mu(x), \qquad (9.4)$$

where $\mathrm{d}\mu(x)$ is the differential of $\mu(x)$.

9.3 The Correlation Sum

Let $\Omega \subset \mathbb{R}^E$, and suppose we have determined N points in Ω. Denote these points by $x(n)$ for $n = 1, 2, \ldots, N$. Denote the E coordinates of $x(n)$ by $x_1(n)$, $x_2(n)$, ..., $x_E(n)$. We can approximate the average pointwise dimension $\langle d \rangle$ defined by (9.4) by averaging $d(x)$ over these points, that is, by computing $(1/N)\sum_{n=1}^N d\big(x(n)\big)$. Since by (9.3) each $d\big(x(n)\big)$ is a limit, then $(1/N)\sum_{n=1}^N d\big(x(n)\big)$ is an average of N limits. An alternative method of estimating (9.4) is to first compute, for each r, the average

$$\langle \mu\big(\Omega(r)\big) \rangle \equiv \frac{1}{N} \sum_{i=1}^N \mu\Big(\Omega\big(x(i),r\big)\Big) \qquad (9.5)$$

and then compute the limit

$$\lim_{r \to 0} \frac{\log\langle \mu\big(\Omega(r)\big) \rangle}{\log r}. \qquad (9.6)$$

Thus, instead of taking the average of N limits, in (9.6) we take the limit of the average of N natural measures. Writing (9.6) in a computationally useful form will lead us to the correlation dimension.

Let $I_0 : \mathbb{R} \to \mathbb{R}$ be the Heaviside step function, defined by

$$I_0(z) = \begin{cases} 1 & \text{if } z \geq 0 \\ 0 & \text{otherwise.} \end{cases} \qquad (9.7)$$

The subscript "0" in I_0 is used to distinguish this step function from one that will be defined in a later chapter. Note that $I_0(0) = 1$. Then

$$I_0\Big(r - dist\big(x(i), x(j)\big)\Big) = \begin{cases} 1 & \text{if } dist\big(x(i), x(j)\big) \leq r \\ 0 & \text{otherwise.} \end{cases} \qquad (9.8)$$

Typically, $dist(x, y)$ is the Euclidean distance; we could alternatively use the L_∞ (sup norm) and define $dist(x, y) = \max_{i=1}^{E} |x_i - y_i|$ [Frank 89]. Define

$$C\big(x(i), r\big) \equiv \sum_{\substack{j=1 \\ j \neq i}}^{N} I_0\Big(r - \big(dist(x(i), x(j))\big)\Big), \qquad (9.9)$$

so $C\big(x(i), r\big)$ is the number of points, distinct from $x(i)$, whose distance from $x(i)$ does not exceed r.[3] Although $C\big(x(i), r\big)$ depends on N, for notational simplicity we write $C\big(x(i), r\big)$ rather than, e.g., $C_N\big(x(i), r\big)$. We have $0 \leq C\big(x(i), r\big) \leq N - 1$. We can approximate $\mu\big(\Omega(x(i), r)\big)$ using

$$\mu\big(\Omega(x(i), r)\big) \approx \frac{C\big(x(i), r\big)}{N - 1}. \qquad (9.10)$$

This is an approximation, and not an equality, since we only have a finite data set; the accuracy of the approximation increases as $N \to \infty$. Define the *correlation sum* $C(r)$ by

$$C(r) \equiv \frac{1}{N(N - 1)} \sum_{i=1}^{N} C\big(x(i), r\big). \qquad (9.11)$$

Although $C(r)$ depends on N, for notational simplicity we write $C(r)$ rather than, e.g., $C_N(r)$. If $0 < dist\big(x(i), x(j)\big) \leq r$, then both (i, j) and (j, i) contribute to $C(r)$.[4] In words, $C(r)$ is the ratio

$$\frac{\text{the number of nonzero distances not exceeding } r}{\text{the total number of nonzero distances}}.$$

Equivalently, $C(r)$ is the fraction of all pairs of distinct points for which the distance between the pair does not exceed r. An equivalent definition of $C(r)$ that invokes the natural measure μ defines $C(r)$ to be the probability that a pair of distinct points, chosen randomly on the attractor with respect to μ, is separated by a distance not exceeding r [Ding 93]. From (9.5), (9.10), and (9.11) we have

$$\langle \mu\big(\Omega(r)\big) \rangle \approx \frac{1}{N} \sum_{i=1}^{N} \mu\big(\Omega(x(i), r)\big) = \frac{1}{N(N - 1)} \sum_{i=1}^{N} C\big(x(i), r\big) = C(r). \quad (9.12)$$

---

[3] Often, e.g., as in [Ding 93], the Heaviside step function is defined such that $I_0(0) = 0$, in which case $\sum_{j \neq i} I_0\left(r - dist(x(i), x(j))\right)$ is the number of points, distinct from $x(i)$, whose distance from $x(i)$ is strictly less than r. The definition in [Grassberger 83] also only counts points satisfying $dist\big(x(i), x(j)\big) < r$. This distinction becomes important when we define the correlation dimension of networks.

[4] Sometimes the correlation sum $C(r)$ is written using the double summation $\sum_{i=1}^{N} \sum_{j=i+1}^{N}$ with the prefactor $2/\big(N(N-1)\big)$ instead of $1/\big(N(N-1)\big)$; with this definition, the double summation counts each pair of distinct points at most once.

Grassberger stressed the importance of excluding $i = j$ from the double sum defining $C(r)$; see [Theiler 90]. However, taking the opposite viewpoint, [Smith 88] wrote that it is essential that the $i = j$ terms be included in the correlation sum, "in order to distinguish the scaling of true noise at small r from fluctuations due to finite N". The argument in [Smith 88] is that, when the $i = j$ terms are omitted, $\log C(r)$ is not bounded as $r \to 0$, and steep descents in $\log C(r)$ may result in large estimates of d_C, which may exceed E. In this chapter, we will use definition (9.9).

Given the set of N points, for $r \geq r_{\max} \equiv \max_{i,j} dist\big(x(i), x(j)\big)$, the distance between each pair of distinct points does not exceed $r_{\max}$, so $C(r) = 1$. For $r < r_{\min} \equiv \min_{i \neq j} dist\big(x(i), x(j)\big)$, the distance between each pair of distinct points is less than $r_{\min}$, so $C(r) = 0$. Hence the useful range of r is the interval $[r_{\min}, r_{\max}]$. If there is a single pair of points at distance $r_{\min}$, then $C(r_{\min}) = 2/[N(N-1)]$, and $\log C(r_{\min}) \approx -2 \log N$. Since $\log C(r_{\max}) = 0$, the range of $\log C(r)$ over the interval $[r_{\min}, r_{\max}]$ is approximately $2 \log N$. However, for box counting with N points, the number of non-empty boxes (i.e., $B_R(r)$ for radius-based boxes and $B_D(s)$ for diameter-based boxes) ranges from 1 to N, so the range of $\log B_R(r)$ or $\log B_D(s)$ is $\log N$, which is only half the range of $\log C(r)$. Similarly, for a single point x, over its useful range the pointwise mass dimension estimate (9.10) ranges from $1/(N-1)$ to 1, so the range of its logarithm is also approximately $\log N$, half the range for $\log C(r)$. The greater range, by a factor of 2, for $\log C(r)$ is "the one advantage" the correlation sum has over the average pointwise dimension [Theiler 90].

Consider an alternate definition $\widetilde{C}(r)$ of the correlation sum, now using the double summation $\sum_{i=1}^{N} \sum_{i=j}^{N}$ which does not exclude the case $i = j$. With this alternative definition, each of the N pairs $\big(x(i), x(i)\big)$ contributes 1 to the double summation. Following [Yadav 10],

$$\widetilde{C}(r) \equiv \frac{1}{N^2} \sum_{i=1}^{N} \sum_{j=1}^{N} I_0\Big(r - dist\big(x(i), x(j)\big)\Big) = \frac{1}{N^2} \sum_{n \in \mathbb{N}} N_n(r), \qquad (9.13)$$

where $N_n(r)$ is the number of points whose distance from point n does not exceed r. For each n, since $N_n(r)$ always counts n itself, then $N_n(r) \geq 1$. For all sufficiently large r we have $N_n(r) = N$ for each n, in which case $\widetilde{C}(r) = 1$. In (9.13), we can interpret $N_n(r)/N$ as the expected fraction of points whose distance from n does not exceed r. Suppose this fraction is independent of n, and suppose we have a probability distribution $P(k, r)$ providing the probability that k of N points are within a distance r from a random point. The expected value (with respect to k) of this probability distribution is the expected fraction of the N points within distance r of a random point, so

$$\frac{N_n(r)}{N} = \sum_{k=1}^{N} k P(k, r), \qquad (9.14)$$

which with (9.13) provides yet another way to write $\widetilde{C}(r)$:

$$\widetilde{C}(r) = \frac{1}{N} \sum_{k=1}^{N} k P(k, r)\,. \tag{9.15}$$

While (9.14) provides the mean (with respect to k) of the distribution $P(k, r)$, to fully characterize the distribution we need its moments. So for $q \in \mathbb{R}$ define the generalized correlation sum

$$\widetilde{C}_q(r) \equiv \frac{1}{N} \sum_{k=1}^{N} k^{q-1} P(k, r)\,, \tag{9.16}$$

which for $q = 2$ yields (9.15). For $q \gg 1$ the terms dominating the sum in (9.16) are terms $k^{q-1}P(k, r)$ for which k is "large" (i.e., many points in the ball of radius r centered at a random point), whereas for $q \ll 0$ the terms dominating the sum in (9.16) are terms $k^{q-1}P(k, r)$ for which k is "small" (i.e., few points in the ball of radius r). Thus studying the behavior of $\widetilde{C}_q(r)$ for $-\infty < q < \infty$ provides information about regions containing many points as well as regions with few points. We will explore this further in Chap. 16 on multifractals.

9.4   The Correlation Dimension

Definition 9.2 [Grassberger 83a, Grassberger 83b, Hentschel 83] If the limit

$$d_C \equiv \lim_{r \to 0} \lim_{N \to \infty} \frac{\log C(r)}{\log r} \tag{9.17}$$

exists, then d_C is called the *correlation dimension* of Ω.   □

Assume for the remainder of this section that d_C exists. Then d_C is independent of the norm used in (9.8) to define the distance between two points. Also, $C(r) \sim r^{d_C}$ as $r \to 0$, where "$\sim$" means "tends to be approximately proportional to" (see Sect. 1.2 for a discussion of this symbol). However, the existence of d_C does not imply that $C(r) \propto r^{d_C}$, since this proportionality may fail to hold, even as $r \to 0$. For $r > 0$, define the function $\Phi : \mathbb{R} \to \mathbb{R}$ by

$$\Phi(r) \equiv C(r)/r^{d_C}\,. \tag{9.18}$$

We cannot conclude that $\Phi(r)$ approaches a constant value as $r \to 0$, but the existence of d_C does imply $\lim_{r \to 0} \log \Phi(r) / \log r = 0$ [Theiler 88].

One great advantage of computing d_C rather than d_B is that memory requirements are greatly reduced, since there is no need to construct boxes covering Ω. With box counting it is also necessary to examine each box to determine if it contains any points; this may be impractical, particularly if r is very small. Another advantage, mentioned in Sect. 9.3 above, is that the range of $\log C(r)$ is twice the range of $\log B_D(s)$ or $\log B_R(r)$. In [Eckmann 85], the correlation dimension was deemed a "quite sound" and "highly successful" means of determining the dimension of a strange attractor, and it was observed that the method "is not entirely justified mathematically, but nevertheless quite sound".[5] In [Ruelle 90] it was noted that "... this algorithm has played a very important role in allowing us to say something about systems that otherwise defied analysis". The importance of d_C was echoed in [Schroeder 91], which noted that the correlation dimension is one of the most widely used fractal dimensions, especially if the fractal comes as "dust", meaning isolated points, sparsely distributed over some region. Lastly, [Mizrach 92] observed that, in the physical sciences, the correlation dimension has become the preferred dimension to use with experimental data.

There are several possible approaches to computing d_C from the N points $x(n)$, $n = 1, 2, \ldots, N$. First, we could pick the smallest r available, call it r_{min}, and estimate d_C by $\log C(r_{min})/\log r_{min}$. The problem with this approach is that the convergence of $\log C(r)/\log r$ to d_C is very slow [Theiler 90]; this is the same problem identified in Sect. 4.2 in picking the smallest s value available to estimate d_B. A second possible approach [Theiler 88] is to take the local slope of the $\log C(r)$ versus $\log r$ curve, using

$$\frac{\mathrm{d} \log C(r)}{\mathrm{d} \log r} = \frac{\left(\frac{\mathrm{d}C(r)}{\mathrm{d}r}\right)}{\left(\frac{C(r)}{r}\right)}. \tag{9.19}$$

A third possible approach [Theiler 88] is to pick two values r_1 and r_2, where $r_1 < r_2$, and use the two-point secant estimate

$$d_C \approx \frac{\log C(r_2) - \log C(r_1)}{\log r_2 - \log r_1}. \tag{9.20}$$

While the third approach avoids the slow logarithmic convergence, it requires choosing r_1 and r_2, both of which should tend to zero. This approach was analyzed in [Theiler 93] and compared to other methods for estimating d_C. (For a network, two-point secant estimates of fractal dimensions were utilized in [Rosenberg 16b, Rosenberg 17c, Wang 11].)

By far, the most popular method for estimating d_C is the *Grassberger–Procaccia* method. The steps are (*i*) compute $C(r)$ for a set of positive radii r_m, $m = 1, 2, \ldots, M$, (*i*) determine a subset of these M values of r_m over which the $\log C(r_m)$ versus $\log r_m$ curve is roughly linear, and (*iii*) estimate d_C as the slope of the best fit line through the points $\left(\log r_m, \log C(r_m)\right)$ over this subset of the M points. As observed in [Theiler 88], there is no clear criterion for how to determine the slope through the chosen subset of the M points. Finding the slope that minimizes the unweighted sum of squares has two drawbacks. First, the

---

[5] The dimension defined by (9.17), which is now universally known as the "correlation dimension", is called the "information dimension" in [Eckmann 85].

least squares model assumes a uniform error in the calculated $\log C(r)$, which is incorrect since the statistics get worse as r decreases. Second, the least squares model assumes that errors in $\log C(r)$ are independent of each other, which is incorrect since $C(r + \epsilon)$ is equal to $C(r)$ plus the fraction of distances between r and ϵ; hence for small ϵ the error at $\log C(r + \epsilon)$ is strongly correlated with the error at $\log C(r)$.

Suppose $C(r_m)$ has been computed for $m = 1, 2, \ldots, M$, where $r_m < r_{m+1}$. In studying the orbits of celestial bodies, [Freistetter 00] used the following iterative procedure to determine the range of r values over which the slope of the $\log C(r_m)$ versus $\log r_m$ curve should be computed. Since the $\log C(r)$ versus $\log r$ curve is not linear in the region of very small and very large r, we trim the ends of the interval as follows. In each iteration we first temporarily remove the leftmost r value (which in the first iteration is r_1) and use linear least squares to obtain d_R, the slope corresponding to $\{r_2, r_3, \ldots, r_M\}$, and the associated correlation coefficient c_R. (Here the *correlation coefficient*, not to be confused with the correlation dimension, is a byproduct of the regression indicating the goodness of fit.) Next we temporarily remove the rightmost r value (which in the first iteration is r_M) and use linear least squares to obtain d_L, the slope corresponding to $\{r_1, r_2, \ldots, r_{M-1}\}$, and the associated correlation coefficient c_L. Since a higher correlation coefficient indicates a better fit (where $c = 1$ is a perfect linear relationship), if $c_R > c_L$ we delete r_L while if $c_L > c_R$ we delete r_M. We begin the next iteration with either $\{r_2, r_3, \ldots, r_M\}$ or $\{r_1, r_2, \ldots, r_{M-1}\}$. The iterations stop when we obtain a correlation coefficient higher than a desired value.

An alternative to the *Grassberger–Procaccia* method is the method of [Takens 85], which is based on maximum likelihood estimation. Suppose we randomly select N points from Ω. Assume that d_C exists and that the function Φ defined by (9.18) takes the constant value k, so $C(r) = kr^{d_C}$. Assume also that the distances between pairs of randomly selected points are independent and randomly distributed such that the probability $P(x \leq r)$ that the distance x between a random pair of points does not exceed r is given by $P(x \leq r) = C(r) = kr^{d_C}$. Let $X(r)$ be the number of pairwise distances less than r, and denote these $X(r)$ distance values by r_i for $1 \leq i \leq X(r)$. The value of d_C that maximizes the probability of finding the $X(r)$ observed distances $\{r_i\}$ is given by the unbiased minimal variance Takens estimator $\widetilde{T}(r)$, defined by Guerrero and Smith [Guerrero 03]

$$\widetilde{T}(r) \equiv \left[\frac{-1}{X(r) - 1} \sum_{i=1}^{X(r)} \log\left(\frac{r_i}{r}\right) \right]^{-1}, \qquad (9.21)$$

so $\widetilde{T}(r)$ uses all distances less than a specified upper bound r, but there is no lower bound on distances. As $r \to 0$ and $X(r) \to \infty$ we have $\widetilde{T}(r) \to d_C$. If we replace $X(r) - 1$ by $X(r)$ (which introduces a slight bias [Guerrero 03]), we can write (9.21) more compactly as

$$T(r) = -1/\langle \log(r_i/r) \rangle \,,$$

where $\langle\ \rangle$ denotes the average over all distances (between pairs of points) not exceeding r. It can be shown that [Theiler 90]

$$T(r) = \frac{C(r)}{\int_0^r \left(\frac{C(x)}{x}\right) \mathrm{d}x} = \frac{\int_0^r \left(\frac{\mathrm{d}C(x)}{\mathrm{d}x}\right) \mathrm{d}x}{\int_0^r \left(\frac{C(x)}{x}\right) \mathrm{d}x} \,. \tag{9.22}$$

As in [Theiler 88], the slightly more unwieldy form of the right-hand side of (9.22) is shown for comparison with (9.19). The estimate

$$\frac{C(r)}{\int_0^r \left(\frac{C(x)}{x}\right) \mathrm{d}x} \tag{9.23}$$

(the middle term in (9.22) above) has been called the Takens-Theiler estimator [Hegger 99]. The advantage of this estimator is that, as $r \to 0$, the Takens-Theiler estimate $T(r)$ approaches d_C much more rapidly than does $\log C(r)/\log r$. However, for some fractal sets $T(r)$ does not converge [Theiler 90]. The non-convergence is explained by considering the function $\Phi(r)$ defined by (9.18). A fractal has "periodic lacunarity" if for some positive number L we have $\Phi(r) = \Phi(Lr)$. For example, the middle-thirds Cantor set exhibits periodic lacunarity with $L = 1/3$. The Takens estimate $T(r)$ does not converge for fractals with periodic lacunarity. Theiler also supplied a condition on $\Phi(r)$ under which $T(r)$ does converge.

Exercise 9.1 Show that the middle-thirds Cantor set exhibits periodic lacunarity with $L = 1/3$. $\square$

To evaluate (9.23), the method of [Hegger 99] can be used. Since $C(r)$ is only evaluated at the discrete points $\{r_1, r_2, r_3, \ldots, r_M\}$, we approximate $\log C(r)$ by the linear function $a_i \log r + b_i$ for $r \in [r_{i-1}, r_i]$, which implies $C(r)$ is approximated by the exponential function $e^{b_i} r^{a_i}$ for $r \in [r_{i-1}, r_i]$. The values a_i and b_i can be computed from the M points $(\log r_i, \log C(r_i))$. Assume $r_i \le r$ for $1 \le i \le M$. Since the integral of a linear function is trivial,

$$\int_{x=0}^r \left(\frac{C(x)}{x}\right) \mathrm{d}x \ \approx \ \sum_{i=2}^M e^{b_i} \int_{r_{i-1}}^{r_i} x^{(a_i-1)} \mathrm{d}x$$

$$= \ \sum_{i=2}^M \frac{e^{b_i}}{a_i} (r_i^{a_i} - r_{i-1}^{a_i}) \,. \tag{9.24}$$

Considering the M points $(\log r_i, \log C(r_i))$ for $i = 1, 2, \ldots, M$, suppose there are indices L and U, where $1 \le L < U \le M$, such that a straight line is a good fit to the points $\big(\log r_i, \log C(r_i)\big)$ for $L \leq i \leq U$. Then, using (9.23) and (9.24), a reasonable estimate of d_C is

$$\frac{C(r_H)}{\sum_{i=L+1}^{U} \frac{e^{b_i}}{a_i}\big(r_i^{a_i} - r_{i-1}^{a_i}\big)}$$

[Hegger 99]. Various other approaches to computing d_C have also been proposed, e.g., [Shirer 97].

Exercise 9.2 Use (9.24) to estimate d_C for the middle-thirds Cantor set for $r \in [0.001, 1]$.   □

When the *Grassberger–Procaccia* method is applied to a time series (Sect. 10.2), as in the calculation of d_C of a strange attractor of a dynamical system, the upper bound

$$d_C \leq 2 \log_{10} N \tag{9.25}$$

applies [Ruelle 90]. To derive this bound, define

$$f(r) \equiv \sum_{i=1}^{N} \sum_{j=i}^{N} I\big(r - dist(x(i), x(j))\big). \tag{9.26}$$

The range of r over which $f(r)$ can be computed is $[r_L, r_U]$, where

$$r_L \equiv \min\{dist\big(x(i), x(j)\big) \mid 1 \leq i, j \leq N, \, i \neq j\}$$

and $r_U \equiv diam(\Omega)$. For time series applications, only part of this range is usable, since the lower part of the range suffers from statistical fluctuations, and the upper part suffers from nonlinearities. Let r_{min} be the smallest usable r for which we evaluate $f(r)$, and let r_{max} be the largest usable r, where $r_L \leq r_{min} < r_{max} \leq r_U$. Suppose $r_{max}/r_{min} \geq 10$, so the range is at least one decade. The correlation dimension d_C is approximately the slope m, where

$$m = \frac{\log_{10} f(r_{max}) - \log_{10} f(r_{min})}{\log_{10} r_{max} - \log_{10} r_{min}}.$$

Since each point is within distance r of itself, then $f(r) \geq N$ for $r > 0$. Hence $f(r_{min}) \geq N$, so $\log_{10} f(r_{min}) \geq 0$. Also, $f(r_{max}) \leq N^2$. Since $r_{max}/r_{min} \geq 10$ then $\log_{10} r_{max} - \log_{10} r_{min} \geq 1$. Hence

$$m = \frac{\log_{10} f(r_{max}) - \log_{10} f(r_{min})}{\log_{10} r_{max} - \log_{10} r_{min}} \leq \log_{10} f(r_{max}) \leq \log_{10} N^2 = 2 \log_{10} N.$$

Thus one should not believe a correlation dimension estimate that is not substantially less than $2 \log_{10} N$ [Ruelle 90].

This bound was used in [Boon 08] to study the visual evoked potential. Unfortunately, numerous studies do violate this bound [Ruelle 90]. If the inequality

$r_{max}/r_{min} \geq 10$ is replaced by the more general bound $r_{max}/r_{min} \geq k$, then we obtain $d_C \leq 2 \log_k N$. Ruelle observed that it seems unreasonable to have the ratio r_{max}/r_{min} be much less than 10.

To conclude this section, we quote these words of caution in [Hegger 99]:

> ... dimensions characterize a set or an invariant measure whose support is the set, whereas any data set contains only a finite number of points representing the set or the measure. By definition, the dimension of a finite set of points is zero. When we determine the dimension of an attractor numerically, we extrapolate from finite length scales ... to the infinitesimal scales, where the concept of dimensions is defined. This extrapolation can fail for many reasons ...

9.5 Comparison with the Box Counting Dimension

In this section we present the proof of [Grassberger 83a, Grassberger 83b] that $d_C \leq d_B$ for a geometric object Ω.[6] We first discuss notation. The pointwise mass measure $\mu\big(\Omega(r)\big)$ was defined in Sect. 9.2 in terms of the box radius r. The correlation sum $C(r)$ given by (9.11), and the correlation dimension d_C given by (9.17), were similarly defined using r. Since $C(r)$ counts pairs of points whose distance is less than r, we could equally well have defined the correlation sum and d_C using using the variable s, rather than r, to represent distance. The motivation for using s rather than r in this section is that we will compute d_C by first covering Ω with diameter-based boxes, and we always denote the linear size of a diameter-based box by s. Therefore, in this section we will denote the correlation sum by $C(s)$; its definition is unchanged.

For a given positive $s \ll 1$, let $\mathcal{B}(s)$ be a minimal covering of Ω by diameter-based boxes of size s, so the distance between any two points in a given box does not exceed s. Let N points be drawn randomly from Ω according to its natural measure and let N_j be the number of points in box $B_j \subset \mathcal{B}(s)$. Discard from $\mathcal{B}(s)$ each box for which $N_j = 0$, and, as usual, let $B_D(s)$ be the number of non-empty boxes in $\mathcal{B}(s)$.

Since $diam(B_j) \leq s$, then the N_j points in B_j yield $N_j(N_j - 1)$ ordered pairs of points such that the distance between the two points does not exceed s. The sum $\sum_{j=1}^{B_D(s)} N_j(N_j - 1)$ counts all ordered pairs $\big(x(i), x(j)\big)$ of points, separated by a distance not exceeding s, such that $x(i)$ and $x(j)$ lie in the same box. Since a given pair $\big(x(i), x(j)\big)$ of points satisfying $dist\big(x(i), x(j)\big) \leq s$ may lie within some box B_j, or $x(i)$ and $x(j)$ may lie in different boxes, from (9.9) and (9.11),

$$C(s) \geq \frac{1}{N(N-1)} \sum_{j=1}^{B_D(s)} N_j(N_j - 1) \tag{9.27}$$

---

[6] In [Grassberger 83b], d_B was called the Hausdorff dimension.

(recall that although $C(s)$ depends on N, for notational simplicity we simply write $C(s)$.) Let $p_j \equiv N_j/N$ be the probability of box B_j, and define

$$\mu_j \equiv \lim_{N \to \infty} p_j. \tag{9.28}$$

Although p_j depends on both N and s, for notational simplicity we write p_j; similarly, although μ_j depends on s, for notational simplicity we write μ_j. Following [Chatterjee 92, Grassberger 83a] and using (9.11) and (9.27),

$$
\begin{aligned}
\lim_{N \to \infty} C(s) &\geq \lim_{N \to \infty} \frac{1}{N(N-1)} \sum_{j=1}^{B_D(s)} N_j(N_j - 1) \\
&= \sum_{j=1}^{B_D(s)} \lim_{N \to \infty} \left(\frac{N_j}{N} \right)^2 \\
&= \sum_{j=1}^{B_D(s)} \lim_{N \to \infty} p_j^2 \\
&= \sum_{j=1}^{B_D(s)} \mu_j^2.
\end{aligned}
\tag{9.29}
$$

Relation (9.29) says that for large N and small s, the correlation sum $C(s)$ is no less than the probability that two points of Ω are contained in some box of size s [Krakovska 95].

A generalization [Peitgen 92] of the correlation sum considers q-tuples $\big(x(1), x(2), \ldots, x(q)\big)$ of distinct points, where $q \geq 2$, such that the *pairwise distance condition* holds: for each pair of points $x(i)$ and $x(j)$ in the q-tuple we have $dist\big(x(i), x(j)\big) \leq s$. With N, $\mathcal{B}(s)$, and $B_D(s)$ as defined as in the previous paragraph, let $Q(N, s, q)$ be the number of q-tuples of the N points such that the pairwise distance condition holds. Then

$$\lim_{N \to \infty} \frac{Q(N, s, q)}{N^q} \propto \sum_{j=1}^{B_D(s)} \mu_j^q. \tag{9.30}$$

The sum on the right-hand side of (9.30) is the probability that q points, randomly drawn from Ω according to its natural measure, fall into any box. From this generalized correlation sum we can define generalized dimensions, to be studied in Chap. 16.

Having shown that we can lower bound the correlation sum by covering Ω with boxes, we now present the proof [Grassberger 83a, Grassberger 83b] that $d_C \leq d_B$. The proof in [Grassberger 83b] assumes $B_D(s) \sim s^{-d_B}$ and $C(s) \sim s^{d_C}$. Since the symbol $\sim$ is ambiguous, we recast these assumptions as Assumptions 9.1 and 9.2 below.

Assumption 9.1 For some positive number d_C and for some function $\Phi_C :$ $\mathbb{R} \to \mathbb{R}$ satisfying

$$\lim_{s \to 0} \log \Phi_C(s)/\log s = 0\,,$$

for all sufficiently small positive s we have

$$\lim_{N \to \infty} C(s) = \Phi_C(s)s^{d_C}\,, \tag{9.31}$$

where $C(s)$ is defined by (9.11). $\square$

Assumption 9.2. For some positive number d_B and for some function $\Phi_B :$ $\mathbb{R} \to \mathbb{R}$ satisfying

$$\lim_{s \to 0} \log \Phi_B(s)/\log s = 0\,,$$

for all sufficiently small positive s we have

$$B_D(s) = \Phi_B(s)s^{-d_B}\,, \tag{9.32}$$

where $B_D(s)$ is the minimal number of diameter-based boxes of size s needed to cover Ω. $\square$

Jensen's inequality says that if f is a convex function, $\mathcal{X}$ is a random variable, and $\mathcal{E}$ denotes expectation, then $\mathcal{E}\big(f(\mathcal{X})\big) \geq f\big(\mathcal{E}(\mathcal{X})\big)$. In particular, for a random variable taking a finite set of values, if $\{\alpha_j\}_{j=1}^J$ are nonnegative weights summing to 1, if $\{x_j\}_{j=1}^J$ are positive numbers, and if $f(z) = z^2$ then

$$\sum_{j=1}^J \alpha_j f(x_j) = \sum_{j=1}^J \alpha_j(x_j)^2 \geq \left(\sum_{j=1}^J \alpha_j x_j\right)^2\,. \tag{9.33}$$

Applying this to the covering of Ω, set $J = B_D(s)$ and $x_j = \mu_j$ and $\alpha_j = 1/J$ for $1 \leq j \leq J$. From (9.33),

$$\frac{1}{B_D(s)} \sum_{j=1}^{B_D(s)} (\mu_j)^2 \geq \left(\frac{1}{B_D(s)} \sum_{j=1}^{B_D(s)} \mu_j\right)^2 = \frac{1}{B_D(s)^2}\,. \tag{9.34}$$

Theorem 9.1 If Assumptions 9.1 and 9.2 hold then $d_C \leq d_B$.

Proof For each s, let $\mathcal{B}(s)$ be a minimal s-covering of Ω and let $B_D(s)$ be the number of boxes in $\mathcal{B}(s)$. Randomly select from Ω, according to its natural measure, the N points $x(1)$, $x(2)$, ..., $x(N)$. Let $\mathbb{N}$ be the set of these N points. For box $B_j \in \mathcal{B}(s)$, let $N_j(s)$ be the number of points of $\mathbb{N}$ that are contained in B_j, so $\sum_{j=1}^{B_D(s)} N_j(s) = N$. From (9.29),

$$\lim_{N \to \infty} C(s) \geq \sum_{j=1}^{B_D(s)} \mu_j^2 = B_D(s)\left(\frac{1}{B_D(s)} \sum_{j=1}^{B_D(s)} \mu_j^2\right) \geq B_D(s)\left(\frac{1}{[B_D(s)]^2}\right) = \frac{1}{B_D(s)}\,, \tag{9.35}$$

where the second inequality follows from (9.34).

By Assumption 9.1, for $s \ll 1$ we have $\lim_{N \to \infty} C(s) = \Phi_C(s)s^{d_C}$. By Assumption 9.2, for $s \ll 1$ we have $B_D(s) = \Phi_B(s)s^{-d_B}$, so $1/B_D(s) = [\Phi_B(s)]^{-1}s^{d_B}$. Hence, from (9.35) for $s \ll 1$ we have

$$\lim_{N \to \infty} C(s) = \Phi_C(s)s^{d_C} \geq 1/B_D(s) = [\Phi_B(s)]^{-1}s^{d_B}.$$

Taking logarithms yields

$$\log \Phi_C(s) + d_C \log s \geq -\log \Phi_B(s) + d_B \log s.$$

Dividing both sides of the inequality by $\log s$, which is negative, and letting $s \to 0$ yields $d_C \leq d_B$.   □

For strictly self-similar fractals such as the Sierpiński gasket, the correlation dimension d_C, the Hausdorff dimension d_H, and the similarity dimension d_S are all equal ([Schroeder 91], p. 216).[7]

---

[7] Schroeder used the term "mass dimension" to refer to what we call, in this book, the "correlation dimension".

Chapter 10

Computing the Correlation Dimension

Nothing makes the earth seem so spacious as to have friends at a distance; they make the latitudes and longitudes.
Henry David Thoreau (1817–1862) American author, poet, and philosopher, best known as the author of "Walden".

In this chapter we investigate several practical issues related to the computation of d_C.

10.1 How Many Points are Needed?

Several researchers have proposed a lower bound on the number of points needed to compute a "good" estimate of d_C of a geometric object $\Omega \subset \mathbb{R}^E$. These bounds were summarized in [Theiler 90], who suggested that the minimal number N of points necessary to get a reasonable estimate of dimension for a d-dimensional fractal is given by a formula of the form $N > ab^d$. Theiler also observed that although it is difficult to provide generically good values for a and b, limited experience indicated that b should be of the order of 10. For example, (9.25) provided the upper bound $d_C \leq 2\log_{10} N$ on d_C, and this inequality also provides a lower bound on N, namely $N \geq 10^{d_C/2}$. In this section we present other lower bounds, which differ markedly. Although this chapter concerns the computation of d_C of a geometric object, in Chap. 11 we will study the "average slope" definition of d_C of a network, and consider the question of how many nodes are required to compute a good estimate of d_C of a rectilinear grid in E dimensions.

Smith's Lower Bound: The lower bound

$$N \geq 42^E \tag{10.1}$$

on the number of data points required to compute d_C to within 5% was provided in [Smith 88]. To derive this bound, consider the ideal case where Ω is

E. Rosenberg, *Fractal Dimensions of Networks*,
https://doi.org/10.1007/978-3-030-43169-3_10

the E-dimensional unit hypercube $[0, 1]^E$, so Ω contains an infinite number of uniformly distributed points, and $d_C = E$. Recalling that with N points $C(r)$ is the fraction of pairs of points separated by a distance at most r, assume that, in the limit of infinite N, the correlation integral $C(r)$ is equal to the probability $P\big(dist(x, y) \le r\big)$ that, for two points x and y randomly chosen from $[0, 1]^E$ and for $r \in (0, 1)$, we have $dist(x, y) \le r$. Let $x = (x_1, x_2, \ldots, x_E)$ and $y = (y_1, y_2, \ldots, y_E)$ be two randomly chosen points in $[0, 1]^E$. For $E = 1$ and $r \in (0, 1)$ we have

$$C(r) = P(|x - y| \le r) = r(2 - r). \tag{10.2}$$

For example, if $r = 1/100$, then $C(r) = (1/100)(199/200) \approx 1/100$. Assume that for $x, y \in [0, 1]^E$ we have $dist(x, y) \equiv \max_{1 \le i \le E} |x_i - y_i|$, i.e., distance is measured using the L_∞ norm. Then $dist(x, y) \le r$ implies $|x_i - y_i| \le r$ for $i = 1, 2, \ldots, E$. For the E-dimensional hypercube $[0, 1]^E$ we have

$$C(r) = P\big(dist(x, y) \le r\big) = [P(|x_i - y_i| \le r)]^E = [r(2 - r)]^E. \tag{10.3}$$

For $r \in (0, 1)$ the correlation dimension d_C can be estimated by the derivative of the $\log C(r)$ versus $\log r$ curve at r. Denoting this derivative by $d_C(r)$,

$$
\begin{aligned}
d_C(r) &\equiv \frac{\mathrm{d}}{\mathrm{d}\log r} \log C(r) = \left(\frac{\mathrm{d}}{\mathrm{d}r} \log C(r)\right)\left(\frac{\mathrm{d}r}{\mathrm{d}\log r}\right) \\
&= \left(\frac{\mathrm{d}}{\mathrm{d}r} \log\left([r(2-r)]^E\right)\right)\left(\frac{\mathrm{d}\log r}{\mathrm{d}r}\right)^{-1} \\
&= E\left(1 - \frac{r}{2-r}\right). \tag{10.4}
\end{aligned}
$$

For $r \in (0, 1)$ we have $0 < d_C(r) < E$. Suppose for a given $\alpha \in (0, 1)$ we want $d_C(r)$ to be at least αE. Since $d_C(r)$ is decreasing in r, let r_{max} be the largest r such that $d_C(r) \ge \alpha E$. From (10.4), r_{max} is the value yielding equality in

$$E\left(1 - \frac{r_{max}}{2 - r_{max}}\right) \ge \alpha E,$$

so $r_{max} = 2(1 - \alpha)/(2 - \alpha)$. Let r_{min} be the smallest usable r; this value must be greater than the mean nearest neighbor distance, and also large enough to overcome the effects of noisy data [Smith 88]. Define $R \equiv r_{max}/r_{min}$. Then

$$r_{min} = \frac{2(1 - \alpha)}{R(2 - \alpha)}. \tag{10.5}$$

If we partition the E-dimensional unit hypercube $[0, 1]^E$ into "small" E-dimensional hypercubes of side length r_{min}, the required number of small hypercubes is $(1/r_{min})^E$. Since r_{min} must be large enough so that each small hypercube contains at least one point, then $(1/r_{min})^E$ is a lower bound on the number N of points required for the estimate of d_C to be at least αd_C. Finally, from (10.5)

$$N \ge (1/r_{min})^E = \left(\frac{R(2 - \alpha)}{2(1 - \alpha)}\right)^E. \tag{10.6}$$

Assuming $\alpha = 0.95$ and $R = 4$, Smith obtained the bound (10.1).   □

Exercise 10.1 Are the choices $\alpha = 0.95$ and $R = 4$ in (10.6) good choices?
□

Exercise 10.2 Referring to (10.2) above, prove the one-dimensional result
that $P(|x - y| \leq r) = r(2 - r)$.   □

Theiler's Lower Bound: It was shown in [Theiler 90] that at least 5^E points
are needed to estimate d_C to within 5%. To derive this bound, much lower than
the bound $N \geq 42^E$, consider again the E-dimensional unit hypercube $[0, 1]^E$.
By (10.4), for $r \ll 1$ we have

$$d_C(r) = E\left(1 - \frac{r}{2 - r}\right) \approx E\left(1 - \frac{r}{2}\right),\tag{10.7}$$

so the relative error $\epsilon(r) \equiv |E - d_C(r)|$ is given by $\epsilon(r) \approx r/2$. Since $C(r)$
varies from a minimal value of $2/[N(N - 1)]$ (if there is a single pair of points at
distance r) to a maximal value of 1, a "natural choice" [Theiler 90] of r is the
value $\widetilde{r}$ for which $C(r) = 1/N$. From (10.3) and $r \ll 1$ we have $C(r) \approx (2r)^E$,
so setting $C(r) = 1/N$ yields

$$\widetilde{r} \approx (1/2)N^{-1/E}\tag{10.8}$$

and $\epsilon(\widetilde{r}) \approx \widetilde{r}/2 = (1/4)N^{-1/E}$ is the relative error. Thus $[4\epsilon(\widetilde{r})]^{-E}$ points are
required to achieve a relative error of $\epsilon(\widetilde{r})$. Setting $\epsilon(\widetilde{r}) = 5\%$ yields $N > 5^E$.
□

Exercise 10.3 In the above derivation of Theiler's bound $N > 5^E$, why is
choosing r to be the value at which $C(r) = 1/N$ a "natural choice"?   □

Krakovská's Lower Bound: Using the same model of the E-dimensional
unit hypercube $[0, 1]^E$, [Krakovska 95] showed that Smith's lower bound of 42^E
is too high, and that Theiler's lower bound of 5^E is too low. For N points
uniformly distributed in an E-dimensional cube with side length 1, the most
probable distance between two points is $r^\star = N^{-1/E}$ [Krakovska 95], and $C(r)$
should not be computed for $r < r^\star$, since that region yields poor statistics.
Since $d_C(r)$, defined by (10.4), is a decreasing function of r, then for $r \geq r^\star$ we
have

$$\begin{aligned} d_C(r) \leq d_C(r^\star) &= E\left(1 - \frac{r^\star}{2 - r^\star}\right) = E\left(1 - \frac{N^{-1/E}}{2 - N^{-1/E}}\right) \\ &= E\left(1 - \frac{1}{2N^{1/E} - 1}\right).\end{aligned}\tag{10.9}$$

Assuming uniformly distributed random points in an E-dimensional unit hypercube, the bound (10.9) provides the N needed to achieve a given accuracy δ, as
follows. From

$$E\left(1 - \frac{1}{2N^{1/E} - 1}\right) = \delta E$$

we obtain

$$N = \left(\frac{2 - \delta}{2 - 2\delta}\right)^E.\tag{10.10}$$

From (10.10) we have $N \to \infty$ as $\delta \to 1$. For 95% accuracy we set $\delta = 0.95$, and we require $(10.5)^E$ points; this estimate lies between Theiler's estimate of 5^E points and Smith's estimate of 42^E points. For the Lorenz attractor, for which $E = 3$ and $d_C \in (2.06, 2.08)$, for $\delta = 0.95$ we have $\delta d_C \in (1.957, 1.976)$. Using Theiler's formula, we need $5^E = 125$ points, and with 125 points the $(\log s, \log C(s))$ curve is very jagged and exhibits no plateau. Using Krakovská's formula, we need $10.5^E = 1158$ points, and the $(\log s, \log C(s))$ curve provides the estimate $d_C \approx 1.96$. Using Smith's formula, we need $42^E = 74{,}088$ points, which provide a much better result than 95% of the true value of d_C.

We can also use (10.9) to determine, with uniformly distributed random points in an E-dimensional hypercube, the amount by which d_C is underestimated when using a finite number of points. For example, with $N = 10^5$ and $E = 5$, the estimated correlation exponent is the fraction $1 - \left(2(10^5)^{1/5} - 1\right)^{-1} \approx 0.947$ of the expected value of 5.

Having considered the issue of a finite rather than infinite number of points, [Krakovska 95] also considered the assumption of a uniform distribution of points. For deterministic systems, since the attractors are more "compressed" than a set of uniformly distributed random points, the same amount of data should, in most cases, yield a better estimate. Let $r^\star$ be the most probable distance between a pair of points on the attractor and define $\nu \equiv d_C(r^\star)$. In practice ν is computed from the slope of the curve of $\log C(r)$ versus $\log r$ over the range of r values for which the slope is reasonably constant. Suppose the attractor is embedded in E-dimensional space, and E is the smallest such value; this implies $E = \lceil d_C \rceil$. Krakovská assumed that, for deterministic systems, corresponding to (10.9) we have

$$\nu = d_C \left(1 - \frac{1}{2N^{1/d_C} - 1}\right), \tag{10.11}$$

so $\nu < d_C$. We approximate the term in parentheses by replacing d_C by E. Since $E > d_C$ this yields

$$\nu > d_C \left(1 - \frac{1}{2N^{1/E} - 1}\right)$$

which we rewrite as

$$d_C < \nu \left(1 + \frac{1}{2\left(N^{1/E} - 1\right)}\right). \tag{10.12}$$

From $\nu < d_C$ and (10.12) we obtain lower and upper bounds on d_C:

$$\nu < d_C < \nu \left(1 + \frac{1}{2\left(N^{1/E} - 1\right)}\right). \qquad \square$$

Eckmann and Ruelle's Lower Bound: The lower bound

$$N > 10^{d_c/2} \tag{10.13}$$

was provided in [Eckmann 92]. To derive this bound, first define

$$F(r) = \sum_{i=1}^{N} \sum_{\substack{j=1 \\ j \neq i}}^{N} I_0\Big(r - \big(dist(x(i), x(j))\big)\Big).\qquad(10.14)$$

By definitions (9.9) and (9.11) we have

$$
\begin{aligned}
C(r) &= \frac{1}{N(N-1)} \sum_{i=1}^{N} C\big(x(i), r\big) \\
&= \frac{1}{N(N-1)} \sum_{i=1}^{N} \sum_{\substack{j=1 \\ j \neq i}}^{N} I_0\Big(r - \big(dist(x(i), x(j))\big)\Big) \\
&= \frac{F(r)}{N(N-1)}.
\end{aligned}
\qquad(10.15)
$$

Setting $r = \Delta$, by definition (9.7) of I_0, for all $x(i)$ and $x(j)$ we have

$$I_0\big(\Delta - dist(x(i), x(j))\big) = 1,$$

so by (10.14) we have $F(\Delta) = N(N-1)$ and by (10.15) we have

$$C(\Delta) = \frac{F(\Delta)}{N(N-1)} = 1.$$

Assume $C(r) \approx kr^{d_C}$ for some constant k. Since $C(\Delta) = 1$ then $k = \Delta^{-d_C}$, which with (10.15) yields

$$F(r) = N(N-1)C(r) \approx N^2 \left(\frac{r}{\Delta}\right)^{d_C}.\qquad(10.16)$$

For statistical reasons we want $F(r) \gg 1$, so by (10.16) we want

$$N^2 \left(\frac{r}{\Delta}\right)^{d_C} \gg 1.\qquad(10.17)$$

In determining the slope of the $\log F(r)$ versus $\log r$ curve, we want $r \ll \Delta$. Define $\rho \equiv r/\Delta$, where $\rho \ll 1$. Taking the logarithm of both sides of (10.17), we obtain

$$2\log N \gg -d_c \log\left(\frac{r}{\Delta}\right) = -d_c \log \rho = d_c \log(1/\rho).\qquad(10.18)$$

Choosing the value $\rho = 10^{-1}$, using base 10 logarithms, and replacing "$\gg$" by "$>$", we obtain $2\log_{10} N > d_c \log_{10}(10)$, which we rewrite as the desired result $N > 10^{d_c/2}$. So if $d_C = 6$, we need at least 1000 points.

Values of ρ larger than 0.1 might be adequate but since d_C is interesting mainly for highly nonlinear systems, such larger values would have to be justified. Rewriting (10.18), an upper bound on d_C is $d_C \leq 2\log_{10} N / \log_{10}(1/\rho)$.

With $N = 1000$ and $\rho = 0.1$ we obtain $d_C \leq 6$. Thus, if the *Grassberger–Procaccia* method yields the estimate $d_C = 6$ for $N = 1000$ points, this estimate is probably worthless, and a larger N is required [Eckmann 92].

The bound (10.13) was used in [Camastra 97] to study electric load forecasting. Using 4320 data points, they computed $d_C \approx 3.9$ using the *Grassberger–Procaccia* method. Since $d_C = 3.9 < 2 \log_{10} N \approx 7.2$, then (10.13) holds. The impact of an insufficient number of points was illustrated in [Camastra 02], who used the *Grassberger–Procaccia* method to determine d_C from a set of N points randomly distributed in the 10-dimensional hypercube $[0,1]^{10}$. Since we expect $d_C = 10$, the lower bound on N is $10^{d_C/2} = 10^5$. The following table from [Camastra 02] provides the computed values of d_C for different values of N (Table 10.1).

Table 10.1: Impact of N on computed d_C

N	Computed d_C
1000	7.83
2000	7.94
5000	8.30
10,000	8.56
30,000	9.11
100,000	9.73

Random sampling statistics were used in [Kember 92] to reduce the number of points required to accurately compute d_C. They observed that the values $I_0\Big(r - dist\big(x(i), x(j)\big)\Big)$ form a population whose values are either 0 or 1. Hence the probability p that $I_0\Big(r - dist\big(x(i), x(j)\big)\Big) = 1$ can be estimated by taking a random sample of the population: the fraction of terms equal to 1 in a random sample with replacement is an unbiased estimator for p. The savings in the number of points required are proportional to $N_\delta(d_C)^2$, where N_δ is the number of points required to estimate d_C to a relative accuracy of δ. Since the bounds of [Smith 88, Theiler 90, Eckmann 92, Krakovska 95] derived above are all of the form $N_\delta \geq \exp(O(d_C))$, then the savings using random sampling are dramatic for large d_C.

Despite the theoretical bounds on the minimal number of points required for an accurate computation of d_C, computational results in [Abraham 86] showed that d_C can be successfully calculated even with as few as 500 data points. For example, for the much-studied Hénon attractor defined by

$$\big(x_{t+1}, y_{t+1}\big) = \big(y_t + 1 - a(x_t)^2, bx_t\big)$$

with $a = 1.4$ and $b = 0.3$, using $N = 500, 1200, 4000$, and $10,000$ points, for the range $s \in [2.5, 5.5]$, they obtained $d_C = 1.28 \pm 0.09, 1.20 \pm 0.04, 1.24 \pm 0.04$, and 1.24 ± 0.02, respectively. Results in [Ramsey 89] found that for simple models, 5000 is a rough lower bound for the number of points needed to achieve reasonable results [Berthelsen 92].

10.2 Analysis of Time Series

A time series is a set of measurements of the same quantity, e.g., heart rate or the price of a stock, taken at regular time intervals. The correlation dimension has been used to analyze time series data in a wide variety of applications. The correlation dimension is the most widely used dimension in applied time series analysis [Martinerie 92, Boon 08], and can determine if the data follows a stochastic process or instead is generated by deterministic chaos.[1] (For stochastic systems, the fractal dimension is equal to the phase-space dimension, while for purely chaotic systems, the fractal dimension is smaller and in general is not an integer [Badii 85].) The reasons for computing the dimension of a time series are [Martinerie 92]:

> Estimating a signal's dimension is often an important step in its dynamical characterization. In ideal cases when large sets of high-quality data are available, the dimension estimate can be used to infer something about the dynamical structure of the system generating the signal. However, even in circumstances where this is impossible, dimension estimates can be an important empirical characterization of a signal's complexity.

To add to this list of reasons, in Sect. 11.3 we will examine how the correlation dimension of a network can be estimated from the time series obtained from a random walk on the network.

We now consider the computation of d_C of the time series, $u_t, t = 1, 2, \ldots, N$, where each u_t is a scalar quantity measured at regular intervals. Let τ be the fixed time interval between measurements, e.g., every 3 min or hourly or every other day. Let E be a given positive integer, called the *embedding dimension*. We create a set of E-dimensional vectors x_t as follows. For $t = 1, 2, \ldots, N - E + 1$, define

$$x_t = (u_t, u_{t+1}, \ldots, u_{t+E-1}) \in \mathbb{R}^E . \tag{10.19}$$

For example, with $E = 3$, we have $x_1 = (u_1, u_2, u_3)$, $x_2 = (u_2, u_3, u_4)$, $x_3 = (u_3, u_4, u_5)$, etc., so the components of the vectors x_t overlap. Figure 10.1 illustrates an embedding with $E = 3$. The embedding of scalar times series data, where the measurements occur with time spacing τ, can be visualized as passing a comb along the data [Boon 08]. The number of teeth of the comb is the

---

[1] See [Bass 99] for an entertaining account of the application, by Farmer and others, of times series analysis to financial markets.

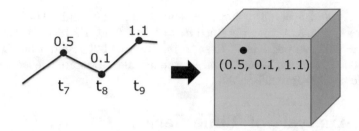

Figure 10.1: Embedding dimension

embedding dimension E, and the distance between the teeth of the comb is τ. When the comb is positioned at the first point in the data, the scalar points touched by the teeth of the comb define the E components of x_1. The coordinates of the next point x_2 are obtained by moving the comb τ time units to the right. The task is to estimate d_C from these $N - E + 1$ vectors, where each vector is a point in $\mathbb{R}^E$.[2]

Following [Barkoulas 98], let X be an open subset of $\mathbb{R}^K$, where K is a positive integer, and suppose that for some function $F : X \to \mathbb{R}^K$ the sequence x_t defined by (10.19) satisfies $x_t = F(x_{t-1})$ for $t = 1, 2, \ldots$. The sequence $\{x_t\}_{t=1}^\infty$ depends on the initial point x_1 but for simplicity of notation we write x_t rather than $x_t(x_1)$. Let $F^2(x_t)$ denote $F\big(F(x_t)\big)$, let $F^3(x_t)$ denote $F\Big(F\big(F(x_t)\big)\Big)$, and in general let $F^j(x_t)$ denote the j-fold composition. A set Ω is *invariant* with respect to the trajectory $\{x_t\}_{t=1}^\infty$ if the following condition holds: if $x_{t_j} \in \Omega$ for some time t_j, then $x_t \in \Omega$ for $t > t_j$. A closed invariant set $\Omega \subset X$ is the *attracting set* for the system $x_t = F(x_{t-1})$ if there exists an open neighborhood Ω_ϵ of Ω such that $\Omega = \{\lim_{t \to \infty} x_t(x_1) \,|\, x_1 \in \Omega_\epsilon\}$. We are interested in the dimension of the attracting set Ω.

In many financial studies, the data $y_t \in \mathbb{R}^K$ represents the market over time. Suppose $y_{t+1} = F(y_t)$ for some function $F : \mathbb{R}^K \to \mathbb{R}^K$. Suppose we observe, at equally spaced time increments, the univariate time series u_t, e.g., u_t might be the price of a single stock over time. Assume $u_t = g(y_t)$, where $g : \mathbb{R}^K \to R$, so the stock price u_t is the output of a dynamical system. We want to recover the market dynamics by analyzing u_t. To do this, let E be the embedding dimension and create, for $t = 1, 2, \ldots$, the E-dimensional vector x_t

---

[2] The choice of τ for a time series can have a dramatic effect on the length and shape of the plateau (the flat region of the $\log C(r)$ versus $\log r$ curve over which d_C is estimated). For some values of τ, a measurable plateau is completely absent even when using high-quality data [Marliere 01].

defined by (10.19). Since $u_t = g(y_t)$ then

$$
\begin{aligned}
x_t &= (u_t, u_{t+1}, u_{t+2}, \ldots, u_{t+E-1}) \\
&= \left(g(y_t), g(y_{t+1}), g(y_{t+2}), \ldots, g(y_{t+E-1}) \right) \\
&= \left(g(y_t), g\big(F(y_t)\big), g\big(F^2(y_t)\big), \ldots, g\big(F^{E-1}(y_t)\big) \right). \\
&\equiv \Phi(y_t).
\end{aligned}
$$

The function $\Phi : \mathbb{R}^K \to \mathbb{R}^E$ is called the *delay coordinate map* [Ding 93].

A theorem of [Takens 85] states that if F and g are smooth, then for $E \geq 2K+1$ the map Φ will be an embedding (a one to one and differentiable mapping [Ding 93]), and an analysis of the time series $\{x_t\}$ gives a correct picture of the underlying dynamics [Barkoulas 98], including the dimension of the attracting set. It actually suffices to choose $E \geq 2d_B + 1$ [Ding 93]. Moreover, it was proved in [Ding 93] that to estimate d_C from a time series, it suffices to have $E \geq d_C$, and the condition $E \geq 2d_B + 1$, while necessary to ensure that the delay coordinate map is one to one, is *not* required to estimate d_C. That is, choosing $E \geq d_C$, the correlation dimension d_C can be estimated whether or not the delay coordinate map is one to one. Based on a random sample of papers published from 1987 to 1992, [Ding 93] found that the assumption that $E \geq 2d_B + 1$ is required to estimate d_C is a widespread misconception.

In theory, if the points x_t for $t = 1, 2, \cdots$ are independent and identically distributed, then the system has infinite dimension. With finite data sets, high dimensionality is indistinguishable from infinite dimensionality [Frank 89]. For a given E, let $d_C(E)$ be the correlation dimension for the series $\{x_t\}$ defined by (10.19). If the data are purely stochastic, then $d_C(E) = E$ for all E, i.e., the data "fills" E-dimensional space. In practice, if the computed values $d_C(E)$ grow linearly with E, then the system is deemed "high-dimensional" (i.e., stochastic). If the data are deterministic, then $d_C(E)$ increases with E until a "saturation" point is reached at which $d_C(E)$ no longer increases with E. In practice, if the computed values $d_C(E)$ stabilize at some value d_C as E increases, then d_C is the correlation dimension estimate.

The steps to estimate d_C from a time series are summarized in [Ding 93] as follows:

(*i*) Construct the sequence $\{x_t\}$, where $x_t \in \mathbb{R}^E$.

(*ii*) Compute the correlation sum $C(r)$ for a set of values of r.

(*iii*) Locate a linear scaling region, for small r, on the $\log C(r)$ versus $\log r$ curve, and compute $d_C(E)$, the slope of the curve over that region.

(*iv*) Repeat this for an increasing range of E values, and if the estimates $d_C(E)$ appear to converge to some value, then this value is the estimate of d_C.

To prevent a spurious estimate of d_C, the following conditions are necessary [Martinerie 92]:

(i) The range of $[r_{min}, r_{max}]$ over which d_C is estimated must be sufficiently large; e.g., by requiring $r_{max}/r_{min} \geq 5$.

(ii) The $\log C(r)$ versus $\log r$ curve must be sufficiently linear over $[r_{min}, r_{max}]$; typically the requirement is that over $[r_{min}, r_{max}]$ the derivative of the $\log C(r)$ versus $\log r$ curve should not exceed 15% of d_C.

(iii) The estimate $d_C(E)$ should be stable as E increases; typically the requirement is that $d_C(E)$ not increase by more than 15% over the range $[E - 4, E]$.

In computing d_C from a time series, it is particularly important to exclude temporally correlated points from the sum (9.11). Since successive elements of a time series are not usually independent, Theiler suggested removing each pair of points whose time indices differ by less than w, with the normalization factor $1/\big(N(N-1)\big)$ appropriately modified. The parameter w is known as the *Theiler window* [Hegger 99].

10.3   Constant Memory Implementation

The correlation dimension is based on computing the distance between each pair of points. While the distances need not be stored, each distance must be computed at least once, and the computational complexity of computing these distances is $\mathcal{O}(N^2)$ [Wong 05]. This complexity is prohibitive for large data sets. An alternative to computing the distance between each pair of points is to use the results of Sect. 9.5, where we showed that the number of pairs whose distance is less than s is approximated by the number of pairs which fall into any box of size s. Using (9.29),

$$\widetilde{C}(s) \approx \sum_{j=1}^{B_D(s)} \left(\frac{N_j(s)}{N} \right)^2. \tag{10.20}$$

We can approximate $\widetilde{C}(s)$ with only a single pass through the N points, so the complexity of computing $\widetilde{C}(s)$ is $\mathcal{O}(N)$. An $\mathcal{O}(N)$ computational cost is acceptable, since each of the N points must be examined at least once, unless sampling is used. The memory required is also $\mathcal{O}(N)$ since for sufficiently small s each point might end up in its own box.

Interestingly, there is a way to reduce the memory requirement without unduly sacrificing speed or accuracy. Drawing upon the results in [Alon 66], it was shown in [Wong 05] how for a given s we can estimate the "second moment" $\sigma^2(s) \equiv \sum_{j=1}^{B_D(s)} [N_j(s)]^2$ in $\mathcal{O}(N)$ time and $\mathcal{O}(1)$ memory, i.e., a constant amount of memory independent of N. The accuracy of estimation is controlled by two user-specified integer parameters: α determines the level of approximation, and β determines the corresponding probability. For any positive integral values of α and β, the probability that the estimate of the second moment has a relative error greater than $\sqrt{16/\alpha}$ is less than $e^{-\beta/2}$.

For each s we generate $\alpha\beta$ 4-tuples of random prime numbers, each of which defines a hash function. Estimating $\sigma^2(s)$ requires one counter per hash function, so there are $\alpha\beta$ counters. Assuming each of the N points lies in $\mathbb{R}^E$, each hash function, which is applied to each of the N points, returns a value of 1 or -1, which is used to increment or decrement the counter associated with the hash function. For each s, the estimate of $\sigma^2(s)$ is computed by first partitioning the $\alpha\beta$ counters into β sets of α counters, squaring the α counter values in each of the β sets, computing the mean for each of the β sets, and finally computing the median of these means.

The method was tested in [Wong 05] using an accuracy parameter $\alpha = 30$ and a confidence parameter $\beta = 5$. A linear regression of the $\left(\log s, \log \sigma^s(s)\right)$ values was used to estimate d_C. The first test of the method considered points on a straight line, and the method estimated d_C values of 1.045, 1.050, and 1.038 for $E = 2, 3, 4$, respectively. On the Sierpiński triangle with $N = 5000$, the method yielded $d_C = 1.520$; the theoretical value is $\log_2 3 \approx 1.585$. A Montgomery County, MD data set with $N = 27{,}282$ yielded $d_C = 1.557$ compared with $d_B = 1.518$, and a Long Beach County, CA data set with $N = 36{,}548$ yielded $d_C = 1.758$ compared with $d_B = 1.732$.

10.4 Applications of the Correlation Dimension

In this section we briefly describe some interesting applications of the correlation dimension. The literature on applications of d_C is quite large, and the set of applications discussed below constitutes a very small sample. In the discussion below, a variety of approaches were used to compute d_C, or some variant of d_C. The first two applications we review also compare d_C and d_B. The discussion below on the hierarchical packing of powders, and on fractal growth phenomena, shows that d_C has been applied to growing clusters of particles; this discussion also sets the stage for our presentation of the *sandbox dimension* in Sect. 10.5 below. Reading this range of applications will hopefully inspire researchers to find new applications of network models in, e.g., biology, engineering, or sociology, and to compute d_C or related fractal dimensions of these networks.

Hyperspectral Image Analysis: In [Schlamm 10], d_C and d_B were compared on 37 hyperspectral images (they also considered the dimension obtained from principal component analysis [Jolliffe 86]).[3] The 37 images were from 11 unique scenes (e.g., desert, forest, lake, urban) and 7 ground sample distances

---

[3] Visible (red, green, and blue), infrared, and ultraviolet are descriptive regions in the electromagnetic spectrum. These regions are categorized based on frequency or wavelength. Humans see visible light (wavelength of 380 nm to 700 nm); goldfish see infrared (700 nm to 1 mm); bumble bees see ultraviolet (10 nm to 380 nm). Multispectral and hyperspectral imagery gives the power to see as humans (red, green, and blue), goldfish (infrared), bumble bees (ultraviolet), and more. This comes in the form of reflected electromagnetic radiation to the sensor. The main difference between multispectral and hyperspectral is the number of bands and how narrow the bands are. Multispectral imagery generally refers to 3 to 10 bands. Hyperspectral imagery consists of much narrower bands (10–20 nm). A hyperspectral image could have hundreds of thousands of bands. http://gisgeography.com/multispectral-vs-hyperspectral-imagery-explained/.

(GSDs).[4] Surprisingly, the d_C estimates of the full images were consistently substantially larger (often by more than a factor of two) than the d_B estimates. They concluded that, compared to the correlation method and principal component analysis, box counting is the best method for full image dimension estimation, as the results are the most reliable and consistent.

Root Systems of Legumes: In [Ketipearachchi 00], d_C and d_B were compared for the root system of six species of legumes (e.g., kidney beans and soybeans). Such analysis assists in understanding the functional differences and growth strategies of root systems among the species, and in screening for genetic improvement. For the d_B calculation, they used box sizes from 2 to 256 pixels, and employed linear regression. For the d_C calculation, they computed the average of 500 pointwise mass dimensions $d(x)$ (Sect. 9.3) centered at points x on the edge of the image. The pointwise dimensions were assumed to satisfy $C(x, r) = kr^{d_C}$ for some constant k, where $C(x, r)$ is the number of pixels in a box or disc of size r centered at x (see (11.3)), and r varies from 3 to 257 pixels. Interestingly, for each species d_C exceeded d_B, as shown in Table 10.2; they noted that they are not the first to observe this relationship for a "natural fractal". These results showed that the six species fall into two groups, with

Table 10.2: Dimensions of legume root systems [Ketipearachchi 00]

Crop	d_B	d_C
Blackgram	1.45	1.52
Cowpea	1.45	1.54
Groundnut	1.61	1.76
Kidneybean	1.61	1.72
Pigeonpea	1.41	1.53
Soybean	1.57	1.68

groundnut, kidney bean, and soybean in the group with the higher fractal dimension. They also observed that d_C is less sensitive than d_B to system size parameters such as root length, and thus d_C is a "suitable alternative" to d_B when evaluating root systems by focusing on branching patterns rather than system size parameters.

Hierarchical Packing of Powders: The correlation dimension can be used to describe the packing of powders in hierarchical structures [Schroeder 91]. Suppose the primary powder particles tend to form clusters (in 3-dimensional space) with density ρ, where $\rho < 1$. Let γ be the ratio of the cluster radius to the primary powder particle radius, where $\gamma > 1$. These first-generation

---

   [4] In a digital photograph, the GSD is the distance between pixel centers measured on the ground. For example, in an image with a one-foot GSD, adjacent pixels image locations are 1 foot apart on the ground. As the GSD decreases, the sampling rate increases to cover the same area on the ground, thereby increasing image resolution. *Rocky Mountain Aerial Surveys* http://rmaerialsurveys.com/?page_id=291.

clusters then cluster together to form second-generation clusters, with the same or similar values of ρ and γ, the second-generation clusters form third-generation clusters, etc. This is illustrated in Fig. 10.2, which shows 7 primary particles in

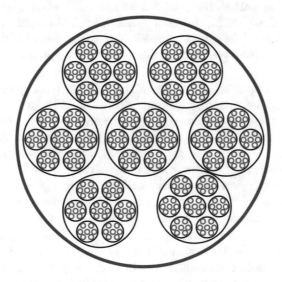

Figure 10.2: Packing of powder particles

each first-generation cluster, 7 first-generation clusters in each second-generation clusters, and 7 second-generation clusters in the third-generation clusters. After t generations, the density of the power is ρ^t and the radius R_t of each generation t cluster is $R_t = \gamma^t$. The mass N_t in each generation t cluster is the product of the cluster density and volume, and the volume is proportional to the cube of the radius, so $N_t \propto \rho^t R_t^3$. Using the identity $\rho^t = (\gamma^t)^{\log \rho / \log \gamma}$, we have $\rho^t = (\gamma^t)^{\log \rho / \log \gamma} = R_t^{\log \rho / \log \gamma}$. For $t \gg 1$,

$$N_t \propto \rho^t R_t^3 = R_t^{\log \rho / \log \gamma} R_t^3 = R_t^{3 + (\log \rho / \log \gamma)} , \qquad (10.21)$$

so $d_C = 3 + (\log \rho / \log \gamma)$ and $3 + (\log \rho / \log \gamma) < 3$, since $\rho < 1$ and $\gamma > 1$.

The β Model: Theoretical aspects of the use of small-angle X-ray scattering to compute the fractal dimension of colloidal aggregates were studied in [Jullien 92]. Derived theoretical results were compared to the results of the following 3-dimensional "random" fractal model, known as the β model. Pick a random positive integer k such that $1 < k < 8$. At $t = 0$, start with the unit cube. At time $t = 1$, enlarge the cube by a factor of 2 in each of the three dimensions, and divide the enlarged cube into $2^3 = 8$ unit cubes. Randomly select k of these 8 cubes and discard the unselected cubes; this completes the $t = 1$ construction. At time $t = 2$, expand the object by a factor of 2 in each of the three dimensions, and, for each of the k cubes selected in step $t = 1$, divide the cube into 8 unit cubes, from which k are randomly selected, and discard the unselected cubes. Each subsequent time t proceeds similarly. At the end of time

t, the original unit cube now has side length 2^t, and we have selected a total of k^t unit cubes. The fractal dimension is $\log(k^t)/\log(2^t) = \log k/\log 2$; compare this to (10.21). As $t \to \infty$, the resulting fractal is not self-similar, since the selection of cubes is random; rather, the fractal is *statistically self-similar*. (Statistically self-similarity is also used in the randomized Sierpiński carpet construction described in Sect. 4.3.) The object generated by the β model construction after some finite t differs from realistic aggregates since the cubes are not necessarily connected.

Brain Electroencephalogram: The d_C of brain electroencephalogram (EEG) patterns produced by music of varying degree of complexity was studied in [Birbaumer 96]. Given a center point x and radius r, the number $M(x, r)$ of points lying in a radius-based box of radius r was computed. Starting with a given r_1 such that $M(x, r_1) = k$, the sequence r_i of radius values was chosen so that $M(x, r_{i+1}) - M(x, r_i) = M(x, r_i)$, yielding $M(x, r_{i+1}) = 2^{i-1}k$. This scheme reduced the problems associated with small radii. For a given x, as r increases, the plot of $\log M(x, r_i)$ versus $\log r_i$ values starts out looking linear, and then levels off. A straight line was fitted to the $\log M(x, r_i)$ versus $\log r_i$ curve for the 10 smallest values of r_i. If the point $\left(\log r_{10}, \log M(x, r_{10})\right)$ has the largest distance (comparing the 10 distances) to the straight line, then the linear fit was recalculated using only the smallest 9 points. If the point $\left(\log r_9, \log M(x, r_9)\right)$ has the largest distance (comparing the 9 distances) to the new straight line, then the linear fit was again recalculated using only the smallest 8 points. This procedure continues until for some r_j, the distance from $\left(\log r_j, \log M(x, r_j)\right)$ to the line determined by the first j points is not the largest distance; the slope of this line is the pointwise dimension for this x. This typically occurred for when j is 5 or 6 or 7. Once the slope has been determined for each x, the correlation dimension was taken to be the median of these pointwise dimensions.

Fat Crystals: The d_C of fat crystals (e.g., of milk fat, palm oil, or tallow) was computed in [Tang 06] by counting the number of particles in a square of size s centered on the image. The box size s ranged from 35% to 100% of the image size. The slope of the best linear fit to the $\log M(s)$ versus $\log s$ curve, obtained using regression, was taken as the estimate of d_C.

Gels and Aerogels: The d_C of gel and aerogel structures, for which the microstructure can be described as a fractal network with the length scale of 10^{-9} to 10^{-7} meters, was calculated in [Marliere 01]. However, this study only counted particles, and did not explicitly consider the network topology. The fractal structure results from an aggregation process in which silica beads form structures whose size is characterized by a fractal dimension of about 1.6. The addition of silica soot "aerosil" strongly modifies the aggregation process, and for sufficiently high concentrations of silica soot the resulting composite is non-fractal.

River Flows: As described in [Krasovskaia 99], river flows can exhibit roughly the same pattern year after year, or they can exhibit irregular behavior by alternating between a few regime types during individual years. In [Krasovskaia 99], d_C was computed from a time series for different values of the embedding dimension. (Although rivers can be modelled as networks, this study did not utilize network models.) They determined that, for the more stable (regular) regimes, $d_C(E)$ quickly converges for small E, leading to a d_C of just over 1. Irregular regimes require $E > 4$ and yield a d_C of over 3.

Astronomy: In astronomy, the correlation sum provides the probability that another galaxy will be found within a given distance of a randomly selected galaxy. The transition to homogeneity, defined as the scale above which the fractal dimension of the underlying point distribution is equal to the ambient dimension of the space in which points are distributed, was studied in [Yadav 10], which provided this description of the role of d_C in astronomy.

> One of the primary aims of galaxy redshift surveys is to determine the distribution of luminous matter in the Universe. These surveys have revealed a large variety of structures starting from groups and clusters of galaxies, extending to superclusters and an interconnected network of filaments which appears to extend across very large scales. We expect the galaxy distribution to be homogeneous on large scales. In fact, the assumption of large-scale homogeneity and isotropy of the Universe is the basis of most cosmological models. In addition to determining the large-scale structures, the redshift surveys of galaxies can also be used to verify whether the galaxy distribution does indeed become homogeneous at some scale. Fractal dimensions of the galaxy matter distribution can be used to test the conjecture of homogeneity.

Medicine: The d_C of the visual evoked potential (VEP), which is a brain electrical activity generated by a visual stimulus such as a checkerboard pattern, was studied in [Boon 08]. The d_C may be a useful diagnostic tool and may find application in the screening and monitoring of acquired color vision deficiencies. In the study of Parkinson's disease, the d_C of the EEG is a sensitive way to discriminate brain electrical activity that is task and disease dependent.[5] In the study of Alzheimer's disease, d_C can also discriminate the EEG arising from dementia from the EEG arising from vascular causes. For the treatment of glaucoma, the d_C of the VEP of patients with glaucoma was lower than for healthy individuals. In the VEP study in [Boon 08], d_C was estimated for the embedding dimensions $E = 1, 2, 3, 4, 5$, and 64 equally spaced values of the correlation distance r were used; practical aspects of applying least squares to determine d_C were also considered.

Human Sleep: Crick and Mitchinson hypothesized that one of the functions of dreams is to erase faulty associations in the brain [Kobayashi 00]. Also, the human brain may have a nighttime mechanism to increase flexibility in order to respond to unexpected stimulations when awake. Much has been learned from computing d_C of the EEG of human sleep patterns. For instance, d_C

---

[5] An electroencephalogram (EEG) is a test that measures brain electrical activity, and records it as wavy lines on paper. Changes in the normal patterns are evidence of certain conditions, such as seizures.

decreases from the "awake" stage to sleep stages 1, 2, 3, and 4, and increases at REM sleep, and d_C during sleep 2 and REM sleep in schizophrenia were significantly lower than in healthy people.[6] Moreover, d_C may relate to the speed of information processing in the brain. The sleep EEG over the entire night was studied in [Kobayashi 00]. The EEG on a tape was converted to 12 bit digital data, and the resulting time series was embedded in phase space, with a maximum embedding dimension of 19. When the change in the slope of $\log C(r)$ versus $\log r$ is less than 0.01 as the embedded dimension is increased, then d_C was defined to be this slope. The mean d_C of the sleep stages were 7.52 for awake, 6.31 for stage 1, 5.54 for stage 2, 3.45 for stage 3, and 6.69 for REM sleep.

Earthquakes: The San Andreas fault in California was studied in [Wyss 04]. They calculated a d_C of 1.0 for the "asperity" region and a d_C of 1.45–1.72 for the "creeping" region. They noted that r should vary by at least one order of magnitude for a meaningful fractal analysis. Their study established a positive correlation between the fractal dimension and the slope of the earthquake frequency-magnitude relationship.

The d_C of earthquakes was also studied in [Kagan 07]. Defining $C(r)$ to be the total number of earthquake pairs at a distance r, the correlation dimension was obtained by estimating $\mathrm{d}\log C(r)/\mathrm{d}\log r$ using a linear fit. Kagan derived formulae to account for various systematic effects, such as errors in earthquake location, the number of earthquakes in a sample, and boundary effects encountered when the size of the earthquake region being studied exceeds r. Without properly accounting for such effects, and also for data errors, practically any value for d_C can be obtained.

Several reasons are provided in [Kagan 07] for utilizing d_C rather than d_B. The distances between earthquakes do not depend, as boxes do, on the system coordinates and grid selection. Distances can be calculated between points on the surface of a sphere, whereas it is not clear how to apply box counting to a large region on the surface of a sphere. Also, since seismic activity in southern California aligns with plate boundaries and the San Andreas fault, issues could arise when boxes are oriented along fault lines. For these reasons, box counting has been employed only in relatively limited seismic regions, like California.

The natural drainage systems in an earthquake-prone region provide clues to the underlying rock formations and geological structures such as fractures, faults, and folds. Analyzing earthquake related data for Northeast India, [Pal 08] computed the correlation dimension d_C^e of earthquake epicenters, the correlation dimension d_C^f of the drainage frequency (total stream segments per unit area, evaluated at the earthquake epicenters), and the correlation dimension d_C^d of the drainage density (total length of stream segments per unit area, evaluated at the earthquake epicenters). They found that d_C^f ranged from 1.10 to 1.64 for the various regions, d_C^d ranged from 1.10 to 1.60, and d_C^e ranged from 1.10 to

---

[6] Schizophrenia is a disorder that affects a person's ability to think, feel, and behave clearly.

1.70, which indicated that d_C^f and d_C^d are very appropriate for earthquake risk evaluation.

Orbit of a Pendulum: The effect of noise in the computation of d_C of a nonlinear pendulum was studied in [Franca 01]. Random noise was introduced to simulate experimental data sets. Their results illustrated the sensitivity of $d_C(E)$ to the embedding dimension E.

Aesthetics of Fractal Sheep: This study [Draves 08] was based on research by Sprott, beginning in 1993, which proposed d_C as a measure of complexity of a fractal image, and examined the relationship of fractal dimension to aesthetic judgment (i.e., which images are more pleasing to the eye). "Fractal sheep" are animated images which are based on the attractors of two-dimensional iterated function systems. The images, used as computer screen-savers, are downloaded automatically (see electricsheep.org) and users may vote for their favorite sheep while the screen-saver is running. The aesthetic judgments of about 20,000 people on a set of 6400 fractal images were analyzed in [Draves 08], and d_C was computed for each frame of a fractal sheep. Since the sheep are animated, d_C varies over time, so an average fractal dimension, averaged over 20 frames, was computed. They found that the sheep most favored have an average fractal dimension between 1.5 and 1.8.

Complexity of Human Careers: In [Strunk 09], d_C was used to compare the career trajectory of a cohort of people whose careers started in the 1970s versus a cohort whose careers started in the 1990s. There were 103 people in the 1970s cohort and 192 people in the 1990s cohort. In each of the first 12 career years of each cohort, twelve measurements were taken for each person; the measurements include, e.g., the number of subordinates, annual income, career satisfaction, and stability of professional relations. The time series of all people in each cohort were aggregated to create a single "quasi time series", and 100 different variants of the time series for each of the two cohorts were generated and calculated separately. Strunk concluded that the first 12 career years of the 1970s cohort has a mean d_C of 3.424, while the first 12 career years of the 1990s cohort has a mean d_C of 4.704, supporting the hypothesis that careers that started later are more complex. Strunk observed that a limitation of d_C is the lack of a test statistic for the accuracy of the estimated d_C, and thus d_C is not a statistically valid test for complexity.

Exercise 10.4 Using the techniques presented in the chapter, estimate the number of people in each cohort required to obtain a statistically significant estimate of d_C. □

Pattern Recognition: The *intrinsic dimension* of a set of points $\Omega \subset \mathbb{R}^E$ is the smallest integer M (where $M \leq E$) such that Ω is contained in an M-dimensional subspace of $\mathbb{R}^E$. Knowing the intrinsic dimension is important in pattern recognition problems since if $M < E$ both memory requirements and algorithm complexity can often be reduced. The intrinsic dimension can be estimated by d_C, but the required number N of points, as determined by the bound (10.13), can be very large. The following method [Camastra 02] estimates the intrinsic dimension of any set Ω of N points, even if N is less than this lower bound.

1. Create a data set Ω' with N points whose intrinsic dimension is known. For example, if Ω' contains N points randomly generated in an E-dimensional hypercube, its intrinsic dimension is E.

2. Compute $d_C(E)$ for the data set Ω'.

3. Repeat Steps 1 and 2, generating a new set Ω', for K different values of E, but always with the same value of N. This generates the set, with K points, $\{(E_1, d_C(E_1)), (E_2, d_C(E_2)), \ldots, (E_K, d_C(E_K))\}$. For example, [Camastra 02] used $K = 18$, with $E_1 = 2$ and $E_{18} = 50$.

4. Let Γ be a curve providing a best fit to these K points. (In [Camastra 02], a multilayer-perceptron neural network was used to generate the curve.) For a point (x, y) on Γ, x is the intrinsic dimension, and y is the correlation dimension.

5. Compute the correlation dimension d_C for the set Ω. If x is the unique value such that (x, d_C) is on the curve Γ, then x is the estimated intrinsic dimension.

The validity of this method rests on two assumptions. First is the assumption that the curve Γ depends on N. Second, since different sets of size N with the same intrinsic dimension yield similar estimates of d_C, the dependence of the curve Γ on the sets Ω' is assumed to be negligible.

Economic Applications: The correlation dimension has been of special interests to economists.

> The correlation dimension, a measure of the relative rate of scaling of the density of points within a given space, permits a researcher to obtain topological information about the underlying system generating the observed data without requiring a prior commitment to a given structural model. If the time series is a realization of a random variable, the correlation dimension estimate should increase monotonically with the dimensionality of the space within which the points are contained. By contrast, if a low correlation dimension is obtained, this provides an indication that additional structures exists in the time series - structure that may be useful for forecasting purposes. In this way, the correlation dimension estimates may prove useful to economists ... Furthermore, potentially the correlation dimension can provide an indication of the number of variables necessary to model a system accurately. [Ramsey 90]

Below are a few selected economic applications of the correlation dimension.

Greek Stock Market: In [Barkoulas 98], d_C was used to investigate the possibility of a deterministic nonlinear structure in Greek stock returns. The data analyzed was the daily closing prices of a value-weighted index of the 30 most marketable stocks on the Athens Stock Exchange during the period 1988 to 1990. As E increased from 1 to 15, the computed $d_C(E)$ rose almost monotonically (with one exception) from 0.816 to 5.842, with some saturation underway by $E = 14$. Based on this and other analysis, [Barkoulas 98] concluded that there was not strong evidence in support of a chaotic structure in the Athens Stock Exchange.

London Metal Exchange: In [Panas 01], d_C was used to study the prices of six metals on the London Metal Exchange. Each metal was studied individually, for $E = 4, 5, \ldots, 10$. Panas observed that if the attractor has an integer d_C, then it is not clear if the dynamical process is chaotic. Therefore, to determine if the system is chaotic, d_C should be accompanied by computing other measures, e.g., the Kolmogorov K-entropy and the Lyapunov exponent [Chatterjee 92].

Thai Stock Exchange: In [Adrangi 04], d_C was used to test for low-dimensional chaos in the Stock Exchange of Thailand. To estimate $d_C(E)$ for a given E and a given set $\{s_j\}$ of decreasing values of s, [Adrangi 04] computed the slopes

$$d_C(E, j) \equiv \frac{\log C_E(s_{j+1}) - \log C_E(s_j)}{\log s_{j+1} - \log s_j},$$

where $C_E(s)$ is the correlation sum for a given E and s. For each value of E, the average of the three highest values of $d_C(E, j)$ was taken to be the estimate of $d_C(E)$.

Gold and Silver Prices: The correlation dimension of gold and silver prices, using daily prices from 1974 to 1986, was studied in [Frank 89]. The range of values for s was $0.9^{33} \approx 0.031$ to $0.9^{20} \approx 0.121$ (gold prices were in United States dollars per fine ounce, and silver prices were in British pence per troy ounce).[7] The correlation dimension $d_C(E)$ was computed for $E = 5, 10, 15, 20, 25$; the results did not substantially change for $E > 25$. For gold, the computed $d_C(E)$ ranged from 1.11 for $E = 5$ to 6.31 for $E = 25$; for silver the range was 2.01 for $E = 5$ to 6.34 for $E = 25$. Using this and other analyses, [Frank 89] concluded that there appears to be some sort of nonlinear process, with dimension between 6 and 7, that generated the observed gold and silver prices.

U.S. Equity Market: In [Mayfield 92], d_C was used to uncover evidence of complex dynamics in U.S. equity markets, using 20,000 observations of the Standard and Poor's (S&P) 500 cash index, taken at 20-s samples. The correlation dimension $d_C(E)$ was computed for $E = 2, 3, \ldots, 10, 15, 20$, and [Mayfield 92] concluded that the underlying market dynamics that determines these S&P index values is either of very high dimension, or of low dimension but with entropy[8] so high that the index cannot be predicted beyond 5 min. [9]

Fractal Growth Phenomena: The correlation dimension has been widely used in the study of fractal growth phenomena, a topic which received considerable attention by physicists starting in the 1980s [Vicsek 89]. Fractal growth phenomena generate such aggregation products as colloids, gels, soot, dust, and

---

[7] See www.goldprice.com/troy-ounce-vs-ounce to learn about troy weights.

[8] Entropy is studied in Sect. 14.1.

[9] The widely held assumption that price is a continuous and differentiable function of time is not obvious is contrary to the evidence, and is in fact contrary to what should be expected, since a competitive price should depend in part on changes in anticipation, which can experience arbitrarily large discontinuities [Mandelbrot 89].

atmospheric and marine snowflakes [Block 91]. Fractal growth processes are readily simulated using *diffusion limited aggregation* (DLA), a model developed in 1981 in which a fractal aggregate of particles, sitting on an E-dimensional lattice, is grown in an iterative process. A walker, added at "infinity" (a faraway position), engages in a random walk until it arrives at a lattice site adjacent to an already occupied site. The walker stops, and this particle is added to the aggregate. The process starts again with another random walker added at infinity. The process is usually started with a seed set of stationary particles onto which a set of moving particle is released. The resulting aggregate is very tenuous, with long branches and considerable space between the branches. There is a large literature on DLA, in particular by Stanley and his students and colleagues; [Stanley 92] is a guide to early work on DLA.

A fractal generated by DLA is an example of one of the two fundamental classes of fractals [Grassberger 93]. For the class encompassing percolation or DLA clusters, the fractal is defined on a lattice, the finest grid used for box counting is this lattice, and each lattice site is either unoccupied or occupied by at most one particle. The other class describes fractals such as strange attractors (Sect. 9.1) for which only a random sample of points on the fractal are known, and we need very many points per non-empty box to compute a fractal dimension.

In the DLA model of [Sorensen 97], one monomer is placed at the center of an E-dimensional sphere. Then other monomers are placed, one at a time, at a starting radius at a random angular position. The starting radius is five lattice units beyond the perimeter of the central cluster. Each monomer follows a random walk until it either joins the central cluster, or wanders past the stopping radius (set to three times the starting radius). Simulation was used to study the number N of monomer (i.e., primary) particles in a cluster formed by aggregation in colloidal or aerocolloidal systems, under the assumption $N = k_0(r_g/a)^{d_C}$, where r_g is the root mean square of monomer positions relative to the center of mass, and the monomer radius a is used to normalize r_g. The focus of [Sorensen 97] was to explore the dependence of the prefactor k_0 on d_C.

A very similar definition of d_C for DLA was given in [Feder 88]; this definition also mirrors (10.21). Let R_0 be the radius of the item (e.g., an atom or molecule) being packed, let R be the radius of the smallest sphere containing a cluster with N items, and let ρ be the density reflecting how densely the items are packed. If as $N \to \infty$ we have

$$N = \rho \left(\frac{R}{R_0} \right)^{d_C} , \qquad (10.22)$$

then d_C is the cluster dimension, also called the mass dimension [Feder 88].[10] For a two-dimensional cluster formed by DLA we have $d_C = 1.71$, and for a three-dimensional cluster we have $d_C = 2.50$ [Feder 88].

---

[10] In this book, we reserve the term "mass dimension" for the correlation dimension of an infinite network, as discussed in Chap. 11.

To simulate diffusion limited cluster aggregation (DLCA), initialize the simulation with a list of monomers. To start the process, remove two monomers from the list. Position one of them at the center, and introduce the second at a random position on the starting radius, as with DLA. This second monomer then follows a random walk until it either joins the monomer at the center or wanders past the stopping radius. If the wandering monomer joins the one at the center, the resulting cluster is placed back into the list. Thus the list in general contains both monomers and clusters. In each successive iteration, two clusters are removed from the list; one is positioned at the center and the other wandered randomly. This process could be stopped after some number of iterations, in which case the list in general contains multiple clusters and monomers. Alternatively, the process could be allowed to continue until there is only a single cluster [Sorensen 97].

Diffusion Limited Deposition (DLP) is a process closely related to DLA. In DLP, particles do not form free-floating clusters, but rather fall onto a solid substrate. Techniques for simulating two-dimensional DLP were presented in [Hayes 94]. Consider a large checkerboard, in which the bottom ("south") boundary (corresponding to $y = 0$) is the substrate, and represents the position of each particle by a cell in the checkerboard. In the simplest DLP, particles are released one at a time from a random x position at the top ("north") boundary of the lattice. Once released, a particle P falls straight down ("south", towards the x axis), stopping when either (i) P is adjacent to another particle (i.e., P and another particle share a common "east" or "south" or "west" boundary), or (ii) P hits the substrate. Once P stops, it never moves again. A variant of this above procedure makes particles move obliquely, as if they were driven by a steady wind.

In simulating DLP, instead of constraining each particle to only drop straight down, or obliquely, we can instead allow a particle to move in all four directions, with each of the four directions being equally likely at each point in time for each particle. A particle stops moving when it becomes adjacent to another particle or hits the substrate. The effect of the "east" and "west" boundary can be eliminated by joining the "east" edge of the checkerboard to the "west" edge, as if the checkerboard was wrapped around a cylinder. This "random walk" DLP generates a forest of fractal trees [Hayes 94]. A straightforward simulation of the "random walk" DEP can require excessive computation time, but two techniques can reduce the time. First, if a particle wanders too far from the growing cluster, pick it up and place it down again closer to the cluster, at a random x position. Second, if a particle is far from the growing cluster, allow it to move multiple squares (rather than just a single square) in the chosen direction. For 10^4 particles, these two techniques reduced the execution time from 30 h (on a 1994 vintage computer) to a few minutes [Hayes 94].

The geometry of the lattice makes a difference: simulations by Meakin on a square (checkerboard) lattice with several million particles generated clusters with diamond or cross shapes aligned with the axes of the lattice, and these shapes do not appear in simulations on a triangular or hexagonal lattice. Thus, as stated in [Hayes 94], "Simulations done with a lattice are more efficient than those without (largely because it is easier to detect collisions), but they are also one step further removed from reality."

Example 10.1 The *gyration dimension* [Kinsner 07] is a variant of d_C that characterizes an asymmetric growing geometric object, such as an object formed by diffusion limited aggregation (DLA). Consider a dynamical process described by $\{\Omega_t\}_{t=1}^{\infty}$, where Ω_t is a finite set of points in $\mathbb{R}^2$. Calling each point a "site", the sequence $\{\Omega_t\}_{t=1}^{\infty}$ might describe a process in which the number of sites increases over time, but the sequence could also describe a cellular automata birth and death process. Defining $N_t \equiv |\Omega_t|$, the *center of gravity* $x_t^{cg} \in \mathbb{R}^2$ of the N_t points in Ω_t at time t is given by $x_t^{cg} \equiv (1/N_t)\sum_{x\in\Omega_t} x$. The center of gravity is also called the "center of mass" or the "centroid". Define

$$r_t \equiv \sqrt{\frac{1}{N_t}\sum_{x\in\Omega_t}[dist(x, x_t^{cg})]^2} \qquad (10.23)$$

to be the radius of gyration at time t of the N_t sites. The center of mass and radius of gyration are "ideally suited" to calculating a fractal dimension for asymmetric objects created by a fractal growth phenomenon such as DLA. For example, consider the asymmetric object (a tree, but treated as a geometric object and not as a network) illustrated in Fig. 10.3. The tree has a root node

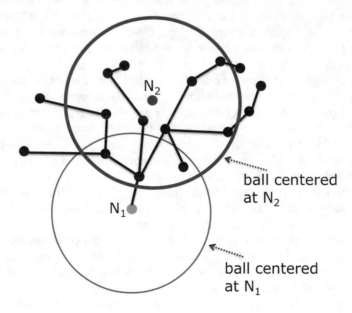

Figure 10.3: Radius of gyration

N_1 (colored green), but balls centered at N_1 do not tend to uniformly cover the tree. In contrast, balls centered at the center of gravity N_2 (colored red) tend to uniformly cover the tree. The center of gravity of the object need not be one of the N given points, as Fig. 10.3 illustrates. If

$$\lim_{t \to \infty} \frac{\log N_t}{\log r_t}$$

exists, then this limit is the *gyration dimension* [Kinsner 07].   □

10.5   The Sandbox Dimension

The *sandbox method* [Tél 89] is a popular method for approximating d_C of a real-world geometric object Ω containing a finite number of points. Randomly choose N points $x(n)$, $n = 1, 2, \ldots, N$, from Ω. Let $B(x(n), r)$ be the box containing each point $y \in \Omega$ such that $dist(y, x(n)) \leq r$, and let $M(x(n), r)$ be the number of points in $B(x(n), r)$. Define

$$M(r) \equiv (1/N) \sum_{n=1}^{N} M(x(n), r). \tag{10.24}$$

Definition 10.1 If for some exponent $d_{sandbox}$ and some range of r we have

$$M(r) \propto r^{d_{sandbox}}, \tag{10.25}$$

then $d_{sandbox}$ is the *sandbox dimension* of Ω.   □

The sandbox dimension was designed for real-world systems of points, e.g., a finite number of points on a lattice, for which the distance between pairs of points does not become arbitrarily small. Thus, unlike the definition of d_C, the definition of $d_{sandbox}$ is not in terms of a limit as $r \to 0$. To see how the sandbox method can be applied to a pixelated image (e.g., of a blood vessel), randomly choose N occupied pixels (a pixel is occupied if it indicates part of the image, e.g., part of the blood vessel) from the full set of occupied pixels. Let $x(n)$ be the coordinates of the center of the n-th occupied pixel. The points $x(n)$ must not be too close to the edge of the image. Let $B(x(n), s)$ be the box centered at $x(n)$ whose side length is s pixels. The value s must be sufficiently small such that $B(x(n), s)$ does not extend beyond the edge of the image. Let $M(x(n), s)$ be the number of occupied pixels in the box $B(x(n), s)$. Define $M(s) = (1/N) \sum_{n=1}^{N} M(x(n), s)$. The scaling $M(s) \propto s^{d_{sandbox}}$ over some range of s defines $d_{sandbox}$.

The sandbox method is sometimes presented (e.g., [Liu 15]) as an extension of the box-counting method. However, Definition 10.1 shows that $d_{sandbox}$ is an approximation of d_C. Indeed, as discussed in Sect. 9.3, in practice the correlation dimension d_C of a geometric object Ω is computed by computing $C(x, r)$ for only a subset of points in Ω, averaging $C(x, r)$ over this finite set to obtain $C(r)$, and finding the slope of the $\log C(r)$ versus $\log r$ curve over a finite range of r.

There are many applications of the sandbox method. It was used in [Hecht 14] to study protein detection. The mass $M(x, r)$ was calculated for all center points

x except for particles that were too close to the periphery of the system. It was used in [Daccord 86] to compute the fractal dimension of the highly branched structures ("fingers") created by DLA on a lattice. In this application, *each* lattice point lying on a finger was chosen as the center of a box (as opposed to choosing only *some* of the lattice points lying on a finger). In [Jelinek 98], the box-counting method, the dilation method (Sect. 4.5), and sandbox method (called the *mass-radius* method in [Jelinek 98]) were applied to 192 images of retinal neurons. Although different spatial orientations or different origins of the box mesh did yield slightly different results, the fractal dimensions computed using these three methods did not differ significantly.

In [Enculescu 02], both d_B and $d_{sandbox}$ of silver nanocrystals were computed. To first calibrate their computer programs, they determined that for colloidal gold, both methods yielded 1.75, the same value determined by previous research. For silver nanoclusters in an alkali halide matrix, they computed $d_{sandbox} = 1.81$ and $d_B = 1.775$; for silver nanocrystals in NaCl they computed $d_{sandbox} = 1.758$ and $d_B = 1.732$; for silver nanocrystals in KBr they computed $d_{sandbox} = 1.769$ and $d_B = 1.772$. Though not mentioned in [Enculescu 02], since $d_{sandbox}$ is an approximation of d_C, and since in theory we have $d_C \leq d_B$ (Theorem 9.1), it is interesting that in two cases (alkali halide and NaCl) they found $d_{sandbox} > d_B$.

As an example of a testimonial to the sandbox method, in an illuminating exchange of letters on the practical difficulties in computing fractal dimensions of geometric objects, Baish and Jain [Corr 01], in comparing the sandbox method to box counting, and to another variant of the Grassberger–Procaccia method for computing d_C, reported that "the sandbox method has been the most robust method in terms of yielding accurate numbers for standard fractal objects and giving a clear plateau for the slope of the power-law plot." The sandbox method also has been a popular alternative to box counting for computing the generalized dimensions of a geometric object (Sect. 16.6) or of a network (Sects. 11.3 and 17.5).

In [Panico 95], a variant of the sandbox method was used to study retinal neurons. Let Ω be the set of N points $x(n)$, $n = 1, 2, \ldots, N$, where $x(n) \in \mathbb{R}^E$. The center of gravity $x_{cg} \in \mathbb{R}^E$ is given by $x_{cg} \equiv (1/N) \sum_{x(n) \in \Omega} x(n)$. The radius of gyration r_g was defined in [Panico 95] by

$$r_g \equiv (1/N) \sum_{x(n) \in \Omega} dist\big(x(n), x_{cg}\big),$$

where $dist\big(x(n), x_{cg}\big)$ is the Euclidean distance from $x(n)$ to x_{cg}, so r_g is the average distance to the center of gravity; this definition differs from (10.23) above. Define Ω_I to be the set of points in Ω within distance r_g of x_{cg}, so Ω_I is the "interior" of Ω. Define $N_I \equiv |\Omega_I|$. For example, suppose in one dimension ($E = 1$) we have $\Omega = \{-3, -2, -1, 1, 2, 3\}$. Then $x_{cg} = 0$ and $r_g = (3 + 2 + 1 + 1 + 2 + 3)/6 = 2$, so $\Omega_I = \{-2, -1, 1, 2\}$. For $r \leq r_g$ and for $x(n) \in \Omega_I$, let $M\big(x(n), r\big)$ be the number of points in Ω within Euclidean distance r of $x(n)$. Define

$$M_I(r) \equiv \frac{1}{N_I} \sum_{x(n) \in \Omega_I} M\big(x(n), r\big).$$

By choosing only centers $x(n)$ in Ω_I, the impact of "edge effects" is mitigated. Then Ω has the *interior sandbox dimension* $d_{sandbox}$ if $M_I(r) \propto r^{d_{sandbox}}$ over some range of r.

One conclusion of [Panico 95] was that this variant of the sandbox method is less sensitive than box counting at discriminating fractal from non-fractal images. For example, for a pixelated skeleton of a river, the linear region of the $\log M_I(r)$ versus $\log r$ plot was substantially smaller for the sandbox method than for box counting. The main result of [Panico 95] was that the calculated fractal dimension d of individual neurons and vascular patterns is not independent of scale, since d reflects the average of the dimension of the interior (this dimension is 2) and the dimension of the border (this dimension approaches 1). Moreover, "Thus, despite the common impression that retinal neurons and vessels are fractal, the objective measures applied here show they are not." The general principle proposed in [Panico 95] is that to accurately determine a pattern's fractal dimension, measurement must be confined to the interior of the pattern.

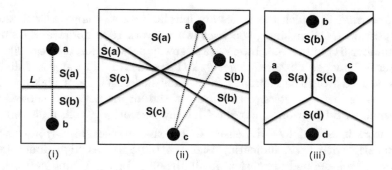

Figure 11.3: Three examples of a Voronoi tessellation

the faster the accumulated area (of the Voronoi regions) increases along the shortest paths at the spatial scale s. Since *area*(i) is added when the distance from node n to node i is reached, $d(n,s)$ can exceed the value 2. The average $\langle d(n,s) \rangle$, where the average is over all nodes and all spatial scales, provides "an interesting quantification of the overall spatial connectivity" of the nodes in $\mathbb{G}$.

Spatially Embedded Random Networks: In [Emmerich 13], d_C was studied for spatially embedded Erdős–Rényi random networks for which nodes are connected to each other with probability $p(r) \sim r^{-\beta}$, where r is the Euclidean distance between the nodes. This model can be viewed as generalizing two previously studied lattice models: the Watts–Strogatz model in which long-range arcs are chosen with the same probability, and the model in [Kleinberg 00] in which arc lengths are chosen from the power-law distribution $p(r) \sim r^{-\beta}$. In [Emmerich 13], d_C was defined by the scaling $C(r) \sim r^{d_C}$, where $C(r)$ is the number of nodes within Euclidean distance r of a random node. In 1 dimension, for $\beta < 1$ the network behaved like an infinite-dimensional network (as in the original Erdős–Rényi network); as β increased d_C became finite for $\beta > 1$, and d_C approached 1 for $\beta > 2$. In 2 dimensions, simulations showed that increasing β causes d_C to change continuously, from $d_C = \infty$ for $\beta < 2$ to $d_C = 2$ for $\beta > 4$.

Real-World Spatially Embedded Networks: In [Gastner 06], the dimension d of an infinite regular E-dimensional lattice was defined by

$$d_C = \lim_{r \to \infty} \frac{\mathrm{d}C_0(n,r)}{\mathrm{d}r},$$

where $C_0(n,r)$, defined by (11.5), is the number of nodes within r hops of a given node n. For any finite network, d_C can be calculated by plotting $\log C_0(n^\star, r)$ against $\log r$ for some central node $n^\star$ and measuring the slope of the initial part of the resulting line. This approach was applied in [Gastner 06] to three real-world spatially embedded networks. To reduce statistical errors, d_C was calculated using $\widetilde{C}(r)$ instead of $C_0(n,r)$, where $\widetilde{C}(r)$ is the average of $C_0(n,r)$ over $n \in \mathbb{N}$, so $\widetilde{C}(r) \approx C(r)$, where $C(r)$ is defined by (11.6). The *Internet* network (7049 nodes and 13,831 arcs and diameter 8) is the March 2003 view

of the Internet, in which the nodes are autonomous systems and the arcs are
direct peering relationships. The *Highway* network (935 nodes and 1337 arcs
and diameter 61) is the U.S. Interstate Highway network for the year 2000; the
diameter is 61 hops. The *Airline* network (176 nodes and 825 arcs and diameter
3) was derived from the February, 2003 published schedule of flights of Delta
Airlines. The low diameters of the Internet and airline networks showed them
to be small-world networks. For the Highway network $d_C \approx 2$, indicating that
this network is essentially 2-dimensional. For the Internet, $\log \widetilde{C}(r)$ grew faster
than linearly with $\log r$, indicating that the Internet has high dimension, or
perhaps no well-defined dimension at all. Results for the airline network were
similar to those for the Internet. The conclusion in [Gastner 06] was that d_C
"distinguishes strongly" between the networks such as the road network, which
is fundamentally two dimensional, and the airline network, which has a much
higher dimension.

Exercise 11.3 For an infinite regular network, the approach in [Gastner 06]
used a limit as $r \to \infty$ to define d_C, yet for a finite network d_C was taken to
be the slope of the initial part of the $\log \widetilde{C}(r)$ versus $\log r$ curve. Are these two
approaches compatible or incompatible?   □

Exercise 11.4 What value of $d(r)$ does (11.11) yield for a node in the interior
of a $K \times K$ rectilinear grid of nodes? What about for a $K \times K \times K$ rectilinear
grid of nodes?   □

Exercise 11.5 If the average node degree of $\mathbb{G}$ is δ, should we expect $d_C = \delta$?
□

Exercise 11.6 How are the branch dimension of [Li 18] and the effective dimension of [Gastner 06] related?   □

11.3   Cluster Growing and Random Walks

In Sect. 10.5 on the sandbox method for geometric fractals, we mentioned that
the sandbox method is also used to compute d_C for a network. In this section,
we discuss the application of the sandbox method to $\mathbb{G}$. We also examine how
d_C can be computed from a random walk in a network. We begin with the
sandbox method.

By Definition 11.1, computing d_C for $\mathbb{G}$ requires using (11.3) to compute
$C(n, r)$ for $n \in \mathbb{N}$ and for a range of r values, and then using (11.6) to compute
$C(r)$. In practice, instead of averaging $C(n, r)$ over all $n \in \mathbb{N}$, we can average
$C(n, r)$ over a randomly selected subset of the nodes in $\mathbb{N}$. This approach
was proposed in [Song 05] and called the *cluster growing method*. Let $\widetilde{\mathbb{N}}$ be a
random subset of $\mathbb{N}$, and let $\widetilde{N} = |\widetilde{\mathbb{N}}|$. In [Costa 07], the nodes in $\widetilde{\mathbb{N}}$ are called
"seed vertices". The cluster growing method of [Song 05] uses $C_0(n, r)$, defined
by (11.5), rather than $C(n, r)$, defined by (11.3). Define

$$C_0(r) = \frac{1}{\widetilde{N}} \sum_{n \in \widetilde{\mathbb{N}}} C_0(n, r). \tag{11.15}$$

If for some c and d_C, and for some range of r, we have

$$C_0(r) \approx c r^{d_C}$$

then d_C is the correlation dimension of $\mathbb{G}$.[3]

Cluster Growing for Spatially Embedded Networks: The cluster growing method (i.e., the sandbox method) was used by Daqing et al. [Daqing 11] to compute d_C for spatially embedded networks. For a given node n, let $\mathbb{N}_k^0(n)$ be the set of all nodes exactly k hops away from n. Define $r_k(n)$ to be the average, over $x \in \mathbb{N}_k^0(n)$, of the Euclidean distance $dist(n,x)$ between nodes x and n, so $r_k(n)$ is the average Euclidean distance from n to a node k hops away from n. Let $M_k(n)$ be the number of nodes at most k hops away from n. Then $M_0(n) = 1$ (since n is 0 hops away from itself) and $M_k(n) = M_{k-1}(n) + |\mathbb{N}_k^0(n)|$ for $k > 1$. This is illustrated in Fig. 11.4, where n is the hollow node in the

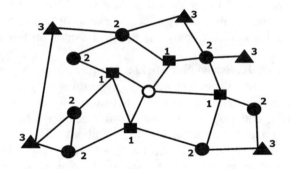

Figure 11.4: $\mathbb{N}_k^0(n)$ for $k = 0, 1, 2$

center. The number next to each node indicates how many hops from n it is. So $|\mathbb{N}_1^0(n)| = 4$ (the squares), $|\mathbb{N}_2^0(n)| = 7$ (the solid circles), and $|\mathbb{N}_3^0(n)| = 5$ (the triangles).

Having computed $r_k(n)$ and $M_k(n)$ for $k = 1, 2, \dots, K$ (for some upper bound K), let $\tilde{n}$ be a different starting node, and repeat these calculations. After doing this for a set $\tilde{\mathbb{N}}$ of starting nodes, for each k we average, over $n \in \tilde{\mathbb{N}}$, the values $r_k(n)$ and $M_k(n)$. This yields a set of averages $\langle r_k \rangle$ and $\langle M_k \rangle$ for $k = 1, 2, \dots, K$. Plotting $\log\langle M_k \rangle$ versus $\log\langle r_k \rangle$, the slope of the linear portion of this curve is the estimate of d_C.

This method was applied in [Daqing 11] to two networks spatially embedded in $\mathbb{R}^2$: an airline network and the Internet. For both networks, the probability

---

[3]In [Song 05], d_C was called the *fractal cluster dimension* of $\mathbb{G}$. Also, in [Song 05], d_C was defined by $C_0(r) \approx r^{d_C}$, but as in [Costa 07]), we interpret $\approx$ to mean $\sim$.

$p(L)$ that the arc length is L follows the power-law distribution $p(L) \sim L^{-\beta}$, and the probability $p(\delta)$ that a node degree is δ follows the power-law distribution $p(\delta) \sim \delta^{-\gamma}$. The airline network has $\beta \approx 3$ and $\gamma \approx 1.8$. The Internet network has $\beta \approx 2.6$ and $\gamma \approx 1.8$. Varying β showed that this parameter significantly impacted the calculated dimension d_C, since for $\beta > 4$, arc lengths were mainly short and $d_C = 2$; for $\beta < 2$, the dimension became infinite; for $2 < \beta < 4$, the dimension increased as β decreased, from $d_C = 2$ to ∞.   □

Cluster Growing of Polymer Networks: In [Sommer 94], cluster growing was used to compute d_C of polymer networks. A *polymer* is a chemical compound comprised of molecules (known as *monomers*) which are bonded together. A polymer can be naturally occurring (e.g., rubber) or man-made (e.g., polytetrafluoroethylene, used in non-stick cooking pans). Viewing each monomer as a logical node, a polymer can be a long chain, or it can be a branched structure, even forming a graph with cycles. Using simulation, [Sommer 94] computed $C_0(r)$ by averaging $C_0(n, r)$ over approximately 100 choices of n (i.e., over 100 monomers) in the central region of the polymer, and obtained $d_C \approx 1.15$. This study anticipated, by a decade or so, the flurry of papers on fractal dimensions of networks.

The Dimension of Railway Networks: In [Benguigui 92], a dimension d was calculated for four railway networks: (*i*) the railway system of the Rhine towns in Germany, (*ii*) the Moscow railway network, (*iii*) the Paris underground (i.e., the "metro"), and (*iv*) the railway network of Paris and its suburbs (an approximately circular region with a radius of about 50 km). For a given r, Benguigui counted the total length $L(r)$ of the railway tracks inside a radius-based box of radius r centered at one specified point in the network. The dimension d was taken to be the slope of the $\log L(r)$ versus $\log r$ curve. For networks (*ii*) and (*iv*), the network seems to have a center node $n^\star$, and each radius-based box was centered at $n^\star$. For networks (*i*) and (*iii*), there is no obvious center node, so several center nodes were used; for each r the mean value is the estimate of $L(r)$. The range of r values used was 2 km–60 km for (*i*), 0.7 km–12 km for (*ii*), 0.25 km–6 km for (*iii*), and 2 km–50 km for (*iv*). The computed values of d are 1.70 for (*i*), 1.70 for (*ii*), 1.80 for (*iii*), and 1.50 for (*iv*). Benguigui suggested that the lower value for (*iv*), for Paris and its suburbs, is easily understood since this network transports people to/from the center of Paris, and is therefore less dense than the other networks.

Exercise 11.7   Consider the railway network illustrated by Fig. 11.5. Assume there are K spokes emanating from the hub, and each spoke has M railway stations, equally spaced. Using the method of [Benguigui 92] described above, what is the fractal dimension of this network?   □

Exercise 11.8   Is it appropriate to call the dimension d calculated in [Benguigui 92] a correlation dimension? If not, how should this dimension be categorized?   □

We now turn to the second topic of this section: computing d_C by simulating a random walk on $\mathbb{G}$. Imagine a random walk on an undirected network $\mathbb{G}$. The random walker starts at some randomly selected node x with node degree δ_x. The walker uses a uniform probability distribution to pick one of the δ_x arcs

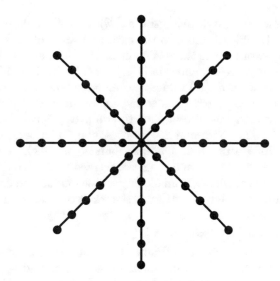

Figure 11.5: A hub and spoke network

incident to x, and traverses this arc, arriving at node y with node degree δ_y. The walker uses a uniform probability distribution to pick one of the δ_y arcs incident to y, and traverses this arc, arriving at node z. If this process continues for a large number of iterations, we generate a time series $n(t)$, $t = 1, 2, \ldots, T$, where $n(1) = x$, $n(2) = y$, $n(3) = z$, etc. As shown in [Lacasa 13, Lacasa 14], this time series can be used to estimate d_C of $\mathbb{G}$. Their method applies to spatially embedded networks in which we associate with each node the coordinate K-tuple $y \in \mathbb{R}^K$. For example, for a road network, $K = 2$, so $y \in \mathbb{R}^2$ identifies the node in a 2-dimensional plane or lattice. For $t = 1, 2, \ldots, T$, let $y_t \in \mathbb{R}^K$ be the coordinates of the node $n(t)$ visited by the random walker at time t. For a given positive integer E, define $x_t \in \mathbb{R}^{KE}$ by $x_t = (y_t, y_{t+1}, \ldots, y_{t+E-1})$. The series $y_t \in \mathbb{R}^K$, $t = 1, 2, \ldots, T$, generates $T - E + 1$ points x_t. For $r > 0$, define the correlation sum

$$C_E(T, r) \equiv \frac{\sum_i \sum_{j \neq i} I(r - dist(y_i, y_j))}{(T - E + 1)(T - E)}.$$

In [Lacasa 13], $dist(\cdot, \cdot)$ was defined using the l_∞ norm. Suppose the limit

$$d_C(E) \equiv \lim_{r \to 0} \lim_{T \to \infty} \frac{\log C_E(T, r)}{\log r} \tag{11.16}$$

exists. If the node coordinates are integers, e.g., if $\mathbb{G}$ is embedded in a lattice, then the limit as $r \to 0$ should be replaced by $r \ll T$. If $d_C(E)$ converges to a value d_C when the embedding dimension E is sufficiently large, then d_C is the correlation dimension of $\mathbb{G}$.

This random walk method was applied in [Lacasa 13] to several networks. For 4000 nodes in a 2-dimensional lattice, where each coordinate of each node is an integer between 1 and 1000, the $d_C(E)$ values increased with E, and $d_C(4) = 1.92 \pm 0.1$, which is close to the expected value of 2. For a 1000 node fully connected network, since there is no natural spatial embedding, each node was labelled by a single real number in $[0, 1]$; they obtained $d_C(E) \approx E$, and the lack of convergence of the $d_C(E)$ values as E increased indicated the network has infinite dimension, as expected (this is the same conclusion obtained from box counting). The method was also applied to two urban networks: San Joaquin, California, with 18,623 nodes and 23,874 arcs, and Oldenberg, Germany, with 6105 nodes and 7035 arcs. San Joaquin, founded in 1920, is a planned city whose road network is gridlike, and $d_C(E)$ converged to 2, with $d_C(3) \approx 1.83$. In contrast, for Oldenberg, which "is an old city whose foundation dates back to the twelfth century and whose road pattern is the result of a self-organized growth", $d_C(E)$ did not converge as E increased, suggesting that the road network of Oldenberg does not have a well defined d_C.

Exercise 11.9 It was argued in [Panico 95] that, to accurately determine a pattern's fractal dimension, measurement must be confined to the interior of the pattern. Should this principle be applied to computing d_C for $\mathbb{G}$?   □

11.4   Overall Slope Correlation Dimension

In this section we present the "average slope" definition of the correlation dimension [Rosenberg 16b] and show its desirable properties for rectilinear grids. (The properties of a two-point approximation to d_C for a geometric object Ω were studied in [Theiler 93].) We begin with the 1-dimensional chain network of K nodes studied in Sect. 11.1. Recalling that $C(r)$ is defined by (11.3), (11.4), and (11.6), consider the "overall slope" $d_C(K)$, defined by

$$d_C(K) \equiv \frac{\log C(\Delta + 1) - \log C(2)}{\log(\Delta + 1) - \log 2} = \frac{\log C(2)}{\log[2/(\Delta + 1)]}, \qquad (11.17)$$

where the equality holds as $C(\Delta + 1) = 1$. This ratio represents the overall slope of the $\log C(s)$ versus $\log s$ curve over the range $s \in [2, \Delta + 1]$. For a 1-dimensional grid of K nodes we have $\Delta = K - 1$. By (11.9) we have $C(2) = 2/K$, so

$$d_C(K) = \frac{\log C(2)}{\log[2/(\Delta + 1)]} = \frac{\log(2/K)}{\log(2/K)} = 1. \qquad (11.18)$$

Thus for each K the overall slope for the 1-dimensional grid of size K is 1, as desired.

Moving to two dimensions, consider a square rectilinear grid $\mathbb{G}$ embedded in $\mathbb{Z}^2$, where $\mathbb{Z}$ denotes the integers. For $x = (x_1, x_2) \in \mathbb{Z}^2$ and $y = (y_1, y_2) \in \mathbb{Z}^2$, we have $dist(x, y) = |x_1 - y_1| + |x_2 - y_2|$. If node n is close to the boundary of $\mathbb{G}$, then the "box" containing all nodes a given distance from n will be truncated, as illustrated by Fig. 11.6. In (i), there are four nodes in $\mathbb{G}$ whose distance to

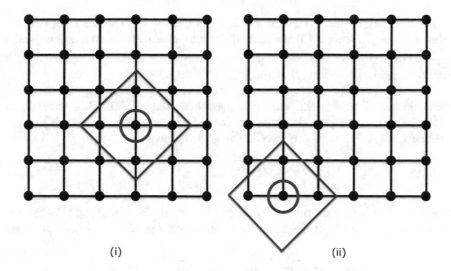

(i)   (ii)

Figure 11.6: Non-truncated and truncated boxes in a 2-dimensional grid

the circled node is 1. In (ii), there are three nodes in $\mathbb{G}$ whose distance to the circled node is 1.

Let E be a positive integer, and consider an undirected rectilinear grid in $\mathbb{Z}^E$, where each edge of the grid contains K nodes. We assert that any definition of the correlation dimension $d_C(K)$ for such a network should satisfy two requirements. The first, which we call the "infinite grid" requirement, is

$$\lim_{K \to \infty} d_C(K) = E. \tag{11.19}$$

Since for a geometric set $\mathcal{X} \subset \mathbb{R}^E$ any reasonable definition of dimension should satisfy the monotonicity property that $\mathcal{Y} \subset \mathcal{X}$ implies $dimension(\mathcal{Y}) \leq dimension(\mathcal{X})$ [Falconer 03], we make the following second assertion. Let $\mathbb{G}_1$ and $\mathbb{G}_2$ be undirected rectilinear grids in $\mathbb{Z}^E$, where each edge of $\mathbb{G}_1$ contains K_1 nodes, and each edge of $\mathbb{G}_2$ contains K_2 nodes. Any definition of the correlation dimension must satisfy the "grid monotonicity" requirement

$$\text{if } K_1 < K_2 \text{ then } d_C(K_1) \leq d_C(K_2). \tag{11.20}$$

From (11.18) we see that, for a 1-dimensional grid, both (11.19) and (11.20) are satisfied by the overall slope definition (11.17). It is instructive to consider some possible definitions of the correlation dimension for a 2-dimensional $K \times K$ rectilinear grid $\mathbb{G}$, and determine if they satisfy (11.19) and (11.20).

Exercise 11.10 For $s = 2, 3, \ldots, 2K - 1$, the slope of the $\log C(s)$ versus $\log s$ curve at s is approximated by

$$F(s) \equiv \frac{\log C(s) - \log C(s-1)}{\log(s) - \log(s-1)}.$$

$F(s)$ is not defined for $s = 1$ and $s = 2$, since $C(1) = 0$. It seems reasonable to define $d_C \equiv F(3)$, since $F(3)$ is a finite difference approximation to the derivative $\frac{\mathrm{d}C(s)}{\mathrm{d}s}$. Demonstrate that $F(3)$ is not a good estimate of d_C of a square $K \times K$ rectilinear grid by showing that $\lim_{K \to \infty} F(3) \approx 2.7$.   □

We now present the results of [Rosenberg 16b] showing that, for a 2-dimensional $K \times K$ rectilinear grid, the overall slope (11.17) has several nice properties. First, we provide an exact closed-form expression for $d_C(K)$.

Theorem 11.1 For a $K \times K$ rectilinear grid we have

$$d_C(K) = \frac{\log[(K^2 + K)/4]}{\log[K - (1/2)]} . \tag{11.21}$$

Proof Each of the four corner nodes of $\mathbb{G}$ (e.g., node v in Fig. 11.7) has two neighbors. Each of the $4(K-2)$ nodes on the boundary of $\mathbb{G}$ that are not corner

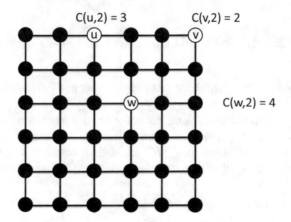

Figure 11.7: Neighbors in a 6×6 grid

nodes (e.g., node u) has three neighbors. Each of the $(K-2)^2$ nodes in the interior of $\mathbb{G}$ (e.g, node w) has four neighbors. Hence from (9.11) we have

$$
\begin{aligned}
C(2) &= \frac{(2)(4) + (3)(4)(K-2) + (4)(K-2)^2}{K^2(K^2-1)} = \frac{4K(K-1)}{K^2(K^2-1)} \tag{11.22} \\
&= \frac{4}{K(K+1)} .
\end{aligned}
$$

Since the diameter of $\mathbb{G}$ is $2(K-1)$, from (11.17) and (11.23) we have

$$d_C(K) = \frac{\log[4/(K^2+K)]}{\log\bigl(2/[2(K-1)+1]\bigr)} = \frac{\log[(K^2+K)/4]}{\log[K-(1/2)]} .   \square$$

Theorem 11.2 below establishes the infinite grid property: the correlation dimension approaches 2 as the grid size approaches infinity. The straightforward proof [Rosenberg 16b], which we omit here, utilizes L'Hôpital's rule.

Theorem 11.2 For a $K \times K$ rectilinear grid we have

$$\lim_{K \to \infty} d_C(K) = 2 \,.$$

Theorem 11.3 For $K \geq 2$ we have

$$d_C(K) < 2 \,.$$

Proof From (11.21),

$$
\begin{aligned}
2 - d_C(K) &= \frac{2\log[K - (1/2)] - \log[K(K+1)/4]}{\log[K - (1/2)]} \\
&= \frac{\log\big([K - (1/2)]^2\big) - \log[K(K+1)/4]}{\log[K - (1/2)]} \,.
\end{aligned}
\tag{11.23}
$$

To prove the result, we will show that the numerator and denominator of (11.23) are positive for $K \geq 2$. We have $\log[K - (1/2)] > 0$ for $K \geq 2$. To show that the numerator of (11.23) is positive for $K \geq 2$, we will show that for $K \geq 2$ we have

$$\frac{4[K - (1/2)]^2}{K^2 + K} > 1 \,. \tag{11.24}$$

This inequality can be written as $4[K^2 - K + (1/4)] > K^2 + K$ which simplifies to $3K^2 - 5K + 1 > 0$. This inequality holds for all sufficiently large K. The roots of $3K^2 - 5K + 1 = 0$ are $(1/6)(5 \pm \sqrt{13})$, and $(1/6)(5 - \sqrt{13}) < (1/6)(5 + \sqrt{13}) < (1/6)(5 + \sqrt{16}) < 2$. It follows that $3K^2 - 5K + 1 > 0$ for $K \geq 2$, which establishes (11.24). $\square$

The next theorem establishes the grid monotonicity property.

Theorem 11.4 The sequence $d_C(K)$ is monotone increasing in K for $K \geq 2$.

Proof We prove the result by treating K as a continuous variable and showing that the first derivative (with respect to K) of $d_C(K)$ is positive. Denote this derivative by $D(K)$. From (11.21) we have

$$D(K) = \frac{f(K)}{g(K)}$$

where

$$f(K) = \frac{\log[K - (1/2)](2K + 1)}{K(K+1)} - \frac{\log[K(K+1)/4]}{K - (1/2)}$$

and $g(K) = \big(\log[K - (1/2)]\big)^2$. Since $g(K) > 0$ for $K \geq 2$, to prove the result it suffices to show that $f(K) > 0$ for $K \geq 2$. The inequality $f(K) > 0$ is equivalent to

$$\frac{[K - (1/2)](2K + 1)}{K(K+1)} > \frac{\log[K(K+1)/4]}{\log[K - (1/2)]} \,. \tag{11.25}$$

Focusing on the left-hand side of (11.25), and employing two partial fractal expansions, we have

$$
\begin{aligned}
\frac{[K-(1/2)](2K+1)}{K(K+1)} &= \frac{2K^2-(1/2)}{K(K+1)} = 2 - \frac{2}{K} + \frac{3/2}{K^2+K} \\
&= 2 - \frac{2}{K} + \left(\frac{3/2}{K} - \frac{3/2}{K+1}\right) \\
&= 2 - \frac{1/2}{K} - \frac{3/2}{K+1} > 2 - \frac{3/2}{K} - \frac{3/2}{K} \\
&= 2\left(1 - \frac{3/2}{K}\right).
\end{aligned}
\tag{11.26}
$$

Focusing on the right-hand side of (11.25), for $K \geq 3$ we have $(K+1)/2 \leq 2K/3$ and $K - (1/2) \geq 3K/4$. For $K \geq 3$ we have

$$
\begin{aligned}
\frac{\log[K(K+1)/4]}{\log[K-(1/2)]} &= \frac{\log K}{\log[K-(1/2)]} + \frac{\log(K+1)}{\log[K-(1/2)]} - \frac{\log 4}{\log[K-(1/2)]} \\
&\leq \frac{\log K}{\log(3K/4)} + \frac{\log(K+1)}{\log(3K/4)} \\
&\quad - \frac{\log 4}{\log(3K/4)} < 2\left(\frac{\log(K+1)}{\log(3K/4)} - \frac{\log 2}{\log(3K/4)}\right) \\
&= 2\left(\frac{\log[(K+1)/2]}{\log(3K/4)}\right) < 2\left(\frac{\log(2K/3)}{\log(3K/4)}\right).
\end{aligned}
\tag{11.27}
$$

From (11.25), (11.26), and (11.27), the inequality $f(K) > 0$ can be shown to hold if $K \geq 3$ and

$$
1 - \frac{3/2}{K} > \frac{\log(2K/3)}{\log(3K/4)}.
$$

After some algebra, this inequality can be rewritten as

$$
\frac{K}{\log(3K/4)} > \frac{3}{2\log(9/8)}.
\tag{11.28}
$$

It is easily verified that (11.28) holds for $K \geq 45$. Numerical computation shows that $d_C(2) > 0$ and $d_C(K)$ is monotone increasing for $2 \leq K \leq 45$. Thus $d_C(K)$ is monotone increasing for $K \geq 2$. $\square$

Figure 11.8 suggests that the convergence of $d_C(K)$ to the limit 2 is very slow. Suppose we seek the value of K for which $d_C(K) = 2 - \epsilon$, where $0 < \epsilon < 2$. From (11.21) we have

$$
\log\left(\frac{K^2+K}{4}\right) = (2-\epsilon)\log[K-(1/2)]
$$

which we rewrite as $K^2 + K = 4[K-(1/2)]^{(2-\epsilon)}$. We approximate $K^2 + K$ by K^2, and $K - (1/2)$ by K, obtaining $K^2 \approx 4K^{(2-\epsilon)}$, which yields $K \approx 4^{1/\epsilon}$. This

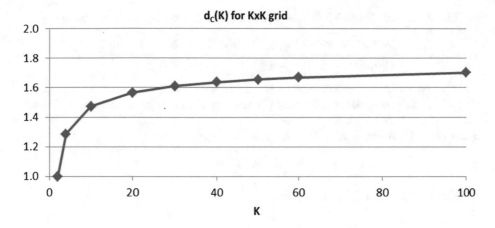

Figure 11.8: Correlation dimension of $K \times K$ grid versus K

estimate is excellent, and shows the very slow convergence. For $\epsilon = 1/5$ we have $K = 4^5 = 1024$ and setting $K = 1024$ in (11.21) yields $d_C(1024) \approx 1.800$; for $\epsilon = 1/50$ we have $K = 4^{50}$ and setting $K = 4^{50}$ in (11.21) yields $d_C(4^{50}) \approx 1.980$.

Exercise 11.11 For a $K \times K$ rectilinear grid, how large must K be to estimate d_C to with 95%? How does this compare to the number of points Krakovská's model (Sect. 10.1) says are needed to estimate d_C to within 95%? □

Now we move to three dimensions, and analyze the overall slope definition (11.17) as applied to a $K \times K \times K$ rectilinear grid.

Theorem 11.5 For a $K \times K \times K$ rectilinear grid $\mathbb{G}$ we have

$$d_C(K) = \frac{\log[(K^3 + K^2 + K)/6]}{\log[(3/2)K - 1]} . \tag{11.29}$$

Proof Each of the 8 corner nodes of $\mathbb{G}$ has 3 neighbors. Each of the 12 edges of $\mathbb{G}$ contains $K - 2$ nodes that are not corner nodes, and each of these $K - 2$ nodes has 4 neighbors. Each of the 6 faces of $\mathbb{G}$ contains $(K - 2)^2$ nodes that do not lie on an edge, and each of these $(K - 2)^2$ nodes has 5 neighbors. Finally, there are $(K - 2)^3$ nodes in the interior of $\mathbb{G}$ (i.e., not lying on a face) and each of these $(K - 2)^3$ nodes has six neighbors. From definition (11.3) and some algebra we have

$$\sum_{n \in \mathbb{G}} C(n, 2) = (8)(3) + (12)(4)(K - 2) + (6)(5)(K - 2)^2 + 6(K - 2)^3$$

$$= 6K^2(K - 1) . \tag{11.30}$$

From definition (9.11) we have

$$C(2) = \frac{6K^2(K - 1)}{K^3(K^3 - 1)} = \frac{6(K - 1)}{K(K - 1)(K^2 + K + 1)} = \frac{6}{K^3 + K^2 + K} . \tag{11.31}$$

The diameter Δ of $\mathbb{G}$ is $3(K-1)$. From (11.17) and (11.31) we have

$$d_C(K) = \frac{\log[(K^3 + K^2 + K)/6]}{\log[3(K-1)+1] - \log 2} = \frac{\log[(K^3 + K^2 + K)/6]}{\log[(3/2)K - 1]} . \qquad \square$$

Theorem 11.6 below establishes the infinite grid property. Just as for Theorem 11.2, the straightforward proof [Rosenberg 16b] of Theorem 11.6, which we omit here, utilizes L'Hôpital's rule.

Theorem 11.6 For a $K \times K \times K$ rectilinear grid we have

$$\lim_{K \to \infty} d_C(K) = 3 .$$

Theorem 11.7 For $K \geq 2$ we have

$$d_C(K) < 3 .$$

Proof From (11.29), we have

$$3 - d_C(K) = \frac{3\log[(3/2)K - 1] - \log[(K^3 + K^2 + K)/6]}{\log[(3/2)K - 1]} . \qquad (11.32)$$

Since the denominator of this fraction is positive for $K \geq 2$, it suffices to show that the numerator is positive for $K \geq 2$. The numerator can be written as

$$\log\left(\frac{6[(3/2)K - 1]^3}{K^3 + K^2 + K} \right) ,$$

and this logarithm is positive if $f(K)$ is positive, where

$$f(K) \equiv 6[(3/2)K - 1]^3 - (K^3 + K^2 + K) . \qquad (11.33)$$

To prove the result we must show that $f(K) > 0$ for $K \geq 2$. We have $f(2) = 34$, $f(3) = 218.25$, and $f(4) = 666$. To show that $f(K) > 0$ for $K > 4$, it suffices to show that $f'(4) > 0$ and $f''(K) > 0$ for $K \geq 4$. We have

$$f'(K) = 27[(3/2)K - 1]^2 - 3K^2 - 2K - 1$$

so $f'(4) = 618$. Also, $f''(K) = (231/2)K - 83$, so $f''(K) > 0$ for $K \geq 1$. $\quad\square$

Theorem 11.4 above shows that, for a two-dimensional grid, $d_C(K)$ is monotone increasing for $K \geq 2$. For a three-dimensional grid, although a plot of $d_C(K)$ shows it to be monotone increasing for $K \geq 2$, we can prove that $d_C(K)$ is monotone increasing for all sufficiently large K.

Theorem 11.8 For all sufficiently large K, $d_C(K)$ is monotone increasing in K.

Proof Define $f(K) = K^3 + K^2 + K$, and let $f'(K)$ denote the first derivative. Let $D(K)$ denote the first derivative of $d_C(K)$ with respect to K. From (11.29), we have

$$D(K) = \frac{\log[(3K/2) - 1] f'(K)/f(K) - (3/2)\log[f(K)/6]/[(3/2)K - 1]}{\left(\log[(3/2)K - 1] \right)^2} .$$

Since the denominator is positive for $K \geq 2$, we need to show that the numerator is positive for all sufficiently large K. After some algebra, the numerator is positive if

$$\log[(3K/2) - 1][3K^3 - (K/3) - (2/3)]$$
$$> (K^3 + K^2 + K)\log[(K^3 + K^2 + K)/6]. \qquad (11.34)$$

Using the Taylor series expansion of $\log(1 + x)$, we have

$$
\begin{aligned}
\log[(K^3 + K^2 + K)/6] &= \log(K/6) + \log\left\{K^2\left(\frac{K^2 + K + 1}{K^2}\right)\right\} \\
&= \log(K/6) + \log(K^2) + \log\left(1 + \frac{K+1}{K^2}\right) \\
&= \log(K/6) + 2\log(K) + \frac{K+1}{K^2} + O\left(\frac{1}{K^2}\right) \quad (11.35)
\end{aligned}
$$

and

$$
\begin{aligned}
\log[(3K/2) - 1] &= \log(3K/2) + \log\left(\frac{(3K/2) - 1}{3K/2}\right) \\
&= \log(3K/2) - \frac{2}{3K} + O\left(\frac{1}{K^2}\right). \quad (11.36)
\end{aligned}
$$

With the expansions (11.35) and (11.36), inequality (11.34) holds if

$$\left(3K^3 + O(K)\right)\left(\log(3K/2) - \frac{2}{3K} + O\left(\frac{1}{K^2}\right)\right)$$
$$> \left(K^3 + O(K^2)\right)\left(\log(K/6) + 2\log(K) + \frac{K+1}{K^2} + O(1/K^2)\right).$$

The highest order term on both sides sides of this inequality is $3K^3 \log K$; the second highest order term on the left-hand side is $3\log(3/2)K^3$, and the second highest order term on the right-hand side is $\log(1/6)K^3$. Hence the inequality holds for all sufficiently large K. $\qquad\square$

Figure 11.9 shows that the convergence of $d_C(K)$ to the limit 3 is also extremely slow (the horizontal axis is $\log_{10} K$). Suppose we seek the value of K for which $d_C(K) = 3 - \epsilon$, where $0 < \epsilon < 3$. From (11.29) we have $\log(K^3 + K^2 + K) - \log 6 = (3-\epsilon)\log[(3/2)K - 1]$. We approximate this equation by $\log K^3 - \log 6 = (3 - \epsilon)\log[(3/2)K]$ which yields $K = (2/3)(81/4)^{1/\epsilon}$. This approximation is also excellent. For $\epsilon = 1$ we have $K = 13.5$ which yields, from (11.29), $d_C(13.5) = 2.060$. For $\epsilon = 1/2$ we have $K \approx 273$ which yields $d_C(273) = 2.502$. For $\epsilon = 1/5$ we have $K \approx 2.270 \times 10^6$ which yields $d_C(2.270 \times 10^6) = 2.800$, so approximately 10^{19} nodes are required to achieve a correlation dimension of 2.8.

Rewriting (11.21) and (11.29) in a more suggestive form, the overall slope results are summarized in Table 11.2. Based on this table, we conjecture that,

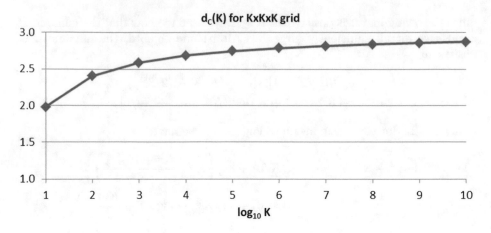

Figure 11.9: Correlation dimension of $K \times K \times K$ grid versus $\log_{10} K$

Table 11.2: Summary of overall slope results

Rectilinear grid	$d_C(K) = \dfrac{\log C(2)}{\log[2/(\Delta + 1)]}$
Chain of K nodes	$d_C(K) = 1$
$K \times K$ grid	$d_C(K) = \dfrac{\log[(K^2 + K)/4]}{\log[(2K - 1)/2]}$
$K \times K \times K$ grid	$d_C(K) = \dfrac{\log[(K^3 + K^2 + K)/6]}{\log[(3K - 2)/2]}$

for any positive integer E, the overall slope correlation dimension $d_C(K, E)$ of a uniform rectilinear grid in $\mathbb{Z}^E$, where each edge of the grid contains K nodes, is given by

$$d_C(K, E) = \frac{\log[(K^E + K^{(E-1)} + \cdots + K^2 + K)/(2E)]}{\log[(EK - (E - 1))/2]} = \frac{\log\left(\frac{K^{E+1} - K}{2E(K-1)}\right)}{\log\left(\frac{EK - E + 1}{2}\right)}. \tag{11.37}$$

Expression (11.37) yields the results in Table 11.2 for $E = 1, 2, 3$.

Exercise 11.12 Use the overall slope approach to derive an expression for d_C for the triangulated hexagon whose perimeter is $6K$ (as illustrated in Fig. 11.10 for $K = 4$). $\square$

An open question is how to extend the overall slope approach to networks without as regular a topology as a rectilinear grid or a triangulated hexagon. Currently there is considerable "art" in the estimation of a fractal dimension,

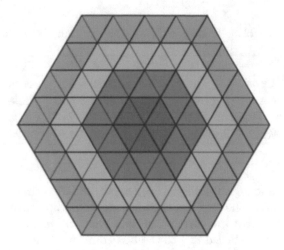

Figure 11.10: Triangulated hexagon

particularly in the selection of the range over which regression is performed. This makes it difficult to compare fractal dimensions computed by different researchers for the same network; the problem is compounded in trying to compare results by different researchers for different networks. The closed-form results derived above for the overall slope correlation dimension of rectilinear grids suggest that it may be possible to obtain similar results for less regular networks. The appropriate modifications of Falconer's "infinite grid" and "grid monotonicity" requirements, which are satisfied for the overall slope correlation dimension, provide reasonable and useful constraints when considering extensions of these results to more general networks.

11.5 Comparing the Box Counting and Correlation Dimensions

In [Wang 13], d_C and d_B were compared for four networks. To compute d_C they first computed $\widetilde{C}(r) \equiv (1/N^2) \sum_{n \in \mathbb{N}} \sum_{x \in \mathbb{N}} I_0(r - dist(x, n))$, where $I_0(z)$ is defined by (9.7). Then d_C was obtained by fitting a straight line to the $\log \widetilde{C}(r)$ versus $\log r$ curve for $r \leq r_{max}$, where r_{max} was manually chosen as the largest value for which the $\log \widetilde{C}(r)$ versus $\log r$ curve appears linear for $r \leq r_{max}$. Table 11.3 provides r_{max} and d_C for the four networks. For these four networks r_{max} is between 5 and 7, which is typical of many of the real-world networks analyzed in this book. However, the plots in [Wang 13] show that for the *E. coli* network, and especially for the *WWW* network, a straight line is

Table 11.3: Results of [Wang 13] for four networks

Network	Nodes	arcs	r_{max}	d_C
E. coli	2859	6890	6	3.50
H. sapiens PIN	1522	2611	5	2.95
S. cerevisiae PIN	1870	2277	7	3.05
World Wide Web	5000	9056	5	4.24

not a good fit to the $\log \widetilde{C}(r)$ versus $\log r$ curve, which suggests that these two networks do not obey the scaling (11.8) and hence do not have a finite d_C.

Table 11.4 compares the correlation dimension d_C calculated in [Wang 13] (given in the column labelled d_C) to the box counting dimension d_B calculated by other researchers. The d_B values calculated by the other researchers are given in the columns labelled "a"–"h"; see [Wang 13] for details, including the caution that the database for the *H. sapiens* PIN network changes as new protein interactions are mapped, and thus comparing d_C with previously calculated d_B values is problematic. Similarly, the *WWW* database is not static. The

Table 11.4: Comparing d_C to various values of d_B

Study	E. coli	H. sapiens	S. cerevisiae	WWW
d_C	3.50	2.95	3.05	4.24
a	3.5	2.3	–	4.1
b	3.5	–	–	–
c	3.5	2.3	–	4.1
d	3.5	2.3	2.1	4.1
e	3.45	–	–	–
f	3.55	5.12	3.41	4.41
g	3.5	2.3	–	4.1
h	3.3	–	2.2	–

computational results in [Wang 13] showed that computing d_C of a network is a viable alternative to computing d_B. Since for geometric objects we have $d_C \leq d_B$ (Theorem 9.1), we might expect that for networks we have the analogous inequality $d_C \leq d_B$. However, Table 11.4 shows that $d_C \leq d_B$ often fails to hold, particularly for the *H. sapiens* and *S. cerevisiae* networks. However, the theoretical result $d_C \leq d_B$ for geometric objects concerns the behavior as $r \to 0$, whereas r is integer valued for the networks of Table 11.4.

The large range of values in each column of Table 11.4 reinforce the message, which was emphasized in the discussions in Sects. 6.5 and 10.4 of applications of d_B and d_C, that excessive focus on the "correct" value of a fractal dimension is of little value. Instead, the practical value of a fractal dimension of a geometric object is typically to detect changes in the object (e.g., the onset of Alzheimer's disease), or differences between objects (e.g., architectural styles of buildings). Similarly, the practical value of a fractal dimension of a network will primarily be to detect changes in the network (e.g., as the network evolves over time), or differences between networks (e.g., differences between road networks in different cities).

Suppose we cover $\mathbb{G}$ with a minimal number $B_R(r)$ of radius-based boxes of radius at most r (Sect. 7.2). Let $N(r)$ be the average, over the $B_R(r)$ boxes, of number of nodes per box. For N points randomly chosen from a geometric object Ω, for large N we have $N \sim B_R(r)N(r)$, which says that N scales as the product of the number of boxes and the average box size [Kim 07b]. This implies that for Ω the scalings $N(r) \sim r^{d_C}$ and $B_R(r) \sim r^{-d_B}$ are equivalent, and that $d_C = d_B$. However, since scale-free networks are small-world, the scaling $N(r) \sim r^{d_C}$ does not hold, and instead we have

$$N(r) \sim e^{r/r_0} \tag{11.38}$$

for some positive characteristic length constant r_0 [Kim 07b]. Thus scale-free networks do not have a finite correlation dimension d_C. As discussed in Sect. 2.4, this behavior occurs since scale-free networks are characterized by a few hubs, and most nodes are only a very few hops away from a hub. By choosing the cluster seeds randomly in the cluster growing method, there is a very high probability of including hubs in the cluster [Song 05, Gallos 07b, Kim 07b, Cohen 08], so nodes with a high degree will be tend to be counted multiple times in the cluster growing method. This biased sampling yields the scaling (11.38). This tendency is absent when using box counting, since each node belongs to only a single box. Similarly, [Costa 07] observed that $d_B = d_C$ holds for a network whose vertices have a "typical" number of connections, but this equality does not hold for scale-free networks. However, the fact that d_C does not exist for scale-free networks does not imply the existence of d_B; that is, a scale-free network is not necessarily fractal [Kim 07b].

Exercise 11.13 In Sect. 6.5 we discussed [Salingaros 12], which used d_B as a measure of which artistic images people found most pleasing. In Sect. 10.4 we discussed [Draves 08], which used d_C to determine the aesthetics of fractal sheep. Is there any published evidence that some real-world network designs or topologies are more aesthetically pleasing, or more relaxing, than other topologies? How could such an experiment be designed? □

11.6 Generalized Correlation Dimension

A generalized correlation dimension of networks was studied in [Eguiluz 03]. For $q \leq N$, let $M_q(r)$ be the number of q-tuples of nodes such that the distance between any two nodes in a q-tuple is less than r. That is, $M_q(r)$ is the cardinality of the set

$$\{(n_{i_1}, n_{i_2}, \ldots, n_{i_q}) \text{ such that } dist(n_{i_j}, n_{i_k}) < r \text{ for } 1 \leq j < k \leq q\}.$$

This definition extends to networks the q-correlation sum defined in [Grassberger 83]. Since the sets are unordered, the number of sets of q nodes is given by the binomial coefficient $\binom{N}{q}$, which is upper bounded by N^q. The q-correlation function is defined as

$$C_q(r) \equiv \frac{M_q(r)}{N^q} , \tag{11.39}$$

which for $N \gg 1$ and $q \ll N$ is approximately the fraction of all sets of q nodes with mutual distance less than r. The q-correlation dimension of $\mathbb{G}$ was defined in [Eguiluz 03] as

$$\lim_{r \to \infty} \frac{1}{q-1} \frac{\log C_q(r)}{\log r} ,$$

so this definition assumes we can grow $\mathbb{G}$ to infinite size. For a finite network $\mathbb{G}$, we can estimate $C_q(r)$ using the approximation [Eguiluz 03]

$$C_q(r) \approx \frac{1}{N} \sum_{n \in \mathbb{N}} \left(\frac{C(n,r)}{N-1} \right)^{q-1} , \tag{11.40}$$

where $C(n,r)$ is defined by (11.3). When $q = 2$ the right-hand sides of (11.40) and (9.11) are identical. In the usual fashion, for a given q the q-correlation dimension of $\mathbb{G}$ is computed by determining a range of r over which $\log C_q(r)$ is approximately linear in $\log r$; the q-correlation dimension estimate is $1/(q-1)$ times the slope of the linear approximation.

Chapter 12

Dimensions of Infinite Networks

Oh, what a tangled web do parents weave
When they think that their children are naive.
Ogden Nash (1902–1971), American poet.

There are two main ways to construct a geometric fractal, or equivalently, two main classes of geometric fractals [Tél 88]. The first class describes a bounded geometric object e.g., the middle-third Cantor set or the Sierpiński triangle, for which the distance between points, or the diameter of replicated self-similar patterns, vanishes. There is no *smallest* scale, and we "zoom in" on the object with a magnification scale that approaches infinity to view the object features which are growing infinitely small. To calculate a fractal dimension (e.g., the box counting dimension d_B, the Hausdorff dimension d_H, or the correlation dimension d_C), we let the box size or radius approach zero.

The second class of fractals described in [Tél 88] describes geometric objects whose size increases to infinity. Such objects often are embedded in some regular lattice for which the inter-lattice spacing remains constant while the size of the object increases to infinity. This class characterizes objects such as aggregates constructed by a growth process, e.g., powder packing or DLA (Sect. 10.4). For such objects, there is no *largest* scale, and we "zoom out" on the object with an ever-increasing window size to see the growing object. This second class also describes networks whose size grows to infinity. A simple example of such a network is the square rectilinear grid in $\mathbb{R}^E$ studied in Sect. 11.4, where we let the side length approach infinity.

The starting point for all our fractal dimensions has been the scaling relationship *bulk* $\sim$ *size*^{d} presented in the beginning of Chap. 1. In this chapter we consider this scaling for Tél's second class, in particular for a sequence $\{\mathbb{G}_t\}_{t=1}^{\infty}$

© Springer Nature Switzerland AG 2020
E. Rosenberg, *Fractal Dimensions of Networks*,
https://doi.org/10.1007/978-3-030-43169-3_12

of networks such that $\mathbb{G}_t$ grows infinitely large, where by "$\mathbb{G}_t$ grows infinitely large" we mean $diam(\mathbb{G}_t) \to \infty$ as $t \to \infty$. For brevity we write $\{\mathbb{G}_t\}$ to denote $\{\mathbb{G}_t\}_{t=1}^{\infty}$.

One possible approach to developing a scaling law for $\{\mathbb{G}_t\}$ is to compute a minimal cover $\mathcal{B}_t(s)$ of $\mathbb{G}_t$ by boxes of size s and study how the number of boxes in $\mathcal{B}_t(s)$ scales as $s \to \infty$ and $t \to \infty$. Another possible approach is to study how the correlation sum of $\mathbb{G}_t$, defined by (9.11), scales as $r \to \infty$ and $t \to \infty$. A third and much simpler approach is to study how the *mass* N_t of $\mathbb{G}_t$, which we define to be the number of nodes in $\mathbb{G}_t$, scales with $diam(\mathbb{G}_t)$. Defining

$$\Delta_t \equiv diam(\mathbb{G}_t), \tag{12.1}$$

the fractal dimension proposed in [Zhang 08] to characterize $\{\mathbb{G}_t\}$ is given by Definition 12.1 below.[1]

Definition 12.1 If

$$d_M \equiv \lim_{t \to \infty} \frac{\log N_t}{\log \Delta_t} \tag{12.2}$$

exists and is finite, then d_M is the *mass dimension* of the sequence $\{\mathbb{G}_t\}$ and $\{\mathbb{G}_t\}$ is *fractal*. If the limit (12.2) is infinite, then $\{\mathbb{G}_t\}$ is *non-fractal*. □

According to Definition 12.1, the sequence $\{\mathbb{G}_t\}$ is fractal if d_M exists and is finite; otherwise, $\{\mathbb{G}_t\}$ is non-fractal. This is a very different meaning of "fractal" than given by Definition 7.5, which says that a given finite network is fractal if d_B exists.

Although the scalings $N_t \sim \Delta_t^{d_M}$ and $C(r) \sim r^{d_C}$ (see (11.8) in Sect. 11.1) are similar, Definition 12.1 defines d_M only for an infinite sequence of networks but not for a single finite network. An advantage of d_M over the correlation dimension d_C is that it is typically much easier to compute the network diameter than to compute $C(n,r)$ for each n and r, as is required to compute $C(r)$ using (11.6).

Exercise 12.1 Can Definition 12.1 be adapted to define d_M for a finite network? Would such a definition be useful? □

In the next two sections we examine three methods of constructing a sequence $\{\mathbb{G}_t\}$ of networks such that $\Delta_t \to \infty$. The constructions are parameterized by a probability $p \in [0,1]$. For the first two constructions studied in Sect. 12.1, $\{\mathbb{G}_t\}$ is (using the terminology of Definition 12.1) fractal. For the third construction, studied in Sect. 12.2, $\{\mathbb{G}_t\}$ is non-fractal, which leads to the definition of the *transfractal* dimension of the sequence $\{\mathbb{G}_t\}$. In the remain-

---

[1]In [Zhang 08] this dimension was denoted by d_B, but we reserve that symbol for the box counting dimension.

ing sections of this chapter we present the pioneering results of Nowotny and Requardt [Nowotny 88] on the fractal dimension of infinite networks.

12.1 Mass Dimension

A procedure was presented in [Zhang 08] that uses a probability p to construct a sequence $\{\mathbb{G}_t\}$ that exhibits a transition from fractal to non-fractal behavior as p increases from 0 to 1. For $p = 0$, the sequence $\{\mathbb{G}_t\}$ does not exhibit the small-world property and has $d_M = 2$, while for $p = 1$ the sequence does exhibit the small-world property and $d_M = \infty$. The construction begins with $\mathbb{G}_0$, which is a single arc, and $p \in [0, 1]$. Let $\mathbb{G}_t$ be the network after t steps. The network $\mathbb{G}_{t+1}$ is derived from $\mathbb{G}_t$.

For each arc in $\mathbb{G}_t$, with probability p we replace the arc with a path of 3 hops (introducing the two nodes c and d, as illustrated by the top branch of Fig. 12.1), and with probability $1 - p$ we replace the arc with a path of 4 hops (introducing the three new nodes c, d, and e, as illustrated by the bottom branch of Fig. 12.1). For $p = 1$, the first three generations of this construction yield the networks of Fig. 12.2.

Colored arcs in the networks for $t = 1$ and $t = 2$ show the construction with $p = 1$: the red arc in the $t = 1$ network becomes the center arc in a chain of three red arcs in the $t = 2$ network, the green arc in the $t = 1$ network becomes the center arc in a chain of three green arcs in the $t = 2$ network, and similarly for the blue arc in the $t = 1$ network. For $p = 0$, the first three generations of this construction yield the networks of Fig. 12.3. This construction builds upon the construction in [Rozenfeld 09] of (u, v) trees.

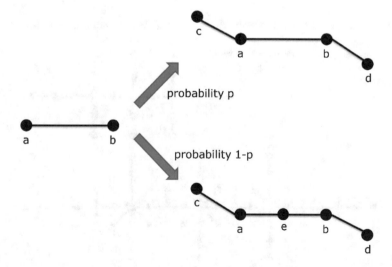

Figure 12.1: Constructing $\mathbb{G}_{t+1}$ from $\mathbb{G}_t$

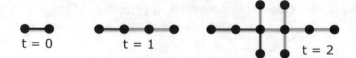

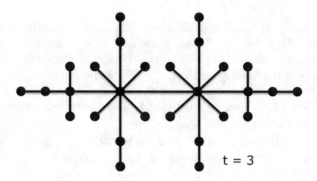

Figure 12.2: Three generations with $p = 1$

Let N_t be the expected number of nodes in $\mathbb{G}_t$, let A_t be the expected number of arcs in $\mathbb{G}_t$, and let Δ_t be the expected diameter of $\mathbb{G}_t$. The quantities N_t, A_t, and Δ_t depend on p, but for notational simplicity we omit that dependence. Since each arc is replaced by three arcs with probability p, and by four arcs with probability $1 - p$, for $t \geq 1$ we have

$$
\begin{aligned}
A_t &= 3pA_{t-1} + 4(1 - p)A_{t-1} = (4 - p)A_{t-1} \\
&= (4 - p)^2 A_{t-2} = \cdots = (4 - p)^t A_0 = (4 - p)^t , \qquad (12.3)
\end{aligned}
$$

where the final equality follows as $A_0 = 1$, since $\mathbb{G}_0$ consists of a single arc.

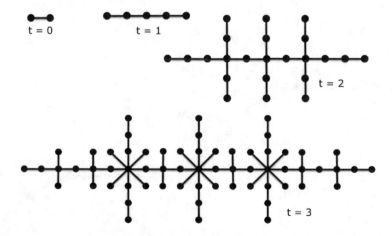

Figure 12.3: Three generations with $p = 0$

Let x_t be the number of new nodes created in the generation of $\mathbb{G}_t$. Since each existing arc spawns two new nodes with probability p and spawns three new nodes with probability $1 - p$, from (12.3) we have

$$x_t = 2pA_{t-1} + 3(1-p)A_{t-1} = (3-p)A_{t-1} = (3-p)(4-p)^{t-1}. \quad (12.4)$$

Since $\mathbb{G}_0$ has 2 nodes, for $t \geq 1$ we have

$$\begin{aligned} N_t &= 2 + \sum_{i=1}^{t} x_i = 2 + \sum_{i=1}^{t}(3-p)(4-p)^{i-1} \\ &= 2 + (3-p)\frac{(4-p)^t - 1}{(3-p)} = (4-p)^t + 1. \end{aligned} \quad (12.5)$$

To compute the diameter Δ_t of $\mathbb{G}_t$, first consider the case $p = 1$. For this case, distances between existing node pairs are not altered when new nodes are added. At each time step, the network diameter increases by two. Since $\Delta_0 = 1$ then $\Delta_t = 2t + 1$. Since $N_t \sim (4-p)^t$, then the network diameter grows as the logarithm of the number of nodes, so $\{\mathbb{G}_t\}$ exhibits the small-world property for $p = 1$. From (12.2) we have $d_M = \infty$.

Now consider the case $0 \leq p < 1$. For this case, the distances between existing nodes are increased. Consider an arc in the network $\mathbb{G}_{t-1}$, and the endpoints i and j of this arc. With probability p, the distance between i and j in $\mathbb{G}_t$ is 1, and with probability $1 - p$, the distance between i and j in $\mathbb{G}_t$ is 2. The expected distance between i and j in $\mathbb{G}_t$ is therefore $p + 2(1-p) = 2 - p$. Since each $\mathbb{G}_t$ is a tree, for $t \geq 1$ we have

$$\Delta_t = p\Delta_{t-1} + 2(1-p)\Delta_{t-1} + 2 = (2-p)\Delta_{t-1} + 2$$

and $\Delta_0 = 1$. This yields [Zhang 08]

$$\Delta_t = \left(1 + \frac{2}{1-p}\right)(2-p)^t - \frac{2}{1-p}. \quad (12.6)$$

From (12.2), (12.5), and (12.6),

$$d_M = \lim_{t \to \infty} \frac{\log N_t}{\log \Delta_t} = \lim_{t \to \infty} \frac{\log[(4-p)^t + 1]}{\log\left[\left(1 + \frac{2}{1-p}\right)(2-p)^t - \frac{2}{1-p}\right]} = \frac{\log(4-p)}{\log(2-p)}, (12.7)$$

so d_M is finite, and $\{\mathbb{G}_t\}$ does not exhibit the small-world property. For $p = 0$ we have $d_M = \log 4 / \log 2 = 2$. Also, $\log(4-p)/\log(2-p) \to \infty$ as $p \to 1$.

Next we present the analysis in [Li 14] of the properties of the family of networks introduced in [Gallos 07a]. The parameters used to construct these networks are two positive integers m and x, where $x \leq m$, and $\alpha \in [0, 1]$. The network $\mathbb{G}_0$ at time $t = 0$ consists of a single arc. The network $\mathbb{G}_{t+1}$ is derived from $\mathbb{G}_t = (\mathbb{N}_t, \mathbb{A}_t)$ as follows.

Step 1. For each arc in $\mathbb{G}_t$, connect each endpoint of the arc to m new nodes. This step creates a total of $2m|\mathbb{A}_t|$ new nodes. This is illustrated, for a single arc, in Fig. 12.4 for $m = 3$. Node u spawns nodes c, e, and g, and node v spawns nodes d, f, and h.

Step 2. Generate a random number p, using the uniform distribution on $[0, 1]$. (Thus p will vary from generation to generation.)

Step 3. We consider two cases, both illustrated in Fig. 12.4 for $m = 3$ and $x = 2$. *Case 1.* If $\alpha \leq p \leq 1$, none of the arcs in $\mathbb{A}_t$ is an arc in $\mathbb{A}_{t+1}$. However, for each $a \in \mathbb{A}_t$ we create x new arcs as follows. Let u and v be the endpoints of a. We select x of the m nodes created in Step 1 that are attached to u, and x of the m nodes created in Step 1 that are attached to v, and connect these two sets of x nodes by x arcs. In Fig. 12.4, arc (u, v) is not preserved, and arcs (c, d) and (e, f) are created. *Case 2.* If $0 \leq p < \alpha$, each $a \in \mathbb{A}_t$ is also an arc in $\mathbb{A}_{t+1}$. Additionally, for each $a \in \mathbb{A}_t$ we create $x - 1$ new arcs as follows. Let i and j be the endpoints of a. We select $x - 1$ of the m nodes created in Step 1 that are attached to u, and $x - 1$ of the m nodes created in Step 1 that are attached to v, and connect these two sets of $x - 1$ nodes by $x - 1$ arcs. In Fig. 12.4, arc (u, v) is preserved, and arc (e, f) is created. Figure 12.5 illustrates three generations

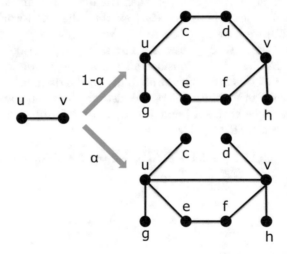

Figure 12.4: $\mathbb{G}_0$ and $\mathbb{G}_1$ for $x = 2$ and $m = 3$

when $\alpha = 0$, $m = 2$, and $x = 2$. Since $\alpha = 0$, we are always in Case 1 of Step 3.

In general, for this construction we have $A_{t+1} = |\mathbb{A}_{t+1}| = (2m + x)A_t$ which implies $A_t = (2m + x)^t$. As for the number of nodes, each arc in $\mathbb{A}_t$ spawns $2m$ new nodes, so

$$N_{t+1} = |\mathbb{N}_{t+1}| = N_t + 2mA_t = N_t + 2m(2m + x)^t,$$

which has the solution

$$N_t = \left(2 - \frac{2m}{2m + x - 1} \right) + \left(\frac{2m}{2m + x - 1} \right)(2m + x)^t. \tag{12.8}$$

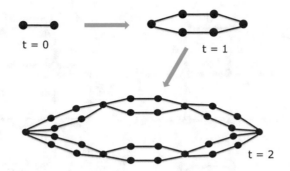

Figure 12.5: $\mathbb{G}_0$ and $\mathbb{G}_1$ and $\mathbb{G}_2$ for $\alpha = 0$ and $x = 2$ and $m = 2$

Let H_t be the hop count between two nodes of $\mathbb{G}_t$. From Fig. 12.4 we see that with probability $1 - \alpha$ we have $H_{t+1} = 3H_p$ and with probability α we have $H_{t+1} = H_t$. Thus $H_{t+1} = (1 - \alpha)3H_t + \alpha H_t = (3 - 2\alpha)H_t$, which implies $\Delta_{t+1} = (3 - 2\alpha)\Delta_t$. Since $\Delta_0 = 1$ then $\Delta_t = (3 - 2\alpha)^t$. By (12.2) the mass dimension d_M of the infinite sequence $\{\mathbb{G}_t\}$ is

$$\frac{\log(2m + x)}{\log(3 - 2\alpha)} . \tag{12.9}$$

Next we examine how the node degree changes each generation. For a given t, let n be any node in $\mathbb{N}_t$ and consider the node degree δ_n. Since n is the endpoint of δ_n arcs, then in Step 1 node n gets $m\delta_n$ new neighbors. In Step 3, if Case 1 occurs, which happens with probability $1 - \alpha$, node n loses each of the δ_n arcs currently incident to n. If Case 2 occurs, which happens with probability α, the $x - 1$ new arcs created are not incident to n. Thus the node degree of n in $\mathbb{G}_{t+1}$ is $\delta_n + m\delta_n - (1 - \alpha)\delta_n = \delta_n(m + \alpha)$, so the degree of each node increases by the factor $m + \alpha$. It was shown in [Song 06, Li 14] that $\mathbb{G}_t$ is scale-free, with the node degree distribution $p_k \sim k^{-\gamma}$, where

$$\gamma = 1 + \frac{\log(2m + x)}{\log(m + \alpha)} . \tag{12.10}$$

12.2 Transfractal Networks

A deterministic recursive construction can be used to create a sequence $\{\mathbb{G}_t\}$ of self-similar networks. Each $\mathbb{G}_t$ is called a (u, v)-*flower*, where u and v are positive integers [Rozenfeld 09]. By varying u and v, both fractal and non-fractal sequences $\{\mathbb{G}_t\}$ can be generated. The construction starts at time $t = 1$ with a cyclic graph (a ring), with $w \equiv u + v$ arcs and w nodes. At time $t + 1$, replace each arc of the time t network by two parallel paths, one with u arcs, and one with v arcs. Without loss of generality, assume $u \leq v$. Figure 12.6 illustrates three generations of a $(1, 3)$-flower. The $t = 1$ network has 4 arcs. To generate the $t = 2$ network, arc a is replaced by the path $\{b\}$ with one arc,

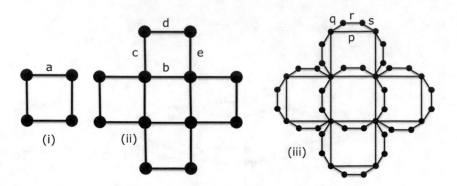

Figure 12.6: Three generations of a $(1,3)$-flower

and also by the path $\{c, d, e\}$ with three arcs; the other three arcs in subfigure (i) are similarly replaced. To generate the $t = 3$ network, arc d is replaced by the path $\{p\}$ with one arc, and also by the path $\{q, r, s\}$ with three arcs; the other fifteen arcs in subfigure (ii) are similarly replaced. The self-similarity of the (u, v)-flowers follows from an equivalent method of construction: generate the time $t + 1$ network by making w copies of the time t network, and joining the copies at the hubs.

Let $\mathbb{G}_t$ denote the (u, v)-flower at time t. The number of arcs in $\mathbb{G}_t$ is $A_t = w^t = (u + v)^t$. The number N_t of nodes in $\mathbb{G}_t$ satisfies the recursion $N_t = wN_{t-1} - w$. With the boundary condition $N_1 = w$ we obtain [Rozenfeld 09]

$$N_t = \left(\frac{w - 2}{w - 1}\right) w^t + \left(\frac{w}{w - 1}\right). \tag{12.11}$$

Consider the case $u = 1$. It can be shown [Rozenfeld 09] that for $(1, v)$-flowers and odd v we have

$$\Delta_t = (v - 1)t + (3 - v)/2$$

while in general, for $(1, v)$-flowers and any v,

$$\Delta_t \sim (v - 1)t. \tag{12.12}$$

Since $N_t \sim w^t$ then $\Delta_t \sim \log N_t$, so $(1, v)$-flowers enjoy the small-world property. By (12.2), (12.11), and (12.12), for $(1, v)$-flowers we have

$$d_M = \lim_{t\to\infty} \frac{\log N_t}{\log \Delta_t} = \lim_{t\to\infty} \frac{\log w^t}{\log t} = \infty, \tag{12.13}$$

so by Definition 12.1 the sequence of $(1, v)$-flowers is non-fractal.

We want to define a new type of fractal dimension that is finite for the sequence of $(1, v)$-flowers and for other non-fractal sequences of networks. For $(1, v)$-flowers, from (12.11) we have

$$N_t \sim w^t = (1 + v)^t$$

as $t \to \infty$ and $\log N_t \sim t \log(1 + v)$. From (12.12) we have $\Delta_t \sim (v - 1)t$ as $t \to \infty$. Since both $\log N_t$ and Δ_t behave like a linear function of t as $t \to \infty$, but with different slopes, let d_E be the ratio of the slopes.[2] Then

$$d_E \equiv \frac{\log(1 + v)}{v - 1}.$$ (12.14)

From (12.14), (12.12), and (12.11), as $t \to \infty$ we have

$$d_E = \frac{t \log(1 + v)}{t(v - 1)} = \frac{\log(1 + v)^t}{t(v - 1)} = \frac{\log w^t}{t(v - 1)} \sim \frac{\log N_t}{\Delta_t},$$ (12.15)

from which we obtain

$$N_t \sim e^{d_E \Delta_t}.$$ (12.16)

Define $\alpha_t \equiv \Delta_{t+1} - \Delta_t$. From (12.16),

$$\frac{N_{t+1}}{N_t} \sim \frac{e^{d_E \Delta_{t+1}}}{e^{d_E \Delta_t}} = e^{d_E \alpha_t}.$$ (12.17)

Writing $N_t = N(\Delta_t)$ for some function $N(\cdot)$,

$$N_{t+1} = N(\Delta_{t+1}) = N(\Delta_t + \alpha_t).$$

From this and (12.17),

$$N(\Delta_t + \alpha_t) \sim N(\Delta_t)e^{d_E \alpha_t}$$ (12.18)

which says that, for $t \gg 1$, when the diameter increases by α_t, the number of nodes increases by a factor which is exponential in $d_E \alpha_t$. As observed in [Rozenfeld 09], in (12.18) there is some arbitrariness in the selection of e as the base of the exponential term $e^{d_E \alpha_t}$ since from (12.14) the numerical value of d_E depends on the logarithm base.

Definition 12.2 [Rozenfeld 09] If (12.18) holds as $t \to \infty$ for a sequence of self-similar graphs $\{\mathbb{G}_t\}$, then d_E is called the *transfinite fractal dimension*. A sequence of self-similar networks whose mass dimension d_M is infinite, but whose transfinite fractal dimension d_E is finite, is called a *transfinite fractal*, or simply *transfractal*. □

The transfractal dimension "usefully distinguishes between different graphs of infinite dimensionality" [Rozenfeld 09]. Thus the sequence of $(1, v)$-flowers is transfractal with transfinite fractal dimension $d_E = \log(1 + v)/(v - 1)$. Finally, consider (u, v)-flowers with $u > 1$. It can be shown [Rozenfeld 09] that $\Delta_t \sim u^t$. Using (12.11),

$$\lim_{t \to \infty} \frac{\log N_t}{\log \Delta_t} = \lim_{t \to \infty} \frac{\log w^t}{\log u^t} = \frac{\log(u + v)}{\log u},$$

---

[2]Having already used d_T, d_B, d_S, d_H, and d_C, we are running out of letters to indicate fractal dimensions. The letter "E" was chosen for the transfinite fractal dimension to denote "exponential", since the usual power-law fractal scaling does not hold.

so

$$d_M = \frac{\log(u+v)}{\log u}.$$

Since d_M is finite, this sequence of networks is fractal, not transfractal, and these networks do not enjoy the small-world property. In [Komjáthy 19], the definition of the transfractal dimension was extended by considering the box counting dimension for infinite sequences of graphs.

12.3   Volume and Surface Dimensions

In this section we study two very early definitions in [Nowotny 88] of the dimension of an infinite network $\mathbb{G}$. Unlike the previous two sections of this chapter, in this section we are not studying a sequence $\{\mathbb{G}_t\}$ of networks such that $\Delta_t \to \infty$; rather, we are studying one network $\mathbb{G}$ whose diameter is infinite. It might be that $\mathbb{G}$ can be viewed as the limit of a sequence $\{\mathbb{G}_t\}$, e.g., an infinite rectilinear grid network embedded in $\mathbb{R}^2$ can be viewed as the limit, as $K \to \infty$, of a $K \times K$ rectilinear grid (as studied in Sect. 11.4), but we do not assume that $\mathbb{G}$ is the limit of some sequence of networks. Since the two network dimensions of [Nowotny 88] are defined in terms of a limit as $r \to \infty$, these dimensions are related to the mass dimension d_M.

The pioneering paper [Nowotny 88] is not well-known, and indeed has barely been mentioned in the bibliographies and reference lists of the bibliographic entries of this book.[3]   The motivation for the two dimensions proposed in [Nowotny 88] is a working philosophy that both physics and the corresponding mathematics are discrete, not continuous, at the "primordial" (i.e., Planck) scale. This discrete primordial substratum consists of elementary cells or modules, interacting with each other over a network of elementary interactions. The term "discrete" in this context does not necessarily imply a lattice or some other countable structure, but rather implies the absence of the usual continuum concepts [Requardt 06].   Restricting the analysis to countably infinite graphs is advantageous, since it facilitates modelling and technical analysis. The network studied in [Nowotny 88] was assumed to be dynamic and highly erratic, where even the distribution of links is dynamical. These networks could be represented using cellular autonoma models; however, such models typically are embedded in fixed and regular geometric arrays. In contrast, the approach of [Nowotny 88] ignored the embedding space and considered an irregular array of nodes and links which dynamically arrange themselves according to some given evolution

---

[3]Their paper was cited in a review [Aaronson 02] of Wolfram's 2002 book "A New Kind of Science". In Chapter 9 of his book, Wolfram conjectured that spacetime is discrete at the Planck scale of about 10^{-33} centimeters or 10^{-43} s. Aaronson noted that this conjecture is not new, and that researchers in quantum gravity have previously proposed that spacetime is a causal network arising from graph updating rules, that particles could arise as 'topological defects' in such a network, and that dimension and other geometric properties can be defined solely in terms of the network's connectivity pattern. Aaronson cited [Nowotny 88] for this last point about dimension.

law. The hope is that, under certain favorable conditions, the system will undergo a series of phase transitions from a disordered chaotic initial state to some kind of macroscopically ordered extended pattern, which can be associated with classical spacetime. From a discrete model, the continuum concepts of ordinary spacetime physics arises via a kind of *geometric renormalization group process* on the much coarser scale corresponding to the comparatively small energies of present high-energy physics. Spacetime possibly emerges via a dynamical process, perhaps resembling a phase transition, from a more chaotic and violent initial phase. The continuum limit is a fixed point of this process, and this idea has some "vague similarities" to the renormalization group process (Chap. 21) of statistical mechanics [Requardt 06]. Computing a dimension for this initial phase, which is discrete but perhaps with complicated interactions, appears to be extremely useful. A "collective geometric" property like dimension suggests that, after some coarse graining and/or scaling, the network displays global smoothness properties which may indicate the transition into a continuum-like macro state [Nowotny 07]. Nowotny and Requardt observed that their work is related to other research in quantum gravity, e.g., to "prophetic" work by Penrose who introduced *spin networks*, which can be viewed as a type of dynamic graph.

As additional motivation for their definitions of dimension, [Nowotny 07] noted that in the renormalization and coarse graining of the discrete primordial network, more and more nodes and arcs are absorbed into the infinitesimal neighborhoods of the emerging points of the continuum. Hence a *local* definition of dimension in the continuum is actually a large scale concept in the network, involving practically infinitely many nodes and their interconnections. They were motivated by wondering what intrinsic global property, independent of some dimension in which the network is embedded, is relevant for the occurrence of such phenomena as phase transitions and critical behavior. They wound up with the answer that the relevant property is the number of new nodes or degrees of freedom seen as the distance from a given node increases.

Assume that $\mathbb{G} = (\mathbb{N}, \mathbb{A})$ is an unweighted and undirected network, where both $\mathbb{N}$ and $\mathbb{A}$ are countably infinite sets. Assume the node degree of each node is finite. For $n \in \mathbb{N}$ and for each nonnegative integer r, define

$$\mathbb{N}(n, r) = \{\, x \in \mathbb{N} \mid dist(n, x) \leq r \,\}, \tag{12.19}$$

so $\mathbb{N}(n, r)$ is the set of nodes whose distance from n does not exceed r. Define

$$M(n, r) \equiv |\mathbb{N}(n, r)| \tag{12.20}$$

$$d_V^i(n) \equiv \liminf_{r \to \infty} \frac{\log M(n, r)}{\log r} \tag{12.21}$$

$$d_V^s(n) \equiv \limsup_{r \to \infty} \frac{\log M(n, r)}{\log r}. \tag{12.22}$$

Definition 12.3 If for some $n \in \mathbb{N}$ we have $d_V^i(n) = d_V^s(n)$, we say that $\mathbb{G}$ has *local volume dimension* $d_V(n)$ at n, where $d_V(n) \equiv d_V^i(n) = d_V^s(n)$. If also

$d_V(n) = d_V$ for each n in $\mathbb{N}$, then we say that the volume dimension of $\mathbb{G}$ is d_V. $\square$

The dimension d_V was called the "internal scaling dimension" in [Nowotny 88], but we prefer the term "volume dimension". If d_V exists then for $n \in \mathbb{N}$ we have $M(n,r) \sim r^{d_V}$ as $r \to \infty$. Note the similarity between the definition of $d_V(n)$ and definition (9.3) of the pointwise mass dimension of a geometric object; the difference is that (12.21) and (12.22) use a limit as $r \to \infty$, while (9.3) uses a limit as $r \to 0$. The volume dimension $d_V(n)$ somewhat resembles the correlation dimension d_C defined by (11.8); the chief difference is that d_C is defined in terms of the correlation sum (9.11), which is an average fraction of mass, where the average is over all nodes in a finite network, while d_V is defined in terms of an infimum and supremum over all nodes. The volume dimension d_V also somewhat resembles the mass dimension d_M defined by (12.2); the chief difference is that d_M is defined in terms of the network diameter, while d_V is defined in terms of the number of nodes within radius r of node n.

The second definition of a network dimension proposed in [Nowotny 88] is based on the boundary (i.e., surface) of $\mathbb{N}(n,r)$, rather than on the entire neighborhood $\mathbb{N}(n,r)$. For $n \in \mathbb{N}$ and for each nonnegative integer r, define

$$\partial\mathbb{N}(n,r) \equiv \{\, x \in \mathbb{N} \mid dist(n,x) = r \,\}, \tag{12.23}$$

so $\partial\mathbb{N}(n,r)$ is the set of nodes whose distance from n is exactly r. For each n we have

$$\partial\mathbb{N}(n,r) = \mathbb{N}(n,r) - \mathbb{N}(n,r-1)$$

and $\partial\mathbb{N}(n,0) = \{n\}$. Define

$$d_U^i(n) \equiv \liminf_{r\to\infty} \left(\frac{\log|\partial\mathbb{N}(n,r)|}{\log r} + 1 \right) \tag{12.24}$$

$$d_U^s(n) \equiv \limsup_{r\to\infty} \left(\frac{\log|\partial\mathbb{N}(n,r)|}{\log r} + 1 \right). \tag{12.25}$$

Definition 12.4 If $d_U^i(n) = d_U^s(n)$, we say that $\mathbb{G}$ has *surface dimension* $d_U(n)$ at n, where $d_U(n) \equiv d_U^i(n) = d_U^s(n)$. If for each n in $\mathbb{N}$ we have $d_U(n) = d_U$, then we say that the surface dimension of $\mathbb{G}$ is d_U. $\square$

This dimension was called the "connectivity dimension" in [Nowotny 88], since "it reflects to some extent the way the node states are interacting with each other over larger distances via the various bond sequences connecting them", but we prefer the term "surface dimension".[4] If d_U exists, then for $n \in \mathbb{N}$ we have $|\partial\mathbb{N}(n,r)| \sim r^{d_U-1}$ as $r \to \infty$. The volume dimension d_V is "rather

---

[4]We cannot use d_S to refer to the surface dimension, since d_S denotes the similarity dimension. Nor can we use d_B to refer to the surface dimension, using "B" to suggest "boundary" or "border", since d_B denotes the box counting dimension. Using "P" for "perimeter" will not work, since d_P is the packing dimension. So, forced to pick something, we choose d_U, where "U" is the second letter in "surface".

a mathematical concept and is related to well-known dimensional concepts in fractal geometry", while the surface dimension d_U "seems to be a more physical concept as it measures more precisely how the graph is connected and how nodes can influence each other". The values of d_V and d_U are identical for "generic" networks, but are different on certain "exceptional" networks [Nowotny 88].

Example 12.1 Turning momentarily to Euclidean geometry in $\mathbb{R}^2$, for a circle with center x and radius r the quantity analogous to $M(n,r)$ is $area(x,r) = \pi r^2$. We have $\lim_{r \to \infty} \log\big(area(x,r)\big)/\log r = 2$. The quantity analogous to $\partial \mathbb{N}(n,r)$ is $area(r) - area(r-\epsilon)$, where ϵ is a small positive number. Since $area(r) - area(r-\epsilon) \approx 2\pi r\epsilon$, we have $\lim_{r \to \infty} \log\big(area(r) - area(r-\epsilon)\big)/\log r = 1$, so this limit and $\lim_{r \to \infty} \log\big(area(x,r)\big)/\log r$ differ by 1, as reflected by the constant 1 added in (12.24). $\square$

Example 12.2 Let $\mathbb{G}$ be the infinite graph whose nodes lie on a straight line at locations $\cdots, -3, -2, -1, 0, 1, 2, 3, \cdots$. Then $M(n,r) = 2r+1$ for each n and r, so $d_V = \lim_{r \to \infty} \log(2r+1)/\log r = 1$. As for the surface dimension, we have $|\partial \mathbb{N}(n,r)| = 2$ for each n and r, so $d_U = \lim_{r \to \infty}[(\log 2/\log r) + 1] = 1$. $\square$

Example 12.3 Let $\mathbb{G}$ be the infinite 2-dimensional rectilinear lattice whose nodes have integer coordinates, and where node (i,j) is adjacent to $(i-1,j)$, $(i+1,j)$, $(i,j-1)$, and $(i,j+1)$. Using the L_1 (i.e., Manhattan) metric, the distance from the origin to node $n = (n_1, n_2)$ is $|n_1| + |n_2|$. For integer $r \geq 1$, the number $|\partial \mathbb{N}(n,r)|$ of nodes at a distance r from a given node n is $4r$ [Shanker 07b], so by (12.20) we have

$$M(n,r) = |\mathbb{N}(n,r)| = 1 + 4\sum_{i=1}^{r} i = 1 + 4r(r+1)/2 = 2r^2 + 2r + 1 \,.$$

Hence

$$d_U = \lim_{r \to \infty}[(\log 4r/\log r) + 1] = 2$$
$$d_V = \lim_{r \to \infty} \log\big(2r^2 + 2r + 1\big)/\log r = 2 \,.   \square$$

Exercise 12.2 Let $\mathbb{G}$ be an infinite rooted tree such that each node spawns k child nodes. Calculate $d_V(n)$ and $d_U(n)$. $\square$

A construction and corresponding analysis in [Nowotny 88] generates an infinite network for which d_V exists but d_U does not exist. Thus the existence of the volume dimension d_V, and its numerical value, do not provide much information about the behavior of $|\partial \mathbb{N}(n,r)|$, although the inequality

$$\limsup_{r \to \infty} \frac{\log |\partial \mathbb{N}(n,r)|}{\log r} \leq d_V(n)$$

is valid for all n. On the other hand, the following result proves that the existence of the surface dimension at n does imply the existence of the volume dimension at n, and the equality of these values.

Theorem 12.1 ([Nowotny 88], Lemma 4.11) If $d_U(n)$ exists and if $d_U(n) > 1$, then $d_V(n)$ exists and $d_V(n) = d_U(n)$.

Proof Let $d = d_U(n)$, and let $\epsilon > 0$ satisfy $d - 1 > \epsilon$. Since $d_U(n)$ exists, then there exists a value $\widetilde{r}$ such that for $r \geq \widetilde{r}$ we have

$$\left| \frac{\log |\partial \mathbb{N}(n,r)|}{\log r} - d + 1 \right| < \epsilon$$

$$\Rightarrow -\epsilon < \frac{\log |\partial \mathbb{N}(n,r)|}{\log r} - d + 1 < \epsilon$$

$$\Rightarrow (d - 1 - \epsilon) \log r < \log |\partial \mathbb{N}(n,r)| < (d - 1 + \epsilon) \log r$$

$$\Rightarrow r^{d-1-\epsilon} < |\partial \mathbb{N}(n,r)| < r^{d-1+\epsilon} .$$

Defining

$$S(r) \equiv \sum_{i=0}^{r} |\partial \mathbb{N}(n,i)| ,$$

for $r \geq \widetilde{r}$ we have

$$M(n,r) = \sum_{i=0}^{r} |\partial \mathbb{N}(n,i)| = S(\widetilde{r}) + \sum_{i=\widetilde{r}+1}^{r} |\partial \mathbb{N}(n,i)| \qquad (12.26)$$

$$\Rightarrow S(\widetilde{r}) + \sum_{i=\widetilde{r}+1}^{r} i^{d-1-\epsilon} \leq M(n,r) \leq S(\widetilde{r}) + \sum_{i=\widetilde{r}+1}^{r} i^{d-1+\epsilon} . \qquad (12.27)$$

Now we provide a lower bound for the sum on the left-hand side and an upper bound for the sum on the right-hand side by replacing them with integrals.

$$\sum_{i=\widetilde{r}+1}^{r} i^{d-1-\epsilon} \geq \int_{\widetilde{r}}^{r} x^{d-1-\epsilon} dx = \left. \frac{x^{d-\epsilon}}{d-\epsilon} \right|_{\widetilde{r}}^{r}$$

$$\sum_{i=\widetilde{r}+1}^{r} i^{d-1+\epsilon} \leq \int_{\widetilde{r}+1}^{r+1} x^{d-1+\epsilon} dx = \left. \frac{x^{d+\epsilon}}{d+\epsilon} \right|_{\widetilde{r}+1}^{r+1} .$$

We now have

$$\log\left(S(\widetilde{r}) + \frac{r^{d-\epsilon} - \widetilde{r}^{d-\epsilon}}{d-\epsilon} \right) \leq \log M(n,r) \leq \log\left(S(\widetilde{r}) + \frac{(r+1)^{d+\epsilon} - (\widetilde{r}+1)^{d+\epsilon}}{d+\epsilon} \right)$$

which implies

$$\log(r^{d-\epsilon}) + \log\left(\frac{S(\widetilde{r})}{r^{d-\epsilon}} + \frac{1}{d-\epsilon}\left(1 - \frac{\widetilde{r}^{d-\epsilon}}{r^{d-\epsilon}} \right) \right) \leq \log M(n,r)$$

$$\leq \log\left((r+1)^{d+\epsilon} \right) + \log\left(\frac{S(\widetilde{r})}{(r+1)^{d+\epsilon}} + \frac{1}{d+\epsilon}\left(1 - \frac{(\widetilde{r}+1)^{d+\epsilon}}{(r+1)^{d+\epsilon}} \right) \right) .$$

Since the arguments of the second logarithm on each side are uniformly bounded for each r, and since

$$\lim_{r \to \infty} \log(r+1)/\log r = 1 ,$$

we can find a value $\hat{r}$, where $\hat{r} > \tilde{r}$, such that whenever $r \geq \hat{r}$ we have

$$d - \epsilon + \frac{\log\left(\frac{S(\tilde{r})}{r^{d-\epsilon}} - \frac{1}{d-\epsilon}\left(1 - \frac{\tilde{r}^{d-\epsilon}}{r^{d-\epsilon}}\right)\right)}{\log r} \geq d - 2\epsilon$$

and

$$(d+\epsilon)\frac{\log(r+1)}{\log r} + \frac{\log\left(\frac{S(\tilde{r})}{(r+1)^{d+\epsilon}} - \frac{1}{d+\epsilon}\left(1 - \frac{(\tilde{r}+1)^{d+\epsilon}}{(r+1)^{d+\epsilon}}\right)\right)}{\log r} \leq d + 2\epsilon .$$

Therefore for $r \geq \hat{r}$ we have

$$\left|\frac{\log M(n,r)}{\log r} - d\right| \leq 2\epsilon$$

which establishes the result. $\square$

The following lemma shows that, under a suitable condition, the convergence of a subsequence of $\log M(n,r)/\log r$ implies convergence of the entire sequence.

Lemma 12.1 ([Nowotny 88], Lemma 4.9) Suppose that for some node $n \in \mathbb{N}$ and for some number $c \in (0,1)$ the subsequence $\{r_m\}$ satisfies the condition $r_m/r_{m+1} \geq c$ for all sufficiently large m. Then

$$\liminf_{m\to\infty} \frac{\log M(n,r_m)}{\log r_m} = \liminf_{r\to\infty} \frac{\log M(n,r)}{\log r} = d_V^i(n) , \qquad (12.28)$$

and similarly for $d_V^s(x)$.

Proof Let $n \in \mathbb{N}$ and $c \in (0,1)$ satisfy the conditions of the lemma. Let r be an arbitrary positive integer, and let m be such that $r_m \leq r \leq r_{m+1}$. Then $\mathbb{N}(n,r_m) \subseteq \mathbb{N}(n,r) \subseteq \mathbb{N}(n,r_{m+1})$. Since $r_m/r_{m+1} \geq c$ then $(r/r_m) \leq (r_{m+1}/r_m) \leq 1/c$ and $(r/r_{m+1}) \geq (r_m/r_{m+1}) \geq c$. Hence

$$\frac{\log M(n,r_m)}{\log r} \leq \frac{\log M(n,r)}{\log r} \leq \frac{\log M(n,r_{m+1})}{\log r}$$

$$\Rightarrow \frac{\log M(n,r_m)}{\log r_m + \log(r/r_m)} \leq \frac{\log M(n,r)}{\log r} \leq \frac{\log M(n,r_{m+1})}{\log r_{m+1} + \log(r/r_{m+1})}$$

$$\Rightarrow \frac{\log M(n,r_m)}{\log r_m + \log(1/c)} \leq \frac{\log M(n,r)}{\log r} \leq \frac{\log M(n,r_{m+1})}{\log r_{m+1} + \log c}$$

$$\Rightarrow \liminf_{r\to\infty} \frac{\log M(n,r)}{\log r} = \liminf_{i\to\infty} \frac{\log M(n,r_m)}{\log r_m} .$$

The same reasoning proves that

$$\limsup_{r\to\infty} \frac{\log M(n,r)}{\log r} = \limsup_{i\to\infty} \frac{M(n,r_m)}{\log r_m} . \qquad \square$$

The following theorem shows that the volume dimension is "pretty much stable under any local changes" [Nowotny 88].

Theorem 12.2 ([Nowotny 88], Lemma 4.10) Let n be a node in $\mathbb{N}$ and let h be a positive number. Then the insertion of arcs between arbitrarily many pairs of nodes (x,y), subject to the constraint that $dist(x,y) \leq h$, does not change $d_V^i(n)$ and $d_V^s(n)$.

Proof Let $\mathbb{G}$ be the original network, with distances $dist(x,y)$ between nodes x and y. Let $\mathbb{G}^+$ be the modified network with arcs between arbitrarily many node pairs (x,y) such that $dist(x,y) \leq h$. Let $dist^+(x,y)$ be the distance in $\mathbb{G}^+$ between nodes x and y. Define $\mathbb{N}^+(n,r) = \{ x \in \mathbb{N} \,|\, dist^+(n,x) \leq r \}$. If for some $x \in \mathbb{N}$ we have $dist(n,x) = \lfloor r/h \rfloor$, then the extra arcs in $\mathbb{G}^+$ imply $dist^+(n,x) \leq \lfloor r/h \rfloor$. Thus

$$\mathbb{N}(n, \lfloor r/h \rfloor) \subseteq \mathbb{N}^+(n, \lfloor r/h \rfloor) . \tag{12.29}$$

On the other hand, suppose that for some $x \in \mathbb{N}$ we have $dist^+(n,x) = \lfloor r/h \rfloor$. Then $dist(n,x) \leq h\lfloor r/h \rfloor$, since the extra arcs in $\mathbb{G}^+$ mean that a single arc in $\mathbb{G}^+$ on the shortest path between nodes n and x corresponds to at most h arcs in $\mathbb{G}$ on the shortest path between n and x. Since $h\lfloor r/h \rfloor \leq r$ we obtain

$$\mathbb{N}^+(n, \lfloor r/h \rfloor) \subseteq \mathbb{N}(n, h\lfloor r/h \rfloor) \subseteq \mathbb{N}(n,r) . \tag{12.30}$$

From (12.29) and (12.30) we have

$$\mathbb{N}(n, \lfloor r/h \rfloor) \subseteq \mathbb{N}^+(n, \lfloor r/h \rfloor) \subseteq \mathbb{N}(n,r) .$$

Defining $M^+(n,r) \equiv |\mathbb{N}^+(n,r)|$, it follows that

$$\frac{\log M(n, \lfloor r/h \rfloor)}{\log \lfloor r/h \rfloor} \leq \frac{\log M^+(n, \lfloor r/h \rfloor)}{\log \lfloor r/h \rfloor} \leq \frac{\log M(n,r)}{\log \lfloor r/h \rfloor} .$$

For all sufficiently large r we have $\lfloor r/h \rfloor > r/2h$ and thus

$$\frac{\log M(n, \lfloor r/h \rfloor)}{\log \lfloor r/h \rfloor} \leq \frac{\log M^+(n, \lfloor r/h \rfloor)}{\log \lfloor r/h \rfloor} \leq \frac{\log M(n,r)}{\log r - \log(2h)}$$

$$\Rightarrow \liminf_{r \to \infty} \frac{\log M^+(n,r)}{\log r} = \liminf_{r \to \infty} \frac{\log M(n,r)}{\log r} ,$$

where the final equality follows from Lemma 12.1. The proof for $\limsup$ is the same. $\square$

Exercise 12.3 How can we reconcile Theorem 12.2 with the Watts–Strogatz results (Sect. 2.2) on the impact of adding some shortcut arcs to a network? $\square$

12.4   Conical Graphs with Prescribed Dimension

Let d be an arbitrary number such that $1 < d \leq 2$. A method to construct a conical graph such that $d_V(n^\star) = d$, where $n^\star$ is a specially chosen node, was provided in [Nowotny 88]. Graphs with higher volume dimension can be

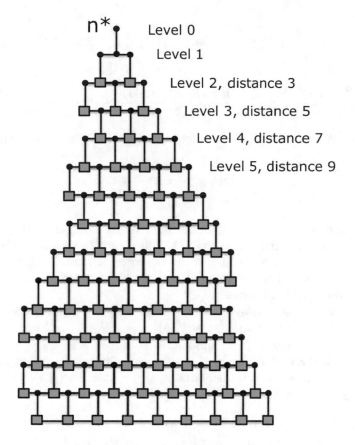

Figure 12.7: Construction of a conical graph for which $d_V(n^\star) = 5/3$

constructed using a nearly identical method. The construction is illustrated in Fig. 12.7 for the choice $d = 5/3$. There are two types of nodes in this figure, small black circles and larger blue squares with a black border. Both types of nodes are nodes in $\mathbb{G}$, but we distinguish these two types to facilitate calculating $d_V(n^\star)$. From the figure we see that boxes connect level m to level $m+1$, where the lower left and lower right corner of each box is a blue square, and the upper left and upper right corner of each box is a black circle. (These boxes have nothing to do with the boxes used to compute the box counting dimension.) For $m \geq 2$, there are $\lfloor (2m-1)^{d-1} \rfloor$ boxes connecting level m to level $m+1$, so the number of blue squares at level m is $1 + \lfloor (2m-1)^{d-1} \rfloor$. Each blue square at level m has distance $2m-1$ to node $n^\star$. Thus, for $m \geq 2$, the set of blue squares at level m is the set $\partial \mathbb{N}(n^\star, 2m-1)$, and

$$|\partial \mathbb{N}(n^\star, 2m-1)| = 1 + \lfloor (2m-1)^{d-1} \rfloor.$$

To compute $d_V(n^\star)$, we first determine $d_U(n^\star)$, the surface dimension at $n^\star$.

Since

$$(2m-1)^{d-1} \le 1 + \lfloor (2m-1)^{d-1} \rfloor \le 1 + (2m-1)^{d-1}$$

then

$$\lim_{m\to\infty} \frac{\log (2m-1)^{d-1}}{\log (2m-1)} \le \lim_{m\to\infty} \frac{\log |\partial \mathbb{N}(n^\star, 2m-1)|}{\log (2m-1)} \le \lim_{m\to\infty} \frac{\log \left(1+(2m-1)^{d-1}\right)}{\log (2m-1)} \, .$$

Since the first and third limits are both equal to $d-1$, then so is the second limit. That is,

$$\lim_{m\to\infty} \frac{\log |\partial \mathbb{N}(n^\star, 2m-1)|}{\log (2m-1)} = d-1 \, . \tag{12.31}$$

At this point we have shown that the limit exists along the subsequence of distances $\{r_m\}_{m=1}^\infty$, where $r_m \equiv 2m-1$. Since convergence of a subsequence does not imply convergence of the entire sequence, it must be shown that the limit exists for all sufficiently large distances, not just along a subsequence of distances. We require a lemma.

Lemma 12.2 ([Nowotny 88], Lemma 4.8) If for some $x, y \in \mathbb{N}$ we have $dist(x,y) < \infty$, then $d_V^i(x) = d_V^i(y)$ and $d_V^s(x) = d_V^s(y)$.

Proof Let $h = dist(x,y)$. If for some node z we have $dist(y,z) \le r-h$, then $dist(x,z) \le dist(x,y) + dist(y,z) \le h + (r-h) = r$, so $\mathbb{N}(y, r-h) \subseteq \mathbb{N}(x,r)$. Similarly, if for some node z we have $dist(x,z) \le r$, then $dist(y,z) \le dist(y,x) + dist(x,z) \le h+r$, so $\mathbb{N}(x,r) \subseteq \mathbb{N}(y, r+h)$. Hence

$$\mathbb{N}(y, r-h) \subseteq \mathbb{N}(x,r) \subseteq \mathbb{N}(y, r+h)$$

$$\Rightarrow \frac{\log M(y, r-h)}{\log r} \le \frac{\log M(x,r)}{\log r} \le \frac{\log M(y, r+h)}{\log r}$$

$$\Rightarrow \frac{\log M(y, r-h)}{\log(r-h) + \log\left(\frac{r}{r-h}\right)} \le \frac{\log M(x,r)}{\log r} \le \frac{\log M(y, r+h)}{\log(r+h) - \log\left(\frac{r+h}{r}\right)}$$

$$\Rightarrow \liminf_{r\to\infty} \frac{\log M(y,r)}{\log r} \le \liminf_{r\to\infty} \frac{\log M(x,r)}{\log r} \le \liminf_{r\to\infty} \frac{\log M(y,r)}{\log r} \, ,$$

so $d_V^i(x) = d_V^i(y)$. Similarly, we obtain $d_V^s(x) = d_V^s(y)$. $\square$

Theorem 12.3 ([Nowotny 88], Section 5.1) For the above conical graph construction, we have $d_V = d$.

Proof For the subsequence $r_m = 2m-1$ the conditions of Lemma 12.1 are satisfied with $c = 1/2$, since $r_m/r_{m+1} = (2m-1)/(2m+1) \ge 1/2$ for $m \ge 2$. By (12.31) and Lemma 12.1 we have

$$\lim_{r\to\infty} \frac{\log |\partial \mathbb{N}(n^\star, r)|}{\log r} = \lim_{m\to\infty} \frac{\log |\partial \mathbb{N}(n^\star, 2m-1)|}{\log (2m-1)} = d-1 \, , \tag{12.32}$$

so by Definition 12.4 we have $d_U(n^\star) = d-1$. Since $d > 1$ it follows from Theorem 12.1 that $d_V(n^\star) = d$. Since the distance between each pair of nodes in the graph is finite, it follows from Lemma 12.2 that for each node n we have $d_V(n) = d$. Thus $d_V = d$. $\quad\square$

Exercise 12.4 Draw the conical graph for $d = 1.01$ and for $d = 1.99$. $\quad\square$

Connections between the above results and some results by Gromov on *geometric group theory* were explored in [Requardt 06]. The application of these techniques provides preliminary answers to such questions as characterizing the set of graphs having the same dimension.

12.5 Dimension of a Percolating Network

A *percolating network* (also called a *percolation network*) is a fundamental model for describing geometrical features of random systems. There are two main types of percolating networks: *site* and *bond*. Consider an infinite lattice in $\mathbb{R}^E$. The lattice need not be rectilinear; other possible geometries are triangle, honeycomb, and diamond. In a site-percolating network (SPN), each lattice point (called a *site*) is occupied at random with probability p. A site is not a node; a site is a position on a lattice, which may be occupied (populated) or empty. Two sites are connected by a *bond* (i.e., an arc) if they are adjacent along a principal direction (e.g., for a rectilinear lattice in $\mathbb{R}^3$, if they are adjacent along the x, y, or z axis.). In a bond-percolating network (BPN), all lattice sites are populated, and bonds between adjacent sites are occupied randomly with probability p. The occupation probability p for sites in a SPN, or for bonds in a BPN, has the same role as the temperature in thermal critical phenomena [Nakayama 94].

As p increases from 0, connected clusters of sites form. As p continues to increase, a critical p_c value is reached. For $p < p_c$, only finite clusters (i.e., clusters of finite size) exist. For $p > p_c$, an infinite cluster (i.e., a cluster of infinite size) is present, as well as finite clusters. The critical value p_c for the BPN is always smaller than the critical value for the SPN, due to the difference in the short range geometrical structure of the two types of lattices: a bond has more nearest neighbors than a site. For example, for a 2-dimensional rectilinear lattice, a given site has 4 adjacent sites, while a given bond is attached to 6 other bonds (3 at each end).

A 1976 paper of de Gennes was credited by Nakayama et al. [Nakayama 94] with sparking interest in the dynamics of a percolating network. That paper posed the following question: an ant parachutes down onto a site on the percolating network and executes a random walk. What is the mean square distance the ant traverses as a function of time?[5] Stanley, in 1977, was the first to recognize that percolating networks exhibit self-similarity and could be characterized by a fractal dimension. Percolating networks have been used to describe a large variety of physical and chemical phenomena, such as gelation processes and conduction in semiconductors. Percolation theory also forms the basis for studies of the flow of liquids or gases through porous media, and has been applied to

---

[5]Let $dist(t)$ be the distance at time t from a random walker to a specified site. The root mean square distance at time t is the square root of the expected value of the square of $dist(t)$.

the stability of computer networks under random failures and under denial of service attacks, to the spread of viruses in computer networks and epidemics in populations, and to the immunization of these networks [Cohen 04].

Let m denote the size of a cluster, i.e., the number of occupied sites in the cluster. (Here "m" suggests "mass".) Although m depends on p, for brevity we write m rather than $m(p)$. Let s_m be the average Euclidean distance between two sites in a cluster of size m. At $p = p_c$ we have, for some d_{pn} (where "pn" denotes "percolating network"),

$$m \propto s_m^{d_{pn}}.$$

The exponent d_{pn} is the "fractal dimension" or "Hausdorff dimension" of the percolating network [Nakayama 94]. For a percolating network in $\mathbb{R}^E$, for $E = 2$ we have $d_{pn} = 91/48$, for $E = 4$ we have $d_{pn} \approx 3.2$, for $E = 5$ we have $d_{pn} \approx 3.7$, and for $E \geq 6$ we have $d_{pn} = 4$. For $E = 3$, the computed values of d_{pn}, as reported in [Nakayama 94], span the range 0.875–2.48.

The chemical distance between two sites u and v in a percolating network is the minimal number of hops between u and v, subject to the constraint that each hop must be an existing bond between connected sites. The *chemical dimension* of a percolating network is d_{ch} if the mass m within chemical distance s_c obeys the scaling

$$m \propto s_c^{d_{ch}}.$$

For 2-dimensional percolating networks, $d_{ch} \approx 1.768$; for 3-dimensional percolating networks, $d_{ch} \approx 1.885$ [Nakayama 94]. Finally, let $chem(L)$ denote the chemical distance between two sites that are separated by the Euclidean distance L. Then

$$chem(L) \propto L^{d_{pn}/d_{ch}},$$

so d_{pn}/d_{ch} is the fractal dimension that relates the chemical distance between two sites in a lattice to the Euclidean distance between two sites [Nakayama 94]. The fractal dimensions of percolating networks were also studied in [Cohen 04, Cohen 08].

Chapter 13

Similarity Dimension of Infinite Networks

Every living being is an engine geared to the wheelwork of the universe. Though seemingly affected only by its immediate surroundings, the sphere of external influence extends to infinite distance.
Nikola Tesla (1856–1943), Serbian-American engineer and physicist

In this chapter we consider self-similar infinite networks and their fractal dimensions. Since self-similarity is such a powerful way to describe a geometric fractal and compute a fractal dimension, extending this theory to networks is very useful. In this chapter we present the method of [Nowotny 88] that uses a self-similar geometric object Ω to create an infinite unweighted graph $\mathbb{G}$ such that Ω and $\mathbb{G}$ have the same fractal dimension.

As we saw in Sect. 5.2, a self-similar geometric object can be constructed by an IFS (Sect. 3.3) which begins with a geometric object Ω_0 and generates Ω_{t+1} by taking K instances of Ω_t, where each copy is scaled by the factor $\alpha < 1$. Clearly, we cannot make K copies of an unweighted network $\mathbb{G}_t$, scale each copy by a factor $\alpha < 1$, and still have an unweighted network. Instead, in [Nowotny 88] the unweighted network $\mathbb{G}_{t+1}$ is generated by making K copies of the unweighted network $\mathbb{G}_t$, without scaling these K copies of $\mathbb{G}_t$, and then interconnecting these K copies of $\mathbb{G}_t$.[1] There are in general many ways to interconnect the K copies of $\mathbb{G}_t$. For example, if we make three copies of the network illustrated in Fig. 13.1(i), two ways of interconnecting the 3 copies are illustrated in Fig. 13.1(ii).

Rather than connecting the K copies in an arbitrary manner, two equivalent methods were described in [Nowotny 88] to construct an infinite sequence $\{\mathbb{G}_t\}$ of self-similar unweighted networks, starting with a given finite network $\mathbb{G}_0$. The network $\mathbb{G}_0$ is called the *network generator*. Let N_0 be the number of nodes in

---

[1]We refer to this as "interconnecting K copies of $\mathbb{G}_t$". Strictly speaking, we make $K - 1$ copies and use the $K - 1$ copies together with the original instance of $\mathbb{G}_t$, but it is convenient to refer to K copies, where the original version of $\mathbb{G}_t$ counts as one of the K copies.

© Springer Nature Switzerland AG 2020
E. Rosenberg, *Fractal Dimensions of Networks*,
https://doi.org/10.1007/978-3-030-43169-3_13

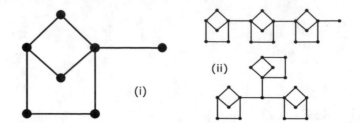

Figure 13.1: A network (i) and two ways to interconnect K copies (ii)

$\mathbb{G}_0$, and let A_0 be the number of arcs in $\mathbb{G}_0$. Assume that a subset of the nodes of $\mathbb{G}_0$ will be used to interconnect two instances of $\mathbb{G}_t$. This subset is called the set of *border nodes* of $\mathbb{G}_0$. If arc a interconnects two instances of $\mathbb{G}_t$, then the endpoints of a must be border nodes of some instance of $\mathbb{G}_0$. For example, if $\mathbb{G}_0$ is a $K \times K$ rectilinear grid as in Fig. 11.6, we might let the set of border nodes be the set of nodes on the perimeter of the $K \times K$ grid, so that two instances of a $K \times K$ grid are connected by an arc connecting a node on the perimeter of one instance to a node on the perimeter of the other instance. Another example of $\mathbb{G}_0$ is illustrated by Fig. 13.2; the four hollow nodes at the corners are the border nodes. Two equivalent ways of generating $\{\mathbb{G}_t\}$ were described

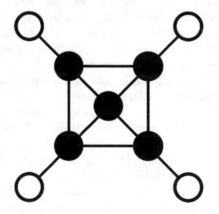

Figure 13.2: $\mathbb{G}_0$ has four border nodes

in [Nowotny 88], where "equivalent" means that the two methods generate the same infinite sequence $\{\mathbb{G}_t\}$, except for the ambiguity mentioned below.

Method 1 ("Construction by Insertion"): Initialize $\mathbb{G}_1 = \mathbb{G}_0$. For $t = 1, 2, \ldots,$ construct $\mathbb{G}_{t+1}$ from $\mathbb{G}_t$ as follows:

1. Replace each node in $\mathbb{G}_t$ by one instance of $\mathbb{G}_0$.

2. For each arc $a = (x, y)$ in $\mathbb{G}_t$, replace a by an arc which connects a border node in the instance of $\mathbb{G}_0$ which replaced node x to a border node in the instance of $\mathbb{G}_0$ which replaced node y.

Method 2 ("Copy and Paste"): Initialize $\mathbb{G}_1 = \mathbb{G}_0$. For $t = 1, 2, \ldots$, construct $\mathbb{G}_{t+1}$ from $\mathbb{G}_t$ as follows:

1. Make N_0 copies of $\mathbb{G}_t$.

2. Create A_0 inter-cluster arcs, as follows: (i) First, arbitrarily associate each node in $\mathbb{G}_0$ with exactly one of the N_0 instances of $\mathbb{G}_t$; let $g(n)$ be the instance of $\mathbb{G}_t$ associated with $n \in \mathbb{G}_0$. (ii) For each arc a in $\mathbb{G}_0$, let u and v be the endpoints of a, and let $g(u)$ and $g(v)$ be the corresponding instances of $\mathbb{G}_t$. Interconnect $g(u)$ and $g(v)$ by creating an arc between some border node $\tilde{u}$ of $g(u)$ and some border node $\tilde{v}$ of $g(v)$; this newly created inter-cluster arc $(\tilde{u}, \tilde{v})$ is an arc in $\mathbb{G}_{t+1}$.

Example 13.1 Consider Fig. 13.3(i), which shows $\mathbb{G}_0$. The four border nodes of $\mathbb{G}_0$ are the hollow nodes (numbered 1–4) on the perimeter; the filled-in node (numbered 5) in the center of $\mathbb{G}_0$ is not a border node. We will construct $\{\mathbb{G}_t\}$

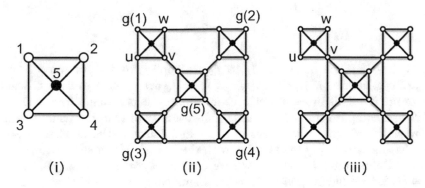

Figure 13.3: Constructing $\mathbb{G}_2$ from $\mathbb{G}_1$

using *Method 2 ("copy and paste")*. Since $N_0 = 5$, we make 5 copies of $\mathbb{G}_0$. Since $\mathbb{G}_0$ has 8 arcs, we create 8 inter-cluster arcs, where each inter-cluster arc interconnects the corresponding two instances of $\mathbb{G}_1$. The inter-cluster arcs must interconnect the instances of $\mathbb{G}_0$ in the same manner as the nodes of $\mathbb{G}_0$ are interconnected. Figure 13.3(ii,iii) show two ways of doing this. In (ii), the interconnections from the cluster instance $g(1)$ on the upper left emanate from different border nodes: an arc from border node u goes to cluster instance $g(3)$, an arc from border node v goes to cluster instance $g(5)$, and an arc from w goes to cluster instance $g(2)$. Alternatively, in (iii), all the interconnections emanate from the single border node v (in practice, the drawback of this alternative is that the failure of node v disconnects $\mathbb{G}_1$). These are just two examples, and other schemes could be used. The end result of Step 2 is a network $\mathbb{G}_1$ with N_0^2 nodes. This construction continues with $t = 2, 3, 4, \cdots$. $\square$

So far we have considered only the construction of $\{\mathbb{G}_t\}$. In the next section we show how a fractal dimension of $\{\mathbb{G}_t\}$ can be determined by relating this construction to the box counting dimension of a geometric object constructed by an IFS.

13.1 Generating Networks from an IFS

A method to generate an infinite sequence of *unweighted* networks from an IFS was described in [Nowotny 88]. The method is illustrated using the Maltese cross fractal of Fig. 13.4. Starting with the unit square (not pictured) whose corner points are $(0,0)$, $(0,1)$, $(1,1)$, and $(1,0)$, we generate subfigure (i) by taking 4 instances, each $1/3$ the size of the original. Letting $x = (x_1, x_2)$, the five squares in (i) are generated by the mappings

$$f_1 : x \to \frac{1}{3}x + (0,0)$$

$$f_2 : x \to \frac{1}{3}x + (0, \frac{2}{3})$$

$$f_3 : x \to \frac{1}{3}x + (\frac{1}{3}, \frac{1}{3}) \qquad (13.1)$$

$$f_4 : x \to \frac{1}{3}x + (\frac{2}{3}, 0)$$

$$f_5 : x \to \frac{1}{3}x + (\frac{2}{3}, \frac{2}{3}).$$

To generate (ii), the mappings f_1 to f_5 are applied to each of the 5 squares in (i). To generate (iii), the mappings f_1 to f_5 are applied to each of the 25 squares in (ii). This process continues indefinitely, generating a fractal Ω.

We can compute the box counting dimension d_B of Ω in the usual manner, by covering the figure with boxes of size s and counting the number of boxes which have a non-empty intersection with the fractal. Using subfigure (i), and recalling Definition 7.4, for $s = 1/3$ we have $B_D(s) = 5$. Using (ii), for $s = 1/9 = (1/3)^2$ we have $B_D(s) = 25 = 5^2$. Using (iii), for $s = 1/27 = (1/3)^3$ we

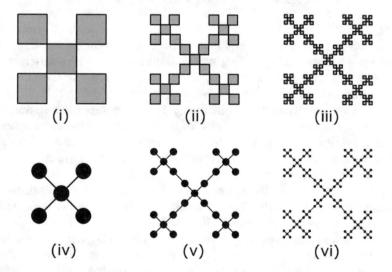

(i)   (ii)   (iii)

(iv)   (v)   (vi)

Figure 13.4: Generating a network from the Maltese cross fractal

have $B_D(s) = 125 = 5^3$. In general, for $s = (1/3)^i$ we have $B_D(s) = 5^i$. Hence $d_B = -\log B_D(s)/\log s = \log 5/\log 3$.

To create a sequence $\{\mathbb{G}_t\}$ of unweighted networks from these geometric pre-fractals, we replace each square by a node, and connect two nodes by an undirected arc if the corresponding squares share an edge or share a corner point. Subfigure (iv) shows the network corresponding to (i), (v) shows the network corresponding to (ii), and (vi) shows the network corresponding to (iii). The sequence $\{\mathbb{G}_t\}$ thus obtained is identical to the sequence obtained by defining the network generator $\mathbb{G}_0$ to be the network of subfigure (iv), defining the border nodes of $\mathbb{G}_0$ to be the four leaf nodes, and applying *Method 1 ("construction by insertion")* or *Method 2 ("copy and paste")* described in the previous section. If we continue this process indefinitely, we obtain an infinite unweighted network, which we denote by $\mathbb{G}_\infty$.

Following [Nowotny 88], we now show that the mass dimension d_M (Definition 12.1) of $\mathbb{G}_\infty$ and the local volume dimension d_V of $\mathbb{G}_\infty$ at the center node (Definition 12.3) are both equal to the box counting dimension of the fractal generated by the IFS defined by f_1, f_2, f_3, f_4, f_5. Let $n^\star$ be the center node in the networks generated by this construction. After the first step of the construction, for the network (iv) all 5 nodes are reachable from $n^\star$ in at most 1 hop, so $M(n^\star, 1) = 5$, where $M(n, r)$ is defined by (12.19) and (12.20). After the second step, for the network (v) all 25 nodes are reachable $n^\star$ in at most 4 hops, so $M(n^\star, 4) = 25$. After the third step, for the network (vi) all 125 nodes are reachable in at most 13 hops, so $M(n^\star, 13) = 125$. Let R_t be the maximal distance from $n^\star$ to any node in the network created after the t-th step. Then $R_1 = 1$ and for $t \geq 2$ we have $R_t = R_{t-1} + 1 + 2R_{t-1} = 1 + 3R_{t-1}$, since it takes R_{t-1} hops from $n^\star$ to reach the boundary of the cluster of the network $\mathbb{G}_{t-1}$ created in step $t-1$, one hop to reach the adjacent copy of $\mathbb{G}_{t-1}$, and $2R_{t-1}$ hops to reach the furthest node in this adjacent copy. This yields $R_t = \sum_{j=0}^{t-1} 3^j = (3^t - 1)/2$. Also, $M(n^\star, R_t) = 5^t$. Hence, for the subsequence R_t for $t = 1, 2, \ldots$, we have

$$\lim_{t \to \infty} \frac{M(n^\star, R_t)}{R_t} = \lim_{t \to \infty} \frac{\log 5^t}{\log[(3^t - 1)/2]} = \frac{\log 5}{\log 3}.$$

For $t \geq 2$ we have

$$\frac{R_t}{R_{t+1}} = \frac{3^t - 1}{3^{t+1} - 1} \geq \frac{3^t - 1}{3^{t+1}} = \frac{1}{3} - \frac{1}{3^{t+1}} \geq \frac{1}{3} - \frac{1}{9}.$$

By Lemma 12.1, the entire sequence converges, so by Definition 12.3 of the local volume dimension we have $d_V(n^\star) = \log 5/\log 3$. Since the number N_t of nodes in $\mathbb{G}_t$ is 5^t, and since the diameter Δ_t of $\mathbb{G}_t$ is $2R_t$, by Definition 12.1 the mass dimension d_M of $\mathbb{G}_\infty$ is $\log 5/\log 3$.

13.2 Analysis of the Infinite Sequence

In Sect. 13.1 we generated the sequence $\{\mathbb{G}_t\}$ of unweighted networks corresponding to an IFS generated by a Maltese cross. In this section we generalize the results of the previous section by presenting the construction of [Nowotny 88] that generates an infinite sequence $\{\mathbb{G}_t\}$ of unweighted self-similar networks corresponding to any IFS satisfying certain conditions. The goal is to relate a fractal dimension of $\mathbb{G}_\infty$ to the similarity dimension or box counting dimension of the geometric object generated by the IFS.

Just as in the Maltese cross example the network generator of Fig. 13.4(iv) was obtained from the IFS generator of Fig. 13.4(i), for the general construction we assume that the network generator $\mathbb{G}_0$ is obtained from an IFS generator Ω_0, and that the IFS satisfies the Open Set Condition (Definition 5.4). Let N_0 be the number of nodes in $\mathbb{G}_0$. It is convenient to analyze the sequence $\{\mathbb{G}_t\}$ assuming it was generated by *Method 2 ("copy and paste")*. We require some conditions on $\mathbb{G}_0$.

Assumption 13.1 One node of $\mathbb{G}_0$, denoted by $n^\star$, is a *center node*. □

Assumption 13.2 For some integer R_0 we have $dist(n^\star, b) = R_0$ for each border node b of $\mathbb{G}_0$. □

Assumption 13.2 says that the distance from the center node $n^\star$ to each border node of $\mathbb{G}_0$ is the same. For example, in Fig. 13.2 we have $R_0 = 2$, while for Fig. 13.3(i) we have $R_0 = 1$.

Assumption 13.3 The distance between each pair of border nodes of $\mathbb{G}_0$ is $2R_0$, and the diameter $diam(\mathbb{G}_0)$ of $\mathbb{G}_0$ is $2R_0$. □

It is possible for a network to satisfy Assumption 13.2 and violate Assumption 13.3, as illustrated by Fig. 13.5, where the distance from $n^\star$ to each of the

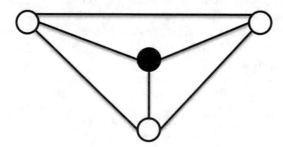

Figure 13.5: A network satisfying Assumption 13.2 but not Assumption 13.3

3 border nodes is 1, and the distance between each pair of border nodes is 1. Define $R_t \equiv \max\{dist(n^\star, n) \mid n \in \mathbb{G}_t\}$, so R_t is the maximal distance from $n^\star$ to any node in $\mathbb{G}_t$.

Lemma 13.1 [Nowotny 88] If Assumptions 13.1–13.3 hold then for $t = 1, 2, \ldots$,

$$R_t = (1/2)[(2R_0 + 1)^t - 1] \tag{13.2}$$

and the diameter $diam(\mathbb{G}_t)$ of $\mathbb{G}_t$ is $2R_t$.

Proof We prove the lemma by induction. Setting $t = 1$, since *Method 2 ("copy and paste")* initializes $\mathbb{G}_1 = \mathbb{G}_0$ then $R_1 = R_0$, so (13.2) holds by Assumption 13.2, and by Assumption 13.3 we have $diam(\mathbb{G}_1) = 2R_1 = 2R_0$. Assume now that the lemma holds for $\mathbb{G}_2$, $\mathbb{G}_3$, ..., $\mathbb{G}_{t-1}$, and consider $\mathbb{G}_t$. Let x be a node in $\mathbb{G}_t$ such that $dist(n^\star, x) = R_t$, and let $\mathcal{P}$ be a shortest path from $n^\star$ to x, so the length of $\mathcal{P}$ is R_t. Starting at $n^\star$, path $\mathcal{P}$ must first go to a border node b of the instance of $\mathbb{G}_{t-1}$ containing $n^\star$; by the induction hypothesis, this segment has length R_{t-1}. From b, path $\mathcal{P}$ must traverse R_0 arcs between instances of $\mathbb{G}_{t-1}$, and $\mathcal{P}$ must traverse R_0 instances of $\mathbb{G}_{t-1}$. The number of hops required to traverse each instance of $\mathbb{G}_{t-1}$ is $diam(\mathbb{G}_{t-1})$, and by the induction hypothesis $diam(\mathbb{G}_{t-1}) = 2R_{t-1}$. Applying (13.2) to G_{t-1},

$$\begin{aligned} R_t &= R_{t-1} + R_0 + (R_0)(2R_{t-1}) \\ &= R_{t-1}(2R_0 + 1) + R_0 \\ &= (1/2)[(2R_0 + 1)^{t-1} - 1](2R_0 + 1) + R_0 \\ &= (1/2)(2R_0 + 1)^t - (1/2)(2R_0 + 1) + R_0 \\ &= (1/2)[(2R_0 + 1)^t - 1] . \quad \square \end{aligned}$$

By Lemma 13.1, since $diam(\mathbb{G}_t) = 2R_t = (2R_0 + 1)^t - 1$, then the diameter increases by about $2R_0 + 1$ in each generation.

Example 13.2 Figure 13.6 illustrates, for $t = 1$, Eq. (13.2) for the very simple case where $\mathbb{G}_0$ is a chain of five nodes and $R_0 = 2$. The hollow nodes, e.g., the

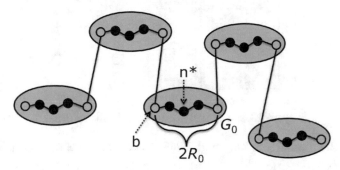

Figure 13.6: $\mathbb{G}_1$ and $\mathbb{G}_2$ for a chain network

node labelled b, are the border nodes of $\mathbb{G}_0$. Since $\mathbb{G}_0$ has 5 nodes and 4 arcs, and since $\mathbb{G}_1 = \mathbb{G}_0$, then $\mathbb{G}_2$ has 5 instances of $\mathbb{G}_1$ and 4 inter-cluster arcs. $\quad \square$

Example 13.3 Figure 13.7 provides a slightly more complicated example where (*i*) illustrates $\mathbb{G}_1$ and (*ii*) illustrates $\mathbb{G}_2$. $\quad \square$

Returning to the general theory, we now consider the correspondence between $\mathbb{G}_t$ and the t-th generation of an IFS that operates on the unit square in $\mathbb{R}^2$. Just as in *Method 1 ("construction by insertion")* and *Method 2 ("copy and paste")* we initialized $\mathbb{G}_1 = \mathbb{G}_0$, assume that for the IFS we initialize $\Omega_1 = \Omega_0$. Consider the Maltese cross, defined by the similarities (13.1). Figure 13.4 (*i*), (*ii*), and (*iii*) illustrate Ω_1, Ω_2, and Ω_3, respectively. To construct a network generator $\mathbb{G}_0$ from Ω_0, we must specify the border nodes. For the network generator $\mathbb{G}_0$ illustrated in Fig. 13.4(*iv*), let each of the four nodes attached to the

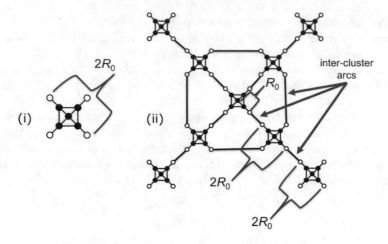

Figure 13.7: Another example of $\mathbb{G}_0$ and $\mathbb{G}_1$

center node be a border node. Figure 13.8(i) illustrates Ω_0, where the hollow boxes correspond to the border nodes of $\mathbb{G}_0$, and (ii) illustrates $\mathbb{G}_0$. We have

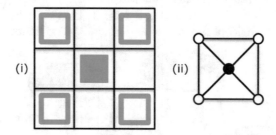

Figure 13.8: Ω_0 and $\mathbb{G}_0$ for the Maltese cross

$R_0 = 1$, so the diameter is $2R_0$, and $2R_0 + 1 = 3$. The IFS contraction factor is

$$\alpha = \frac{1}{2R_0 + 1} = \frac{1}{3}.$$

Referring back to Fig. 13.4, since each square in (i), (ii), and (iii) touches some other square, then the networks in (iv), (v), and (vi) are connected; more generally, each G_t is connected. Since $diam(\mathbb{G}_t) = 2R_t = (2R_0 + 1)^t - 1$ and $R_0 = 1$, the diameter increases by about $2R_0 + 1 = 3$ in each generation.

The IFS generator Ω_0 must be defined carefully. For example, considering the $\mathbb{G}_0$ given in Fig. 13.4(iv), suppose we want to specify an IFS generator Ω_0 which yields $\mathbb{G}_0$. Figure 13.9 illustrates two invalid choices of Ω_0. Figure 13.9(i) is invalid since it yields a disconnected $\mathbb{G}_0$ (only boxes of Ω_0 meeting at an edge or corner generate an arc of $\mathbb{G}_0$). Figure 13.9(ii) is invalid since there are no small squares located along the perimeter of the figure; recalling that $\Omega_1 = \Omega_0$, none of the instances of Ω_1 will share a common edge or corner when Ω_2 is constructed, so the corresponding graph $\mathbb{G}_2$ will be disconnected.

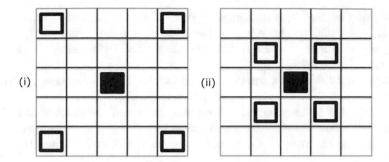

Figure 13.9: Two invalid choices of Ω_0

Example 13.4 For another example of the representation of $\mathbb{G}_0$ by a geometric set Ω_0, consider Fig. 13.10. Define an IFS by letting Ω_0 be as shown in

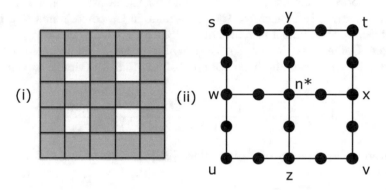

Figure 13.10: Example for which $2R_0 + 1 \neq 1/\alpha$

Fig. 13.10(i). In the IFS terminology of Sect. 3.3, we have $K = 21$ since there are 21 small boxes in (i), and the contraction ratio is $\alpha = 1/5$, so the geometric fractal Ω produced by the IFS has similarity dimension $d_S = \log 21/\log 5$.

To construct a network generator $\mathbb{G}_0$ from Ω_0, we must specify the border nodes. Consider the network generator $\mathbb{G}_0$ illustrated by Fig. 13.10(ii), where each node corresponds to a box in (i). Since the distance from $n^\star$ to s, t, u, and v is 4, and the distance from $n^\star$ to w, x, y, and z is 2, by Assumption 13.2 all eight of these nodes cannot be border nodes. Suppose we choose only s, t, u, and v to be the border nodes of $\mathbb{G}_0$. Then Assumptions 13.2 and 13.3 are satisfied with $R_0 = 4$. From (13.2) the diameter of $\mathbb{G}_t$ increases by approximately $2R_0 + 1 = 9$ in each generation. Since

$$\frac{1}{\alpha} = 5 \neq 9 = 2R_0 + 1\,,$$

the pre-fractal sequence $\{\Omega_t\}$ and the network sequence $\{G_t\}$ scale by different factors. $\quad\square$

Returning to the general theory, to prove that the similarity dimension d_S of the IFS is equal to the volume dimension d_V of the infinite graph $\mathbb{G}_\infty \equiv \lim_{t\to\infty} \mathbb{G}_t$, examples like Fig. 13.10 are eliminated by making the following assumption [Nowotny 88].

Assumption 13.4 The contraction ratio α for the IFS satisfies $\alpha = 1/(2R_0+1)$. □

For example, Assumption 13.4 is satisfied for the Maltese cross of Fig. 13.4, since $2R_0+1 = 3$ and the IFS has contraction factor $\alpha = 1/3$. Assumption 13.4 is not satisfied for Example 13.4 with border nodes s, t, u, and v, since $2R_0+1 = 9$ but $1/\alpha = 5$.

We can now prove, under appropriate conditions, the main result of this section: the equality of the volume dimension d_V of $\mathbb{G}_\infty$ and the similarity dimension of the geometric object Ω generated by the IFS.[2]

Theorem 13.1 [Nowotny 88] If Assumptions 13.1–13.4 hold, then $d_S(\Omega) = d_V(\mathbb{G}_\infty)$.

Proof Suppose Assumptions 13.1–13.4 hold for the network generator $\mathbb{G}_0$ and the corresponding IFS generator Ω_0. Each node in $\mathbb{G}_0$ corresponds to a box in Ω_0. Let N_0 be the number of nodes in $\mathbb{G}_0$. Since $\mathbb{G}_0$ has N_0 nodes, then Ω_0 has N_0 boxes. Define $\beta \equiv 2R_0 + 1$, so $\beta = 1/\alpha$. Since the IFS is assumed to satisfy the Open Set Condition, by Assumption 13.4 and (5.8) the similarity dimension of Ω is

$$d_S(\Omega) = \frac{-\log N_0}{\log\big(1/(2R_0+1)\big)} = \frac{\log N_0}{\log(2R_0+1)} = \frac{\log N_0}{\log \beta} \ .$$

Consider the sequence of radius values $\{R_t\}_{t=1}^\infty$ defined by (13.2). For $t \geq 1$ we have

$$\frac{R_t}{R_{t+1}} = \frac{\beta^{t+1}-1}{\beta^{t+2}-1} > \frac{\beta^{t+1}-1}{\beta^{t+2}} = \frac{1}{\beta} - \frac{1}{\beta^{t+2}} > \frac{1}{\beta} - \frac{1}{\beta^2} = \alpha(1-\alpha) \ .$$

Since $0 < \alpha(1-\alpha) < 1$ then the constant c required by Lemma 12.1 exists.

By definition, R_t is the maximal distance from the center node $n^\star$ to any node of $\mathbb{G}_t$. The graph $\mathbb{G}_t$ contains $N_t = N_0^t$ nodes. Recalling Definition 12.3, the local volume dimension $\mathbb{G}_\infty(n^\star)$ of $\mathbb{G}$ at $n^\star$ is

---

[2]From Sect. 5.2 we know that for this IFS, which satisfies the open set condition, we have $d_S(\Omega) = d_B(\Omega)$. Although [Nowotny 88] invoked only the box counting dimension, we prefer to invoke to the similarity dimension, since it emphasizes that Ω is generated by an IFS.

$$\lim_{t\to\infty}\frac{\log M\left(n^{\star},R_t\right)}{\log R_t} = \lim_{t\to\infty}\frac{\log N_0^t}{\log R_t} = \lim_{t\to\infty}\frac{\log N_0^t}{\log[(\beta^t-1)/2]}$$

$$= \lim_{t\to\infty}\frac{\log N_0^t}{\log\beta^t+\log[(1-\beta^{-t})/2]}$$

$$= \lim_{t\to\infty}\frac{\log N_0^t}{\log\beta^t} = \frac{\log N_0}{\log\beta} = \frac{\log N_0}{\log(1/\alpha)} = d_S(\Omega)\,.$$

Having established convergence for the subsequence R_t of radius values, by Lemma 12.1 we conclude that for the full sequence we have

$$\lim_{r\to\infty}\frac{\log M(n^{\star},r)}{\log r} = d_S(\Omega)\,,$$

so $\mathbb{G}_\infty$ has local volume dimension $d_S(\Omega)$ at $n^{\star}$; that is, $d_V(n^{\star}) = d_S(\Omega)$. Since $\mathbb{G}_\infty$ is connected, by Lemma 12.2 for $n\in\mathbb{N}$ we have $d_V(n) = d_S(\Omega)$; that is, $\mathbb{G}_\infty$ has volume dimension $d_S(\Omega)$. $\square$

The example of Fig. 13.10, for which $2R_0+1 \neq 1/\alpha$, demonstrates that Theorem 13.1 can fail to hold when Assumption 13.4 fails to hold. On the other hand, [Nowotny 88] observed that we may have $d_S(\Omega) = d_V(\mathbb{G}_\infty)$ even when $\mathbb{G}_0$ has no center node. For example, equality holds when Ω_0 is the generator for the Sierpiński triangle, illustrated in Fig. 3.4. Moreover, [Nowotny 88] observed that it is difficult to give a general proof of the equality $d_S(\Omega) = d_V(\mathbb{G}_\infty)$ for arbitrary self-similar sets. Figure 13.11(i) provides another example of an Ω_0, without a center box, for which $d_S(\Omega) = d_V(\mathbb{G}_\infty)$. The corresponding $\mathbb{G}_0$ is

(i) (ii)

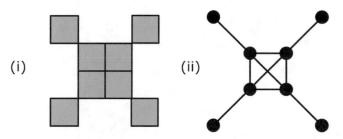

Figure 13.11: $\mathbb{G}_0$ has no center node

shown in Fig. 13.11(ii). Since the 4 inner squares in (i) intersect on some edge, the corresponding 4 inner nodes of $\mathbb{G}_0$ are fully interconnected, as shown in (ii).

Exercise 13.1 Example 13.4 shows that picking s, t, u, and v to be the border nodes of $\mathbb{G}_0$ yields $5 = 1/\alpha \neq 2R_0+1 = 9$. Can we instead pick w, x, y, and z to be the border nodes of $\mathbb{G}_0$? $\square$

Exercise 13.2 Figure 13.12 (left) illustrates a network generator $\mathbb{G}_0$, which has 7 nodes (of which 4 are border nodes), 8 arcs, and $R_0 = 2$. With this $\mathbb{G}_0$, there are many ways to construct $\mathbb{G}_2$ using *Method 2 ("copy and paste")*. One way is illustrated in Fig. 13.12 (right). Can we apply Theorem 13.1 to compute the dimension of $\mathbb{G}_\infty$?

probability. We call the set $\{p_j(s) \mid B_j \in \mathcal{B}(s)\}$ the *coarse-grained probability distribution at resolution s.* Define

$$H(s) \equiv - \sum_{B_j \in \mathcal{B}(s)} p_j(s) \log p_j(s) \, . \tag{14.4}$$

Definition 14.1 The information dimension d_I of the probability distribution is given by

$$d_I \equiv - \lim_{s \to 0} \frac{H(s)}{\log s} \, , \tag{14.5}$$

assuming this limit exists.   □

Since by (14.1) the numerator $H(s)$ is the entropy of the distribution, then this limit, which we call the *information dimension*, is the rate at which the information scales as the precision of measurement is increased [Farmer 82]. Expressed slightly differently, since by (14.5) for small s we have $H(s) \approx d_I \log(1/s)$, we may view d_I as telling how fast the information necessary to specify a point on the attractor increases as s decreases [Farmer 83]. The information dimension d_I is a property of any random variable with a probability measure defined on it [Farmer 82a]. Farmer observed that the information dimension was originally defined by Balatoni and Rényi in 1956, but in a context unrelated to dynamical systems.[4]

In the beginning of this chapter, we motivated the study of d_I by showing that $H(s) = -E \log s$ bits are needed to specify the position of a point in the unit hypercube in $\mathbb{R}^E$ to an accuracy of s, where $s < 1$. This corresponds to the special case of (14.4) for which each of the $(1/s)^E$ boxes of size s in the E-dimensional unit cube is equally likely, with probability $p_j = s^E$. For then we have $|\mathcal{B}(s)| = s^{-E}$ and

$$- \sum_{B_j \in \mathcal{B}(s)} p_j(s) \log p_j(s) = - \sum_{B_j \in \mathcal{B}(s)} s^E \log s^E = - \log s^E = -E \log s \, .$$

Moreover, we can recast (14.5) in a form reminiscent of the definition (1.1) of dimension as the way bulk scales with size [Theiler 90]. Since $H(s)$ is the entropy, the total number of available states is $2^{H(s)}$, and $2^{-H(s)}$ is the average "bulk" of each state when the accuracy is s. With this interpretation, d_I expresses how the logarithm of the average "bulk" of each state scales with $\log s$.

For the study of chaotic dynamical systems, the information dimension is more useful than the box counting dimension, because the information dimension has a direct physical interpretation and is easier to compute. For

---

[4]In [Farmer 82], the information dimension was defined as $\lim_{s \to 0} \frac{-H(s)}{s}$, where s rather than $\log s$ is the denominator. This does not appear to be a misprint, since [Farmer 82] then stated that if d_I is known, the entropy in a measurement of resolution s is approximately $d_I s$. In another 1982 paper [Farmer 82a], d_I was defined using (14.5), which uses $\log s$. These are very early papers in the history of the information dimension, and thus appear to reflect the evolution of the definition of d_I. Several references providing probabilistic definitions of dimension were supplied in [Farmer 82], which offered the rule of thumb that definitions involving probability are usually equivalent to the information dimension. The definition (14.5) is the same definition of d_I used in [Theiler 90]; it is now the standard definition, and the one we adopt.

a chaotic dynamical system, most information about initial conditions is lost in a finite amount of time. Once this information is lost, what remains is statistical information, i.e., a probability measure. The natural probability measure of chaotic attractors is not smooth but rather has structure on all scales, and hence can be termed a *fractal measure* [Farmer 82]. In a dynamical system, d_I can be computed in a manner similar to the computation of d_B [Bais 08, Farmer 82, Halley 04]. Suppose the dynamical motion is confined to some set $\Omega \subset \mathbb{R}^E$ equipped with a natural invariant measure μ. We cover Ω with a set $\mathcal{B}(s)$ of boxes of size s such that for each box we have $p_j > 0$ and that for any two boxes B_i and B_j we have $\mu(B_i \cap B_j) = 0$ (i.e., the intersection of each pair of boxes has measure zero). Define the probability p_j of box B_j by

$$p_j \equiv \frac{\mu(B_j)}{\mu(\Omega)}\,.$$

For example, we can define p_j to be the fraction of time spent by the dynamical system in box B_j [Ruelle 90]. To estimate the p_j values, suppose the system is in equilibrium, and each measurement of the system determines that the state of the system is in a given box B_j. Given N observations from Ω, we discard any box in $\mathcal{B}(s)$ containing none of the N points. For the remaining boxes we estimate $p_j(s)$ by $N_j(s)/N$, where $N_j(s)$ is the number of points contained in box B_j. We compute

$$H(s) \equiv - \sum_{B_j \in \mathcal{B}(s)} \big(N_j(s)/N\big) \log \big(N_j(s)/N\big)$$

and repeat this process for different box sizes. If the $H(s)$ versus $\log(1/s)$ curve is approximately linear over some region of s, the slope of the linear approximation over that region is an estimate of d_I.

Since the computation of both d_B and d_I begins by covering Ω by a set of boxes of size s, then the complexity of computing d_I is the same as the complexity of computing d_B. However, when the distribution has a long tail, reflecting many very improbable boxes, computing d_I may be more efficient than computing d_B, since computing d_B requires counting *each* box containing at least one of the N points, while if $p_j \approx 0$, then the impact on d_I of ignoring box B_j is negligible.

Theorem 14.2 $d_I \leq d_B$.

Proof [Grassberger 83b] After discarding each box in $\mathcal{B}(s)$ containing none of the N points, define $B(s) \equiv |\mathcal{B}(s)|$. Consider the probability distribution $\{\widetilde{p}_j(s) \mid B_j \in \mathcal{B}(s)\}$ for which all the $p_j(s)$ values are equal, having the common value $1/B(s)$. From (14.4), the entropy $\widetilde{H}(s)$ of this distribution is

$$\widetilde{H}(s) = - \sum_{B_j \in \mathcal{B}(s)} \widetilde{p}_j(s) \log \widetilde{p}_j(s) = \sum_{j=1}^{B(s)} \big(1/B(s)\big) \log B(s) = \log B(s)\,.$$

By Theorem 14.1, the entropy is maximal for a uniform probability distribution, so the entropy $H(s)$ of any probability distribution $\{p_j(s) \mid B_j \in \mathcal{B}(s)\}$ satisfies $\widetilde{H}(s) \geq H(s)$, where equality holds if $\{p_j(s) \mid B_j \in \mathcal{B}(s)\}$ is the uniform distribution. Thus

$$d_B = -\lim_{s \to 0} \frac{\log B(s)}{\log s} = \lim_{s \to 0} \frac{\widetilde{H}(s)}{\log(1/s)} \geq \lim_{s \to 0} \frac{H(s)}{\log(1/s)} = d_I, \qquad (14.6)$$

and strict equality holds if $\{p_j(s) \mid B_j \in \mathcal{B}(s)\}$ is the uniform distribution. $\quad\square$

We can therefore view the box counting dimension as a special case of the information dimension for which each box has the same probability.

Exercise 14.3 Recall Definition 9.2 of the correlation dimension d_C, where the correlation integral is defined by (9.12) as $C(r) = \frac{1}{N} \sum_{i=1}^{N} \mu\left(\Omega\big(x(i), r\big)\right)$. A particularly simple approach to estimating d_C is to pick some $x(n)$ and estimate d_C by the local pointwise dimension $d\big(x(n)\big)$, where $d(x)$ is given by (9.3). However, it was stated in [Donner 11] that since $d(x)$ does not consider two-point correlations, then $d(x)$ provides a local estimate of the information dimension d_I and not a local estimate of d_C. Justify or refute this statement. $\quad\square$

Example 14.2 The "shades of grey" construction [Farmer 82, Farmer 82a], based on the classical middle-third Cantor set construction, generates a probability distribution whose information dimension d_I is less than the box counting dimension d_B of the support of the distribution. The construction starts with the unit interval $[0, 1]$ and a given probability $p \in (0, 1)$. In generation 1, we assign probability p to the middle third of the interval $[0, 1]$, and assign probability $(1 - p)/2$ to each of the outer thirds of $[0, 1]$. This generates the probability density function $Q_1(x)$ defined on $[0, 1]$. We then apply the same construction in each successive generation, by distributing the probability within each third with the same ratio. For example, in generation two, we partition $[0, 1/3]$ using the probabilities $(1 - p)/2$, p, and $(1 - p)/2$ for the first, second, and third sub-interval, respectively, so the probability assigned to $[0, \frac{1}{9}]$ is $(1-p)^2/2^2$, the probability assigned to $[\frac{1}{9}, \frac{2}{9}]$ is $(p)(1 - p)/2$, and the probability assigned to $[\frac{2}{9}, \frac{1}{3}]$ is $(1 - p)^2/2^2$. The partitioning of $[\frac{2}{3}, 1]$ yields the same probabilities as the partitioning of $[0, \frac{1}{3}]$. The partitioning of $[\frac{1}{3}, \frac{2}{3}]$ into three thirds yields the probability $p(1 - p)/2$ for the first and last thirds, and p^2 for the middle third. These nine probabilities, each associated with an interval of length $\frac{1}{9}$, define the probability density function $Q_2(x)$ defined on $[0, 1]$. Continuing this process indefinitely, we obtain a limiting distribution $Q_\infty(x)$ defined on $[0, 1]$; a formula for $Q_\infty(x)$ was given in [Farmer 82a]. Farmer referred to this construction as the "shades of grey" construction, since we can view the probability density as proportional to the shade of grey, where a very light grey has low probability and a very dark grey has high probability.

Since at each stage of this construction any sub-interval $[a, b] \subset [0, 1]$ (where $a < b$) has nonzero probability, then for all t the support of the probability density function $Q_t(x)$ is $[0, 1]$, for which $d_B = 1$. However, the information

dimension corresponding to the limiting probability density function $Q_\infty(x)$ is
[Farmer 82a]

$$d_I = \frac{p \log \frac{1}{p} + (1-p) \log \frac{2}{1-p}}{\log 3} \, .$$

(14.7)

Recalling that the classical triadic Cantor set has $d_B = \log 2 / \log 3$, there is a
value of p for which $d_I = \log 2 / \log 3$ if for some p the numerator of (14.7) equals
$\log 2$. Figure 14.2 plots $p \log \frac{1}{p} + (1-p) \log \frac{2}{1-p} - \log 2$ versus p. There are two

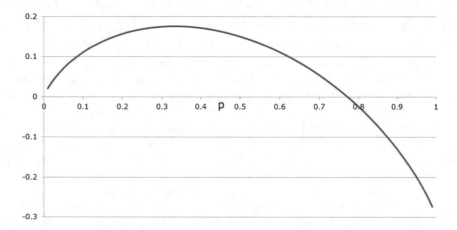

Figure 14.2: Farmer's shades of grey example

values of p for which the curve crosses the horizontal axis. One of these is $p = 0$;
this "unstable fixed point" [Farmer 82a] corresponds to the classical Cantor set.
The second value is approximately 0.773. For this value, the "shades of grey"
limiting distribution $Q_\infty(x)$, whose support is $[0,1]$, has $d_I \approx \log 2 / \log 3$, which
is the box counting dimension of the Cantor set.   □

14.3   Information Dimension Bounds the Correlation Dimension

One of the classical results in the theory of fractal dimensions of geometric
objects is that $d_C \leq d_I$. In this section we present the proof, by Grassberger
and Procaccia [Grassberger 83, Grassberger 83a], of this result. Assume we
have a set $\mathbb{N}$ of N points from a time series generated by a dynamical system
and lying on the strange attractor $\Omega \subset \mathbb{R}^E$ of the dynamical system. Recalling
definition (11.4), define

$$C_N(s) = \frac{1}{N^2} \sum_{x \in \mathbb{N}} \sum_{y \in \mathbb{N}} I\big(s - dist(x,y)\big)$$

and

$$C(s) = \lim_{N \to \infty} C_N(s). \tag{14.8}$$

Cover Ω by a set $\mathcal{B}(s)$ of E-dimensional hypercubes of side length s. Let $N_j(s)$ be number of points in $\mathbb{N}$ contained in box $B_j \in \mathcal{B}(s)$, and let $p_j(s) = \lim_{N \to \infty} N_j(s)/N$. Remove from $\mathcal{B}(s)$ any box B_j for which $p_j = 0$, and define $B(s) \equiv |\mathcal{B}(s)|$. From the definition of $C(s)$ we have, up to a factor of order unity,

$$C(s) \approx \sum_{B_j \in \mathcal{B}(s)} p_j^2. \tag{14.9}$$

This approximation replaces the number of pairs of points separated by a distance less than s by the number of pairs of points which fall into the same box of size s; the error in this approximation should be independent of s and thus should not affect the estimation of d_C ([Grassberger 83], p. 197). The following assumption implies the existence of d_I as defined by (14.5); it is stronger than the assumption that d_I exists.

Assumption 14.1 For some positive constants k_I and d_I and for all sufficiently small box sizes s we have

$$H(s) = k_I - d_I \log s. \quad \square \tag{14.10}$$

We require a lemma, also utilized in [Grassberger 83], which credits [Feller 71] for the proof. Let x_k, $1 \le k \le K$ be positive numbers, and let δ_k, $1 \le k \le K$ be nonnegative numbers such that $\sum_{k=1}^{K} \delta_k = 1$. The classical geometric inequality [Duffin 67] is

$$\sum_{k=1}^{K} \delta_k x_k \ge \prod_{k=1}^{K} x_k^{\delta_k}, \tag{14.11}$$

where equality holds if $x_1 = x_2 = \cdots = x_K$. Letting $y_k = \delta_k x_k$, inequality (14.11) can be rewritten as

$$\sum_{k=1}^{K} y_k \ge \prod_{k=1}^{K} \left(\frac{y_k}{\delta_k}\right)^{\delta_k}, \tag{14.12}$$

for any nonnegative numbers y_k and any nonnegative numbers δ_k whose sum is 1. Here $(y_k/\delta_k)^{\delta_k} = 1$ if $\delta_k = 0$. For brevity, we will write $\sum_{k=1}^{K}$ and $\prod_{k=1}^{K}$ as $\sum$ and $\prod$, respectively.

Lemma 14.4 Let ω_k, $1 \le k \le K$ be positive numbers. Then

$$\log\left(\frac{\sum \omega_k^2}{\sum \omega_k}\right) \ge \frac{\sum \omega_k \log \omega_k}{\sum \omega_k}. \tag{14.13}$$

Proof Examining the right-hand side of (14.13), we have

$$\frac{\sum \omega_k \log \omega_k}{\sum \omega_k} = \frac{\sum \log \omega_k^{\omega_k}}{\sum \omega_k} = \frac{\log \left(\prod \omega_k^{\omega_k} \right)}{\sum \omega_k} = \log \left\{ \left(\prod \omega_k^{\omega_k} \right)^{1/\sum \omega_k} \right\}$$

$$= \log \left(\prod_k \omega_k^{\omega_k / \sum \omega_k} \right).$$

Hence, to prove (14.13), it suffices to prove that

$$\frac{\sum \omega_k^2}{\sum \omega_k} \geq \prod_k \omega_k^{\omega_k / \sum \omega_k}.$$

For $1 \leq k \leq K$, let $\delta_k = \omega_k / \sum \omega_k$. By (14.12),

$$\sum \omega_k^2 \geq \prod (\omega_k^2 / \delta_k)^{\delta_k}$$

$$= \left\{ \prod (1/\delta_k)^{\delta_k} \right\} \left\{ \prod \left(\omega_k^2 \right)^{\delta_k} \right\}$$

$$= \left\{ \prod \left(\frac{\sum \omega_k}{\omega_k} \right)^{\omega_k / \sum \omega_k} \right\} \left\{ \prod \left(\omega_k^2 \right)^{\omega_k / \sum \omega_k} \right\}$$

$$= \left\{ \prod \left(\sum \omega_k \right)^{\omega_k / \sum \omega_k} \right\} \left\{ \prod \omega_k^{-\omega_k / \sum \omega_k} \right\} \left\{ \prod \omega_k^{2\omega_k / \sum \omega_k} \right\}$$

$$= \left(\sum \omega_k \right) \prod_k \omega_k^{\omega_k / \sum \omega_k}. \tag{14.14}$$

Dividing both sides of (14.14) by $\sum \omega_k$ completes the proof. $\square$

The proof of Theorem 14.3 below requires Assumption 14.1 above, and also Assumption 14.2, which is embedded in the following proof of Theorem 14.3.

Theorem 14.3 [Grassberger 83, Grassberger 83a] Under Assumptions 14. 1 and 14.2 we have $d_C \leq d_I$.

Proof By (14.10), for all sufficiently small s we have

$$H(2s) - H(s) = -d_I \big(\log(2s) - \log s \big) = -d_I \log 2.$$

From $C(s) \sim s^{d_C}$, for all sufficiently small s we have

$$\log \left(\frac{C(s)}{C(2s)} \right) = \log \left(\frac{s^{d_C}}{(2s)^{d_C}} \right) = -d_C \log 2.$$

Suppose the following inequality holds:

$$\log \left(\frac{C(s)}{C(2s)} \right) \geq H(2s) - H(s). \tag{14.15}$$

Then

$$\log \left(\frac{C(s)}{C(2s)} \right) = -d_C \log 2 \geq H(2s) - H(s) = -d_I \log 2$$

which implies $d_I \geq d_C$. The remainder of the proof is devoted to showing that (14.15) holds.

Having covered Ω by a set $\mathcal{B}(s)$ of boxes of size s, we want this covering to be nested inside a covering of Ω by boxes of size $2s$. So let $\mathcal{F}(2s)$ be a covering of Ω by E-dimensional hypercubes of side length $2s$. Let $Y_k(s)$ be the number of points in $\mathbb{N}$ contained in box $F_k \in \mathcal{F}(2s)$, and let $P_k(s) = \lim_{N \to \infty} Y_k(s)/N$. Remove from $\mathcal{F}(2s)$ any box F_k for which $P_k = 0$, and let $B(2s)$ be the number of boxes in $\mathcal{F}(2s)$. Denote the $B(s)$ boxes of size s by B_j, $j = 1, 2, \ldots, B(s)$, so j always refers to a "small" box. Denote the $B(2s)$ boxes of size $2s$ by F_k, $k = 1, 2, \ldots, B(2s)$, so k always refers to a "large" box. Since the covering by small boxes is nested inside the covering by large boxes, and since each box lies in $\mathbb{R}^E$, then each large box $F_k \in \mathcal{F}(2s)$ contains 2^E small boxes. The set of all $B(2s)$ non-empty large boxes of size $2s$ in $\mathcal{F}(2s)$ contains a total of $B(s) = 2^E B(2s)$ small boxes of size s.

Recall that, for $j = 1, 2, \ldots, B(s)$, p_j is the probability that a point in $\mathbb{N}$ lies in small box B_j, and $p_j > 0$. Also, for $k = 1, 2, \ldots, B(2s)$, P_k is the probability that a point in $\mathbb{N}$ lies in large box F_k, and $P_k > 0$. Let $J(k)$ be the set of indices j such that small box B_j is contained in large box F_k. Then for each k we have $\sum_{j \in J(k)} p_j = P_k$. For each k and each $j \in J(k)$, define $\omega_j \equiv p_j / P_k$. Then for each k we have $\sum_{j \in J(k)} \omega_j = 1$ and

$$\sum_{j=1}^{B(s)} \omega_j = \sum_{k=1}^{B(2s)} \sum_{j \in J(k)} \omega_j = \sum_{k=1}^{B(2s)} 1 = B(2s). \qquad (14.16)$$

With these definitions, we present the final assumption made in [Grassberger 83, Grassberger 83a].

Assumption 14.2 The random variables ω_j are distributed independently of the random variables P_k. $\qquad \square$

In the context of strange attractors of dynamical systems, which is the subject of [Grassberger 83, Grassberger 83a], Assumption 14.2 says that locally the attractor looks the same in regions where it is denser (i.e., where P_k is large) as it does in regions where it is sparser (i.e., where P_k is small). As to the validity of this assumption, it was stated in [Grassberger 83a] that "Although we cannot further justify this assumption, it seems to us very natural." Continuing the proof of Theorem 14.3, by (14.9) we have

$$C(s) \approx \sum_{j=1}^{B(s)} p_j^2 = \sum_{j=1}^{B(s)} (\omega_j P_k)^2 = \sum_{k=1}^{B(2s)} P_k^2 \sum_{j \in J(k)} \omega_j^2 \qquad (14.17)$$

and

$$C(2s) \approx \sum_{k=1}^{B(2s)} P_k^2 \,. \qquad (14.18)$$

Assume that the approximations in (14.17) and (14.18) hold as equalities. Then

$$\frac{C(s)}{C(2s)} = \frac{\sum_{k=1}^{B(2s)} P_k^2 \sum_{j \in J(k)} \omega_j^2}{\sum_{k=1}^{B(2s)} P_k^2} . \tag{14.19}$$

By Assumption 14.2, $\sum_{j \in J(k)} \omega_j^2$ is independent of k, for $k = 1, 2, \ldots, B(2s)$. Define $\alpha \equiv \sum_{j \in J(k)} \omega_j^2$. Then

$$\sum_{j=1}^{B(s)} \omega_j^2 = \sum_{k=1}^{B(2s)} \sum_{j \in J(k)} \omega_j^2 = \sum_{k=1}^{B(2s)} \alpha = B(2s)\,\alpha ,$$

so for $k = 1, 2, \ldots, B(2s)$ we have

$$\alpha = \sum_{j \in J(k)} \omega_j^2 = \frac{\sum_{j=1}^{B(s)} \omega_j^2}{B(2s)} = \frac{\sum_{j=1}^{B(s)} \omega_j^2}{\sum_{j=1}^{B(s)} \omega_j} , \tag{14.20}$$

where the final equality follows from (14.16). From (14.19) and (14.20) we have

$$\frac{C(s)}{C(2s)} = \frac{\sum_{k=1}^{B(2s)} P_k^2 \sum_{j \in J(k)} \omega_j^2}{\sum_{k=1}^{B(2s)} P_k^2} = \frac{\sum_{k=1}^{B(2s)} P_k^2\, \alpha}{\sum_{k=1}^{B(2s)} P_k^2} = \alpha = \frac{\sum_{j=1}^{B(s)} \omega_j^2}{\sum_{j=1}^{B(s)} \omega_j} . \tag{14.21}$$

Turning now to the information dimension, by Assumption 14.2 the sum $\sum_{j \in J(k)} \omega_j \log \omega_j$ is independent of k. Define $\beta \equiv \sum_{j \in J(k)} \omega_j \log \omega_j$. Then

$$\sum_{j=1}^{B(s)} \omega_j \log \omega_j = \sum_{k=1}^{B(2s)} \sum_{j \in J(k)} \omega_j \log \omega_j = \sum_{k=1}^{B(2s)} \beta = B(2s)\,\beta ,$$

so for $k = 1, 2, \ldots, B(2s)$ we have

$$\beta = \sum_{j \in J(k)} \omega_j \log \omega_j = \frac{\sum_{j=1}^{B(s)} \omega_j \log \omega_j}{B(2s)} = \frac{\sum_{j=1}^{B(s)} \omega_j \log \omega_j}{\sum_{j=1}^{B(s)} \omega_j} \tag{14.22}$$

where the final equality follows from (14.16). Since $p_j = \omega_j P_k$ for $j \in J(k)$ and $\sum_{j \in J(k)} \omega_j = 1$, from (14.1) we have

$$
\begin{aligned}
H(2s) - H(s) &= -\sum_{k=1}^{B(2s)} P_k \log P_k + \sum_{j=1}^{B(s)} p_j \log p_j \\
&= -\sum_{k=1}^{B(2s)} P_k \log P_k + \sum_{k=1}^{B(2s)} P_k \sum_{j \in J(k)} \omega_j \left(\log \omega_j + \log P_k \right) \\
&= -\sum_{k=1}^{B(2s)} P_k \log P_k + \sum_{k=1}^{B(2s)} P_k \sum_{j \in J(k)} \omega_j \log P_k \\
&\quad + \sum_{k=1}^{B(2s)} P_k \sum_{j \in J(k)} \omega_j \log \omega_j \\
&= \sum_{k=1}^{B(2s)} P_k \sum_{j \in J(k)} \omega_j \log \omega_j = \sum_{k=1}^{B(2s)} P_k \beta = \beta = \frac{\sum_{j=1}^{B(s)} \omega_j \log \omega_j}{\sum_{j=1}^{B(s)} \omega_j},
\end{aligned}
$$

so

$$
H(2s) - H(s) = \frac{\sum_{j=1}^{B(s)} \omega_j \log \omega_j}{\sum_{j=1}^{B(s)} \omega_j}. \tag{14.23}
$$

By (14.13), we have $\log \alpha \geq \beta$. Hence by (14.21), (14.22), and (14.23) we have

$$
\log \left(\frac{C(s)}{C(2s)} \right) = \log \left(\frac{\sum_{j=1}^{B(s)} \omega_j^2}{\sum_{j=1}^{B(s)} \omega_j} \right) = \log \alpha \geq \beta = \frac{\sum_{j=1}^{B(s)} \omega_j \log \omega_j}{\sum_{j=1}^{B(s)} \omega_j} = H(2s) - H(s).
$$

Thus (14.15) holds, which concludes the proof of Theorem 14.3. $\square$

The computational results in [Grassberger 83b] found d_B, d_C, and d_I to be very close in the systems they studied, except for the Feigenbaum map and the Mackey–Glass delay equation.[5]

14.4 Fixed Mass Approach

Another way of calculating the information dimension from a set of N points is to use the method of [Badii 85]. This method is fundamentally different than all the other methods we have so far examined, which can be described as *fixed size* methods. For example, to calculate d_B or d_I, the methods we have studied so far cover Ω by boxes of a given size and count the number of non-empty boxes (if computing d_B), or the number of points in each non-empty box (if computing d_I). Similarly, to calculate d_C we count the number of points whose distance from a center node does not exceed a given radius. These methods are *fixed size* methods, since the box size s or radius r is fixed when we count boxes or points. In contrast, for a *fixed mass* method, e.g., [Badii 85], we compute the

---

[5]In [Grassberger 83b], d_B was called the "Hausdorff dimension".

average distance from a point to its k-th nearest neighbor. By varying k we can estimate a fractal dimension of Ω.

To describe the fixed mass approach, suppose we have N points $\{x_j\}_{j=1}^{N} \subset \Omega$, randomly chosen with respect to the natural measure. Let $dist(x_j, k)$ be the smallest distance such that k of the N points (not including x_j itself) are within distance $dist(x_j, k)$ of x_j. That is, $dist(x_j, k)$ is the distance from x_j to the k-th closest neighbor of x_j. We compute $dist(x_j, k)$ for each x_j, and define

$$dist(k) = \frac{1}{N} \sum_{j=1}^{N} dist(x_j, k) .$$

Then $dist(k)$ exhibits the scaling [Theiler 90, Badii 85]

$$dist(k) \sim k^{1/d} \text{ as } N \to \infty$$

for some d. Rewriting this as $k \sim dist(k)^d$ and comparing $k \sim dist(k)^d$ to (1.1), we can view d as a dimension. Moreover, the scaling of $dist(k)$ with k provides a way to compute d_I [Hegger 99]. Define

$$\langle \ln dist(k) \rangle \equiv \frac{1}{N} \sum_{j=1}^{N} \ln dist(x_j, k) .$$

Then

$$\exp\left(\langle \ln dist(k) \rangle\right) = \exp\left(\sum_{j=1}^{N} \ln\left(dist(x_j, k)^{1/N} \right) \right)$$

$$= \prod_{j=1}^{N} dist(x_j, k)^{1/N} = \sqrt[N]{\prod_{j=1}^{N} dist(x_j, k)}$$

so $\exp\left(\langle \ln dist(k) \rangle\right)$ is the geometric mean length scale for which k nearest neighbors are found from the N points. A "rather robust" [Hegger 99] estimate of d_I is given by the derivative

$$d_I = \lim_{(k/N) \to 0} \frac{\mathrm{d} \ln (k/N)}{\mathrm{d} \langle \ln dist(k/N) \rangle} .$$

Computational considerations with this approach were discussed in [Hegger 99]. In particular, finite sample corrections are required if k is small. It was observed in [Chatterjee 92] that the fixed mass approach is generally computationally more efficient than the fixed size approach.

14.5 Rényi and Tsallis Entropy

We have shown that d_I, defined by (14.5), quantifies how the Shannon entropy $H(s)$ scales with the box size. We have also seen that the Shannon entropy is not the only entropy that can be defined; in Sect. 14.1 we considered the Hartley entropy, which can be viewed as the Shannon entropy for the special case of a uniform distribution, and showed that d_B quantifies how the Hartley entropy scales with the box size. We now return to the Shannon entropy (14.1) and show how it can be generalized.

Formula (14.1) applies the weight p_j to the information measure $\log p_j$. Yet there are other ways to apply weights to the $\log p_j$ terms. A general definition of a weighted mean, using the probabilities p_j, is a function of the form

$$f^{-1}\left(\sum_{j=1}^{J} p_j f(-\log p_j)\right),$$

where f is a continuous, strictly monotonic, and invertible function (often referred to as the *Kolmogorov–Nagumo function* [Jizba 04]), and f^{-1} denotes the inverse of f. For $f(x) = cx$, where $c \neq 0$, we obtain the Shannon entropy. It can be shown [Jizba 04] that the appropriate axioms on the additivity of information for independent events can be satisfied for only one other function f, given by $f(x) = [2^{(1-q)x} - 1]/\beta$, where $q \neq 1$ and $\beta \neq 0$. Recalling that in this chapter log denotes $\log_2$, for this f we have $f^{-1}(y) = [\log(\beta y + 1)]/(1 - q)$. Since

$$f(-\log p) = \frac{2^{-(1-q)\log p} - 1}{\beta} = \frac{p^{q-1} - 1}{\beta}$$

then

$$f^{-1}\left(\sum_{j=1}^{J} p_j f(-\log p_j)\right) = f^{-1}\left(\sum_{j=1}^{J} p_j \frac{(p_j^{q-1} - 1)}{\beta}\right)$$

$$= \frac{1}{1-q}\log\left(\sum_{j=1}^{J} p_j^q - \sum_{j=1}^{J} p_j + 1\right) = \frac{1}{1-q}\log\left(\sum_{j=1}^{J} p_j^q\right). \quad (14.24)$$

The right-hand side of (14.24) is called the *Rényi entropy of order q* [Rényi 70], and is denoted by H_q. That is,

$$H_q \equiv \frac{1}{1-q}\log\left(\sum_{j=1}^{J} p_j^q\right). \quad (14.25)$$

Exercise 14.4 Show that, for $q \neq 1$, the functions $f(x) = [2^{(1-q)x} - 1]/\beta$ and $f^{-1}(y) = [\log(\beta y + 1)]/(1-q)$ are inverse functions. □

Exercise 14.5 Under what conditions does H_q vanish? □

For $q = 0$ we have $H_0 = \log J$, which is the Hartley entropy of Sect. 14.1. For $q = 1$ we obtain the indeterminate form $0/0$; applying L'Hôpital's rule yields

$$H_1 \;=\; \lim_{q \to 1} H_q = - \lim_{q \to 1} \frac{\mathrm{d}}{\mathrm{d}q} \log \sum_{j=1}^{J} p_j^q = - \lim_{q \to 1} \frac{\sum_{j=1}^{J} p_j^q \log p_j}{\sum_{j=1}^{J} p_j^q} \quad (14.26)$$

$$=\; -\sum_{j=1}^{J} p_j \log p_j \,,$$

so H_1 is the Shannon entropy (14.1).

If $p_j = 1/J$ for all j, then for $q \neq 1$ we have $H_q = [1/(1-q)] \log[J(1/J)^q] = \log J$. For a probability distribution with $J = 2$ and the two probabilities p and $1 - p$, Fig. 14.3 plots H_q versus q for $-15 \leq q \leq 15$. The four curves shown are for $p = 0.001$, $p = 0.1$, $p = 0.3$, and 0.5. Since $J = 2$, the curve for $1 - p$ is identical to the curve for p.

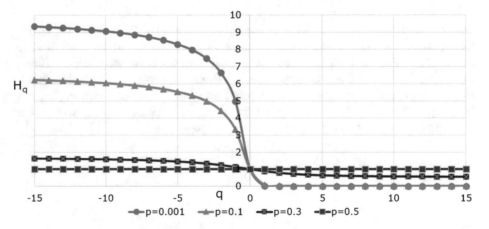

Figure 14.3: H_q versus q for four values of p

Theorem 14.4 For $q \neq 1$,

$$\frac{\mathrm{d}H_q}{\mathrm{d}q} \leq 0 \,.$$

Proof From (14.25), for $q \neq 1$ we have

$$\frac{\mathrm{d}H_q}{\mathrm{d}q} = \frac{G}{(1-q)^2} \,,$$

where

$$G = \frac{(1-q) \sum_j p_j^q \log p_j}{F} + \log F$$

and

$$F = \sum_j p_j^q \,.$$

To prove the theorem we must show that $G \leq 0$. It suffices to show that $\widetilde{G} \leq 0$, where

$$\widetilde{G} = \frac{(1-q)\sum_j p_j^q \ln p_j}{F} + \ln F$$

For $j = 1, 2, \ldots, J$, define $z_j = p_j^q/F$. Then $\sum_j z_j = 1$ and

$$
\begin{aligned}
\widetilde{G} &= (1-q)\sum_j \left(\frac{p_j^q}{F}\right) \ln p_j + \ln F = (1-q)\sum_j z_j \ln p_j + \sum_j z_j \ln F \\
&= \sum_j z_j \ln p_j - \sum_j z_j \ln p_j^q + \sum_j z_j \ln F = \sum_j z_j \ln p_j - \sum_j z_j (\ln p_j^q - \ln F) \\
&= \sum_j z_j \ln p_j - \sum_j z_j \ln z_j = \sum_j z_j \ln \left(\frac{p_j}{z_j}\right) \leq 0,
\end{aligned}
$$

where the final inequality follows by Lemma 14.2 (the Gibbs inequality).   □

If the probability density is uniform, so $p_j = 1/J$ for each j, then $z_j = 1/J$ for each j, so in the above proof we have $\widetilde{G} = 0$, which implies that the derivative of H_q is zero. This is to be expected, since we have already shown that if $p_j = 1/J$ for each j, then for all q we have $H_q = \log J$.

The connections between Rényi entropy and thermodynamics were explored in [Baez 11]. The "most convincing" uses of Rényi entropy in physics seem to involve the limiting cases $\lim_{q\to 0} H_q(s)$ and $\lim_{q\to\infty} H_q(s)$, known as the "max-entropy" and "min-entropy", respectively, since $H_q(s)$ is a non-increasing function of q. These limiting cases arise in the study of the work value of information and in the thermodynamic meaning of negative entropy. For a physical system in thermal equilibrium, if we suddenly divide the temperature by q, where $q > 0$, then the maximal amount of work the system can do as it moves to equilibrium at the new temperature, divided by the change in temperature, is equal to the Rényi entropy of the system in its original state.

For another perspective on the Rényi entropy, recall that Axioms 14.5–14.7 of Sect. 14.1 yield the Shannon entropy. Suppose we require that a measure of information must satisfy Axioms 14.5 and 14.6 but not Axiom 14.7, which is the branching assumption on the joint entropy of non-independent subsystems. Unlike the other assumptions, it is unclear whether Axiom 14.7 should be required of a measure of information. We *do* require that if we observe outcomes of two independent events, the total information measure should be the sum of the information received from the two events. This implies that the information conveyed by receiving a symbol (i.e., message) with probability p_j is proportional to $\log p_j$. If we weight, by taking an arithmetic mean of the p_j values, all the information conveyed when reading J symbols, we obtain the Shannon entropy.

But suppose we now replace Axiom 14.7 by the less stringent condition (14.3), which specifies that the entropy of independent systems should be additive. These conditions yield the Rényi entropies [Beck 09]. Just as the Shannon entropies are additive for independent systems, the Rényi entropies are additive

for independent systems for any q. However, Axiom 14.7 does not hold for
the Rényi entropy, and there is no way to express the total Rényi entropy of
a combined system of non-independent subsystems as a simple function of the
Rényi entropies of the interacting subsystems. Another drawback of the Rényi
entropies is that they are neither convex nor concave, since the second derivative
with respect to p_k can be positive or negative. This poses a serious problem in
formulating a generalized statistical mechanics [Beck 09]. Thus various other
entropy measures have been proposed as alternatives to the Rényi entropy. In
1988, Tsallis defined the family of entropies

$$\frac{1}{q-1}\left(1 - \sum_k p_k{}^q\right).$$ (14.27)

They are concave for $q > 0$, convex for $q < 0$, and yield the Shannon entropy as
$q \to 1$. Despite the fact that they are not additive for independent subsystems,
they are a better candidate for generalizing statistical mechanics than the Rényi
entropy [Beck 09]. The Rényi, Tsallis, and several other entropy formulations
were compared in [Beck 09].

Exercise 14.6 Show that when $q = 1$ the Tsallis entropy is equal to the
Shannon entropy.   □

14.6 Applications of the Information Dimension

Below we briefly describe several applications of d_I. However, as described
in [Rosenberg 17d], in the first two applications an incorrect definition of d_I was
used.

Schizophrenia: In [Zhao 16], d_I was used to study schizophrenia. Three-
dimensional magnetic resonance images (MRIs) were obtained from 38 people,
19 with schizophrenia, and 19 without. Using a 3-dimensional grid of voxels
(a *voxel* is a value in a 3-dimensional grid, just as a pixel is a value in a 2-
dimensional grid), d_I was computed for different sub-cortical areas of the brain,
especially the hippocampus (which contributes to the encoding, consolidation,
and retrieval of memory, and the representation of the temporal context). The
reason for computing d_I, rather than some other fractal dimension, was that d_I
has been shown to have a higher "inter-class correlation" [Goni 13]. The plots in
[Zhao 16] of $\log H(s)$ versus $\log s$ show that $H(s)$ does not scale linearly for large
box sizes, and instead oscillates; it was noted that reason for this oscillation is
not well understood. To estimate d_I, points that deviated from the regression
line were repeatedly excluded, and a new regression line computed, until most
data points could be fitted. The results of this study included a finding that d_I
is significantly lower in the left and right hippocampus and in the left thalmus in
patients with schizophrenia, compared to healthy patients. However, this study
used an incorrect definition of d_I, namely

$$d_I = -\lim_{s \to 0} \frac{\log H(s)}{\log s}$$

which is incorrect since the numerator should be $H(s)$ and not $\log H(s)$. This is not a single "typo" or misprint, since $\log H(s)$ was used throughout the paper. **Asthma:** In [Boser 05], d_I was used to study differences between the lungs of three groups of deceased human non-smokers: fatal asthma (people who died from asthma), non-fatal asthma, and non-asthma control. From lung images of 1280×960 pixels, d_I was calculated using 9 box sizes, ranging from about 10 pixels to about 100 pixels. The calculated values were $d_I = 1.83$ for the non-asthma control group, $d_I = 1.76$ for the non-fatal asthma group, and $d_I = 1.72$ for the fatal asthma group. A statistical analysis showed that d_I for the fatal asthma group and non-fatal asthma group were significantly lower than d_I for the non-asthma group. However, the difference in d_I for the fatal asthma group and non-fatal asthma group was not significantly significant. It was observed in [Boser 05] that an advantage of using d_I rather d_B is that d_I is less likely to result in a false detection of multifractal characteristics (Chap. 16). It was also observed that the exact interpretation of a fractal dimension, when used to quantify structure or change in structure, is less important than the ability of a fractal dimension to discriminate between states. However, [Boser 05] also used an incorrect definition of d_I, namely

$$\log H(s) \approx -d_I \log s.$$

The fact that incorrect definitions of d_I were used in [Zhao 16] and [Boser 05] does not necessarily negate their findings. However, it does mean that their findings are valid only with respect to their definition of d_I, and other researchers may obtain very different conclusions when using the correct definition of d_I [Rosenberg 17d].

Soil Drainage: In [Martin 08], d_I was used to study soil drainage networks. Soil samples from a square lattice of 256 locations in a vineyard in Central Spain were analyzed to determine the concentrations of potassium and phosphorous. From the concentrations, a probability distribution of 256 values was obtained for potassium, and similarly for phosphorous. These distributions yielded the values $d_I = 1.973$ for potassium and $d_I = 1.964$ for phosphorous. The closeness of these values means that "both minerals have a very similar probability to be in contact with soil water during saturated flow". Moreover, $d_I - d_B$ is "an indicator that might reflect the likelihood of leaching or dispersion of soil chemicals". In particular, for soil contaminants, $d_I - d_B$ "would measure the risk of exporting pollution to the surrounding areas".

Stability of a Suspension Bridge: In [Lazareck 01], d_I was computed for the calculated instability in a model of the Tacoma Narrows suspension bridge in Tacoma, Washington, which collapsed in 1940 due to excessive oscillations. A plot relating the amount of oscillation to the wind velocity was divided into $J = 1/s$ boxes of size s. Defining $p_j = N_j/N$, where N_j is the number of points in cell j and N is the total number of points, d_I was calculated for three cases: a low wind speed of 20 mph (the bridge is stable, $d_I = 1.3055$), a critical wind speed of 35 mph (at which the bridge oscillates continuously at the same amplitude, $d_I = 1.4453$), and a high wind speed of 42 mph (at which the bridge collapses, $d_I = 1.9293$).

Ecogenomics: Ecogenomics is the application of advanced molecular technologies to study the responses of organisms to environmental challenges in their natural settings. In [Chapman 06], the box counting dimension d_B, the information dimension d_I, and the correlation dimension d_C were computed to study the white-spot syndrome virus, a lethal pathogen of the Pacific white shrimp. This fractal analysis was done for both one-dimensional and two-dimensional images. For four different types of shrimp tissue, the ranking of the tissues by fractal dimension was found to be the same whether the ranking used d_B, d_I, or d_C. Since for both one-dimensional and two-dimensional images the inequality $d_I > d_B$ was typically found to hold, the conclusion in [Chapman 06] was that the images are not "true fractals", since then the relation $d_B \geq d_I$ would hold.[6]

Stock Prices: The stock prices of 60 companies in the Dow Jones Industrial Index were analyzed in [Mohammadi 11]. Results showed that d_B was more sensitive to outliers than d_C or d_I, but the effect of outliers decreased as the number of observations increases. Two studies using local fractal dimension for the detection of outliers were cited in [Mohammadi 11].

Microvascular Network of Chicks: The CAM (chorioallantoic membrane) of the avian chick was studied in [Arlt 03]. The CAM is the respiratory organ of the chick embryo, and the microvascular network (the vessels with a diameter less than $30\,\mu$m) lies on the CAM in a two-dimensional area. In [Arlt 03], the three dimensions d_B, d_I, and d_C of the microvascular network were computed as a measure of its complexity. (So this is yet another example, just prior to the adoption of network-based methods for computing fractal dimensions of a network, of computing fractal dimensions of a network by treating it as a geometric object.) One set of results computed the average d_B, d_I, and d_C over four different CAMs, yielding $d_B \approx 1.8606 \pm 0.0093$, $d_I \approx 1.853 \pm 0.018$, and $d_C \approx 1.844 \pm 0.021$, in agreement with the theoretical result $d_B \geq d_I \geq d_C$.

Being 2-dimensional, the fractal dimension of the microvascular network lies between 1 and 2. The value 2 would be optimal for the exchange of oxygen. On the other hand, the CAM also must transport nutrients necessary for cell growth, and a dimension less than 2 is needed so that "the flow of the blood and the metabolism can be guided".

The influence of noise on the fractal dimensions was also studied. The images were 1091μm $\times$ 751μm and any spot of size up to 19.2μm^2 (20 pixels) was considered to be noise in an image. For 19 images, [Arlt 03] computed $\delta(B) \equiv d_B(\text{noise}) - d_B(\text{no noise})$ and similarly computed $\delta(I)$ for d_I and $\delta(C)$ for d_C. The "best" dimension is the one yielding $\min\{\delta(B), \delta(I), \delta(C)\}$. The motivation in [Arlt 03] for this definition of "best" is the *stability* property (Sect. 5.1) that

---

[6]The proof in Sect. 14.3 of Theorem 14.2 shows that the inequality $d_B \geq d_I$ holds in the limit as the box size vanishes. Since a real-world image has finite resolution, it necessarily can only approximate a fractal, and this inequality may not hold.

should be satisfied by a dimension function: If $\mathcal{X}$ and $\mathcal{Y}$ are subsets, then $dim(\mathcal{X} \cup \mathcal{Y}) = \max\big(dim(\mathcal{X}), dim(\mathcal{Y})\big)$. Treating $\mathcal{X}$ as the image with noise and $\mathcal{Y}$ as the image with noise removed, $\mathcal{X} \cup \mathcal{Y} = \mathcal{X}$. The dimension which best satisfies this stability property was taken to be the preferred dimension. Since computational results showed that $\delta(B) > \delta(I) > \delta(C)$, the conclusion in [Arlt 03] was that the correlation dimension has the best performance.

Exercise 14.7 In the above discussion of the study in [Arlt 03] of the fractal dimension of the microvascular network, the finding that $\delta(B) > \delta(I) > \delta(C)$ was explained as follows:

> The box dimension is computed with a fixed grid. The value of the fractal dimension differs depending on the position of the grid. To compensate [for] this error, the information dimension takes into account that the density of the image (points per area) differs. Therefore, the probability to find a point in a grid is [considered]. The correlation dimension does not need a grid. Only the distances of all points are used for calculation. The grid-based calculations of the box and information dimensions worsen the result of the fractal dimension.

Based on the material in this book on the computation of d_B, d_C, and d_I, to what extent do you agree or disagree with this explanation? □

Chapter 15

Network Information Dimension

A thousand fibers connect us with our fellow men; and among those fibers, as sympathetic threads, our actions run as courses, and they come back to us as effects.
Herman Melville (1819–1891), American novelist, poet, and writer of short stories, best known for the whaling novel "Moby-Dick" (1851).

In this chapter we discuss the information dimension of a network. We begin by considering the information dimension of a node. We then present several definitions of the entropy of a network. For geometric objects, we know from Sect. 14.2 that d_I quantifies how the entropy scales with the box size. Similarly, the information dimension d_I of a network quantifies how the network entropy scales with the box size.

15.1   Entropy of a Node and a Network

The local dimension (Sect. 11.2) of node $n \in \mathbb{N}$ can be used to define the entropy and the local information dimension of n [Bian 18]. Define

$$\Delta_n \equiv \max_{\substack{x \in \mathbb{N} \\ x \neq n}} dist(n, x) \,,$$

so Δ_n is the maximal distance from n to any other node. For integer r between 1 and Δ_n, define $\Delta_n(r) = \lceil \Delta_n/r \rceil$. By definition (11.12), for $j = 1, 2, \ldots, \Delta_n(r)$, the local dimension $d(n, j)$ of node n at radius j is given by

$$d(n, j) = \frac{j \, M^0(n, j)}{M(n, j)} \,,$$

© Springer Nature Switzerland AG 2020
E. Rosenberg, *Fractal Dimensions of Networks*,
https://doi.org/10.1007/978-3-030-43169-3_15

where $M^0(n, j)$ is the number of nodes exactly j hops from n and $M(n, j)$ is the number of nodes within j hops of n. For $j = 1, 2, \ldots, \Delta_n(r)$, define

$$p_j(n, r) \equiv \frac{d(n, j)}{\sum_{j=1}^{\Delta_n(r)} d(n, j)}$$

and define the entropy $I_n(r)$ for node n at radius r by

$$I(n, r) \equiv - \sum_{j=1}^{\Delta_n(r)} p_j(n, r) \log p_j(n, r).$$

The node information dimension $d_I(n)$ was defined in [Bian 18] by

$$d_I(n) \equiv - \lim_{r \to 0} I(n, r) / \log r.$$

Since r assumes only integer values, $\lim_{r \to 0} I(n, r) / \log r$ does not exist, so we take $d_I(n)$ to be the slope of the $I(n, r)$ versus $\log r$ plot, taken over a range of r for which the plot is sufficiently linear. In [Bian 18], $d_I(n)$ was compared to other node metrics that measure node importance.

Shannon Entropy of a Network

One way to define the entropy of a network is to apply Shannon's formula

$$H \equiv - \sum_k p_k \log p_k \qquad (15.1)$$

with appropriately defined probabilities. It is useful to apply the term *Shannon entropy of a network* when the entropy of $\mathbb{G}$ is computed using Shannon's formula. There are many ways to define the probabilities $\{p_k\}$ for a network. Regardless of how the network probability distribution is defined, Shannon network entropy is maximized when the probability distribution is uniform; i.e., when all the probabilities are equal. One way to define the network probability distribution is to let K be the set of values for which $\delta_n = k$ for some $n \in \mathbb{N}$; i.e., K is the set of values of k such that at least one node has degree k. For $k \in K$, let N_k be the number of nodes with degree k, and define $p_k \equiv N_k/N$, so p_k is the probability that a node has degree k and $\sum_{k \in K} p_k = 1$. When the probabilities are defined this way, the Shannon network entropy is related to the resilience of networks to attacks, and to lethality in protein interaction networks [Costa 07].

An interesting consequence of the Shannon network entropy, when $p_k \equiv N_k/N$ and N_k is the number of nodes with degree k, was presented in [Bonchev 05]. Consider N nodes and two different graphs on these N nodes. The first graph, call it $\mathbb{G}_1$, has no arcs, so each node is isolated. The second graph, call it $\mathbb{G}_2$, is K_N, the complete graph on N nodes, so each node has degree $N - 1$. For $\mathbb{G}_1$ we have $p_0 = 1$ since each node has degree 0, so by (15.1) we have $H = 0$. For $\mathbb{G}_2$ we have $p_{N-1} = 1$ since each node has degree $N - 1$, so we again have $H = 0$. It is unsettling that both graphs have the same entropy, which is zero.[1] This issue was considered in [Bonchev 05]:

---

[1]This example also appears in Gell-Mann's book *The Quark and the Jaguar*.

The logic of the above arguments is flawless. Yet, our intuition tells us that the complete graph is more complex than the totally disconnected graph. There is a hidden weak point in the manner the Shannon theory is applied, namely how symmetry is used to partition the vertices into groups. One should take into account that symmetry is a simplifying factor, but not a complexifying one. A measure of structural or topological complexity must not be based on symmetry. The use of symmetry is justified only in defining compositional complexity, which is based on equivalence and diversity of the elements of the system studied.

This argument led [Bonchev 05] to consider other ways to define network entropy. A general scheme described in [Bonchev 05] represents the graph by K elements (e.g., nodes, arcs, or distances) and assigns a weight w_k to element k. The probability of k is $p_k \equiv w_k / \sum_k w_k$, and these probabilities can be used in (15.1). As an example, let the elements be the N nodes, and set $w_n = \delta_n$, where δ_n is the degree of node n. With this choice,

$$p_n \equiv \frac{\delta_n}{\sum_{n \in \mathbb{N}} \delta_n},\tag{15.2}$$

and the entropy H of the complete graph K_N is $\log N$. This approach to defining a network probability distribution was also utilized in [ZhangLi 16]. It is convenient to refer to the Shannon entropy (15.1), when computed using the probabilities defined by (15.2), as the *Shannon degree entropy*.

A third way [ZhangLi 16] to define a network probability distribution is to use the node betweenness values $node\_bet(n)$ defined by (2.5). With this approach,

$$p_n \equiv \frac{node\_bet(n)}{\sum_{n \in \mathbb{N}} node\_bet(n)}.\tag{15.3}$$

Since there are N nodes, there are N probabilities, and $\sum_{n=1}^{N} p_n = 1$. It is convenient to refer to the Shannon entropy (15.1), when computed using the probabilities defined by (15.3), as the *Shannon node betweenness entropy*.

Tsallis Entropy of a Network

The entropy of $\mathbb{G}$ can also be defined using the Tsallis entropy (14.27), introduced in Sect. 14.5. The Tsallis network entropy H_T (where "T" denotes "Tsallis") defined in [ZhangLi 16] is

$$H_T \equiv \sum_{n \in \mathbb{N}} \frac{p_n^{q_n} - p_n}{1 - q_n},\tag{15.4}$$

where p_n is defined by (15.2), and where, for $n \in \mathbb{N}$,

$$q_n \equiv 1 + \left(\max_{i \in \mathbb{N}} node\_bet(i) \right) - node\_bet(n).\tag{15.5}$$

We have $q_n > 0$ for $n \in \mathbb{N}$. In contrast to definition (14.27) of the Tsallis entropy of a probability distribution which utilizes a single value of q, the Tsallis network entropy defined by (15.4) utilizes the set of N values $\{q_n\}_{n \in N}$.

The value of the Tsallis entropy H_T is illustrated by considering networks (i)–(iv) in Figs. 15.1 and 15.2 and Table 15.1, all from [ZhangLi 16]. In each network, each arc cost is 1 except as otherwise indicated. Networks (i) and (ii) have the same Shannon degree entropy, and networks (iii) and (iv) have the same Shannon node betweenness entropy. The Tsallis entropy H_T has a different value for the four networks, illustrating that H_T can sometimes provide a better way to characterize networks than the Shannon entropy.

Table 15.1: Network entropies for four networks

	network (i)	network (ii)	network (iii)	network (iv)
Shannon degree entropy	1.7918	1.7918	1.7721	1.6434
Shannon node betweenness entropy	1.7918	1.5955	1.5763	1.5763
Tsallis entropy H_T	1.7918	1.6308	1.5475	1.4564

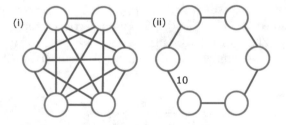

Figure 15.1: Networks (i) and (ii)

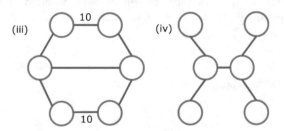

Figure 15.2: Networks (iii) and (iv)

Exercise 15.1 Derive all the values in Table 15.1   □

Entropy of Randomized Network Ensembles

A different approach defines the entropy of an ensemble of networks to be the logarithm of the number of networks with a given structural characteristic [Anand 09, Bianconi 07, Bianconi 09]. Two examples of given structural characteristics are a given number of nodes, and a given degree sequence. As the number of networks in the ensemble decreases, the complexity of the ensemble increases, and the entropy of the ensemble decreases. This is modelled by starting with a large collection of randomized networks and successively adding constraints that reflect the characteristics of real networks. Each time we add a constraint, the entropy of the resulting network ensemble decreases, and the difference between the entropies in two successive ensembles quantifies how restrictive is the constraint that was added.

We start with an ensemble $\mathbb{G}(N, A)$ of random networks defined only by the number N of nodes and the number A of arcs. Then we impose on $\mathbb{G}(N, A)$ a given node degree sequence $\{\delta_n\}$; this lowers the size of the ensemble and decreases the entropy. Then we further impose, for example, a given community structure, or a given cluster coefficient (see (2.3)), further lowering the size of the ensemble and decreasing the entropy. While the results in [Anand 09, Bianconi 07, Bianconi 09] estimated the entropy of randomized network ensembles, the computation of d_I was not considered.

The analysis in [Bianconi 07] was based upon defining a set of partition functions. The partition function of an ensemble counts the number of undirected and unweighted networks, defined by their node-node adjacency matrices, that satisfy all the constraints defining that ensemble. Recall from Sect. 2.1 that the node-node adjacency matrix M of a network is defined by $M_{ij} = 1$ if an arc exists between nodes i and j, and $M_{ij} = 0$ otherwise. Hence $\sum_{i<j} M_{ij}$ is the total number A of arcs. Define $I(\cdot, \cdot)$ by $I(x, y) = 1$ if $x = y$ and $I(x, y) = 0$ otherwise (this is the Kronecker delta function). The analysis starts with the ensemble of networks with N nodes and A arcs. Let $\sum_{\{M_{ij}\}}$ denote the sum over all possible $N \times N$ adjacency matrices. For $1 \le i, j \le N$, let h_{ij} be variables; these variables are called "auxiliary fields" in statistical mechanics. The partition function Z_0 for the first ensemble is defined by

$$Z_0 \equiv \sum_{\{M_{ij}\}} I\Big(A, \sum_{i<j} M_{ij}\Big) \exp\Big(\sum_{i<j} h_{ij} M_{ij}\Big),$$

so $I\big(A, \sum_{i<j} M_{ij}\big)$ is 1 if the number of arcs in the network is A, and 0 otherwise. The number X_0 of networks in this ensemble is given by

$$X_0 = Z_0|_{h_{ij}=0,\ 1 \le i,j \le N}.$$

For this first ensemble, X_0 is immediately available, since X_0 is the number of ways to select A arcs from $N(N-1)/2$ arcs, so

$$X_0 = \binom{N(N-1)/2}{A}.$$

The entropy per node of this ensemble is $(1/N) \log X_0$. Now we impose the node degree constraints. Since δ_i is the node degree of node i, imposing the

N node degree constraints means imposing the constraints $\sum_{j=1}^{N} M_{ij} = \delta_i$ for $i = 1, 2, \ldots, N$. The partition function Z_1 for this second ensemble is defined by

$$Z_1 \equiv \sum_{\{M_{ij}\}} \prod_i I\left(\delta_i, \sum_{j=1}^{N} M_{ij}\right) \exp\left(\sum_{i<j} h_{ij} M_{ij}\right).$$

The partition functions for the ensembles with additional constraints are defined similarly. The next step is to approximate the partition function Z_1; details can be found in [Bianconi 07].

The Entropy of Dynamic Social Networks

The entropy of dynamic social networks was studied in [Zhao 11]. Consider an arbitrary node n (e.g., a phone user) and let g be an arbitrary connected subgraph of the static network $\mathbb{G}$. To indicate that an interaction is occurring among the nodes in g at time t we write $g(t) = 1$. The history $hist_t$ is the set of all groups existing at time t', for $t' < t$. Let $p(g(t) = 1 \mid hist_t)$ be the probability that $g(t) = 1$ given the history $hist_t$. The entropy H of the network characterizes the logarithm of the typical number of different group configurations that can be expected at time t, and $H \equiv -\sum_g p(g(t) = 1 \mid hist_t) \log p(g(t) = 1 \mid hist_t)$. By analyzing the call sequence of subscribers of a major European mobile service provider (634 million calls among 6.2 million subscribers), they showed that H depends on the time of day in a typical weekday, with H varying from 0 to about 0.11. They also showed that the duration of phone calls differs significantly from the duration of face-to-face interactions, and this can be quantified using H.

Network Entropy and Cancer Detection

Cancer cells cause a fundamental dynamic re-wiring of the cellular interaction network, ultimately impairing normal cell physiology. Studies of yeast and *C. elegans* have shown that genes that contribute most to the network entropy are more likely to be essential genes for the organism. The role of network entropy in cancer was studied in [West 12], using weighted networks in which arc weights reflect correlations in gene expression.[2] The weights were normalized to define a random walk on the network. The protein interaction network (PIN) was treated as a backbone topological structure, and phenotype-specific [3] gene expression data (in particular, the correlations in gene expression over the disease phenotype) were used to approximate the interaction probabilities. With this model, cancer cells differ from normal cells due to different arc weights on the same underlying network.

The PIN studied in [West 12] had $N = 10{,}720$ nodes and $A = 152{,}889$ documented arcs (interactions). Let $\mathbb{N}(i)$ be the neighbors of gene i in the PIN

---

[2]Gene expression is the process by which genetic instructions are used to synthesize gene products. These products are usually proteins, which go on to perform essential functions, e.g., as enzymes, hormones, and receptors. MedicineNet.com.

[3] The phenotype is the appearance or characteristic of an individual, which results from the interaction of the person's genetic makeup and his or her environment. By contrast, the genotype is merely the genetic constitution (genome) of an individual. MedicineNet.com.

and let w_{ij} be the weight of arc (i, j). The weight w_{ij} depends on the gene expression between genes i and j. Define the stochastic matrix $\{p_{ij}\}_{i,j=1}^{N}$ by $p_{ij} \equiv w_{ij}/\sum_{k \in \mathbb{N}(i)} w_{ik}$, so $\sum_{j \in \mathbb{N}(i)} p_{ij} = 1$. These p_{ij} values, which define a random walk on the network, can be used to define an "information (or probability) flux" p_{ij}^{s} between any two nodes i and j separated by a path in $\mathbb{G}$ of s hops. The "total probability flux" E_{ij} between i and j was defined by

$$E_{ij} = \gamma \sum_{s=1}^{\infty} (1/s!) p_{ij}^{s},$$

where γ is a normalization factor, and where the weight $1/s!$ suppresses the contributions of longer paths. Finally, let t be the "temperature" and define the stochastic matrix

$$K_{ij}(t) = \frac{1}{e^{t}-1} \sum_{s=1}^{\infty} (t^{s}/s!) p_{ij}^{s}.$$

For large t, the matrix $K(t)$ approximates a solution of a certain heat diffusion equation, and the interpretation of $K(t)$ on a network is related to the analysis of the PageRank algorithm of Brin and Page (the founders of Google). The matrix $K_{ij}(t)$ requires a sum over all s, but $K_{ij}(t)$ can be well approximated by a sum up to path lengths of five hops; for larger s the computation times were excessive. Letting Q denote the number of nonzero elements of the matrix $K(t)$, the network entropy $H(t)$ was defined by

$$H(t) \equiv -\frac{1}{\log Q} \sum_{i,j} K_{ij}(t) \log K_{ij}(t).$$

15.2 Defining the Network Information Dimension

In this section we consider several definitions of the information dimension d_I of a network $\mathbb{G}$. We first discuss a definition based on a random walk [Argyrakis 87]. Then we discuss the definition of [Wei 14a] and show that this definition does not lead to a unique value of d_I, since different minimal s-coverings of $\mathbb{G}$ can yield different values of d_I. Finally, we present the definition of [Rosenberg 17a] which eliminated the ambiguity in [Wei 14a] by requiring that the minimal s-covering of $\mathbb{G}$ must also maximize an appropriately defined network entropy. Computing a maximal entropy minimal s-covering of $\mathbb{G}$ is no more computationally burdensome than computing a minimal s-covering of $\mathbb{G}$ using, e.g., any of the heuristics discussed in Chap. 8.

In [Argyrakis 87], random walks were used to define a network information dimension.[4] Consider a particle performing a random walk of T steps on a

---

[4]Recall that random walks were used, in Sect. 11.3, to compute the correlation dimension of $\mathbb{G}$.

percolating cluster (Sect. 12.5). Let $\mathbb{N}$ be the set of distinct occupied sites visited at least once on the walk, and define $N \equiv |\mathbb{N}|$. Then N is a measure of the overall spread or range of the particle motion, but N is insensitive to the number of times the particle visited the same site. For $n \in \mathbb{N}$, let $v(n)$ be the number of times that site n has been visited in the walk. Since $\mathbb{N}$ only contains sites visited at least once, $v(n) \geq 1$ and $\sum_{n \in \mathbb{N}} v(n) = T$. For $n \in \mathbb{N}$, the probability p_n that the particle visits site n is given by $p_n \equiv v(n)/T$. Define $H \equiv - \sum_{n \in \mathbb{N}} p_n \log p_n$ and $d_I \equiv H / \log T$. Then d_I is the information dimension of the random walk. If each site has the same probability $1/N$ of being visited, then $H = \log N$ and $d_I = \log N / \log T$, which is within a constant factor of the spectral dimension, also known as the fracton dimension (Sect. 20.1).

Computational results in [Argyrakis 87] simulated a lattice in a two-dimensional square topology, where each potential lattice site in the square topology is occupied with probability p, where p ranges from p_c to 1, and where $p_c \approx 0.593$ is the critical percolation threshold.[5] Setting $p = 1$ generates a perfect lattice (a lattice in which each site is occupied). The largest cluster was identified, and all calculations were made on this cluster. The numerical results showed that $d_I \approx 0.89$ for a perfect lattice, and $d_I \approx 0.62$ for a percolating lattice corresponding to $p = p_c$.

In [Ramirez 19], the property of d-summability of a compact set was used to define the d-summable information dimension of a network. However, since the d-summable information dimension is a function of an unknown parameter ν, computing this dimension requires estimating ν using nonlinear regression.

In Sect. 15.1 we discussed three ways of defining the probabilities required to compute the Shannon entropy (15.1) of $\mathbb{G}$. A fourth way [ZhangLi 16] to define the probabilities is to first cover $\mathbb{G}$ by diameter-based boxes of size s or radius-based boxes of radius r (Sect. 7.2). Let $\mathcal{B}$ be the set of boxes used to cover $\mathbb{G}$, and let N_j be the number of nodes in box $B_j \in \mathcal{B}$. Since each box in the covering contains at least one node, we obtain a set of positive probabilities by defining

$$ p_j \equiv \frac{N_j}{\sum_{n \in \mathbb{N}} N_j} \, . \tag{15.6} $$

Defining the probabilities this way is the natural extension of the definition in [Grassberger 83a, Grassberger 83b] of the probabilities used to compute the correlation dimension of Ω (Sect. 9.5). Since N_j is the "mass" of box B_j, it is convenient to refer to the Shannon entropy (15.1), when computed using the probabilities defined by (15.6), as the *Shannon mass entropy*.

The Shannon mass entropy was used by Wei et al. [Wei 14a] to define d_I for $\mathbb{G}$. Suppose, for a range of s, we have computed a minimal s-covering $\mathcal{B}(s)$. For $B_j \in \mathcal{B}(s)$, let $N_j(s)$ be the number of nodes of $\mathbb{G}$ in box $B_j \in \mathcal{B}(s)$. Define

---

[5] Percolation threshold is a mathematical concept related to percolation theory, which is the formation of long-range connectivity in random systems. Below the threshold a giant connected component does not exist; above the threshold, there exists a giant component of the order of system size. https://everipedia.org/wiki/lang_en/Percolation_threshold/.

$p_j(s) \equiv N_j(s)/N$, and define $H(s)$ by (14.4). The information dimension d_I of $\mathbb{G}$ was defined in [Wei 14a] by

$$d_I \equiv - \lim_{s \to 0} \frac{H(s)}{\log s} = \lim_{s \to 0} \frac{\sum_{B_j \in \mathcal{B}(s)} p_j(s) \log p_j(s)}{\log s}. \tag{15.7}$$

Although it was clearly recognized in [Wei 14a] that box counting applies to both weighted and unweighted networks (box counting for weighted networks was studied in Sect. 8.2), the definition of d_I in [Wei 14a] was given only for unweighted networks, and their computational results considered only unweighted networks. However, defining $p_j(s) \equiv N_j(s)/N$ and using (15.7) to compute d_I is equally applicable to weighted networks, and [Wen 18] used (15.7) to compute d_I for several weighted networks. Since (15.7) uses the Shannon mass entropy, it will later be useful to refer to d_I, as defined by (15.6) and (15.7), as the *Shannon mass information dimension* of $\mathbb{G}$.

Although both [Wei 14a] and [Wen 18] used (15.7) to compute d_I, for both unweighted and weighted networks letting $s \to 0$ is neither meaningful nor computationally useful. For an unweighted network, we cannot let $s \to 0$ since the distance between each pair of nodes is at least 1. Moreover, we cannot use the value $s = 1$ in (15.7), since then the denominator of the fraction is zero, while for $s \geq 2$ we have $H(s) > 0$ and $\log s > 0$, which implies $-H(s)/\log s < 0$ and thus, from (15.7), $d_I < 0$. The fact that s cannot become arbitrarily small was recognized in [Wei 14a], which called (15.7) a "theoretic formulation". Similarly, letting $s \to 0$ is not meaningful or computationally useful for weighted networks. These problems with (15.7) can be eliminated by using Definition 15.1 below, which is a computationally useful definition of d_I that is applicable to both weighted and unweighted networks. Moreover, by quantifying how $H(s)$ scales with $\log(s/\Delta)$, rather than how $H(s)$ scales with $\log s$, this definition has the same functional form as the definition of d_I in [Grassberger 83a] and [Grassberger 83b]. We require, as is typical in studying fractal dimensions, the existence of a "regime" or "plateau" over which the desired scaling relation holds approximately.

Definition 15.1 The weighted network $\mathbb{G}$ has the Shannon mass information dimension d_I if for some constant c and for some range of s we have

$$H(s) \approx -d_I \log \left(\frac{s}{\Delta} \right) + c, \tag{15.8}$$

where $H(s)$ is defined by (15.1), $p_j(s) = N_j(s)/N$ for $B_j \in \mathcal{B}(s)$, and $\mathcal{B}(s)$ is a minimal s-covering of $\mathbb{G}$. □

Using this definition, the following example shows that $d_I = 3$ for a 3-dimensional cubic rectilinear lattice. The analysis has the obvious extension to an E-dimensional rectilinear lattice of hypercubes, for any positive integer E.

Example 15.1 [Rosenberg 17a] Let $\mathbb{G}(L)$ be a 3-dimensional cubic rectilinear lattice of L^3 nodes, where L is the number of nodes on an edge of the square (this network, the familiar primitive Bravais lattice in three dimensions, was also studied in Sect. 11.4). The diameter Δ of $\mathbb{G}(L)$ is $3(L-1)$. Suppose $L = 2^K$ for a given positive integer K. For $M = 1, 2, \ldots, K - 1$, we can cover $\mathbb{G}(2^K)$ using copies of $\mathbb{G}(2^M)$. The number of copies required is $2^{3(K-M)}$, and the diameter of $\mathbb{G}(2^M)$ is $3(2^M - 1)$. Using $2^{3(K-M)}$ copies of $\mathbb{G}(2^M)$ approximates, for large M, a minimal covering of $\mathbb{G}(2^K)$ by boxes of size $s = 3(2^M - 1) + 1$. Since the number of nodes in $\mathbb{G}(2^M)$ is 2^{3M} then $p_j = 2^{3M}/2^{3K} = 2^{3(M-K)}$. With $s = 3(2^M - 1) + 1$ we have

$$\frac{s}{\Delta} = \frac{(3)(2^M - 1) + 1}{(3)(2^K - 1)} \approx 2^{M-K}. \tag{15.9}$$

From (14.4) we have

$$H(s) = -2^{3(K-M)} \left(2^{3(M-K)} \log 2^{3(M-K)} \right) = -3(M - K) \log 2 \tag{15.10}$$

which is positive for $M < K$. From (15.9) and (15.10) we have $H(s)/\log\left(s/\Delta\right) \approx -3$. By (15.8), choosing $c = 0$, for $\mathbb{G}(L)$ we have $d_I \approx 3$. $\square$

Example 15.2 [Rosenberg 17a] Figure 15.3 illustrates a *hub and spoke with a tail* network with N nodes. The spokes are the $N - K$ nodes $K + 1$, $K + 2$, ..., $K + N$ that are connected to node K. The diameter Δ of this network is K.

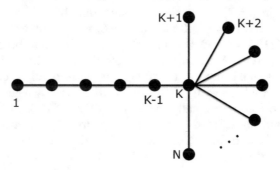

Figure 15.3: *Hub and spoke with a tail* network

The coverings $\mathcal{B}(s)$ for $s = 3$ and $s = 5$ are illustrated in Fig. 15.4 for $K = 6$. When $s = 3$, three boxes are required; one has all $N - K$ spokes and 2 other

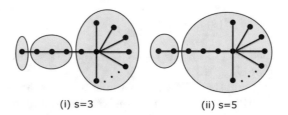

(i) s=3   (ii) s=5

Figure 15.4: Coverings for $K = 6$, for $s = 3$ and $s = 5$

nodes, one has 3 nodes, and the last one has 1 node. When $s = 5$, two boxes are required; one has all $N - K$ spokes and 4 other nodes, and one has 2 nodes.

In general, to compute the covering $\mathcal{B}(s)$ for a positive integer s we must consider two cases. (*Case 1*). $K \leq s - 1$. For this case, one box contains all N nodes. The probability assigned to this box is 1. (*Case 2*). $K \geq s$. For this case, the first box contains $N - K + s - 1$ nodes, namely, the $N - K$ spokes and $s - 1$ other nodes. The probability assigned to this box is $(N - K + s - 1)/N$. This box leaves $K - s + 1$ nodes uncovered, so $\lfloor (K - s + 1)/s \rfloor$ boxes, each with s nodes, are required. The probability assigned to each of these $\lfloor (K - s + 1)/s \rfloor$ boxes is s/N. Finally, if $x \equiv (K - s + 1) - s\lfloor (K - s + 1)/s \rfloor > 0$, a last box covers the remaining x nodes, and the probability of this box is x/N. For example, let $N = 100$, $K = 10$, and $s = 3$. The first box covers the 90 spokes and 2 other nodes; this box has probability $92/100$ and leaves 8 nodes uncovered. These 8 nodes form a chain, so a box of size 3 contains 3 nodes. Using $\lfloor 8/3 \rfloor = 2$ boxes, each with probability $3/100$, we cover 6 nodes, leaving 2 nodes to be covered by the last box, which has probability $2/100$. From (14.4), for *Case 2* we have

$$
\begin{aligned}
H(s) \;=\; & -\left(\frac{N - K + s - 1}{N}\right) \log\left(\frac{N - K + s - 1}{N}\right) \\
& -\left(\left\lfloor \frac{K - s + 1}{s} \right\rfloor\right) \frac{s}{N} \log \frac{s}{N} \\
& -\left(\frac{(K - s + 1) - s\lfloor (K - s + 1)/s \rfloor}{N}\right) \\
& \quad \log\left(\frac{(K - s + 1) - s\lfloor (K - s + 1)/s \rfloor}{N}\right).
\end{aligned}
\tag{15.11}
$$

The number of nodes in the last box (if needed) is less than s. Assume $s \ll N$. Then the third term in (15.11) has the form $(x/N)\log(x/N)$, where $x = (K - s + 1) - s\lfloor (K - s + 1)/s \rfloor < s \ll N$, and by L'Hôpital's rule we ignore this term. Assuming further that $s \ll K$ and $s \ll N - K$, we have

$$
\begin{aligned}
H(s) \;\approx\; & -\left(\frac{N - K + s - 1}{N}\right) \log\left(\frac{N - K + s - 1}{N}\right) - \left(\left\lfloor \frac{K - s + 1}{s} \right\rfloor\right) \frac{s}{N} \log \frac{s}{N} \\
\approx\; & -\left(\frac{N - K}{N}\right) \log\left(\frac{N - K}{N}\right) - \frac{K}{N} \log \frac{s}{N} \\
=\; & -\left(\frac{N - K}{N}\right) \log\left(\frac{N - K}{N}\right) - \frac{K}{N} \log\left(\frac{s}{K} \cdot \frac{K}{N}\right) \\
=\; & -\left[\left(\frac{N - K}{N}\right) \log\left(\frac{N - K}{N}\right) + \frac{K}{N} \log \frac{K}{N}\right] - \frac{K}{N} \log \frac{s}{K}.
\end{aligned}
\tag{15.12}
$$

Setting

$$
c = -\left[\left(\frac{N - K}{N}\right) \log\left(\frac{N - K}{N}\right) + \frac{K}{N} \log \frac{K}{N}\right],
$$

from (15.12) we have $H(s) \approx -(K/N)\log(s/K) + c$ for $s \ll K$ and $s \ll N - K$. Since $\Delta = K$, by (15.8) we have $d_I = K/N$. When $N = K + 1 \gg 1$, the network is a long chain (i.e., a rectilinear lattice in one dimension), for which $d_I = 1$.

When $N \gg K = 2$, the network is a pure hub and spoke network, for which $d_I = 0$.

To determine the accuracy of the approximation $d_I = K/N$ for this example, we compute $H(s)$ for $N = 1000$ and for five values of K, namely, $K = 10, 100,$ $500, 900,$ and 990. For each value of K, we compute $H(s)$ using (15.11), and then fit a line through the nine points $\left(-\log(s/K), H(s)\right)$ for $s = 2, 3, \ldots, 10$; by (15.8), the slope of the line is the estimate of d_I for this value of K. Figure 15.5 (left) shows that, for each of the five values of K, the entropy $H(s)$ is a linear function of $-\log(s/K)$ over the range $s = 2, 3, \ldots, 10$. Figure 15.5 (right), which plots the computed d_I versus K/N for each of the five values of K, shows that K/N is an excellent approximation to d_I for the *hub and spoke with a tail* network. $\square$

Exercise 15.2 Determine d_I of the network of Fig. 15.6, which is constructed by connecting three copies of the network of Fig. 15.3 to a superhub node. $\square$

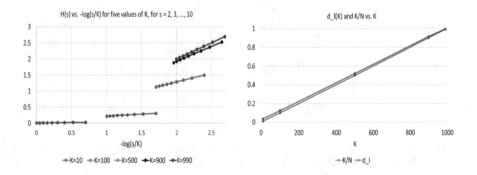

Figure 15.5: Results for the *hub and spoke with a tail* network

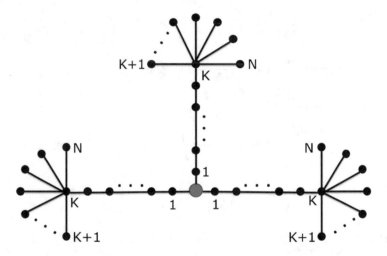

Figure 15.6: Three hubs connected to a superhub

15.3   Comparison with the Box Counting Dimension

We proved in Sect. 14.2 that for a probability distribution we have $d_I \leq d_B$, where d_B is the box counting dimension of the support of the distribution. In this section we explore whether the inequality $d_I \leq d_B$ holds for an undirected, unweighted network $\mathbb{G}$ when d_I is the Shannon information dimension defined by Definition 15.1 and d_B is the network box counting dimension defined by Definition 7.5.

Let $\{p_1, p_2, \ldots, p_J\}$ be positive numbers whose sum is 1. Define the entropy $H \equiv -\sum_{j=1}^{J} p_j \log p_j$. Theorem 14.1 above proved that $H \leq \log J$ and that equality holds when $p_j = 1/J$. When the network probabilities are the mass probabilities (15.6) then in general these probabilities are not equal, and so in general we cannot expect that the Shannon mass entropy $H(s)$, defined by (15.1) and (15.6), is equal to $\log B(s)$. Figure 15.7 illustrates the inequality $H(s) < \log B(s)$ for Example 15.2, the *hub and spoke with a tail* network studied in Sect. 15.2, with $K = 500$.

We now turn to the inequality $d_I \leq d_B$. For geometric fractals, this result is Theorem 14.4. We cannot apply the proof of Theorem 14.4 to $\mathbb{G}$, since we cannot take the limit as $s \to 0$. In [Wei 14a], d_B and the Shannon mass information dimension d_I were computed for four unweighted networks, and for all four networks $d_I > d_B$, with $(d_I - d_B)/d_B$ ranging from 3% to 17%. Yet

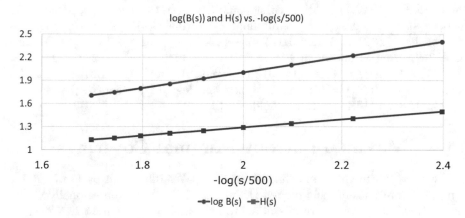

Figure 15.7: $\log B(s)$ and $H(s)$ for the *hub and spoke with a tail* network

$d_I > d_B$ does not always hold for a network: consider again the *hub and spoke with a tail* network of Example 2, with $K = 500$. Figure 15.7 shows that $H(s)$ and $\log B(s)$ are linear in $-\log(s/K)$ for $2 \leq s \leq 10$. Fitting a line to these points yields $d_I \approx 0.52$ and $d_B \approx 0.99$, so $d_I < d_B$.

Suppose the approximate equalities in (15.8) and (7.1) hold as strict equalities over the range $L \leq s \leq U$, where $L < U$. Then we can derive a sufficient condition for the inequality $d_I \le d_B$ to hold for $\mathbb{G}$, as follows. For integers s_1 and s_2 such that $L \le s_1 < s_2 \le U$,

$$d_I = \frac{H(s_1) - H(s_2)}{\log s_2 - \log s_1} \quad \text{and} \quad d_B = \frac{\log B(s_1) - \log B(s_2)}{\log s_2 - \log s_1} . \quad (15.13)$$

Both fractions have the same denominator, which is positive, so $d_I \le d_B$ holds if

$$H(s_1) - H(s_2) \le \log B(s_1) - \log B(s_2) . \quad (15.14)$$

Example 15.3 Consider the five node network of Fig. 15.8. Since $\Delta = 3$

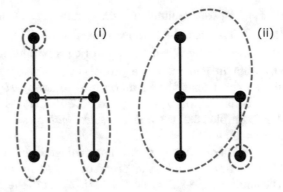

Figure 15.8: Network for which $d_I > d_B$: minimal coverings for $s = 2$ and $s = 3$

then (15.8) and (7.1) hold as strict equalities over the range $[L, U] = [2, 3]$. Choose $s_1 = 2$ and $s_2 = 3$. A minimal 2-covering $\mathcal{B}(s)$ is shown in Fig. 15.8 (*i*); we have $B(2) = 3$ and $H(2) = -\left(\frac{2}{5}\log\frac{2}{5} + \frac{2}{5}\log\frac{2}{5} + \frac{1}{5}\log\frac{1}{5}\right)$. A minimal 3-covering is shown in Fig. 15.8 (*ii*); we have $B(3) = 2$ and $H(3) = -\left(\frac{4}{5}\log\frac{4}{5} + \frac{1}{5}\log\frac{1}{5}\right)$. From (15.13) we have $d_B = 1$ and $d_I \approx 1.37$, so $d_I > d_B$. Also, $H(s_1) - H(s_2) = (4/5)\log 2 \approx 0.24$ and $\log B(s_1) - \log B(s_2) = \log 3 - \log 2 \approx 0.18$, so (15.14) does not hold. $\square$

15.4   Maximal Entropy Minimal Coverings

We have shown that the definition of d_I in [Wei 14a] given by (15.7) is not computationally useful, and offered Definition 15.1 as a computationally viable alternative. However, there is still a problem with Definition 15.1. Consider again Example 15.3 and the minimal coverings illustrated by Fig. 15.8. Using Definition 15.1 with the range $s \in [2, 3]$ we obtain

$$d_I = \frac{H(2) - H(3)}{\log 3 - \log 2} = \left(\frac{4}{5}\right)\left(\frac{\log 2}{\log 3 - \log 2}\right) \approx 1.37 .$$

Yet there is another minimal 3-covering, shown in Fig. 15.9. Using this covering,

$$d_I = \frac{H(2) - H(3)}{\log 3 - \log 2} = \frac{\frac{3}{5}\log 3 - \frac{2}{5}\log 2}{\log 3 - \log 2} \approx 0.94 .$$

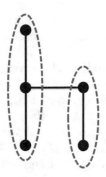

Figure 15.9: Maximal entropy minimal 3-covering

Thus the Shannon mass entropy and hence the Shannon mass information dimension d_I are not uniquely determined by a minimal covering of $\mathbb{G}$. The following definition provides a way to eliminate the ambiguity in the choice of s-coverings used to compute d_I.

Definition 15.2 [Rosenberg 17a] The minimal s-covering $\mathcal{B}(s)$ of $\mathbb{G}$ is a *maximal entropy* minimal s-covering if the entropy $H(s)$ of $\mathcal{B}(s)$, defined by (15.1) and (15.6), is no less than the entropy of any other minimal s-covering. $\qquad \square$

For Example 15.3, the minimal 3-covering in Fig. 15.8 (*ii*) has entropy $-\left(\frac{4}{5}\log\frac{4}{5} + \frac{1}{5}\log\frac{1}{5}\right) \approx 0.217$ while the maximal entropy minimal covering in Fig. 15.9 has entropy $-\left(\frac{3}{5}\log\frac{3}{5} + \frac{2}{5}\log\frac{2}{5}\right) \approx 0.292$.

The justification for this maximal entropy definition is the argument of [Jaynes 57] that the significance of Shannon's information theory is that there is a unique, unambiguous criterion for the uncertainty associated with a discrete probability distribution, and this criterion agrees with our intuitive notion that a flattened distribution possesses more uncertainty than a sharply peaked distribution. Jaynes argued that when making physical inferences with incomplete information about the underlying probability distribution, we must use that probability distribution which maximizes the entropy subject to the known constraints. The distribution that maximizes entropy is the unique distribution that should be used "for the positive reason that it is maximally noncommittal with regard to missing information, instead of the negative one that there was no reason to think otherwise". Accordingly, we replace Definition 15.1 of d_I with the following new definition.

Definition 15.3 [Rosenberg 17a] The network $\mathbb{G}$ has the information dimension d_I if for some constant c and for some range of s,

$$H(s) \approx -d_I \log\left(\frac{s}{\Delta}\right) + c, \qquad (15.15)$$

where $H(s)$ is defined by (15.1) and (15.6), $p_j(s) = N_j(s)/N$ for $B_j \in \mathcal{B}(s)$, and $\mathcal{B}(s)$ is a maximal entropy minimal s-covering. $\qquad \square$

Applying Definition 15.3 to the network of Example 15.3 yields $d_I = 0.94$. Procedure 15.1 below shows how any box-counting method can easily be modified to generate a maximal entropy minimal s-covering for $\mathbb{G}$. Aside from using (15.1) to compute $H(s)$ from $\mathcal{B}(s)$, computing a maximal entropy minimal covering is no more computationally burdensome than computing a minimal covering.

Procedure 15.1 Let $\mathcal{B}_{\min}(s)$ be the best s-covering obtained over all executions of whatever box-counting method is utilized. Suppose we have executed box counting some number of times, and stored $\mathcal{B}_{\min}(s)$ and $H^{\star}(s) \equiv H\big(\mathcal{B}_{\min}(s)\big)$, so $H^{\star}(s)$ is the current best estimate of the maximal entropy of a minimal s-covering. Now suppose we execute box counting again, and generate a new s-covering $\mathcal{B}(s)$ using $B(s)$ boxes. Let $H \equiv H\big(\mathcal{B}(s)\big)$ be the entropy associated with $\mathcal{B}(s)$. If $B(s) < B_{\min}(s)$ holds, or if $B(s) = B_{\min}(s)$ and $H > H^{\star}(s)$ hold, set $\mathcal{B}_{\min}(s) = \mathcal{B}(s)$ and $H^{\star}(s) = H$.   □

The remainder of this section presents computational results [Rosenberg 17a] for four networks. The box-counting method used was the *Random Sequential Node Burning* heuristic (Sect. 8.3).

Example 15.4 Figure 15.10 illustrates an Erdős–Rényi random network, with 100 nodes, 250 arcs, and diameter 6. Table 15.2 and Fig. 15.11 show the results after 1000 executions of *Random Sequential Node Burning*. Over the range

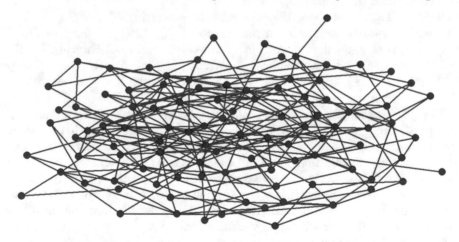

Figure 15.10: *Erdős–Rényi* network

Table 15.2: Results for the *Erdős–Rényi* network

r	$B_{\min}(s)$	$H_{\min}(s)$	$H_{\max}(s)$
1	31	3.172	3.246
2	9	1.676	1.951
3	2	0.168	0.680
4	1	0	0

$1 \leq r \leq 3$ (which, using $s = 2r + 1$, translates to a box size range of $3 \leq s \leq 7$), the best linear fits of $\log B_{\min}(s)$ and $H_{\max}(s)$ versus $\log s$ yielded $d_B = 3.17$ and $d_I = 2.99$, so $d_I < d_B$. The plot of $H_{\max}(s)$ versus $\log s$ was substantially

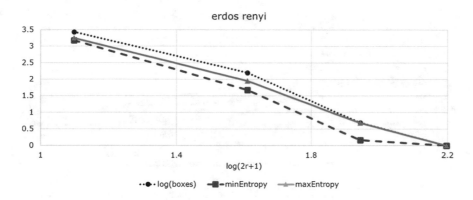

Figure 15.11: Results for the *Erdős–Rényi* network

more linear than the plot of $H_{\min}(s)$ versus $\log s$, so the use of maximal entropy minimal coverings enhanced the ability to compute d_I. Redoing this analysis with only 100 executions of *Random Sequential Node Burning* yielded the quite different results $d_B = 2.71$ and $d_I = 2.72$, so the number of executions of *Random Sequential Node Burning* should be sufficiently high so that the results have converged.   □

Example 15.5 The *jazz* network, with 198 nodes, 2742 arcs, and diameter 6, is a collaboration network of jazz musicians [Gleiser 03]. Table 15.3 and Fig. 15.12 show the results after 1000 executions of *Random Sequential Node Burning*. Over the range $1 \leq r \leq 3$, the best linear fits of $\log B_{\min}(s)$ and $H_{\max}(s)$ versus

Table 15.3: Table of results for the *jazz* network

r	$B_{\min}(s)$	$H_{\min}(s)$	$H_{\max}(s)$
1	20	2.185	2.185
2	4	0.291	0.831
3	2	0.032	0.658
4	1	0	0

$\log s$ yielded $d_B = 2.75$ and $d_I = 1.87$, so $d_I < d_B$.   □

Example 15.6 This network, the anonymized giant component of Lada Adamic's social network, has 350 nodes, 3492 arcs, and diameter 8. Table 15.4 and Fig. 15.13 show the results for the giant component after 2000 executions of *Random Sequential Node Burning*. Over the range $1 \leq r \leq 3$, the best linear fits of $\log B_{\min}(s)$ and $H_{\max}(s)$ versus $\log s$ yielded $d_B = 3.23$ and $d_I = 2.83$, so $d_I < d_B$. For this network we again see that the $H_{\max}(s)$ versus $\log s$ plot is substantially more linear than the $H_{\min}(s)$ versus $\log s$ plot.   □

Example 15.7 This network, illustrated in Fig. 15.14, is a 1998 communications network with 174 nodes, 557 arcs, and diameter 7. Table 15.5 and Fig. 15.15 show the results after 2000 executions of *Random Sequential Node Burning*.

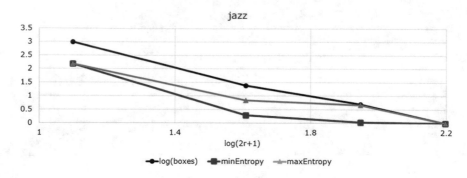

Figure 15.12: Results for the *jazz* network

Table 15.4: Table of results for the *social* network

r	$B_{\min}(s)$	$H_{\min}(s)$	$H_{\max}(s)$
1	62	3.116	3.116
2	12	0.749	1.703
3	4	0.141	0.711
4	1	0	0

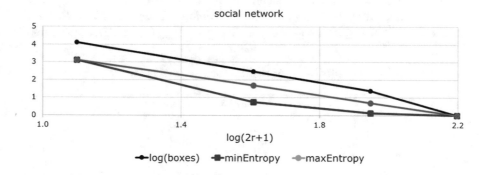

Figure 15.13: Results for *Lada Adamic's social network*

Over the range $1 \leq r \leq 3$, the best linear fits of $\log B_{\min}(s)$ and $H_{\max}(s)$ versus $\log s$ yielded $d_B = 3.25$ and $d_I = 3.28$, so, by a very narrow margin, we have $d_I > d_B$. For this example also, the plot of $H_{\max}(s)$ versus $\log s$ is more linear than the plot of $H_{\max}(s)$ versus $\log s$, validating the usefulness of computing the maximal entropy minimal covering.   □

Summarizing the comparison of d_B and d_I of these four networks, $d_I < d_B$ for three of the four; for one network, d_I was very slightly larger than d_B. For three of the four networks, the use of maximal entropy minimal coverings also had the beneficial effect of producing a more linear plot of $H(s)$ versus $\log s$.

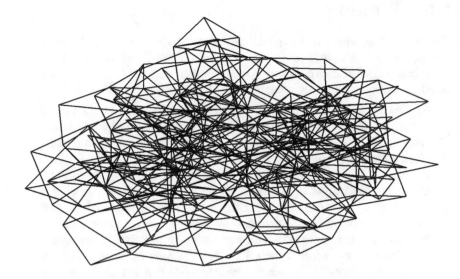

Figure 15.14: *Communications* network

Table 15.5: Table of results for the *communications* network

r	$B_{\min}(s)$	$H_{\min}(s)$	$H_{\max}(s)$
1	50	3.670	3.670
2	14	1.926	2.131
3	3	0.466	0.868
4	1	0	0

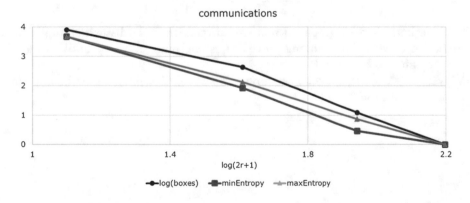

Figure 15.15: Results for the *communications* network

15.5   Binning

When dealing with event distributions best plotted on a double logarithmic scale, e.g., the box counting scaling $B(s) \propto s^{-d_B}$, care must be taken in the proper *binning* of the experimental data. With binning, a range of values is treated as a single value. As described in [Adami 02]:

> Say we are interested in the probability distribution $P(n)$ of an event distribution over positive integer values of n. We conduct N trials, resulting in a data set [with $Q(n)$ observations for each value of n.] For ranges of n where the expected or observed number of events for each n is much higher than 1, normally no binning is required. However, for ranges of n where $Q(n)$ or $P(n)$ is small, binning is necessary to produce both statistically significant data points, and intuitively correct graphical representations. A constant bin size has several drawbacks: One must guess and choose an intermediate bin size to serve across a broad range of parameter space, and the shape and slopes of the curve (even in a double log plot) are distorted. These disadvantages can be overcome by using a variable bin size.

The *data threshold variable bin size method* [Adami 02] begins by selecting a threshold value T. Starting with $n = 1$ and proceeding to higher values of n, no binning is done until a value of n is found for which $Q(n) < T$. When such a value n_1 is found, subsequent $Q(n)$ values are added to $Q(n_1)$ until the sum of these values exceed T, that is, until we reach some $n_2 > n_1$ for which

$$\sum_{n=n_1}^{n_2} Q(n) \geq T.$$

We replace the $n_2 - n_1 + 1$ values of n, and their corresponding $Q(n)$ values, by the single point

$$\left(\frac{n_1 + n_2}{2}, \frac{\sum_{n=n_1}^{n_2} Q(n)}{n_2 - n_1 + 1} \right).$$

This method yields a graphical representation with little distortion and good predictive power. This binning procedure is continued from n_2 until no more data remains to be binned.

Exercise 15.3 Should the *data threshold variable bin size method* be applied when computing d_I?   □

Chapter 16

Generalized Dimensions and Multifractals

But I'll tell you what hermits realize. If you go off into a far, far forest and get very quiet, you'll come to understand that you're connected with everything.
Alan Watts (1915–1973), British-born philosopher, theologian, and lecturer, author of many books on the philosophy and psychology of religion, and Zen Buddhism in particular.

In the preceding two chapters, we studied the information dimension d_I of a probability distribution and of a network. However, in general a single fractal dimension does not suffice to quantify the fractal characteristics of a probability distribution, and instead an entire spectrum of fractal dimensions is required; when this is true, the distribution is said to be *multifractal*. Multifractal behavior was first observed in the 1970s by Mandelbrot in studies of fluid turbulence, and the theory was greatly advanced in 1983 by Grassberger, Hentschel, and Procaccia in their studies of nonlinear dynamical systems [Stanley 88]. By the late 1980s, multifractals had become an extremely active area of investigation for physicists and chemists, with particular interest in the behavior of complex surfaces and interfaces, and in fluid flow in porous media. This chapter presents the theory of multifractal distributions, and the next chapter applies this theory to networks.

The definition in Sect. 9.2 of the correlation dimension d_C started with a probability measure μ. Let Ω be the support of μ, and assume Ω is bounded.

© Springer Nature Switzerland AG 2020
E. Rosenberg, *Fractal Dimensions of Networks*,
https://doi.org/10.1007/978-3-030-43169-3_16

Recall that by (9.3) the pointwise dimension $d(x)$ of μ at $x \in \Omega$ is

$$d(x) \equiv \lim_{r \to 0} \frac{\log \mu\big(\Omega(x,r)\big)}{\log r} , \qquad (16.1)$$

where $\Omega(x,r)$ is the set of points in Ω whose distance from x does not exceed r. The correlation dimension d_C is the average of $d(x)$ over all $x \in \Omega$, and d_C does not capture the variation of $d(x)$ with x. For a multifractal measure μ we are interested in how $d(x)$ varies with x. We study this behavior by defining, for $\alpha \geq 0$, the set $\Omega(\alpha) \equiv \{x \in \Omega \,|\, d(x) = \alpha\}$. This yields the partition $\Omega = \cup_{\alpha \geq 0}\Omega(\alpha)$. If each $\Omega(\alpha)$ is a fractal set, we can determine its fractal dimension, e.g., using box counting. If μ is a multifractal distribution, then the sets $\{\Omega(\alpha)\}_{\alpha \geq 0}$ do not all have the same fractal dimension, but rather there is a range of fractal dimensions.

As an example [Stanley 88] of how a multifractal distribution can occur, consider a random process for which each realization is the product of eight numbers, where each number is randomly chosen, with equal probability, to be either $\frac{1}{2}$ or 2. For example, one realization might generate the eight numbers $(\frac{1}{2}, \frac{1}{2}, 2, \frac{1}{2}, \frac{1}{2}, \frac{1}{2}, 2, 2)$ whose product is $2^3(\frac{1}{2})^5$. There are $2^8 = 256$ possible realizations of this random product. If we simulate this process, and take the average of J realizations of such 8-tuples, then as $J \to \infty$ the average value of the product is $[(2 + \frac{1}{2})/2]^8 = (\frac{5}{4})^8$. One might expect that a small number of realizations yields an average product close to $(\frac{5}{4})^8$, since random sampling procedures typically provide an approximately correct value even when only a small fraction of the possible configurations have been sampled. However, for this process, small random samples yield an average that is not close to $(\frac{5}{4})^8$, due to a small number of low probability configurations (e.g., when all of the eight numbers are 2), which significantly bias the average. To obtain an average of $(\frac{5}{4})^8$, we must actually take the average of about 256 realizations. Multifractal analysis facilitates the identification and characterization of such low probability regions.

Exercise 16.1 For the above example, where a realization is the product of 8 numbers, and where each number is equally like to be 2 or $\frac{1}{2}$, generate a plot showing how the average converges with the number of realizations.   □

Another example in [Stanley 88] of multifractal behavior is illustrated in Fig. 16.1.

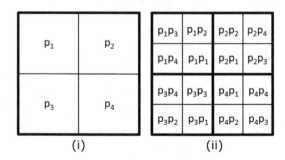

Figure 16.1: Two generations in a multifractal construction

Let p_1, p_2, p_3, and p_4 be nonnegative probabilities whose sum is 1. The four quadrants of the square are assigned these probabilities, as in subfigure (i). Each of these generation 1 quadrants is then divided into four generation 2 quadrants, and the probabilities p_1 to p_4 are randomly assigned to each of the second generation quadrants within each first generation quadrant, as in subfigure (ii). If we continue this for t generations, we obtain 4^t quadrants, where each generation t quadrant is associated with a product p of t probabilities, and each probability is p_1, p_2, p_3, or p_4. If we associate with each possible p value a color, or a shade of gray, the 4^t colors or shades of gray produce a nonuniform pattern. Although this object is not self-similar, we can employ multifractal analysis to characterize those generation t quadrants for which the product of the t probabilities are identical.

16.1 Multifractal Cantor Set

We have already encountered a multifractal in Farmer's "shades of gray" construction in Example 14.2, although that discussion did not use the term "multifractal". We now revisit this construction and present the analysis in [Tél 88] of this multifractal. Let p_1 and p_2 be positive numbers such that $2p_1 + p_2 = 1$. Starting at time $t = 0$ with the unit interval $[0, 1]$, at time $t = 1$ we subdivide $[0, 1]$ into thirds. We assign probability p_1 to the first third and last third, and assign probability p_2 to the middle third, as illustrated in Fig. 16.2. At time $t = 2$, we subdivide into thirds each of the three segments created at

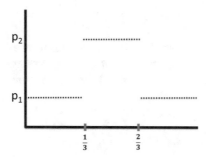

Figure 16.2: Generation 1 of the multifractal Cantor set

time $t = 1$, and compute the probabilities for the $3^2 = 9$ segments by the same rule applied at time $t = 1$. That is, both of the segments with probability p_1 are divided into thirds, and the probabilities of the three segments are $p_1 p_1$, $p_1 p_2$, and $p_1 p_1$. The segment with probability p_2 is also divided into thirds, so the probabilities of these three segments are $p_2 p_1$, $p_2 p_2$, and $p_2 p_1$. Figure 16.3 shows, for $p_1 = 0.2$ and for $t = 1, 2, 3, 4$, the probabilities of each segment generated at time t. The probabilities of the $t = 1$ segments are 0.2, 0.6, and 0.2. Each of the two $t = 1$ segments with probability 0.2 generates, at time $t = 2$, three segments with probabilities $(0.2)(0.2)$, $(0.2)(0.6)$, and $(0.2)(0.2)$, and the $t = 1$ segment with probability 0.6 generates three segments at time $t = 2$ with probabilities $(0.6)(0.2)$, $(0.6)(0.6)$, and $(0.6)(0.2)$. After the time t

Figure 16.3: Four generations of the multifractal Cantor set for $p_1 = 0.2$

construction we have 3^t segments, each of length $(1/3)^t$, and for each t the sum of the 3^t probabilities is 1. Each of the time t probabilities takes one of the $t+1$ values

$$P_m = p_1{}^m p_2{}^{t-m}, \tag{16.2}$$

where m is an integer between 0 and t. Since there are 3^t segments at time t, but only $t + 1$ possible values for the probability assigned to a segment, then multiple segments have the same probability, and the number of segments with probability P_m is

$$N_m = \binom{t}{m} 2^m. \tag{16.3}$$

Since the total number of segments is 3^t, we have

$$\sum_{m=0}^{t} \binom{t}{m} 2^m = 3^t; \tag{16.4}$$

alternatively, (16.4) follows immediately from the binomial expansion

$$(x + y)^t = \sum_{m=0}^{t} \binom{t}{m} x^m y^{t-m} \tag{16.5}$$

with $x = 2$ and $y = 1$.

If $p_2 = 0$, then for each t each nonzero probability has the common value $(1/2)^t$, so we have a uniform distribution. If $p_1 = p_2 = 1/3$, then for each t the probability is constant, with the common value $(1/3)^t$, over the entire interval $[0, 1]$. To better discern the structure of the probabilities, Fig. 16.4 shows the natural logarithm of each of the $3^4 = 81$ probabilities computed in generation 4 with $p_1 = 0.2$. The density of the asymptotic probability distribution function

Figure 16.4: Logarithm of generation 4 probabilities for $p_1 = 0.2$

obtained after an infinite number of steps is discontinuous everywhere on $[0, 1]$ [Tél 88]. For a given t, if we treat each time t segment as a box of size $s = (1/3)^t$, then the value P_m associated with a given segment can be considered a box probability.

16.2 Generalized Dimensions

In the previous section on the multifractal Cantor set, we derived a set of box probabilities P_m and the number N_m of segments in the unit interval whose probability is P_m. In this section we introduce the concept of *generalized dimensions*, which are computed from the box probabilities. For this discussion, we use a more general setting [Theiler 88], but will return to the multifractal Cantor set to illustrate the concepts.

Let Ω be a compact subset of $\mathbb{R}^E$ and let μ be a measure defined on subsets of Ω. For a given $s > 0$, let $\mathcal{B}(s)$ be an s-covering of Ω (Definition 5.1). Let μ_j be the measure of box $B_j \in \mathcal{B}(s)$, and let $p_j = \mu_j/\mu(\Omega)$ be the normalized probability of box B_j. The probabilities p_j depend on the box size s, but for simplicity of notation we write p_j rather than $p_j(s)$. Discarding from $\mathcal{B}(s)$ any box for which $p_j = 0$, we assume $p_j > 0$ for each j, and let $B(s)$ be the number of such boxes. For a dynamical system, p_j is the probability that a random point on the attractor lies in box B_j, and p_j can be estimated from a finite sample by the fraction of points that lie in box B_j. For brevity, let $\sum_j$ denote $\sum_{j=1}^{B(s)}$. For $q \in \mathbb{R}$, define

$$\langle p^{q-1} \rangle \equiv \sum_j p_j^q \ . \tag{16.6}$$

Since $\sum_j p_j^q = \sum_j (p_j^{q-1}) p_j$ then $\langle p_j^{q-1} \rangle$ is the weighted average, using the weights (i.e., the probabilities) $\{p_j\}$, of the terms $\{p_j^{q-1}\}_{j=1}^{B(s)}$. For $q \in \mathbb{R}$ and $q \neq 1$,

$$\frac{1}{q-1} \log \sum_j p_j^q = \frac{1}{q-1} \log \langle p^{q-1} \rangle = \log \left(\langle p^{q-1} \rangle^{\frac{1}{q-1}} \right)$$

and $\langle p^{q-1} \rangle^{\frac{1}{q-1}}$ can be viewed as a generalized average probability per box. If $q = 2$, then

$$\langle p^{q-1} \rangle^{\frac{1}{q-1}} = \langle p \rangle = \sum_j p_j p_j$$

is the ordinary weighted average of the probabilities; if $q = 3$, then

$$\langle p^{q-1} \rangle^{\frac{1}{q-1}} = \langle p^2 \rangle^{\frac{1}{2}} = \sqrt{\sum_j p_j^2 p_j}$$

is a root mean square [Theiler 90].

Definition 16.1 For $q \in \mathbb{R}$ and $q \neq 1$, the *generalized dimension of order q* is

$$D_q \equiv \frac{1}{q-1} \lim_{s \to 0} \frac{\log \langle p^{q-1} \rangle}{\log s} \ , \tag{16.7}$$

assuming the limit exists. For $q = 1$, the *generalized dimension of order 1* is

$$D_1 \equiv \lim_{q \to 1} D_q \ . \quad \square \tag{16.8}$$

The values $\{D_q \mid q \in \mathbb{R}\}$ are the *generalized dimensions* of the measure μ, and they quantify how the generalized average probability per box scales with the box size. Generalized dimensions were introduced in [Grassberger 83, Hentschel 83]. For $q = 0$ we have

$$D_0 = -\lim_{s \to 0} \frac{\log \sum_j p_j^0}{\log s} = -\lim_{s \to 0} \frac{\log B(s)}{\log s}$$

so D_0 is the box counting dimension d_B of Ω (Definition 4.6). For $q = 1$, interchanging the order of limits and applying L'Hôpital's rule yields

$$D_1 \equiv \lim_{q \to 1} D_q = \lim_{q \to 1} \frac{1}{q-1} \lim_{s \to 0} \frac{\log \sum_j p_j^q}{\log s} = \lim_{s \to 0} \frac{1}{\log s} \lim_{q \to 1} \frac{\log \sum_j p_j^q}{q-1} \tag{16.9}$$

$$= \lim_{s \to 0} \frac{1}{\log s} \lim_{q \to 1} \frac{\sum_j p_j^q \log p_j}{\sum_j p_j^q} = \lim_{s \to 0} \frac{1}{\log s} \frac{\sum_j p_j \log p_j}{\sum_j p_j} = d_I$$

so D_1 is the information dimension d_I of μ (Definition 14.1). While the generalized dimensions for $q = 0$, 1, and 2 are the well-known and first "observed" generalized dimensions, the whole spectrum D_q, $q \in \mathbb{R}$ of generalized dimensions plays a non-trivial role in the study of chaos and dynamical systems [Barbaroux 01].

Exercise 16.2 Under what conditions is it valid to interchange the order of the limits as was done in (16.9) to show that $D_1 = d_I$?   □

Exercise 16.3 Show that if the probability measure μ is uniform on Ω, i.e., if all the p_j values are the same, then D_q is independent of q.   □

Exercise 16.4 Prove that D_q (as a function of q) is continuous at $q = 1$.   □

Exercise 16.5 For the multifractal Cantor set, prove that if $p_1 = 1/3$, then $D_q = 1$ for all q.   □

Exercise 16.6 Prove that $\lim_{q\to\infty} \langle p^{q-1}\rangle^{\frac{1}{q-1}} = \max_j p_j$ and that $\lim_{q\to-\infty} \langle p^{q-1}\rangle^{\frac{1}{q-1}} = \min_j p_j$.   □

Suppose the probability measure is nonuniform on Ω. From Exercise 16.6 we have $\lim_{q\to\infty} \langle p^{q-1}\rangle^{\frac{1}{q-1}} = \max_j p_j$. Interchanging limits, and remembering that the p_j values depend on the box size s, we obtain

$$
\begin{aligned}
D_\infty &\equiv \lim_{q\to\infty} D_q = \lim_{q\to\infty} \frac{1}{q-1} \lim_{s\to 0} \frac{\log \langle p^{q-1}\rangle}{\log s} \\
&= \lim_{s\to 0} \frac{1}{\log s} \lim_{q\to\infty} \frac{1}{q-1}\log \langle p^{q-1}\rangle = \lim_{s\to 0} \frac{\log \max_j p_j}{\log s}
\end{aligned}
\tag{16.10}
$$

so D_∞ is the generalized dimension of the most dense region of Ω. Similarly, since $\lim_{q\to-\infty} \langle p^{q-1}\rangle^{\frac{1}{q-1}} = \min_j p_j$, by interchanging limits we obtain

$$
\begin{aligned}
D_{-\infty} &\equiv \lim_{q\to-\infty} D_q = \lim_{q\to-\infty} \frac{1}{q-1} \lim_{s\to 0} \frac{\log \langle p^{q-1}\rangle}{\log s} \\
&= \lim_{s\to 0} \frac{1}{\log s} \lim_{q\to-\infty} \frac{1}{q-1}\log \langle p^{q-1}\rangle = \lim_{s\to 0} \frac{\log \min_j p_j}{\log s}
\end{aligned}
\tag{16.11}
$$

so $D_{-\infty}$ is the generalized dimension of the least dense region of Ω.

For the multifractal Cantor set, Fig. 16.5 plots D_q versus q for $q \in [-5, 5]$ and for five values of p_1. Although D_q is defined in terms of a limit as $s \to 0$, for Fig. 16.5 we used the approximation

$$
D_q \approx \frac{1}{q-1} \frac{\log \langle p^{q-1}(s)\rangle}{\log s}
$$

with $s = 3^{-4}$. For the multifractal Cantor set, the support of μ is the unit interval, for which $d_B = D_0 = 1$; however, $D_q > 1$ when $q < 0$. We will show later that for any multifractal measure, D_q is a non-increasing function of q.

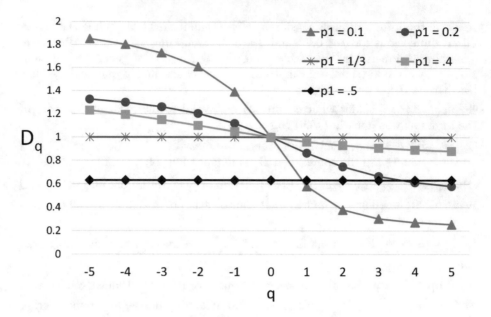

Figure 16.5: Generalized dimensions of the multifractal Cantor set

For box size s, define

$$Z_q(s) \equiv \sum_{B_j \in \mathcal{B}(s)} [p_j(s)]^q .$$

Let the box sizes U and L be given, where $U > L$. For $q \neq 1$ we can estimate D_q using the two-point "secant" approximation [Prigarin 13]

$$D_q(L,U) \equiv \frac{\log Z_q(U) - \log Z_q(L)}{(q-1)\big(\log U - \log L\big)} = \frac{1}{(q-1)\log(U/L)} \log\left(\frac{Z_q(U)}{Z_q(L)}\right). \quad (16.12)$$

Define

$$H(s) \equiv - \sum_{B_j \in \mathcal{B}(s)} p_j(s) \log p_j(s)$$

so $H(s)$ is the entropy at box size s. Setting $q = 1$, the two-point secant estimate of D_1 is [Prigarin 13]

$$D_1(L,U) \equiv \frac{H(L) - H(U)}{\log(U/L)} . \quad (16.13)$$

An analogous two-point secant approximation to D_q for networks [Rosenberg 17c] is studied in Chap. 17.

Exercise 16.7 Using (16.12), derive (16.13).   □

16.3 Spectrum of the Multifractal Cantor Set

Another way to obtain the generalized dimensions defined by (16.7) was provided in [Tél 88]. We illustrate this alternative approach using the multifractal Cantor set presented in Sect. 16.1. Recall that the box probability P_m is given by (16.2) and the number N_m of boxes with probability P_m is given by (16.3). We are interested in the value m for which, as $t \to \infty$, all the boxes with probability P_m contribute the most to the total probability. That is, we want to determine the $m^\star$ which maximizes $N_m P_m$ as $t \to \infty$.

Assume, temporarily, that $p_2 > p_1$. Since $p_2 = 1 - 2p_1$ then $p_1 < 1/3$. By (16.2), the probability P_m is maximized for $m = 0$, in which case $P_0 = p_2^t$, but there is only a single such box (the "middle box" of each interval), since from (16.3) we have $N_0 = 1$. Thus $N_0 P_0 = p_2^t$, and $p_2^t \to 0$ as $t \to \infty$. At the other extreme, for $m = t$ we have $N_t = 2^t$, so such boxes are plentiful. However, $P_t = p_1^t$ and $N_t P_t = (2p_1)^t \to 0$ as $t \to \infty$, so the contribution from these boxes also vanishes.

Without assuming $p_2 > p_1$, we maximize $N_m P_m$ utilizing Stirling's approximation $\log t! \approx t \log t$ for large t (Sect. 23.4). From (16.2) and (16.3),

$$\log N_m P_m = \log t! - \log m! - \log(t - m)! + m \log 2 + m \log p_1$$
$$+ (t - m) \log p_2 \approx t \log t - m \log m - (t - m) \log(t - m)$$
$$+ m \log 2 + m \log p_1 + (t - m) \log p_2 . \tag{16.14}$$

Treating m as a continuous variable, expression (16.14) is maximized over m at the value $m_1 \equiv 2p_1 t$. The reason for the subscript "1" in m_1 will be evident shortly. It is easily verified that expression (16.14) vanishes when $m = m_1$. Thus for large t and for $m = m_1$ we have $\log N_m P_m \to 0$ and $N_{m_1} P_{m_1} \to 1$ as $t \to \infty$. We consider only those boxes with $m = m_1$ and ignore the other boxes, since their contribution to the total probability is negligible.

From (16.3) and again using Stirling's approximation, we obtain

$$\log N_{m_1} = -(2p_1 \log p_1 + p_2 \log p_2) t \tag{16.15}$$

so N_{m_1} grows exponentially with t. Letting $s = (1/3)^t$, where s is the box size, we can determine the dimension f_1 such that

$$N_{m_1} = s^{-f_1} . \tag{16.16}$$

We obtain

$$f_1 = \frac{2p_1 \log p_1 + p_2 \log p_2}{\log(1/3)} . \tag{16.17}$$

These N_{m_1} boxes cover a fractal subset of the interval $[0, 1]$ and f_1 is the dimension of this subset. Since N_{m_1} is a number of boxes, and s is a box size, we might be tempted to call f_1 a box counting dimension. However, as discussed in [Evertsz 92], f_1 is not a box counting dimension, which applies to covering a given set by a collection $\mathcal{B}(s)$ of boxes, and letting $s \to 0$. For the binomial distribution we are considering, a box (i.e., an interval) with a given P_m value contains contains smaller sub-boxes with different P_m values. We will return to this point later.

Exercise 16.8 Prove that (16.14) is maximized over m when $m = 2p_1 t$.   □

To uncover the multifractal characteristics of the probability distribution obtained from the multifractal Cantor set, for $q \in \mathbb{R}$ consider the q-th power of the box probabilities, and the sum

$$\sum_{m=0}^{t} N_m P_m^q = (2p_1^q + p_2^q)^t,$$

where the equality follows directly from (16.2), (16.3), and (16.4). When $q = 1$ we have $(2p_1^q + p_2^q)^t = 1$ for all t (in each generation, the sum of all the segment probabilities is 1), and the dominant term in $\sum_{m=0}^{t} N_m P_m^q$ corresponds to the value of m that maximizes $N_m P_m$, as considered above. In general, for $q \in \mathbb{R}$, the dominant term in $\sum_{m=0}^{t} N_m P_m^q$ corresponds to the value of m that maximizes $N_m P_m^q$. To maximize $N_m P_m^q$ we maximize $\log N_m P_m^q$. Using Stirling's approximation to approximate $\log N_m P_m^q$, the value of m maximizing $\log N_m P_m^q$ is

$$m_q \equiv \left(\frac{2p_1^q}{2p_1^q + p_2^q} \right) t. \tag{16.18}$$

The reason for introducing the subscript "1" in m_1 is now evident; it indicates $q = 1$. Indeed, setting $q = 1$ in (16.18) yields $m_1 = 2p_1 t$, as obtained above. Analogous to what was shown above with $q = 1$, for $q \in \mathbb{R}$ we have $\sum_{m=0}^{t} N_m P_m^q \sim N_{m_q} P_{m_q}^q$ as $t \to \infty$. We leave the proof of this as an exercise.

Exercise 16.9 Use Stirling's approximation to show that for large t the value of m maximizing $\log N_m P_m^q$ is given by (16.18), and that for large t we have $N_{m_q} P_{m_q}^q \approx 1$.   □

Equation (16.18) can be rewritten as

$$\frac{m_q}{t} = \frac{1}{1 + \frac{1}{2}\left(p_2/p_1\right)^q}. \tag{16.19}$$

Suppose $p_2 > p_1$ and let t be fixed. From (16.19) we have $\lim_{q \to \infty} m_q/t = 0$, so $\lim_{q \to \infty} m_q = 0$ and (16.2) yields $\lim_{q \to \infty} P_{m_q} = p_2^t$, corresponding to the most probable box; there is only one such box. Similarly, $\lim_{q \to -\infty} m_q/t = 1$, so $\lim_{q \to -\infty} m_q = t$ and $\lim_{q \to -\infty} P_{m_q} = p_1^t$, corresponding to the least probable box; there are 2^t such boxes. As q increases over the range $(-\infty, \infty)$, different regions of the interval $[0, 1]$ constitute the set of boxes for which $N_m P_m^q$ is maximal. For each integer value of t there are only $t + 1$ possible value of m_q, so we cannot say, when treating m as a discrete variable, that each q maps to a unique m_q.

Above, we determined the fractal dimension f_1, given by (16.17), such that $N_{m_1} = s^{-f_1}$. Generalizing this, we seek the fractal dimension f_q such that

$$N_{m_q} = s^{-f_q} . \tag{16.20}$$

In words, f_q is the dimension of the subset of $[0,1]$ yielding the dominant contribution to the sum $\sum_{m=0}^{t} N_m P_m^q$ [Tél 88]. As with f_1, the dimension f_q is not a box counting dimension. To compute f_q, recall that $s = (1/3^t)$. Consider first the limiting cases $q = \infty$ and $q = -\infty$. Assume for the moment that $p_2 > p_1$. Since $\lim_{q \to \infty} m_q = 0$ and $N_0 = 1$, from (16.20) we have $1 = (1/3^t)^{-d_\infty}$, so $d_\infty = 0$. For the other limiting case, since $\lim_{q \to -\infty} m_q = t$ and $N_t = 2^t$, from (16.20) we have $2^t = (1/3^t)^{-d_{-\infty}}$, so $d_{-\infty} = \log 2 / \log 3$. Having dispensed with these limiting cases, from (16.3), (16.19), (16.20), and Stirling's approximation we obtain for $q \in \mathbb{R}$ the closed-form result

$$f_q = \left[\frac{2p_1^q \log p_1^q + p_2^q \log p_2^q}{2p_1^q + p_2^q} - \log\left(2p_1^q + p_2^q\right) \right] \frac{1}{\log(1/3)} . \tag{16.21}$$

Equation (16.21), which is independent of t, holds for any positive p_1 and p_2 satisfying $2p_1 + p_2 = 1$. When $p_1 = p_2 = 1/3$, for $q \in \mathbb{R}$ we obtain $f_q = 1$. When $p_2 = 0$ and $p_1 = 1/2$, for $q \in \mathbb{R}$ we obtain $f_q = \ln 2 / \ln 3 \approx 0.631$, which is the box counting dimension of the classical middle-third Cantor set. Numerical results for the multifractal Cantor construction for $p_1 = 0.1$ are provided in Table 16.1.

Table 16.1: Three values of f_q for the multifractal Cantor set

q	m_q	P_{m_q}	N_{m_q}	f_q
-6	5	10^{-5}	32	0.631
1	1	0.04096	10	0.582
6	0	0.32765	1	9.4×10^{-5}

Exercise 16.10 Derive equation (16.21). ☐

For the multifractal Cantor set, Fig. 16.6 illustrates the behavior of f_q as a function of p_1, for various values of q. Along the horizontal axis, p_1 ranges from 0.02 to 0.32. Separate curves are shown for $q = -5, -3, -1, 0, 1, 3, 5$.

Definition 16.2 The set $\{f_q \mid q \in \mathbb{R}\}$ is called the *spectrum* of the probability distribution. If for some $f \in \mathbb{R}$ we have $f_q = f$ for $q \in \mathbb{R}$, then the probability distribution is *monofractal*; otherwise, the probability distribution is *multifractal*. ☐

Expressed differently, the distribution is multifractal if the set $\{f_q \mid q \in \mathbb{R}\}$ contains more than one distinct element. Equation (16.21) implies that, as $t \to \infty$, if $p_2 \in (0,1)$ and $p_2 \neq 1/3$, then the probability distribution $\{P_m\}_{m=0}^{t}$

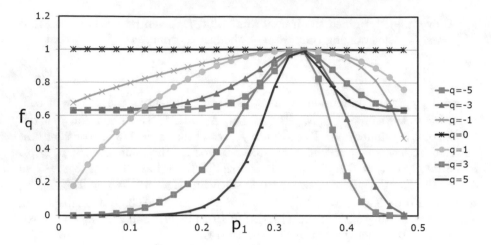

Figure 16.6: f_q versus p_1 for the multifractal Cantor set

is a multifractal distribution. However, for each t the support of the distribution $\{P_m\}_{m=0}^t$ is the interval $[0,1]$ for which $d_B = 1$.

Exercise 16.11 The basis for the computation of f_q for the multifractal Cantor set is the assumption that $\sum_{m=0}^t N_m P_m^q$ is dominated by the single term $N_{m_q} P_{m_q}$. Perform some calculations to quantify how good an approximation this is. □

Continuing the analysis of the multifractal Cantor set, we now define a new exponent which relates the box size to the corresponding probability P_{m_q}. This exponent, denoted by α_q and called the *crowding index*, is defined for the multifractal Cantor set by

$$P_{m_q} = [(1/3)^t]^{\alpha_q} .\qquad(16.22)$$

From (16.2), (16.18), and (16.22), a little algebra yields

$$\alpha_q = \left(\frac{2p_1^q \log p_1 + p_2^q \log p_2}{2p_1^q + p_2^q}\right)\left(\frac{-1}{\log 3}\right)$$

which we rewrite as

$$\alpha_q = \left(\frac{2(p_1/p_2)^q \log p_1 + \log p_2}{2(p_1/p_2)^q + 1}\right)\left(\frac{-1}{\log 3}\right) .\qquad(16.23)$$

From (16.23), for $p_2 > p_1$ we have $\alpha_\infty \equiv \lim_{q\to\infty} \alpha_q = -\log p_2/\log 3$ and $\alpha_{-\infty} \equiv \lim_{q\to-\infty} \alpha_q = -\log p_1/\log 3$ so $\alpha_\infty < \alpha_{-\infty}$ and $\alpha_\infty < \alpha_q < \alpha_{-\infty}$ for all q [Tél 88].

Exercise 16.12 Derive (16.23) from (16.22). □

Using (16.21) we can plot f_q as a function of q for $q \in \mathbb{R}$, and using (16.23) we can plot α_q as a function of q. It is common in the multifractal literature to also plot f_q as a function of α_q; this plot is known as the $f(\alpha)$ spectrum.[1] To compute the $f(\alpha)$ spectrum, define the function $f(\cdot)$ by $f(\alpha_q) \equiv f_q$, so the $f(\alpha)$ spectrum is the set $\{(\alpha_q, f_q)\}_{q \in R}$ of ordered pairs. Figure 16.7 plots the $f(\alpha)$ spectrum for the multifractal Cantor set for $p_1 = 0.2$ and for $p_1 = 0.4$.

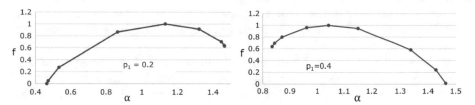

Figure 16.7: $f(\alpha)$ spectrum for the multifractal Cantor set

Exercise 16.13 For the multifractal Cantor set, create a plot similar to Fig. 16.7 for $p_1 = 0.3$.   □

We showed above for the multifractal Cantor set that for $q \in \mathbb{R}$ we can compute an exponent f_q which is the dimension of the fractal subset of $[0, 1]$ providing the dominant contribution to the sum $\sum_{m=0}^{t} N_m P_m^q$. Since f_q is a dimension, it relates the box size, which at time t is $(1/3)^t$, to the number N_{m_q} of boxes. We also mentioned that f_q is not a box counting dimension. To further explore this, we require two more definitions. Let μ be a probability measure with support $\mathcal{X}$, and let $x \in \mathcal{X}$. For $r > 0$, let $B(x, r)$ be the closed ball of radius r centered at x.

Definition 16.3 For $r \in (0, 1)$, the *coarse-grained Hölder exponent* is

$$\alpha(x, r) \equiv \frac{\log \mu(B(x, r))}{\log r}.   \square \tag{16.24}$$

From (16.22), we see that $\alpha(x, r)$ looks like the crowding index. Letting $r \to 0$ in (16.24), we obtain the *local Hölder exponent*.

Definition 16.4 [Falconer 03] The *local Hölder exponent* of μ at x is

$$\alpha(x) \equiv \lim_{r \to 0} \alpha(x, r).   \square \tag{16.25}$$

The local Hölder exponent is also known as the *singularity index* or the *singularity strength* or the *local dimension* of μ at x [Falconer 03]. Comparing (16.25) with (9.3), the local Hölder exponent is another name for the pointwise dimension. For $c \geq 0$, let $\mathcal{X}(c)$ be the set of all points in $\mathcal{X}$ for which $\alpha(x) = c$. It can be shown [Evertsz 92] that for a class of multifractal measures, which includes the binomial measure, $f(c)$ is the Hausdorff dimension of $\mathcal{X}(c)$.

---

[1] While (16.20) implies that $f(\alpha) > 0$ for all α, in [Mandelbrot 90] the theory is extended to allow $f(\alpha) < 0$ for some α values. Roughly, positive values of $f(\alpha)$ define and describe a typical distribution of a random fractal measure, and negative values of $f(\alpha)$ describe the fluctuations one may expect in a finite size sample.

16.4   Spectrum of a Multifractal

In Sect. 16.3 we defined the dimension f_q and the crowding index α_q for the multifractal Cantor set, and provided closed-form expressions for f_q and α_q. In this section we generalize the results of Sect. 16.3 by defining f_q and α_q for any multifractal distribution. We will present the theory in two steps, first the discrete version, and then the continuous version, which replaces each finite sum by an integral.

Following [Frigg 10], let Ω be a bounded subset of $\mathbb{R}^E$. Let μ be a probability measure defined on Ω such that $\mu(\Omega) = 1$. For example, for a dynamical system, Ω could be a box sufficiently large such that the probability is zero that the dynamical system leaves Ω. Let $\mathcal{B}(s)$ be an s-covering of Ω (Definition 5.1). As usual, let $B(s) = |\mathcal{B}(s)|$. We require that the intersection of any two boxes in $\mathcal{B}(s)$ has measure zero. By discarding any box $B_j \in \mathcal{B}(s)$ for which $\mu(B_j) = 0$, we can assume that $\mu_j \equiv \mu(B_j) > 0$ for each B_j. We can approximate μ_j by p_j, the fraction of observed points in box j. (Although B_j and μ_j and p_j depend on s, for notational simplicity we ignore that dependence.) For example, considering a time series $\{x_t\}$ for a dynamical system, we can define $k_j(t)$ to be the number of times that the finite sequence $\{x_1, x_2, \cdots, x_t\}$ visits box B_j, and approximate μ_j by $p_j \equiv k_j(t)/\sum_{B_j \in \mathcal{B}(s)} k_j(t)$ for large t.

Definition 16.5 For $q \in \mathbb{R}$, the *partition function* for the measure μ is

$$Z_q(s) \equiv \sum_{B_j \in \mathcal{B}(s)} \mu_j^q . \qquad \square \qquad (16.26)$$

The partition function is a central concept in thermodynamics and statistical mechanics. For a given s, since $0 < \mu_j < 1$ for each j then $Z_q(s)$ is strictly decreasing in q. We have $Z_0(s) = B(s)$, the number of boxes, and $Z_1(s) = \sum_{B_j \in \mathcal{B}(s)} \mu_j = 1$.

Example 16.1 For the multifractal Cantor set we have $\Omega = [0, 1]$, and $\mathcal{B}(s)$ is the set of the 3^t intervals of length $s = (1/3)^t$ at time t. The probability μ_j of box B_j is P_m, where P_m is given by (16.2), and $Z_q(s) = \sum_{m=0}^{t} N_m P_m^q$, where N_m is given by (16.3). $\square$

Definition 16.6 [Riedi 95, Schroeder 91] Define the function $\tau : \mathbb{R} \to \mathbb{R}$ by

$$\tau(q) \equiv \limsup_{s \to 0} \frac{\log Z_q(s)}{-\log s} . \qquad (16.27)$$

The values $\tau(q)$, $q \in \mathbb{R}$ are *singularity exponents* of μ. $\square$

Since $-\log s > 0$ for $s < 1$ then $\tau(q) \geq 0$ when $Z_q(s) \geq 1$, which occurs when $q \leq 1$; similarly, $\tau(q) \leq 0$ when $Z_q(s) \leq 1$, which occurs when $q \geq 1$. From (16.7), for $q \neq 1$ we have

$$D_q = \frac{\tau(q)}{1 - q} \qquad (16.28)$$

and $D_q \geq 0$, while for $q = 1$ using L'Hôpital's rule we have

$$D_1 = \lim_{q \to 1} D_q = \lim_{q \to 1} \lim_{s \to 0} \frac{1}{(1-q)} \frac{\log Z_q(r)}{(-\log s)} = \lim_{s \to 0} \left(\frac{1}{\log s} \sum_{B_j \in \mathcal{B}(s)} \mu_j \log \mu_j \right) \quad (16.29)$$

Exercise 16.14 Under what conditions is final equality in (16.29) valid? □

Exercise 16.15 If μ is the uniform distribution, what is the value of $\tau(q)$? □

There are many subtle mathematical points in the theory of multifractals, and one arises in relation to definition (16.27). This definition "turns out to be unsatisfactory for reasons of convergence as well as for an undesired dependence on coordinates", and the difficulties are "imperceptibly hidden" in the negative q domain [Riedi 95]. This was illustrated in [Riedi 95] for a binomial measure by showing that $\tau(q) = \infty$ for $q < 0$. Having shown that $\tau(q)$ is an imperfect definition of a singularity exponent, [Riedi 95] proposed a revised theory. However, in this book, we will continue to use (16.27), since that definition is easily presented and is still widely used.

For a given q, as $s \to 0$ the partition function $Z_q(s)$ can be approximated by $\sum_{B_j \in \widetilde{\mathcal{B}}_q(s)} \mu_j^q$ for some subset $\widetilde{\mathcal{B}}_q(s) \subset \mathcal{B}(s)$ such that the probability of each box in $\widetilde{\mathcal{B}}_q(s)$ is approximated by a constant value which we denote by μ_q. Let $B_q(s)$ be the number of boxes in the set $\widetilde{\mathcal{B}}_q(s)$. As $s \to 0$ we have

$$Z_q(s) \approx \sum_{B_j \in \widetilde{\mathcal{B}}_q(s)} \mu_j^q \approx B_q(s)[\mu_q]^q. \quad (16.30)$$

Assume for now that the approximation (16.30) is valid; later in this section we will justify this approximation.

Just as we defined the dimension f_q, the crowding index α_q, and the spectrum $f(\alpha)$ for the multifractal Cantor set, we can define them for any multifractal measure, and describe the multifractal measure in terms of local Hölder exponents $\alpha(x)$ and interwoven sets of Hausdorff dimension $f(\alpha)$, as follows [Chhabra 89]. For each q, the $B_q(s)$ boxes in $\widetilde{\mathcal{B}}_q(s)$ cover a fractal subset of Ω. Let f_q denote the Hausdorff dimension of this fractal subset. Then

$$B_q(s) \sim s^{-f_q} \text{ as } s \to 0. \quad (16.31)$$

Since $Z_0(s) = B(s)$, then f_0 is the box counting dimension d_B of Ω, the support of the multifractal distribution μ. It follows that $f_0 \geq f_q$ for $q \in \mathbb{R}$, since Ω is the union, over $\{q \geq 0\}$, of the subset of Ω with Hausdorff dimension f_q. The plot of f_q versus q increases until it reaches the maximal value f_0 and then decreases.

Definition 16.7 [Chhabra 89] The *crowding index* α_q is defined by

$$\mu_q(s) \sim s^{\alpha_q} \text{ as } s \to 0. \quad \square \quad (16.32)$$

The crowding index is also known as the *scaling index* [Theiler 90] or the *singularity strength*. As higher powers of q select denser regions, α_q is monotone non-increasing in q [Tél 88]. From (16.30), (16.31), and (16.32), as $s \to 0$,

$$Z_q(s) \approx B_q(s)[\mu_q]^q \sim s^{-f_q}(s^{\alpha_q})^q = s^{q\alpha_q - f_q}. \tag{16.33}$$

For $q \neq 1$, from (16.7), (16.26), and (16.33), as $s \to 0$,

$$D_q = \frac{1}{q-1}\lim_{s \to 0}\frac{\log\langle\mu^{q-1}\rangle}{\log s} = \frac{1}{q-1}\lim_{s \to 0}\frac{\log Z_q(s)}{\log s}$$

$$\approx \frac{1}{q-1}\lim_{s \to 0}\frac{\log\left(B_q(s)[\mu_q]^q\right)}{\log s} \sim \frac{1}{q-1}\lim_{s \to 0}\frac{\log s^{q\alpha_q - f_q}}{\log s} = \frac{q\alpha_q - f_q}{q-1}. \tag{16.34}$$

From Definition 16.1 for $q \neq 1$ we have $Z_q(s) \sim s^{(q-1)D_q}$ as $s \to 0$, and from (16.34) we have $Z_q(s) \sim s^{q\alpha_q - f_q}$ as $s \to 0$, so the generalized dimensions satisfy $(q-1)D_q = q\alpha_q - f_q$, hence

$$D_q = \frac{q\alpha_q - f_q}{q-1}. \tag{16.35}$$

From (16.33) and (16.35), we have

$$Z_q(s) \sim s^{(q-1)D_q} \text{ as } s \to 0. \tag{16.36}$$

If $q = 1$, then (16.26) implies $Z_1(s) = 1$ for all s; similarly, if $q = 1$, then (16.36) implies $Z_1(s)$ is independent of s as $s \to 0$. Together, the sets $\{D_q\}_{q \in \mathbb{R}}$ and $\{\alpha_q\}_{q \in \mathbb{R}}$, or the sets $\{f_q\}_{q \in \mathbb{R}}$ and $\{\alpha_q\}_{q \in \mathbb{R}}$, characterize the multifractal distribution μ.

Exercise 16.16 Prove that if for each s we have $\mu_j = 1/B(s)$, then $D_q = d_B$ for all q.   □

Recall from Sect. 11.4 that for $\mathcal{X} \subset \mathbb{R}^E$ and $\mathcal{Y} \subset \mathbb{R}^E$ any reasonable definition of dimension should satisfy the monotonicity property

$$\mathcal{Y} \subset \mathcal{X} \text{ implies dimension}(\mathcal{Y}) \leq \text{dimension}(\mathcal{X}). \tag{16.37}$$

Since $\widetilde{\mathcal{B}}_q(s) \subset \Omega$ then for D_q to be a "valid" dimension, D_q should not exceed the dimension of Ω. Since $f_0 \geq f_q \geq 0$ for $q \in \mathbb{R}$, then (16.35) implies $\lim_{q \to \infty} D_q = \alpha_\infty$ and $\lim_{q \to -\infty} D_q = \alpha_{-\infty}$. For the multifractal Cantor set example of Sect. 16.1, and for $p_2 > p_1$, from (16.23) we have $\alpha_{-\infty} = -\log p_1/\log 3$, which can be made arbitrary large by making p_1 sufficiently small. Thus D_q can be arbitrarily large when q is sufficiently negative; in particular, D_q can exceed the box counting dimension or the Hausdorff dimension of Ω. Thus the "generalized dimension" D_q does not really deserve to be called a "dimension", since D_q does not satisfy the monotonicity property (16.37) [Tél 88].

From (16.7), (16.6), and (14.25),

$$D_q \equiv \frac{1}{q-1}\lim_{s \to 0}\frac{\log\langle p^{q-1}\rangle}{\log s} = \frac{1}{q-1}\lim_{s \to 0}\frac{\log\sum_j p_j^q}{\log s} = \lim_{s \to 0}\frac{-H_q(s)}{\log s}. \tag{16.38}$$

Hence the generalized dimensions are intimately related to the Rényi entropy (Sect. 14.5). Suppose we cover Ω with boxes of size s, so the probability that the system is in box B_j is $\mu_j = \mu(B_j)$. For $q > 0$, the amount of information obtained when we learn which single box the system occupies is given by the Rényi entropy $H_q(s)$, defined by (14.25). Thus the generalized dimension D_q quantifies how the Rényi entropy scales as $s \to 0$.

We next show that the generalized dimensions are non-increasing in q. A condensed proof of this result is presented in [Grassberger 83]. Theorem 16.1 assumes that D_q exists and is differentiable at $q \neq 0$. However, it is possible for discontinuities to exist in the D_q curve (e.g., this occurs for the logistics map $x_{t+1} = 4x_t(1 - x_t)$), and this corresponds to a phase transition in equilibrium statistical mechanics [Chhabra 89].

Theorem 16.1 If D_q exists and is differentiable at $q \neq 0$, then

$$\frac{\mathrm{d}}{\mathrm{d}q} D_q \leq 0 \,.$$

Informal "proof". Without loss of generality, assume that log is the natural logarithm. Recalling (16.7) and (16.26), for $s > 0$ and each j define $z_j(s) \equiv p_j^q(s)/Z_q(s)$, so $\sum_{B_j \in \mathcal{B}(s)} z_j(s) = 1$. For $q \neq 1$ the derivative of D_q with respect to q is

$$\frac{\mathrm{d}}{\mathrm{d}q} D_q = \frac{\mathrm{d}}{\mathrm{d}q} \left\{ \frac{1}{q-1} \lim_{s \to 0} \frac{\ln Z_q(s)}{\ln s} \right\}$$

$$= \lim_{s \to 0} \frac{1}{\ln s} \frac{\mathrm{d}}{\mathrm{d}q} \left\{ \frac{\ln Z_q(s)}{q-1} \right\}$$

$$= \lim_{s \to 0} \frac{1}{\ln s} \frac{1}{(q-1)^2} \left\{ \frac{(q-1)}{Z_q(s)} \sum_{B_j \in \mathcal{B}(s)} p_j^q(s) \ln p_j(s) - \ln Z_q(s) \right\}$$

$$= \lim_{s \to 0} \frac{1}{\ln s} \frac{1}{(q-1)^2} \left\{ (q-1) \sum_{B_j \in \mathcal{B}(s)} z_j(s) \ln p_j(s) - \sum_{B_j \in \mathcal{B}(s)} z_j(s) \ln Z_q(s) \right\}$$

$$= \lim_{s \to 0} \frac{1}{\ln s} \frac{1}{(q-1)^2}$$

$$\times \left\{ \sum_{B_j \in \mathcal{B}(s)} z_j(s) \ln p_j^q(s) - \sum_{B_j \in \mathcal{B}(s)} z_j(s) \ln p_j(s) - \sum_{B_j \in \mathcal{B}(s)} z_j(s) \ln Z_q(s) \right\}$$

$$= \lim_{s \to 0} \frac{1}{\ln s} \frac{1}{(q-1)^2} \sum_{B_j \in \mathcal{B}(s)} z_j(s) \ln \left(\frac{z_j(s)}{p_j(s)} \right) \,. \tag{16.39}$$

(Note that the second equality above implicitly assumes that we may interchange the order of two limits, namely the derivative with respect to q, and the limit as $s \to 0$.) By Lemma 14.2 we have

$$\sum_{B_j \in \mathcal{B}(s)} z_j(s) \ln \left(\frac{z_j(s)}{p_j(s)} \right) \geq 0$$

and for $s < 1$ we have $\ln s < 0$, so by (16.39) we have $\frac{\mathrm{d}}{\mathrm{d}q} D_q \leq 0$. $\square$

Following [Halsey 86], we now justify the approximation (16.30). Consider the s-covering $\mathcal{B}(s)$ of Ω. For box $B_j \in \mathcal{B}(s)$, define the *crowding index* $\alpha(j)$ by

$$p_j = s^{\alpha(j)} . \tag{16.40}$$

Note that $\alpha(j)$ is the crowding index associated with box $B_j \in \mathcal{B}(s)$, not to be confused with definition (16.32) of α_q. Since $p_j < 1$ then for $s < 1$ we have $\alpha(j) > 0$. Replace the $B(s)$ values of $\alpha(j)$ by a continuous density function, by assuming that the number of times α lies in the range $[\alpha, \alpha + d\alpha]$ is $\rho(\alpha) s^{-f(\alpha)} d\alpha$, where $\rho(\alpha)$ and $f(\alpha)$ are continuous functions. Then

$$Z_q(s) = \sum_{B_j \in \mathcal{B}(s)} p_j^q = \sum_{B_j \in \mathcal{B}(s)} s^{q\alpha(j)} \approx \int_{\alpha=0}^{\infty} s^{q\alpha} \rho(\alpha) s^{-f(\alpha)} d\alpha$$

$$= \int_{\alpha=0}^{\infty} \rho(\alpha) s^{q\alpha - f(\alpha)} d\alpha . \tag{16.41}$$

Assume $\rho(\alpha) > 0$ for all α. For $s \ll 1$, using Laplace's *steepest descent* method (Sect. 23.3), the integral (16.41) can be approximated by another integral whose limits of integration are a neighborhood of the point minimizing $q\alpha - f(\alpha)$. Let α_q be the value of the crowding index α such that α_q yields a minimum of the exponent $q\alpha - f(\alpha)$, and such that there is no other minimum in any sufficiently small neighborhood of α_q. Then α_q must satisfy the extremal conditions

$$\frac{\mathrm{d}}{\mathrm{d}\alpha} \{q\alpha - f(\alpha)\} \Big|_{\alpha=\alpha_q} = 0 \tag{16.42}$$

and

$$\frac{\mathrm{d}^2}{\mathrm{d}\alpha^2} \{q\alpha - f(\alpha)\} \Big|_{\alpha=\alpha_q} > 0 . \tag{16.43}$$

The first order optimality condition (16.42) implies

$$f'(\alpha_q) = q . \tag{16.44}$$

The second order optimality condition (16.43) implies $f''(\alpha_q) < 0$, which in turn implies that $f(\alpha)$ is strictly concave at α_q ([Peitgen 92] p. 941; [Halsey 86] p. 1145), ensuring that α_q is an isolated minimizer of $q\alpha - f(\alpha)$. Define $\beta \equiv -\log s$, so $\beta > 0$ for $s \ll 1$. Using (16.41) and applying (23.6) (from Sect. 23.3) with $g(\alpha) \equiv q\alpha - f(\alpha)$, as $s \to 0$ (equivalently, as $\beta \to \infty$),

$$
\begin{aligned}
Z_q(s) &\approx \int_{\alpha=0}^{\infty} \rho(\alpha) s^{q\alpha - f(\alpha)} d\alpha = \int_{\alpha=0}^{\infty} \rho(\alpha) e^{-\beta[q\alpha - f(\alpha)]} d\alpha \\
&\approx \left(\rho(\alpha_q) \sqrt{\frac{2\pi}{\beta g''(\alpha_q)}} \right) e^{-\beta g(\alpha_q)} \sim e^{-\beta g(\alpha_q)} \sim s^{\min_\alpha \{q\alpha - f(\alpha)\}}.
\end{aligned}
$$

Thus as $s \to 0$ we obtain the scaling $Z_q(s) \sim s^{\min_\alpha \{q\alpha - f(\alpha)\}}$. Comparing this scaling with (16.36) yields

$$
(q-1)D_q = \min_\alpha \{q\alpha - f(\alpha)\} . \tag{16.45}
$$

Since by (16.28) we have $\tau(q) = (q-1)D_q$, we can rewrite (16.45) as

$$
\tau(q) = \min_\alpha \{q\alpha - f(\alpha)\} , \tag{16.46}
$$

so $\tau(q)$ is the Legendre transform of $f(\alpha)$ (see Sect. 23.5 for a short tutorial on the Legendre transform). Equation (16.46) immediately implies that $\tau(q)$ is concave function of q for $q \in \mathbb{R}$, since the pointwise infimum of an arbitrary collection of linear functions (each linear function being $q\alpha - f(\alpha)$ for a given value of α) is a concave function (Exercise 16.17). Moreover, since the Legendre relation is symmetric, $f(\alpha)$ is the Legendre transform of $\tau(q)$:

$$
f(\alpha) = \min_q \{q\alpha - \tau(q)\}
$$

and $f(\alpha)$ is a concave function of α ([Theiler 90], p. 1061).[2]

Theorem 16.2 $\max_\alpha f(\alpha) = D_0$.

Proof. The function $f(\alpha)$ attains its maximal value $f^\star$ at the point $\alpha^\star$ if for some value of q the linear function $q\alpha - \tau(q)$ and $f(\cdot)$ have the same value at $\alpha^\star$ and the linear function has slope 0. Since q is the slope of the linear function $q\alpha - \tau(q)$ then $q = 0$ is the desired value. When $q = 0$ the linear function $q\alpha - \tau(q)$ has the constant value $-\tau(0)$, so $f^\star = f(\alpha^\star) = -\tau(0)$. From (16.28) we have $\tau(0) = -D_0$, so $f^\star = D_0$. $\square$

The geometry is illustrated in Fig. 16.8. The five lines $L1$–$L5$ are all tangent to the $f(\alpha)$ curve, and the slope of each line is a different value of q. If the slope of line $L1$ is q_1, the slope of line $L2$ is q_2, etc., then $q_1 > q_2 > q_3 = 0 > q_4 = q_5$.

Exercise 16.17 Let $\mathcal{I}$ be an arbitrary (finite or countable or uncountable) set, and for $i \in \mathcal{I}$ let $f_i : \mathbb{R}^E \to \mathbb{R}$ be concave. Prove that $F \equiv \liminf_{i \in \mathcal{I}} f_i$ is concave. $\square$

Exercise 16.18 Let $\mathcal{I}$ be an arbitrary (finite or countable or uncountable) set, and for $i \in \mathcal{I}$ let $f_i : \mathbb{R}^E \to \mathbb{R}$ be convex. Prove that $F \equiv \limsup_{i \in \mathcal{I}} f_i$ is convex. $\square$

---

[2] As will be discussed later in this section, despite the numerous appearances in the multifractal literature of the assertion that f is concave, the concavity of f holds only under additional assumptions.

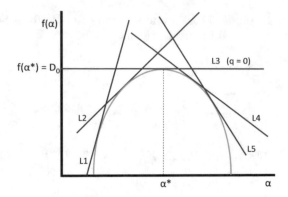

Figure 16.8: Geometry of the Legendre transform of $f(\alpha)$

It was observed in [Olsen 95] that "a number of claims [by theoretical physicists and mathematicians] have been made on the basis of heuristics and physical intuition". The purpose of Olsen's lengthy (114 pages) and highly technical paper is to "determine to what extent rigorous arguments can be provided for this theory". Similarly, it was observed in [Barbaroux 01] that although generalized dimensions had appeared fairly frequently in the physics literature, the main knowledge about these dimensions was "mostly heuristic", and [Barbaroux 01] cited a 1998 paper observing that generalized dimensions "were never examined with real mathematical rigor". To conclude this section we present some equivalent definitions of generalized dimensions that were given in [Barbaroux 01]. We will not present the proofs in [Olsen 95] and [Barbaroux 01], some of which are quite lengthy. Rather our intent is to provide, in preparation for our discussion in Chap. 17 of multifractal networks, more information on generalized dimensions, and an appreciation of how heuristic and "intuitive" reasoning may turn out to be incorrect.

The various definitions in [Barbaroux 01] and [Olsen 95] of generalized dimensions begin with a metric space $\mathcal{X}$ and Borel measure μ which assigns a positive probability to each Borel set S, where $S \subset \mathbb{X}$. Roughly speaking, a Borel set is a set that can be constructed from open or closed sets by repeatedly taking countable unions and intersections[3]; a formal definition can be found in [Royden 68]. The support of μ, denoted by supp μ, is the set of points $x \in \mathbb{R}$ for which every open neighborhood of x has positive measure. For $x \in \mathcal{X}$ and $r > 0$, let $B(x, r)$ be the closed ball with center x and radius r. The *upper local dimension* and *lower local dimension* of μ at x [Olsen 95, Nasr 02] are

$$\overline{d}(x) \equiv \limsup_{r \to 0} \frac{\log \mu\big(B(x, r)\big)}{\log r} \qquad \underline{d}(x) \equiv \liminf_{r \to 0} \frac{\log \mu\big(B(x, r)\big)}{\log r} \, ;$$

---

[3] http://mathworld.wolfram.com/BorelSet.html.

compare these definitions with Definition 9.1 of the pointwise mass dimension, and with Definition 16.3 of the coarse-grained Hölder exponent. For $\alpha \geq 0$, define

$$\underline{\mathcal{X}}(\alpha) \equiv \{x \in \operatorname{supp} \mu \mid \alpha \leq \underline{d}(x)\} \qquad \overline{\mathcal{X}}(\alpha) \equiv \{x \in \operatorname{supp} \mu \mid \alpha \geq \overline{d}(x)\}$$
$$\mathcal{X}(\alpha) \equiv \underline{\mathcal{X}}(\alpha) \cap \overline{\mathcal{X}}(\alpha).$$

This decomposes the (perhaps fractal) set $\operatorname{supp} \mu$ into a family $\{\mathcal{X}(\alpha) \mid \alpha \geq 0\}$ of "subfractals". The idea of such a multifractal decomposition dates back to 1972 and 1974 papers by Mandelbrot on turbulent flows.[4]

The "main problem" of multifractal theory is to estimate the size of $\mathcal{X}(\alpha)$. This is accomplished, in part, by introducing the function $f(\alpha) \equiv d_H\big((\mathcal{X}(\alpha))$, where $d_H\big((\mathcal{X}(\alpha))$ is the Hausdorff dimension of $\mathcal{X}(\alpha)$. This function was first explicitly defined in [Halsey 86]. It was proved in [Halsey 86] that $\{f(\alpha)\}_{\alpha \geq 0}$ and the generalized dimensions $\{D_q\}_{q \in \mathbb{R}}$ can be derived from each other, and are thus equivalent from a physical and heuristic point of view, but, as shown in [Olsen 95], not from a mathematically rigorous point of view. Also, despite the fact that $f(\alpha)$ is often asserted to be a concave function of α (several citations of this assertion are provided earlier in this section), this assertion is debunked in [Olsen 95] as another piece of "multifractal folklore"; conditions under which $f(\alpha)$ is concave were studied in [Olsen 95] and Chapter 17 of [Falconer 03].

We now present some results on equivalent definitions of dimensions [Barbaroux 01]. While all the definitions below are readily extended to measures on $\mathbb{R}^E$, they are presented in [Barbaroux 01], and hence here, only for the real line $\mathbb{R}$. For brevity, we present only results for $q \neq 1$; see [Barbaroux 01] for the $q = 1$ analysis. For $q \neq 1$ and $\epsilon > 0$, define

$$I_\mu(q, \epsilon) = \int_{\operatorname{supp} \mu} \mu\big([x - \epsilon, x + \epsilon)\big)^{q-1} \mathrm{d}\mu(x),$$

$$\tau_\mu^+(q) = \limsup_{\epsilon \downarrow 0} \frac{\log I_\mu(q, \epsilon)}{-\log \epsilon},$$

$$D_\mu^+(q) = \limsup_{\epsilon \downarrow 0} \frac{\log I_\mu(q, \epsilon)}{(q-1)\log \epsilon},$$

with $\tau_\mu^-(q)$ and $D_\mu^-(q)$ defined the same way but using $\liminf$. The quantities $D_\mu^+(q)$ and $D_\mu^-(q)$ are called the *upper and lower generalized fractal dimensions* of μ. For $q < 1$ we have $D_\mu^+(q) = [1/(1-q)]\tau_\mu^+(q)$ and $D_\mu^-(q) = [1/(1-q)]\tau_\mu^-(q)$; for $q > 1$ we have $D_\mu^+(q) = [1/(1-q)]\tau_\mu^-(q)$ and $D_\mu^-(q) = [1/(1-q)]\tau_\mu^+(q)$ (the $+$ and $-$ signs are reversed). The function $\tau_\mu^+(q)$ is a convex function of q for $q \in \mathbb{R}$, and $\tau_\mu^+(q)$, $\tau_\mu^-(q)$, $D_\mu^+(q)$, and $D_\mu^-(q)$ are all non-increasing functions of q for $q \in \mathbb{R}$.

---

[4]The early history of multifractals was reviewed in [Cawley 92], which in particular cited the analysis of [Grassberger 83, Grassberger 83a, Hentschel 83] which is based on Rényi entropies of order $q \geq 0$, and the extension in [Grassberger 85] to $q \in \mathbb{R}$, along with a "slightly sharper notion of generalized dimensions which seems to have anticipated the later partition function formalism".

To discretize these definitions [Barbaroux 01], for $\epsilon \in (0,1)$, and for integer j, define the interval $I(j,\epsilon) \equiv [j\epsilon, (j+1)\epsilon)$. Define $p_j \equiv \mu\big(I(j,\epsilon)\big)$. For $q \neq 1$, define $Z_\mu(q,\epsilon) \equiv \sum_{p_j>0} p_j^q$. The *upper generalized Rényi dimensions* of μ are

$$ RD_\mu^+(q) \equiv \limsup_{\epsilon \downarrow 0} \frac{\log Z_\mu(q,\epsilon)}{(q-1)\log \epsilon} $$

with the *lower generalized Rényi dimensions* of μ defined the same way but using lim inf. It was proved in [Barbaroux 01] that if $q > 0$ and $q \neq 1$, then all the dimensions are equal:

$$ D_\mu^-(q) = RD_\mu^-(q) \ \text{ and } \ D_\mu^+(q) = RD_\mu^+(q). $$

However, for $q \leq 0$ we obtain inequalities, but not equalities:

$$ D_\mu^-(q) \leq RD_\mu^-(q) \ \text{ and } \ D_\mu^+(q) \leq RD_\mu^+(q). \tag{16.47} $$

The fact that strict inequality can occur in (16.47) means that considering the grid $[j\epsilon, (j+1)\epsilon)$ is "too crude to encode totally the $D_\mu(q)$ for $q < 0$" [Barbaroux 01]. Thus the region $q \leq 0$ "seems to be a totally different (and interesting) regime" compared to $q > 0$.

16.5   Computing the Multifractal Spectrum and Generalized Dimensions

In this section we consider the computation of the generalized dimensions and the singularity spectrum of a multifractal measure μ with support Ω. Considering first the computation of the generalized dimensions, for $q \neq 1$ the generalized dimension D_q is typically computed, using (16.27) and (16.28), by computing $Z_q(s)$ for a range of s values for $s \ll 1$, plotting $\big(1/(q-1)\big)\log Z_q(s)$ versus $\log s$, and computing the slope of the line through these points. It is well-known that numerical instabilities can occur if $q < 0$; this is known as the *scattering* problem [Feeny 00].

Example 16.2 Suppose we have 1000 points from Ω, and that these points can be covered by 10 boxes, with 100 points per box. Then each $p_j = 1/10$, so $Z_q(s) = (10)(100/1000)^q = 10^{-q+1}$, so for $q = -5$ we have $Z_q(s) = 10^6$. Now suppose that we have 1001 points from Ω, and that these points can be covered by 11 boxes, with 100 points in each of 10 boxes, and 1 point (an outlier) in the last box. Then $Z_q(s) = (10)(100/1001)^q + (1/1001)^q \approx 10^{-q+1} + 10^{-3q}$. For $q = -5$ the last term dominates, yielding $Z_q(s) \approx 10^{15}$. Thus an outlier point can lead to large variations in $Z_q(s)$. $\square$

Example 16.2 above illustrates that for some q and boxes i and j we might have $p_i^q \gg p_j^q$. If this happens, numerical errors can arise when adding p_i^q and p_j^q as required to compute $Z_q(s)$, due to the finite precision of floating point arithmetic. The use of sorting prevents the contribution of small numbers from being lost when added to much larger numbers. However, since sorting has time complexity $\mathcal{O}(N \log N)$, [Feeny 00] recommended not using sorting. Conclusions in [Roberts 05] were that (i) estimating D_q for $q \leq 0$ is "nonsense", that estimates of D_q for small positive q are "sensitive", and that D_q is robust only for $q \geq 1$; (ii) for very large q, D_q is determined by only the very densest regions of the multifractal and so is not representative of the whole fractal, and (iii) of all the generalized dimension of order $q \geq 1$, the information dimension D_1 is the dimension most representative of the fractal as a whole.

As discussed in Chap. 6, with a straightforward implementation of the box-counting method, the memory and computational complexity both grow faster than $\mathcal{O}(N)$. The method of [Block 90] reduces the memory requirement to $\mathcal{O}(N)$ and has a computational complexity of only $\mathcal{O}(N \log N)$. Assuming the data lies in $\mathbb{R}^E$, this method further requires the data to lie in $\mathcal{Q}$, the unit hypercube in $\mathbb{R}^E$. This can be accomplished as follows. Let Ω be the set of N points $\{x(n)\}_{n=1}^{N}$, where $x(n) \in \mathbb{R}^E$. For $i = 1, 2, \cdots, E$, let $x_i(n)$ be the i-th coordinate of $x(n)$. Let γ be any positive number, and define

$$\Delta \equiv \gamma + \max_{1 \leq i \leq E} \left\{ \max_{x(n) \in \Omega} x_i(n) - \min_{x(n) \in \Omega} x_i(n) \right\}.$$

For any two points x and y in the set $(1/\Delta)\Omega$ we have $\max_{1 \leq i \leq E} |x_i - y_i| < 1$. Define

$$\alpha \equiv \min_{x \in (1/\Delta)\Omega} \min_{1 \leq i \leq E} x_i .$$

Adding $\max\{-\alpha, 0\}$ to each coordinate of each point in Ω will ensure that each coordinate is now nonnegative, and this translation does not change the value of Δ. For $n = 1, 2, \cdots, N$ and each coordinate i, define

$$\widetilde{x}_i(n) \equiv (1/\Delta) \left(x_i(n) + \max\{-\alpha, 0\} \right) . \tag{16.48}$$

Each $\widetilde{x}(n)$ lies inside the unit cube $\mathcal{Q}$. Moreover, since $\gamma > 0$ then $\widetilde{x}_i(n) < 1$ for each i and n. Thus we can, without loss of generality, assume that the N points $\{x(n)\}_{n=1}^{N}$ lie in the unit hypercube $\mathcal{Q}$. Define

$$\delta \equiv \min_{\substack{u,v \in \Omega \\ u \neq v}} dist(u, v) ,$$

so δ is the smallest distance between any two of the N points in Ω. Then $\delta < 1$ and for N large we expect $\delta \ll 1$.

Let $M > 1$ be a fixed positive integer. We decompose $\mathcal{Q}$ into boxes of size s_m, for $m = 0, 1, \cdots, M$, as follows. For $m = 0, 1, \cdots, M$, define

$$k_m \equiv \lfloor \delta^{-m/M} \rfloor \tag{16.49}$$

and define $s_m \equiv 1/k_m$ to be the side length of the boxes in stage m. Since k_m is an integer, for each m we have decomposed the cube $\mathcal{Q}$ into $(k_m)^E$ boxes of size s_m. Note that $k_0 = 1$, that $k_M = \lfloor \delta^{-1} \rfloor > 1$, that

$$1 = k_0 \leq k_1 \leq \cdots \leq k_M ,$$

and therefore

$$1 = s_0 \geq s_1 \geq \cdots \geq s_M\,.$$

Each box size s_m satisfies $s_m \geq \delta$, since $\delta < 1$ and $m \leq M$ imply

$$s_m = \frac{1}{k_m} = \frac{1}{\lfloor \delta^{-m/M} \rfloor} \geq \frac{1}{\delta^{-m/M}} = \delta^{m/M} \geq \delta\,.$$

The box sizes s_m are nearly evenly spaced on a logarithmic scale, since

$$\log s_m - \log s_{m+1} = \log k_{m+1} - \log k_m \approx \log \delta^{-(m+1)/M} - \log \delta^{-m/M}$$
$$= (-1/M)\log \delta\,,$$

and $(-1/M)\log \delta$ is independent of m.

Let $\mathcal{Z}_m$ be the set of E-dimensional vectors $(z_1, z_2, \cdots, z_E)$ such that each z_j is an integer between 0 and $k_m - 1$. Since a box size of s_m generates $(k_m)^E$ boxes in the partition of the cube $\mathcal{Q}$, we can label each box with the appropriate vector in $\mathcal{Z}_m$. Define the function $g_m : \mathcal{Z}_m \to \mathbb{R}$ by

$$g_m(z) \equiv \sum_{i=1}^{E} (k_m)^{i-1} z_i\,. \tag{16.50}$$

For $z \in \mathcal{Z}_m$ we have $g_m(z) \geq 0$ and $g_m(z) \leq (k_m)^E - 1$, since

$$\sum_{i=1}^{E} (k_m)^{i-1} z_i \;\leq\; \sum_{i=1}^{E} (k_m)^{i-1}(k_m - 1) = \sum_{i=1}^{E} (k_m)^i - \sum_{i=1}^{E} (k_m)^{i-1}$$
$$= \frac{k_m[(k_m)^E - 1]}{k_m - 1} - \frac{(k_m)^E - 1}{k_m - 1} = (k_m)^E - 1\,. \tag{16.51}$$

Since $x_i(n) < 1$ for each i and n, then $\lfloor k_m x_i(n) \rfloor \leq k_m - 1$, so for $n = 1, 2, \cdots, N$ we have

$$\big(\lfloor k_m x(n) \rfloor\big) \equiv \big(\lfloor k_m x_1(n) \rfloor, \lfloor k_m x_2(n) \rfloor, \cdots, \lfloor k_m x_E(n) \rfloor\big) \in \mathcal{Z}_m\,.$$

Hence $\big(\lfloor k_m x(n) \rfloor\big)$ is in the domain of g_m, and for $x \in \mathcal{Q}$ we have

$$g_m\big(\lfloor k_m x(n) \rfloor\big) = \sum_{i=1}^{E} (k_m)^{i-1} \lfloor k_m x_i(n) \rfloor\,. \tag{16.52}$$

It follows that for $x \in \mathcal{Q}$ we have $g_m\big(\lfloor k_m x(n) \rfloor\big) \leq (k_m)^E - 1$. Thus, rather than associate each of the N points with an integer vector in $\mathbb{R}^E$, we map each point to a *box address*, which does not exceed $(k_m)^E - 1$, and the addresses are stored in a linear array of size N.

Example 16.3 To illustrate the steps discussed so far, consider the following six points in $\mathbb{R}^2$ given in Table 16.2. Then $\alpha = -7$. Choosing $\gamma = 3$, we obtain $\Delta = 3 + [x_1(2) - x_1(1)] = 3 + [11 - (-7)] = 21$.

Table 16.2: Six points $x(n)$ to be mapped to the unit square

$x(1) = (-7, -4)$	$x(2) = (11, -7)$	$x(3) = (0, 7)$
$x(4) = (8, 5)$	$x(5) = (11, 5)$	$x(6) = (11, 11)$

Applying the mapping (16.48), we obtain the points $\widetilde{x}(n)$ given in Table 16.3 and displayed in Fig. 16.9.

Table 16.3: The mapped points $\widetilde{x}(n)$ inside the unit square

$\widetilde{x}(1) = (0, \frac{1}{7})$	$\widetilde{x}(2) = (\frac{6}{7}, 0)$	$\widetilde{x}(3) = (\frac{1}{3}, \frac{2}{3})$
$\widetilde{x}(4) = (\frac{5}{7}, \frac{4}{7})$	$\widetilde{x}(5) = (\frac{6}{7}, \frac{4}{7})$	$\widetilde{x}(6) = (\frac{6}{7}, \frac{6}{7})$

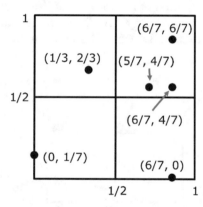

Figure 16.9: The six points in the unit square

The two points closest to each other are $\widetilde{x}(4)$ and $\widetilde{x}(5)$, and $\delta = ||\widetilde{x}(4) - \widetilde{x}(5)|| = 1/7$.

Choose $M = 2$, so m takes the values 0 or 1 or 2. We have $k_m = \lfloor \delta^{-m/M} \rfloor$, so $k_0 = 1$ (the entire cube is one box); $k_1 = \lfloor \delta^{-1/2} \rfloor = \lfloor \sqrt{7} \rfloor = \lfloor 2.64 \rfloor = 2$ (the cube is subdivided into four equal-sized boxes); and $k_2 = \lfloor \delta^{-1} \rfloor = 7$ (the cube is subdivided into 49 equal-sized boxes). Using (16.50), for g_1 we obtain

$$g_1\left(\left\lfloor k_1\left(0,\frac{1}{7}\right)\right\rfloor\right) = \lfloor(2)(0)\rfloor + 2\left\lfloor(2)\left(\frac{1}{7}\right)\right\rfloor = 0$$

$$g_1\left(\left\lfloor k_1\left(\frac{6}{7},0\right)\right\rfloor\right) = \left\lfloor(2)\left(\frac{6}{7}\right)\right\rfloor + 2\lfloor(2)(0)\rfloor = 1$$

$$g_1\left(\left\lfloor k_1\left(\frac{1}{3},\frac{2}{3}\right)\right\rfloor\right) = \left\lfloor(2)\left(\frac{1}{3}\right)\right\rfloor + 2\left\lfloor(2)\left(\frac{2}{3}\right)\right\rfloor = 2$$

$$g_1\left(\left\lfloor k_1\left(\frac{5}{7},\frac{4}{7}\right)\right\rfloor\right) = \left\lfloor(2)\left(\frac{5}{7}\right)\right\rfloor + 2\left\lfloor(2)\left(\frac{4}{7}\right)\right\rfloor = 3$$

$$g_1\left(\left\lfloor k_1\left(\left(\frac{6}{7}\right),\left(\frac{4}{7}\right)\right)\right\rfloor\right) = \left\lfloor(2)\left(\frac{6}{7}\right)\right\rfloor + 2\lfloor(2)(4/7)\rfloor = 3$$

$$g_1\left(\left\lfloor k_1\left(\frac{6}{7},\frac{6}{7}\right)\right\rfloor\right) = \left\lfloor(2)\left(\frac{6}{7}\right)\right\rfloor + 2\left\lfloor(2)\left(\frac{6}{7}\right)\right\rfloor = 3\,.$$

The function g_1 cannot assume a value higher than 3 since by (16.51) for $m = 1$ we have

$$g_1(z) = \sum_{i=1}^{E}(k_m)^{i-1}z_i \le (k_1)^E - 1 = 2^2 - 1 = 3\,. \qquad \square$$

Continuing the presentation of the method of [Block 90], label the four subdivisions of the unit cube Q in $\mathbb{R}^2$ as A, B, C, D, as illustrated in Fig. 16.10. In the partition of Q we must be explicit about the boundaries of the cubes.

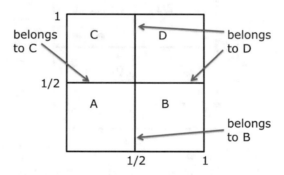

Figure 16.10: Specifying to which box the boundaries belong

We assign the set of points $x = (x_1, x_2)$ for which $x_1 = \frac{1}{2}$ and $x_2 < \frac{1}{2}$ to B. We assign the set of points $x = (x_1, x_2)$ for which $x_1 = \frac{1}{2}$ and $\frac{1}{2} \le x_2 < 1$ to D. We assign the set of points $x = (x_1, x_2)$ for which $x_2 = \frac{1}{2}$ and $x_1 < \frac{1}{2}$ to C. We assign the set of points $x = (x_1, x_2)$ for which $x_2 = \frac{1}{2}$ and $\frac{1}{2} \le x_1 < 1$ to D. The top border of the cube (points for which $x_2 = 1$) is not contained in the cube, so these points are not in C or D. Similarly, the right-hand border of the cube (points for which $x_1 = 1$) is not contained in the cube, so these points are not in B or D.

We now show that g_1 maps each point in A to the value 0. To see this, pick $x \in A$. Then $0 \le x_1 < \frac{1}{2}$ and $0 \le x_2 < \frac{1}{2}$, so $x_1 = \frac{1}{2} - \epsilon_1$, where $0 < \epsilon_1 \le \frac{1}{2}$ and

$x_2 = \frac{1}{2} - \epsilon_2$, where $0 < \epsilon_2 \leq \frac{1}{2}$. Then $g_1(x) = \lfloor (2)(\frac{1}{2} - \epsilon_1) \rfloor + 2\lfloor (2)(\frac{1}{2} - \epsilon_2) \rfloor = 0$. Similarly, we can show that g_1 maps each point in B to the value 1, and each point in C to the value 2. Finally, suppose $x \in D$. Then $\frac{1}{2} \leq x_1 < 1$, and $\frac{1}{2} \leq x_2 < 1$, so $x_1 = 1 - \epsilon_1$, where $0 < \epsilon_1 \leq \frac{1}{2}$ and $x_2 = 1 - \epsilon_2$, where $0 < \epsilon_2 \leq \frac{1}{2}$. Then $g_1(x) = \lfloor (2)(1 - \epsilon_1) \rfloor + 2\lfloor (2)(1 - \epsilon_2) \rfloor = 3$.

In general, for N points $x(n)$ in $\mathbb{R}^E$, and for arbitrary M and m satisfying $0 \leq m < M$, the value $g_m(\lfloor k_m x(n) \rfloor)$ is an integer between 0 and $(k_m)^E - 1$, and each of the $(k_m)^E$ boxes in the decomposition of $\mathcal{Q}$ is mapped to a unique integer between 0 and $(k_m)^E - 1$. Thus the number of points $x(n)$ which map to a given integer z, where $0 \leq z < (k_m)^E - 1$, is exactly the number of points in the corresponding box in the decomposition of $\mathcal{Q}$ which are mapped by g_m to z. So we need only compute $g_m(\lfloor k_m x(n) \rfloor)$ for each of the N points to know the weight of each box.

Returning now to the method of [Block 90], let m be fixed. Compute $g_m(\lfloor k_m x(n) \rfloor)$ for $n = 1, 2, \cdots, N$, and order these N values in ascending order. Let $\hat{g}_n$ be the sorted values; for simplicity of notation, we write $\hat{g}_n$ rather than $g_m(\lfloor k_m x(n) \rfloor)$. Then

$$\hat{g}_1 \leq \hat{g}_2 \leq \cdots \leq \hat{g}_N.$$

For example, suppose $N = 9$, $E = 2$ (the data lie in $\mathbb{R}^2$), $\delta = \frac{1}{8}$, $M = 12$, and $m = 4$. From (16.49) we have $k_m = 8^{4/12} = 2$, so the unit cube is covered by 4 boxes. Suppose four of the 9 points are in box 2, two points are in box 3, three points are in box 4, and no points are in box 1. Then $(\hat{g}_1, \hat{g}_2, \cdots, \hat{g}_N) = (2, 2, 2, 2, 3, 3, 4, 4, 4)$. For a given q, the following pseudocode computes the partition function $Z(q, m) \equiv \sum_{j=1}^{B(s_m)} [p_j(s_m)]^q$. The counter c is the number of points mapped by g_m to the value v.

```
1    initialize: n = 2, v = ĝ₁, c = 1, and Z(q,m) = 0;
2    while (n ≤ N) {
3        if (ĝₙ = v)  { c = c + 1 }
4        else  { Z(q,m) = Z(q,m) + (c/N)^q, v = ĝₙ,  and c = 1 }
5        n = n + 1
6    }
```

Once $Z(q, m)$ has been computed for $m = 0, 1, \cdots, M$, we can compute D_q, e.g., using regression. As for the computational complexity of this method, the computation of δ (the minimal distance between any two points) is $\mathcal{O}(N^2)$. However, this computation is only done once, since it is independent of M and m. The N values $\hat{g}_n$, $n = 1, 2, \cdots, N$ must be sorted for each m, where m ranges from 0 to M, and each sort has complexity $\mathcal{O}(N \log N)$. As for memory requirements, besides the storage needed for the N points $x(n)$, an array of length N is needed for the $\hat{g}_n$ values, and an array of length $\log N$ is needed as temporary storage for the sorting method. This concludes the presentation of the method of [Block 90].

Computation of the Singularity Spectrum

We now consider the computation of the singularity spectrum $\{f(\alpha)\}_{\alpha\geq 0}$. In the multifractal literature, the generalized dimensions $\{D_q\}_{q\in\mathbb{R}}$ were proposed earlier than the singularity spectrum, and the generalized dimensions are easier to compute than the singularity spectrum, which accounts for their wide usage [Chhabra 89]. Since $\{D_q\}_{q\in\mathbb{R}}$ has in the past been easier to evaluate than $\{f(\alpha)\}_{\alpha\geq 0}$ for measures arising from real or computer experiments, usually $f(\alpha)$ has been computed by the Legendre transform of the $\tau(q)$ curve. However, a major issue arises when trying to calculate $f(\alpha)$ from $\tau(q)$. The issue is that for a given q the computation of D_q involves finding a range of s over which the plot of $\log Z_q(s)$ versus $\log s$ is approximately linear. The uncertainty (as quantified by the error bar) in the estimate of D_q depends on q. In particular, the D_q value for $q\ll 0$ may have a large error bar since such a D_q corresponds to a low probability event. Hence determining error bars for the Legendre transform of the $\tau(q)$ curve is "hazardous". Therefore it is desirable to calculate the multifractal spectrum directly from experimental data, without having to use the Legendre transform.

One might imagine that $\{f(\alpha)\}_{\alpha\geq 0}$ can be calculated in a straightforward manner from (16.31) and (16.32), as follows. First cover the support of the measure with boxes of size s, and compute the $p_j(s)$ values. Then compute α_j using (16.32). Let $B(\alpha_j,s)$ be the number of boxes that possess the singularity strength α_j. The slope of the $\log B(\alpha_j,s)$ versus $\log s$ plot yields, by (16.31), an estimate of $f(\alpha_j)$. However, for finite data such a procedure yields inaccurate estimates due to poor convergence to asymptotic values. For example, Mandelbrot found that for the binomial measure the entire $f(\alpha)$ curve visibly overshoots the true values even with 10^{15} boxes, due to the scale-dependent prefactors in (16.32).

The method proposed in [Chhabra 89] directly computes $\{f(\alpha)\}_{\alpha\geq 0}$ for a special class of measures that arise from multiplicative processes, e.g., the multiplicative process yielding Farmer's "shades of gray" construction of Example 14.2. For a range of box sizes s, cover Ω with a set $\mathcal{B}(s)$ of boxes of size s and compute, as usual, the box probabilities $p_j(s)$. Discard all boxes with $p_j(s)=0$. For $q\in\mathbb{R}$, define

$$\mu_j(q,s) \equiv \frac{[p_j(s)]^q}{\sum_{B_j\in\mathcal{B}(s)}[p_j(s)]^q}.$$

As usual, setting $q\gg 1$ amplifies the denser regions of the measure, setting $q\ll 0$ amplifies the less dense regions, and setting $q=1$ replicates the original measure. Then

$$f(q) = \lim_{s\to 0}\frac{\sum_{B_j\in\mathcal{B}(s)}\mu_j(q,s)\log\mu_j(q,s)}{\log s} \tag{16.53}$$

and the average value of the singularity strength is

$$\alpha(q) = \lim_{s\to 0}\frac{\sum_{B_j\in\mathcal{B}(s)}\mu_j(q,s)\log p_j(s)}{\log s}, \tag{16.54}$$

so (16.53) and (16.54) implicitly relate the Hausdorff dimension f and the average singularity strength α to q, without the need to compute the Legendre transform of the $\tau(q)$ curve.

To conclude this section, we note that statistical tests can be used to confirm the presence of multifractality. In applying the correlation dimension to Chinese stock markets, [Jiang 08] used a bootstrap approach to test whether the "width" $W(\alpha) \equiv \alpha_{max} - \alpha_{min}$ of the spectrum exceeds the width that would be produced by chance. They reshuffled the time series S times to remove any potential temporal correlation, and ran the same multifractal analysis as for the original unshuffled data. For each set of shuffled data they computed $W_S(\alpha)$, the width of the spectrum for the shuffled data. The statistic $(1/S)(\#[W(\alpha) \leq W_S(\alpha)])$, where $\#[W(\alpha) \leq W_S(\alpha)]$ is the number of times that $W(\alpha)$ does not exceed $W_S(\alpha)$, was used to determine if the multifractal behavior is statistically significant.

16.6 The Sandbox Method for Geometric Multifractals

As discussed in Sect. 16.4, to compute the generalized dimensions D_q of a geometric object Ω using (16.7), we first cover Ω by a set $\mathcal{B}(s)$ of boxes of size s. To compute $\sum_{B_j \in \mathcal{B}(s)} p_j^q(s)$ we need to estimate $p_j(s)$, the probability of box $B_j \in \mathcal{B}(s)$. From a given set $\mathbb{N}$ of N points in Ω, we estimate $p_j(s)$ by $N_j(s)/N$, where $N_j(s)$ is the number of points in box B_j. Suppose for some box we have $p_j(s) \approx 0$. As observed in Sect. 16.5, for $q \ll 0$ the term $p_j^q(s)$ in (16.7) can become large, dominating the partition sum $\sum_{B_j \in \mathcal{B}(s)} p_j^q(s)$ [Ariza 13, Fernandez 99]. Thus the partition sum is very sensitive to the estimate of $p_j(s)$, which in turn is very sensitive to the placement of the box grid. The *sandbox method* [Tél 89] is a popular method for approximating D_q. This method, originally used to study geometric multifractals obtained by simulating diffusion limited aggregation on a lattice [Vicsek 89, Vicsek 90, Witten 81], "drastically reduces the undesired effect of almost empty boxes, which is a limitation of the box counting method" [DeBartolo 04]. We have already encountered the sandbox method, in Sect. 10.5, as a way of approximating the correlation dimension. In this section we build upon Sect. 10.5 by showing how the sandbox method can be used to compute the generalized dimensions of Ω.

Consider a lattice $\mathcal{L} \subset \mathbb{R}^E$. Let L be the linear size, e.g., side length, of $\mathcal{L}$. Each position on $\mathcal{L}$ is either empty or occupied; an occupied position is called a *point* (e.g., a point might be a particle in a diffusion limited aggregate). Let N be the number of points on $\mathcal{L}$. Let δ be the spacing between adjacent lattice positions (e.g., between adjacent horizontal and vertical positions on a lattice in $\mathbb{R}^2$). Using the box counting approach to computing D_q, we would first compute a set $\mathcal{B}(s)$ of boxes, of linear size s, covering the N particles, such that each box in $\mathcal{B}(s)$ contains at least one of the N particles. This is illustrated by Fig. 16.11

for a lattice $\mathcal{L}$ in $\mathbb{R}^2$, where the lattice positions are the black circles, the $N = 9$ points (occupied positions) are blue squares, the side length of $\mathcal{L}$ is $L = 5\delta$, and the nine points in $\mathcal{L}$ are covered by two boxes of side length $s = 2\delta$.

The sandbox method does not use a covering of $\mathcal{L}$ by boxes of size s. Instead, the sandbox method randomly chooses $\widetilde{N}$ of the N points as centers, and considers a radius-based box of radius r centered on each of these $\widetilde{N}$ points. These boxes may overlap. The term "sandbox" connotes a set of sandpiles, where the "mass" of each sandpile of a given radius is the number of points (e.g., grains of sand) in the pile. For each r, the average sandpile mass is used to compute the sandbox dimension.

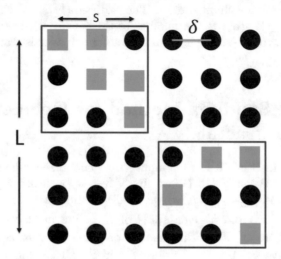

Figure 16.11: Nine points covered by two boxes

For a geometric multifractal for which D_q exists, by (16.26) and (16.36) we have, as $s \to 0$,

$$Z_q(s) = \sum_{B_j \in \mathcal{B}(s)} p_j^q(s) = \sum_{B_j \in \mathcal{B}(s)} p_j(s)[p_j(s)]^{q-1} \sim s^{(q-1)D_q} \ .$$

To approximate $Z_q(s)$, let $\widetilde{\mathbb{X}}$ be a randomly chosen subset of the N points and define $\widetilde{N} \equiv |\widetilde{\mathbb{X}}|$. Recalling that $M(x, r)$, defined by (12.19) and (12.20), is the number of points whose distance from x does not exceed r, define

$$avg\big(p^{q-1}(r)\big) \equiv \frac{1}{\widetilde{N}} \sum_{x \in \widetilde{\mathbb{X}}} \left(\frac{M(x, r)}{N} \right)^{q-1} , \qquad (16.55)$$

where the notation $avg\big(p^{q-1}(r)\big)$ is chosen to make it clear that this average uses equal weights of $1/\widetilde{N}$. The essence of the sandbox method is the approximation, for $r \ll L$,

$$avg\big(p^{q-1}(r)\big) \sim (r/L)^{(q-1)D_q} . \qquad (16.56)$$

Note that $\sum_{B_j \in \mathcal{B}(s)} p_j^q(s)$ is a sum over the set of non-empty grid boxes, and the weight applied to $[p_j(s)]^{q-1}$ is $p_j(s)$. In contrast, $avg(p^{q-1}(r))$ is a sum over a randomly selected set of sandpiles, and the weight applied to $(M(x,r)/N)^{q-1}$ is $1/\widetilde{N}$. Since the $\widetilde{N}$ sandpile centers are randomly chosen from the N points, the sandpiles may overlap. Because the sandpiles may overlap, and the sandpiles do not necessarily cover all the N points, in general $\sum_{x \in \widetilde{\mathbb{X}}} M(x,r) \neq N$, and we cannot regard the values $\{M(x,r)/N\}_{x \in \widetilde{\mathbb{X}}}$ as a probability distribution.

Definition 16.8 [Tél 89] For $q \neq 1$, the *sandbox dimension function* of order q is the function of r defined for $\delta \leq r \ll L$ by

$$D_q^{sandbox}(r/L) \equiv \frac{1}{q-1} \frac{\log avg(p^{q-1}(r))}{\log(r/L)} \; . \qquad \square \qquad (16.57)$$

For a given $q \neq 1$ and lattice size L, the sandbox dimension function does not define a single sandbox dimension, but rather a range of sandbox dimensions, depending on r. It is not meaningful to define $\lim_{r \to 0} D_q^{sandbox}(r/L)$, since r cannot be smaller than the spacing δ between lattice points. In practice, for a given $q \neq 1$ and L, a single value $D_q^{sandbox}$ of the sandbox dimension of order q is typically obtained by computing $D_q^{sandbox}(r/L)$ for a range of r values (where r is at least two orders of magnitude smaller than L and two orders of magnitude greater than the pixel size [Fernandez 99]) and finding the slope of the $\log avg(p^{q-1}(r))$ versus $\log(r/L)$ curve. The estimate of $D_q^{sandbox}$ is $1/(q-1)$ times this slope.

If for $q \neq 1$ the fractal has sandbox dimension $D_q^{sandbox}$, then over some range $r_{min} \leq r \leq r_{max}$, where $\delta < r_{min} < r_{max} < L$, we have

$$avg(p^{q-1}(r)) \sim \left(\frac{r}{L}\right)^{(q-1)D_q^{sandbox}} . \qquad (16.58)$$

From Sect. 16.4 we know that $D_0 = d_B$. Setting $q = 0$, from (16.58) we have

$$avg(p^{-1}(r)) \sim \left(\frac{r}{L}\right)^{-D_0^{sandbox}} ,$$

so $D_0^{sandbox}$ is obtained by plotting

$$\log \left(\frac{1}{\widetilde{N}} \sum_{x \in \widetilde{X}} \left(\frac{M(x,r)}{N} \right)^{-1} \right)$$

against $\log(r/L)$.

To define the sandbox dimension function of order q for $q = 1$ we follow the derivation of [DeBartolo 04], which began by defining the sandbox dimension of order q as

$$D_q^{sandbox} \equiv \frac{1}{q-1} \lim_{r \to 0} \frac{\log avg(p^{q-1}(r))}{\log r} . \qquad (16.59)$$

This is a different definition than Definition 16.8 of the sandbox dimension, since (16.59) takes a limit as $r \to 0$ but there is no limit taken in (16.57). With this caveat, we continue with the derivation in [DeBartolo 04] for $q = 1$. In (16.58), set $q = 1 + dq$, where dq is the differential. Then

$$
\begin{aligned}
avg\big(p^{1+dq-1}(r)\big) &= avg\big(p^{dq}(r)\big) = avg\big(e^{dq\,\log p(r)}\big) \sim \left(\frac{r}{L}\right)^{(1+dq-1)D_{1+dq}^{sandbox}} \\
&= \left(\frac{r}{L}\right)^{dq\,D_{1+dq}^{sandbox}}.
\end{aligned} \tag{16.60}
$$

Since $|dq| \ll 1$, the first order Taylor series approximation $e^x \approx 1 + x$ yields

$$
avg\big(e^{dq\,\log p(r)}\big) \approx avg\big(1 + dq\,\log p(r)\big) = 1 + dq\,avg\big(\log p(r)\big). \tag{16.61}
$$

From (16.60) and (16.61),

$$
1 + dq\,avg\big(\log p(r)\big) \sim \left(\frac{r}{L}\right)^{dq\,D_{1+dq}^{sandbox}}. \tag{16.62}
$$

Taking the logarithm of both sides of (16.62) and using the approximation $\log(1 + \alpha) \approx \alpha$ for $|\alpha| \ll 1$,

$$
dq\,avg\big(\log p(r)\big) \sim dq\,D_{1+dq}^{sandbox}\,\log(r/L)
$$

which yields

$$
D_{1+dq}^{sandbox} = \lim_{r \to 0} \frac{avg\big(\log p(r)\big)}{\log(r/L)}
$$

and finally,

$$
D_1^{sandbox} \equiv \lim_{dq \to 0} D_{1+dq}^{sandbox} = \lim_{r \to 0} \frac{avg\big(\log p(r)\big)}{\log(r/L)}. \tag{16.63}
$$

In practice, $D_q^{sandbox}(r/L)$ is a good approximation to D_q[Tél 89, DeBartolo 04], and the sandbox method has been quite popular. For geometric multifractals with known theoretical dimension, the sandbox method provides better estimates of the generalized dimensions D_q than the estimates based on box counting; however, the sandbox method requires considerably more computing time [Tél 89]. Also, although the box counting method is "impractical" due to its significant bias for small samples, the sandbox method, while outperforming box counting, is "not reliable" for estimating D_q when $q < 0$ [Kamer 13]. We are warned in [Daxer 93], who cited [Vicsek 89], that the sandbox method may yield incorrect results if the object exhibits inhomogeneity, and therefore the sandbox method should be used with caution.

Comparing (16.55) with the definitions in Sects. 9.2 and 9.4 shows that the sandbox method, which averages a set of pointwise mass dimensions with random centers, is a variant of the correlation dimension. Indeed, [Kamer 13] observed that the sandbox method was inspired by correlation method of

[Grassberger 83b], and in [Daxer 93] the terms "sandbox method" and "mass-radius method" were used synonymously. However, the sandbox method is sometimes viewed as a variant of the box counting dimension, e.g., by [Amarasinghe 15], who computed three different fractal dimensions of electrical discharges from a set of photographic images. The estimated fractal dimensions d_B, $D_0^{sandbox}$, and d_C for long laboratory sparks were 1.20 ± 0.06, 1.66 ± 0.05 and 1.52 ± 0.12, respectively.

The sandbox dimension was used to study generalized number systems [Farkas 08], and used to relate the amount of branching of a fungus to the fungal secretions [Jones 99]. To compare the fat infiltration in two types of meat (Iberian pork and White pork), [Serrano 13] used the sandbox method to compute D_q for $q \in [-10, 10]$ and to compute the spectrum $f(\alpha)$; the spectrum for Iberian pork was found to be wider and rounder than for White pork. To study river networks, [DeBartolo 04] applied the sandbox method for $q \in [-10, 10]$, using at most 100 sandpiles (i.e., $\widetilde{N} \leq 100$) and $r \in [0.004, 0.252]$. For each network, a polynomial approximation of the mass exponent function $\tau(q)$ was calculated (using a fifth to seventh order polynomial) and used to compute the multifractal spectrum $f(\alpha)$. One conclusion of [DeBartolo 04] was that while $f(\alpha)$ is a measure of the geometric complexity of a river network, more work is needed to link $f(\alpha)$ to the physical characteristics of the network.

The sandbox method was applied in [Tél 89] to the "growing asymmetric Cantor set", whose points are positive integers. The linear size of C_t is 4^t, and the number of particles in C_t is 3^t. An exact expression can be derived for D_q [Tél 89]:

$$D_q = \frac{1}{q-1}\left(q + \frac{\log \frac{-1 + \sqrt{1 + 4(3/4)^q}}{2}}{\log 2}\right).$$

For $t = 8$ and using $\widetilde{N} = 75$ (i.e., 75 sandpiles), for $q = -8, -3, 3$, and 8 the sandbox dimension $D_q^{sandbox}(r/L)$ closely approximated (to within 3%) D_q over a wide range of r/L values.

The sandbox method as proposed in [Tél 89] is a "fixed size method" which counts the mass inside a box of a given radius or dimension. (Recall that a "fixed mass method" (Sect. 14.4) determines the smallest radius from a center node to include a given mass.) Based on several studies which reported fixed mass methods to be superior to fixed size methods, [Kamer 13] proposed the following fixed mass version of the sandbox method. Let $\mathbb{X}$ be a given set of N points in $\mathbb{R}^E$. The *barycenter* of $\mathbb{X}$, also called the *centroid* of $\mathbb{X}$, is the point $y \in \mathbb{R}^E$ defined by $y_i \equiv (1/N)\sum_{x \in \mathbb{X}} x_i$, where x_i is the i-th coordinate of x. For a given q, consider the computation of the estimate $D_q^{sandbox}$ of D_q. The sandbox method as described above uses a randomly selected set of points as sandbox centers. The method of [Kamer 13] employed the alternative approach of using a test to determine if $x \in \mathbb{X}$ should be included in the set $\widetilde{\mathbb{X}}$ used to estimate $D_q^{sandbox}$ using (16.55); i.e., if x should be the center of a sandbox. Let k be a positive integer. For $x \in \mathbb{X}$, the method first computes, (*i*) the set $\mathbb{X}(x, k)$ of the k closest neighbors in $\mathbb{X}$ to x, (*ii*) the barycenter $y(x, k)$

of $\mathbb{X}(x,k)$, and (*iii*) the distance $r(x,k) \equiv \max\{dist(x,z) \,|\, z \in \mathbb{X}(x,k)\}$. If $dist\big(y(x,k),x\big) < dist\big(y(x,k),z\big)$ for $z \in \mathbb{X}(n,k)$, then we include x in $\widetilde{\mathbb{X}}$ with the corresponding radius $r(x,k)$. In words, we accept x as one of the sandbox centers if the distance from the barycenter $y(x,k)$ to x is less than the distance from $y(x,k)$ to each of the k closest neighbors of x.

Example 16.4 To illustrate, in one dimension, this test for choosing sandbox centers, let $\mathbb{X} = \{-8, -1, 0, \frac{1}{4}, 1, 7\}$. For $x = 0$ and $k = 3$, the three closest neighbors of x are -1, 1, and $\frac{1}{4}$. Then $y(0,3) = \frac{1}{12}$, and $y(0,3)$ is closer to x than to -1, 1, or $\frac{1}{4}$, so we include x in $\widetilde{\mathbb{X}}$. For $x = -1$ and $k = 3$, the three closest neighbors of x are 0, $\frac{1}{4}$, and 1, and $y(-1,3) = \frac{5}{12}$. Since $y(-1,3)$ is closer to $\frac{1}{4}$ than to -1, we do not include x in $\widetilde{\mathbb{X}}$.   □

With this test for choosing sandbox centers, the points in $\widetilde{X}$ tend to reside in dense areas of the fractal. Since this might bias the computed dimension, [Kamer 13] incorporated a "non-overlapping criterion", which is related to the packing dimension (Sect. 5.4), modified to use boxes of varying size.

The sandbox method was used in [Berthelsen 92] to estimate the fractal dimension of human DNA, by representing nucleic acid sequences as random walks in $\mathbb{R}^2$. The sandbox dimension $D_0^{sandbox}$ was calculated for embedding dimensions of 2, 4, 6, and 8. The sandbox radii ranged from 2 to 26 units, in increments of 2, so the range was only 1.3 decades. As for the size of $\widetilde{N}$, 1% of the points were randomly sampled, the same fraction used in [Tél 89]. For an embedding dimension of 2, [Berthelsen 92] obtained a mean $D_0^{sandbox}$ of 1.395; for an embedding dimension of 4 the mean was 1.68, indicating that a two-dimensional embedding is insufficient. Embeddings exceeding 4 did not yield a significant increase in $D_0^{sandbox}$, indicating that a four-dimensional embedding was sufficient.

The generalized dimensions D_q (obtained by box counting) and the sandbox dimensions $D_q^{sandbox}$ were compared in [Fernandez 99] for cat and turtle retinal neurons, and for diffusion limited aggregates (DLAs). Results showed that, for both neurons and DLAs, $D_{-\infty} < D_\infty$, contrary to the expected non-increasing behavior of D_q (Theorem 16.1). In particular, for neurons [Fernandez 99] found that D_q can be increasing in q for $q < 0$, and that such anomalous behavior has been observed by other researchers. However, for both neurons and DLAs, the sandbox dimensions $D_q^{sandbox}$ behaved as expected. It was observed in [Fernandez 99] that it commonly suffices to pick $\widetilde{N}$, the number of sandbox centers, to be less than $N/10$. However, this study used $\widetilde{N} = N/20$, and increasing $\widetilde{N}$ did not change the results.

Exercise 16.19 In Sect. 16.6 we presented the recommendation in [Panico 95] that the calculations of the fractal dimension of an object should use measurements only from the interior of the object. Does this principle have applicability when calculating $D_q^{sandbox}$?   □

16.7 Particle Counting Dimension

The mass distribution of a growing geometric object defined on a lattice in $\mathbb{R}^2$ was studied in [Tél 87], which showed that in the limit as the object size gets very large, the object exhibits multifractal behavior. Although the growing object considered in [Tél 87] was not a network, their analysis is of particular interest since it resembles the analysis in Chap. 12 of infinite networks. The geometric object considered in [Tél 87] starts with a single particle and gets infinitely large. Let Ω_t be the object at time t. At time $t = 0$ the object Ω_0 consists of a single particle. At time $t > 1$, the object Ω_t is obtained by adding, to the four corners of Ω_{t-1}, the twice enlarged (in each of the two orthogonal directions) copy of Ω_{t-1}. Figure 16.12 illustrates Ω_0, Ω_1, and Ω_2. For example,

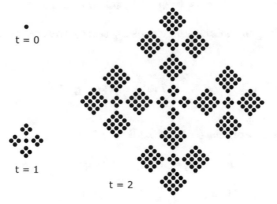

Figure 16.12: Two generations of the Tél-Viscek multifractal

the 1×1 array in Ω_0 becomes a 2×2 array in the corners of Ω_1, and each 2×2 array of particles in Ω_1 becomes a 4×4 in the corners of Ω_2. Let N_t be the number of particles in (i.e., the mass of) Ω_t. For $t \geq 1$, we can view Ω_t as comprised of four "corner" clusters with a "central" cluster in the middle. The mass of each of the four corner clusters of Ω_t is $4N_{t-1}$, and the mass of the center cluster of Ω_t is N_{t-1}. Hence $N_t = (4)(4)N_{t-1} + N_{t-1} = 17N_{t-1}$. For $t \geq 1$, the linear size of the center cluster of Ω_t is $1/5$ the size of Ω_t and the size of each of the four corner clusters of Ω_t is $2/5$ the size of Ω_t. Even though Ω_t contains dense blocks of size 2^t, the size of each such block relative to the size of Ω_t does not exceed $(2/5)^t$, which vanishes as $t \to \infty$.

Let δ be the distance (in each of the orthogonal directions) between adjacent lattice points. Let $t \gg 0$ be fixed, and let L be the linear size of Ω_t (e.g., the maximal distance between any two points in Ω_t). Since t is fixed, define $N \equiv N_t$. To uncover the multifractality of Ω_t, pick a box size s such that $\delta \ll s \ll L$ and cover Ω_t with a set $\mathcal{B}(s)$ of $B(s)$ boxes of size s. Let $N_j(s)$ be the mass of box $B_j \in \mathcal{B}(s)$, so $\sum_{j=1}^{B(s)} N_j(s) = N$. For box $B_j \subset \mathcal{B}(s)$, define the *mass index* α_j of B_j by

$$N_j(s) \sim N \left(\frac{s}{L}\right)^{\alpha_j} \text{ for } s/L \ll 1. \tag{16.64}$$

If not all boxes have the same mass index, then the growing structure is a *geometric multifractal*. The set of all boxes with the same mass index is a subset of $\mathcal{B}(s)$. Let $B_\alpha(s)$ be the number of boxes with mass index α. The centers of the $B_\alpha(s)$ boxes with mass index α are on a fractal set of dimension $f(\alpha)$, so

$$B_\alpha(s) \sim \left(\frac{s}{L}\right)^{-f(\alpha)} \text{ for } s/L \ll 1.$$

Using (16.64), the generalized dimensions D_q satisfy the scaling relation (16.36), thus

$$\Psi_q(N, s, L) \equiv \sum_{j=1}^{B(s)} N_j^q \sim \sum_{j=1}^{B(s)} N^q (s/L)^{(q-1)D_q}, \tag{16.65}$$

where $f(\alpha)$ and D_q are related by (16.23), so

$$(q-1)D_q = q\alpha(q) - f(\alpha(q)). \tag{16.66}$$

and $f'(\alpha(q)) = q$; see (16.44).

To calculate D_q we use (16.65) together with the following self-similarity argument. As noted above, the linear size of each corner cluster of Ω_t is twice the linear size of the central cluster of Ω_t, and the mass of each corner cluster of Ω_t is four times the mass of the center cluster of Ω_t. The fraction of the total mass of Ω_t contained in each corner cluster is $4/((4 \cdot 4) + 1) = 4/17$, and the fraction of the total mass of Ω_t contained in the center cluster is $1/17$. The ratio of the linear dimension of each corner cluster of Ω_t to the linear dimension of Ω_t is $2/((2 \cdot 2) + 1) = 2/5$, and the ratio of the linear dimension of the center cluster to the linear dimension of Ω_t is $1/((2 \cdot 2) + 1) = 1/5$. By reducing the large cluster Ω_t, which has size L, by the factors 5 and 5/2 we obtain two structures, the first corresponding to the central square of linear size $L/5$ with mass $N/17$, and the second structure corresponding to the four larger squares of linear size $2L/5$, each with mass $4N/17$. Thus we have the self-similarity relationship [Tél 87]

$$\Psi_q(N, s, L) = \Psi_q\left(\frac{N}{17}, \frac{s}{(1/5)}, L\right) + 4\Psi_q\left(\frac{4N}{17}, \frac{s}{(2/5)}, L\right). \tag{16.67}$$

Using (16.65) and defining $x \equiv s/L$, Eqs. (16.65) and (16.67) yield

$$N^q x^{(q-1)D_q} = \left(\frac{N}{17}\right)^q (5x)^{(q-1)D_q} + 4\left(\frac{4N}{17}\right)^q \left(\frac{5x}{2}\right)^{(q-1)D_q}$$

which simplifies to an implicit equation for D_q:

$$\left(\frac{1}{17}\right)^q 5^{(q-1)D_q} + 4\left(\frac{4}{17}\right)^q \left(\frac{5}{2}\right)^{(q-1)D_q} = 1. \tag{16.68}$$

Exercise 16.20 Numerically calculate D_q for each nonzero integer between -10 and 10. How can we calculate D_q for $q = 1$? □

Exercise 16.21 Show that $\lim_{q \to -\infty} D_q = \log 17 / \log 5 \approx 1.760$, and $\lim_{q \to \infty} D_q = \log(17/4)/\log(5/2) \approx 1.579$. □

Exercise 16.22 It was observed above that even though Ω_t contains dense blocks of size 2^t, the size of each such block relative to the size of Ω_t does not exceed $(2/5)^t$, which vanishes as $t \to \infty$. This observation was used in [Tél 87] to conclude that Ω_t can be considered a fractal for large t. Justify this conclusion. □

Exercise 16.23 Consider the curve (q, D_q). Using (16.68), what can we determine about the shape of this curve (e.g., is it convex or concave)? □

16.8 Applications of Multifractals

Some applications of multifractals were described in Sect. 16.6 above. In this final section of this chapter, we describe a few more applications. Multifractals are of particular interests to physicists, since the partition function $Z_q(s)$ is formally equivalent to the partition function in the classical theory of thermodynamics, $\tau(q)$ is analogous to the free energy, and $f(\alpha)$ is analogous to the entropy [Stanley 88].

DNA: There is by now a fairly large literature on the fractal dimension of DNA; e.g., [Berthelsen 92] was briefly discussed in Sect. 16.6. Three methods to compute the fractal dimension of DNA were compared in [Oiwa 02]. A strand of DNA was represented as a random walk in two dimensions, where the horizontal axis represents the DNA walk for the nucleic acids thymine and adenine, and the vertical axis represents the walk for guanine and cytosine. When thymine is the next acid in the strand, the walk moves one step to the right, when adenine is next, the walk moves one step to the left. Similarly, when guanine is next, the walk moves one step up, and when cytosine is next, the walk moves one step down. Thus the walk can be represented by a rectilinear network, which in general contains cycles. (However, the fractal dimension calculations described below do not take advantage of the network representation.) We can visualize the results of box counting, for a given box size, by a histogram in which the height of each histogram bar is the number of steps of the random walk contained within the box.

The generalized dimensions D_q of the random walk were estimated in [Oiwa 02] by three techniques: the box counting dimension, the moving box method, and the sandbox method. The box-counting method covers the walk with a grid of boxes of size s, where $s = 1$ corresponds to the size of one base pair (i.e., one pair of nucleic acids). For a given box size s, and for a given box j, define $p_j(s)$ to be the ratio

$$\frac{\# \text{ of random walk steps in the box}}{\# \text{ of steps in the entire random walk}}.$$

If we perform this covering for a set $\{s_i\}$ of box sizes, then D_q is the slope of the line through the points $(\log s_i, \log \sum_j p_{jq}(s_i))$. For a given s, positive values of q put more weight on boxes for which $p_j(s)$ is large (i.e., many random walk steps in the box), while negative values of q put more weight on boxes for which $p_j(s)$ is small (i.e., few random walk steps in the box). The *moving box method* is a variant of box counting which covers the walk with a set of boxes, without having to define a grid. This method reduces error by allowing boxes to be moved in space based upon the geometry of the object. With the *sandbox* method, D_q was taken to be the average slope of the points $(\log s_i, \log \sum_j p_j^q(s_i))$. Results in [Oiwa 02] showed that the three methods give consistent results, and that the sandbox method yielded results of the same quality as the moving box method.

Diffusion Limited Aggregation: As explained in [Stanley 88], the notion of the surface of a DLA cluster is not well defined. For example, consider a large number (e.g., 10^6) random walkers which randomly hit the same object. A picture showing the surface sites hit by at least one of the random walkers appears very different than a picture showing the surface sites hit by at least 50 of the random walkers. Define the "hit probability" p_i to be the probability that surface site i is the next site to be hit by a random walker. The p_i values can be used to form a probability distribution function $\mu(x)$, where $\mu(x)\mathrm{d}x$ is the number of p_i in the range $[x, x + \mathrm{d}x]$. This probability distribution, like all probability distributions, is characterized by its moments, which can be analyzed using multifractal techniques [Kadanoff 90].

Soil: To characterize the variation of soil properties, including electrical conductivity, soil pH, and organic matter content, [Caniego 05] calculated D_q and the $f(\alpha)$ spectrum over the range $q \in [-4, 4]$. The probability measure over an interval $[a, b]$, obtained by a transect (a narrow cut or slice), was divided into 2^i boxes of size $s_i = (b - a)2^{-i}$ in iteration i (this was called "dyadic scaling" in [Caniego 05]). The soil property was sampled at points along the transect, and these values were used to calculate the box probabilities, which then were used to compute D_q. Linear regression was used for the computation of both $f(\alpha)$ and the D_q. The results showed that, while both D_q and $f(\alpha)$ are useful for characterizing soil variability, the D_q values displayed sharper distinction between transects than did $f(\alpha)$.

Population Studies: The generalized dimensions D_q and the spectrum $f(\alpha)$ were used in [Orozco 15] to characterize the spatial distribution pattern of the population in all of Switzerland, and for three regions (Alps, Plateau, and Jura) in Switzerland. It was observed in [Orozco 15] that "The distribution of the

Swiss population has certainly been shaped by the socio-economic history and the complex geomorphology of the country. These factors have rendered a highly clustered, inhomogeneous and very variable population distribution at different spatial scales".

Other interesting applications of multifractal analysis include forecasting sudden floods [Ferraris 03] and the effect of ocean turbulence on the mating encounters of millimeter-scale aquatic crustaceans [Seuront 14]. There are alternatives to the use of multifractals to describe an object for which a single number, e.g., the box counting dimension, does not suffice. For example, the Fourier fractal dimension, based on the Fourier transform of a mass distribution defined over a bounded subset of $\mathbb{R}^E$, was used in [Florindo 11] to characterize the texture of digital images.

16.9 Generalized Hausdorff Dimensions

To conclude this chapter, we present the *generalized Hausdorff dimensions* of a geometric object. This discussion, important in its own right, also provides the foundation for the presentation, in Chap. 18, of the generalized Hausdorff dimensions of a network [Rosenberg 18a].

Generalizing the the Hausdorff dimension d_H (Sect. 5.1), [Grassberger 85] defined the generalized Hausdorff dimension of a compact set $\Omega \subset \mathbb{R}^E$. Recall from Definition 5.1 that, for $s > 0$, an s-covering of Ω is a finite collection of J sets $\{X_1, X_2, \cdots, X_J\}$ that cover Ω (i.e., $\Omega \subseteq \cup_{j=1}^J X_j$) such that for each j we have $diam(X_j) \leq s$. Let $\mathcal{C}(s)$ be the set of all s-coverings of Ω and let s_j be the diameter of box B_j in the s-covering $\{B_1, B_2, \cdots, B_J\}$ of Ω. Let μ be a given measure concentrated on Ω, where μ is possibly distinct from the Hausdorff measure $v^\star(\cdot, d_H(\Omega))$ defined by (5.1). Define $\mu_j \equiv \mu(B_j)$, so μ_j is the measure of box B_j. For $q \in \mathbb{R}$, define

$$v^\star(\Omega, q, d) \equiv \begin{cases} \limsup_{s \to 0} \inf_{\mathcal{C}(s)} \left(\sum_{j=1}^J \mu_j (s_j^d/\mu_j)^{1-q} \right)^{1/(1-q)} & \text{if } q \neq 1, \\ \limsup_{s \to 0} \inf_{\mathcal{C}(s)} \exp \left(\sum_{j=1}^J \mu_j \log(s_j^d/\mu_j) \right) & \text{if } q = 1. \end{cases} \tag{16.69}$$

As in (5.1), we take the infimum since the goal is to cover Ω with small sets X_j as efficiently as possible. The generalized Hausdorff dimension $D_q^H(\Omega)$ of Ω is defined as the (unique) value of d for which

$$v^\star(\Omega, q, d) \equiv \begin{cases} \infty & \text{for } d < D_q^H(\Omega) \\ 0 & \text{for } d > D_q^H(\Omega). \end{cases} \tag{16.70}$$

If there is no ambiguity about the set Ω, for brevity we write D_q^H rather than $D_q^H(\Omega)$. In [Grassberger 85], D_q^H was denoted by $D^{(\alpha)}$. We prefer the notation D_q^H, which uses the superscript H to denote "Hausdorff" and avoids introducing the new symbol α. Setting $q = 0$ in (16.69) yields

$$m(\Omega, d, 0) = \limsup_{s \to 0} \inf_{\mathcal{C}(s)} \sum_{j=1}^{J} s_j^d$$

so $D_0^H = d_H$.

Compared to the generalized dimensions $\{D_q\}_{q \in \mathbb{R}}$, the generalized Hausdorff dimensions $\{D_q^H\}_{q \in \mathbb{R}}$ defined by (16.70) have not enjoyed widespread popularity. For example, three well-known books [Feder 88, Peitgen 92, Schroeder 91], which devote considerable attention to multifractals, discuss D_q but not D_q^H. It is not surprising that D_q^H is rarely considered in computing fractal dimensions of a geometric object Ω, since equal-diameter boxes are typically employed for covering a geometric object. Indeed, Grassberger observed in [Grassberger 85] that in all applications of which he is aware, s-coverings with $s_j \leq s$ can be replaced with s-coverings with $s_j = s$, and such coverings yield the fractal dimensions $\{D_q\}_{q \in \mathbb{R}}$.

Chapter 17

Multifractal Networks

My power flurries through the air into the ground
My soul is spiraling in frozen fractals all around
And one thought crystallizes like an icy blast
I'm never going back
The past is in the past

Lyrics from the song "Let It Go", from the 2013 movie "Frozen". Music
and lyrics by Kristen Anderson-Lopez and Robert Lopez.

In Chap. 16 we presented the theory of generalized dimensions for a geometric object Ω, and discussed several methods for computing the generalized
dimensions of Ω. In this chapter we draw upon the content of Chap. 16 to
show how generalized dimensions can be defined and computed for a complex
network $\mathbb{G}$.

17.1 Lexico Minimal Coverings

In Chap. 15, we showed that the definition proposed in [Wei 14a] of the
information dimension d_I of a network $\mathbb{G}$ is ambiguous, since d_I is computed
from a minimal s-covering of $\mathbb{G}$, for a range of s, and there are in general
multiple minimal s-coverings of $\mathbb{G}$, yielding different values of d_I. Using the
maximal entropy principle of [Jaynes 57], we resolved the ambiguity for each
s by maximizing the entropy over the set of minimal s-coverings. We face the
same ambiguity when using box counting to compute D_q, since the different
minimal s-coverings of $\mathbb{G}$ can yield different values of D_q. In this section we
illustrate this indeterminacy and show how it can be resolved by computing
lexico minimal summary vectors, and this computation requires only negligible
additional work beyond what is required to compute a minimal covering.

To begin, recall from Chap. 16 that, for $q \in \mathbb{R}$, the partition function obtained from an s-covering $\mathcal{B}(s)$ of a geometric object Ω is

© Springer Nature Switzerland AG 2020
E. Rosenberg, *Fractal Dimensions of Networks*,
https://doi.org/10.1007/978-3-030-43169-3_17

$$Z_q\big(\mathcal{B}(s)\big) = \sum_{B_j \in \mathcal{B}(s)} [p_j(s)]^q \tag{17.1}$$

and for $q \neq 1$ the generalized dimension D_q of Ω is given by

$$D_q = \frac{1}{q-1} \lim_{s \to 0} \frac{\log Z_q\big(\mathcal{B}(s)\big)}{\log s}.$$

Considering now network $\mathbb{G}$, let $\mathcal{B}(s)$ be an s-covering of $\mathbb{G}$. For $B_j \in \mathcal{B}(s)$, define $p_j(s) \equiv N_j(s)/N$, where $N_j(s)$ is the number of nodes in B_j. The partition function of $\mathbb{G}$ corresponding to $\mathcal{B}(s)$ is also $Z_q\big(\mathcal{B}(s)\big)$, as given by (17.1). However, since the box size s is always a positive integer, for $q \neq 1$ we cannot define the generalized dimension D_q of $\mathbb{G}$ to be

$$\frac{1}{q-1} \lim_{s \to 0} \frac{\log Z_q\big(\mathcal{B}(s)\big)}{\log s}.$$

Instead, we want D_q of $\mathbb{G}$ to satisfy, over some range of s,

$$D_q \approx \frac{1}{q-1} \frac{\log Z_q\big(\mathcal{B}(s)\big)}{\log s}. \tag{17.2}$$

This was the approach used by [Wang 11], who used radius-based boxes to cover $\mathbb{G}$, and who for $q \neq 1$ defined D_q to be the generalized dimension of $\mathbb{G}$ if over some range of r and for some constant c,

$$\log Z_q\big(\mathcal{B}(r)\big) \approx (q-1)D_q \log(r/\Delta) + c, \tag{17.3}$$

where $Z_q\big(\mathcal{B}(r)\big)$ is the average partition function value obtained from a covering of $\mathbb{G}$ by radius-based boxes of radius r. (A randomized box counting heuristic was used in [Wang 11], and $Z_q\big(\mathcal{B}(r)\big)$ was taken to be the average partition function value, averaged over some number of executions of the randomized heuristic.) However, definition (17.3) is ambiguous, since different minimal coverings can yield different values of $Z_q\big(\mathcal{B}(r)\big)$ and hence different values of D_q.

Example 17.1 Consider the *chair* network of Fig. 17.1, which shows two minimal 3-coverings and a minimal 2-covering. Choosing $q = 2$, for the covering $\widetilde{\mathcal{B}}(3)$ from (17.1) we have $Z_2\big(\widetilde{\mathcal{B}}(3)\big) = (\frac{3}{5})^2 + (\frac{2}{5})^2 = \frac{13}{25}$, while for $\widehat{\mathcal{B}}(3)$ we have $Z_2\big(\widehat{\mathcal{B}}(3)\big) = (\frac{4}{5})^2 + (\frac{1}{5})^2 = \frac{17}{25}$. For $\mathcal{B}(2)$ we have $Z_2\big(\mathcal{B}(2)\big) = 2(\frac{2}{5})^2 + (\frac{1}{5})^2 = \frac{9}{25}$. If we use $\widetilde{\mathcal{B}}(3)$, then from (17.3) and the range $s \in [2,3]$ we obtain $D_2 = \big(\log \frac{13}{25} - \log \frac{9}{25}\big)/(\log 3 - \log 2) = 0.907$. If instead we use $\widehat{\mathcal{B}}(3)$ and the same range of s we obtain $D_2 = \big(\log \frac{17}{25} - \log \frac{9}{25}\big)/(\log 3 - \log 2) = 1.569$. Thus the method of [Wang 11] can yield different values of D_2 depending on the minimal covering selected.   □

To devise a computationally efficient method for selecting a unique minimal covering, first consider the maximal entropy criterion presented in Chap. 15. A

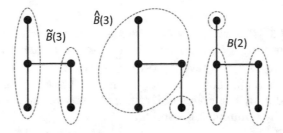

Figure 17.1: Two minimal 3-coverings and a minimal 2-covering for the chair network

maximal entropy minimal s-covering is a minimal s-covering for which the entropy $-\sum_j p_j(s) \log p_j(s)$ is largest. It is well-known that entropy is *maximized* when all the probabilities are equal. Theorem 17.1 below, concerning a continuous optimization problem, proves that the partition function is *minimized* when the probabilities are equal. For integer $J \geq 2$, let $\mathcal{P}(q)$ denote the continuous optimization problem

$$\text{minimize } \sum_{j=1}^{J} p_j^q \text{ subject to } \sum_{j=1}^{J} p_j = 1 \text{ and } p_j \geq 0 \text{ for each } j \,.$$

Theorem 17.1 For $q > 1$, the solution of $\mathcal{P}(q)$ is $p_j = 1/J$ for each j, and the optimal objective function value is J^{1-q}.

Proof [Rosenberg 17b] If $p_j = 1$ for some j, then for each other j we have $p_j = 0$ and the objective function value is 1. Next, suppose each $p_j > 0$ in the solution of $\mathcal{P}(q)$. Let λ be the Lagrange multiplier associated with the constraint $\sum_{j=1}^{J} p_j = 1$. The first order optimality condition for p_j is $q p_j^{q-1} - \lambda = 0$. Thus each p_j has the same value, which is $1/J$. The corresponding objective function value is $\sum_{j=1}^{J} (1/J)^q = J^{1-q}$. This value is less than the value 1 (obtained when exactly one $p_j = 1$) when $J^{1-q} < 1$, which holds for $q > 1$. For $q > 1$ and $p > 0$, the function p^q is a strictly convex function of p, so $p_j = 1/J$ is the unique global (as opposed to only a local) solution of $\mathcal{P}(q)$. Finally, suppose that for some solution of $\mathcal{P}(q)$ we have $p_j \in [0, 1)$ for each j, but the number of positive p_j is J_0, where $J_0 < J$. The above arguments show that the optimal objective function value is J_0^{1-q}. Since for $q > 1$ we have $J_0^{1-q} > J^{1-q}$, any solution of $\mathcal{P}(q)$ must have $p_j > 0$ for each j. $\square$

Applying Theorem 17.1 to $\mathbb{G}$, minimizing the partition function $Z_q\big(\mathcal{B}(s)\big)$ over all minimal s-coverings of $\mathbb{G}$ yields a minimal s-covering for which all the probabilities $p_j(s)$ are approximately equal. Since $p_j(s) = N_j(s)/N$, having the box probabilities almost equal means that all boxes in the minimal s-covering have approximately the same number of nodes. The following definition of an (s, q) minimal covering, for use in computing D_q, is analogous to the definition in [Rosenberg 17a] of a maximal entropy minimal covering, for use in computing d_I.

Definition 17.1 [Rosenberg 17b] For $q \in \mathbb{R}$, the covering $\mathcal{B}(s)$ of $\mathbb{G}$ is an (s, q) minimal covering if (i) $\mathcal{B}(s)$ is a minimal s-covering and (ii) for any other minimal s-covering $\widetilde{\mathcal{B}}(s)$ we have $Z_q\big(\mathcal{B}(s)\big) \leq Z_q\big(\widetilde{\mathcal{B}}(s)\big)$. $\square$

Procedure 17.1 below shows how an (s, q) minimal covering can be computed, for a given s and q, by a simple modification of whatever box-counting method is used to compute a minimal s-covering.

Procedure 17.1 Let $\mathcal{B}_{\min}(s)$ be the best s-covering obtained over all executions of whatever box-counting method is utilized. Suppose we have executed box counting some number of times, and stored $\mathcal{B}_{\min}(s)$ and $Z_q^{\star}(s) \equiv Z_q\big(\mathcal{B}_{\min}(s)\big)$, so $Z_q^{\star}(s)$ is the current best estimate of the minimal value of the partition function. Now suppose we execute box counting again, and generate a new s-covering $\mathcal{B}(s)$ using $B(s)$ boxes. Let $Z_q \equiv Z_q\big(\mathcal{B}(s)\big)$ be the partition function value associated with $\mathcal{B}(s)$. If $B(s) < B_{\min}(s)$ holds, or if $B(s) = B_{\min}(s)$ and $Z_q < Z_q^{\star}(s)$ hold, set $\mathcal{B}_{\min}(s) = \mathcal{B}(s)$ and $Z_q^{\star}(s) = Z_q$. $\square$

Let s be fixed. Procedure 17.1 shows that we can easily modify any box-counting method to compute the (s, q) minimal covering for a given q. However, this approach to eliminating ambiguity in the computation of D_q is not attractive, since it requires computing an (s, q) minimal covering for each value of q for which we wish to compute D_q. Moreover, since there are only a finite number of minimal s-coverings, as q varies we will eventually rediscover some (s, q) minimal coverings, that is, for some q_1 and q_2 we will find that the (s, q_1) minimal covering is identical to the (s, q_2) minimal covering. An elegant alternative to computing an (s, q) minimal covering for each s and q, while still achieving our goal of unambiguously computing D_q, is to use lexico minimal summary vectors.

For a given box size s, we summarize an s-covering $\mathcal{B}(s)$ by the point $x(s) \in \mathbb{R}^J$, where $J \equiv B(s)$, where $x_j(s) = N_j(s)$ for $1 \leq j \leq J$, and where $x_1(s) \geq x_2(s) \geq \cdots \geq x_J(s)$. We say "summarize" since $x(s)$ does not specify all the information in $\mathcal{B}(s)$; in particular, $\mathcal{B}(s)$ specifies exactly which nodes belong to each box, while $x(s)$ specifies only the number of nodes in each box. We use the notation $x(s) = \sum \mathcal{B}(s)$ to mean that $x(s)$ summarizes the s-covering $\mathcal{B}(s)$ and that $x_1(s) \geq x_2(s) \geq \cdots \geq x_J(s)$. For example, if $N = 37$, $s = 3$, and $B(3) = 5$, then we might have $x(s) = \sum \mathcal{B}(3)$ for $x(s) = (18, 7, 5, 5, 2)$. However, we cannot have $x(s) = \sum \mathcal{B}(3)$ for $x(s) = (7, 18, 5, 5, 2)$ since the components of $x(s)$ are not ordered correctly. If $x(s) = \sum \mathcal{B}(s)$, then each $x_j(s)$ is positive, since $x_j(s)$ is the number of nodes in box B_j. We call $x(s) = \sum \mathcal{B}(s)$ a *summary vector* of $\mathcal{B}(s)$. By "$x(s)$ is a summary vector" we mean $x(s)$ summarizes some $\mathcal{B}(s)$.

Let $x \in \mathbb{R}^K$ for some positive integer K. Let $right(x) \in \mathbb{R}^{K-1}$ be the point obtained by deleting the first component of x. For example, if $x = (18, 7, 5, 5, 2)$, then $right(x) = (7, 5, 5, 2)$. Similarly, we define $right^2(x) \equiv right\big(right(x)\big)$, so $right^2(7, 7, 5, 2) = (5, 2)$. Let $u \in \mathbb{R}$ and $v \in \mathbb{R}$ be numbers. We say that $u \succeq v$ (in words, u is *lexico* greater than or equal to v) if ordinary inequality holds, that is, $u \succeq v$ if $u \geq v$. (We use *lexico* instead of the longer *lexicographically*.)

Thus $6 \succeq 3$ and $3 \succeq 3$. Now let $x \in \mathbb{R}^K$ and $y \in \mathbb{R}^K$. We define lexico inequality recursively. We say that $y \succeq x$ if either (i) $y_1 > x_1$ or (ii) $y_1 = x_1$ and $right(y) \succeq right(x)$. For example, for $x = (9, 6, 5, 5, 2)$, $y = (9, 6, 4, 6, 2)$, and $z = (8, 7, 5, 5, 2)$, we have $x \succeq y$ and $x \succeq z$ and $y \succeq z$.

Definition 17.2 Let $x(s) = \sum \mathcal{B}(s)$. Then $x(s)$ is *lexico minimal* if (i) $\mathcal{B}(s)$ is a minimal s-covering and (ii) if $\widetilde{\mathcal{B}}(s)$ is a minimal s-covering distinct from $\mathcal{B}(s)$ and $y(s) = \sum \widetilde{\mathcal{B}}(s)$, then $y(s) \succeq x(s)$. $\square$

Theorem 17.2 For each s there is a unique lexico minimal summary vector.

Proof [Rosenberg 17b] Pick s. Since there are only a finite number of minimal s-coverings of $\mathbb{G}$, and the corresponding summary vectors can be ordered using lexicographic ordering, then a lexico minimal summary vector exists. To show that there is unique lexico minimal summary vector, suppose that $x(s) = \sum \mathcal{B}(s)$ and $y(s) = \sum \widetilde{\mathcal{B}}(s)$ are both lexico minimal. Define $J \equiv B(s)$, so J is the number of boxes in both $\mathcal{B}(s)$ and in $\widetilde{\mathcal{B}}(s)$. If $J = 1$, then a single box covers $\mathbb{G}$, and that box must be $\mathbb{G}$ itself, so the theorem holds. So assume $J > 1$. If $x_1(s) > y_1(s)$ or $x_1(s) < y_1(s)$, then $x(s)$ and $y(s)$ cannot both be lexico minimal, so we must have $x_1(s) = y_1(s)$. We apply the same reasoning to $right(x(s))$ and $right(y(s))$ to conclude that $x_2(s) = y_2(s)$. If $J = 2$ we are done; the theorem is proved. If $J > 2$, we apply the same reasoning to $right^2(x(s))$ and $right^2(y(s))$ to conclude that $x_3(s) = y_3(s)$. Continuing in this manner, we finally obtain $x_J(s) = y_J(s)$, so $x(s) = y(s)$. $\square$

For $x(s) = \sum \mathcal{B}(s)$ and $q \in \mathbb{R}$, define

$$Z(x(s), q) \equiv \sum_{B_j \in \mathcal{B}(s)} \left(\frac{x_j(s)}{N} \right)^q. \tag{17.4}$$

Lemma 17.1 Let α, β, and γ be positive numbers. Then $(\alpha + \beta)/(\alpha + \gamma) > 1$ if and only if $\beta > \gamma$.

Theorem 17.3 Let $x(s) = \sum \mathcal{B}(s)$. If $x(s)$ is lexico minimal, then $\mathcal{B}(s)$ is (s, q) minimal for all sufficiently large q.

Proof [Rosenberg 17b] Pick s and let $x(s) = \sum \mathcal{B}(s)$ be lexico minimal. Let $\widetilde{\mathcal{B}}(s)$ be a minimal s-covering distinct from $\mathcal{B}(s)$, and let $y(s) = \sum \widetilde{\mathcal{B}}(s)$. By Theorem 17.2, $y(s) \neq x(s)$. Define $J \equiv B(s)$. Consider the ratio

$$F \equiv \frac{Z(y(s), q)}{Z(x(s), q)} = \frac{\sum_{j=1}^{J} (y_j(s)/N)^q}{\sum_{j=1}^{J} (x_j(s)/N)^q}.$$

Let k be the smallest index such that $y_j(s) \neq x_j(s)$. For example, if $x(s) = (8, 8, 7, 3, 2)$ and $y(s) = (8, 8, 7, 4, 1)$, then $k = 4$. Since $x_j(s) \leq x_k(s)$ for $k \leq j \leq J$, for $q > 1$ we have

$$\sum_{j=k}^{J} \left(\frac{x_j(s)}{x_k(s)} \right)^q < \sum_{j=k}^{J} \left(\frac{x_j(s)}{x_k(s)} \right) \leq J - k + 1. \tag{17.5}$$

By Lemma 17.1, $F > 1$ if and only if $\sum_{j=k}^{J} \left(y_j(s)/N\right)^q > \sum_{j=k}^{J} \left(x_j(s)/N\right)^q$. Since $\mathcal{B}(s)$ and $\widetilde{\mathcal{B}}(s)$ are minimal s-coverings, and since $x(s)$ is lexico minimal, then $y_k(s) > x_k(s)$. We have

$$\frac{\sum_{j=k}^{J} \left(y_j(s)/N\right)^q}{\sum_{j=k}^{J} \left(x_j(s)/N\right)^q} = \frac{N^q\left(x_k(s)\right)^{-q} \sum_{j=k}^{J} \left(y_j(s)/N\right)^q}{N^q\left(x_k(s)\right)^{-q} \sum_{j=k}^{J} \left(x_j(s)/N\right)^q}$$

$$= \frac{\sum_{j=k}^{J} \left(y_j(s)/x_k(s)\right)^q}{\sum_{j=k}^{J} \left(x_j(s)/x_k(s)\right)^q}$$

$$> \frac{\left(y_k(s)/x_k(s)\right)^q}{\sum_{j=k}^{J} \left(x_j(s)/x_k(s)\right)^q} \geq \frac{\left(y_k(s)/x_k(s)\right)^q}{J - k + 1},$$

where the final inequality holds by (17.5). Since $y_k(s) > x_k(s)$ then for all sufficiently large q we have $F > 1$. $\square$

Analogous to Procedure 17.1, Procedure 17.2 below shows how, for a given s, the lexico minimal $x(s)$ can be computed by a simple modification of whatever box-counting method is used to compute a minimal s-covering.

Procedure 17.2 Let $\mathcal{B}_{\min}(s)$ be the best s-covering obtained over all executions of whatever box-counting method is utilized. Suppose we have executed box counting some number of times, and stored $\mathcal{B}_{\min}(s)$ and $x_{\min}(s) = \sum \mathcal{B}_{\min}(s)$, so $x_{\min}(s)$ is the current best estimate of a lexico minimal summary vector. Now suppose we execute box counting again, and generate a new s-covering $\mathcal{B}(s)$ using $B(s)$ boxes. Let $x(s) = \sum \mathcal{B}(s)$. If $B(s) < B_{\min}(s)$ holds, or if $B(s) = B_{\min}(s)$ and $x_{\min}(s) \succeq x(s)$ hold, set $\mathcal{B}_{\min}(s) = \mathcal{B}(s)$ and $x_{\min}(s) = x(s)$. $\square$

Procedure 17.2 shows that the only additional steps, beyond the box-counting method itself, needed to compute $x(s)$ are lexicographic comparisons, and no evaluations of the partition function $Z_q\big(\mathcal{B}(s)\big)$ are required. Theorem 17.2 proved that $x(s)$ is unique. Theorem 17.3 proved that $x(s)$ is "optimal" (i.e., (s,q) minimal) for all sufficiently large q. Thus an attractive way to unambiguously compute D_q is to compute $x(s)$ for a range of s and use the $x(s)$ summary vectors to compute D_q, using Definition 17.3 below.

Definition 17.3 For $q \neq 1$, the network $\mathbb{G}$ has the generalized dimension D_q if for some constant c, for some positive integers L and U satisfying $2 \leq L < U \leq \Delta$, and for each integer $s \in [L, U]$,

$$\log Z(x(s), q) \approx (q - 1)D_q \log(s/\Delta) + c, \qquad (17.6)$$

where $Z(x(s), q)$ is defined by (17.4) and where $x(s) = \sum \mathcal{B}(s)$ is lexico minimal. $\square$

Other than the minor detail that (17.3) is formulated using radius-based boxes rather than diameter-based boxes, (17.6) is identical to (17.3). Definition 17.3 modifies the definition of D_q in [Wang 11] to ensure uniqueness of D_q. Since $x(s)$ summarizes a minimal s-covering, using $x(s)$ to compute D_q, even when q is not sufficiently large as required by Theorem 17.3, introduces no error in the computation of D_q. By using the summary vectors $x(s)$ we can unambiguously compute D_q, with negligible extra computational burden.

Example 17.1 (Continued) Consider again the chair network of Fig. 17.1. Choose $q = 2$. For $s = 2$ we have $x(2) = \sum \mathcal{B}(2) = (2, 2, 1)$ and $Z(x(2), 2) = \frac{9}{25}$. For $s = 3$ we have $x(3) = \sum \widetilde{\mathcal{B}}(3) = (3, 2)$ and $Z(x(3), 2) = \frac{13}{25}$. Choose $L = 2$ and $U = 3$. From Definition 17.3 we have $D_2 = \log(13/9)/\log(3/2) = 0.907$. $\square$

Definition 17.4 If for some constant d we have $D_q = d$ for $q \in \mathbb{R}$, then $\mathbb{G}$ is *monofractal*; if $\mathbb{G}$ is not *monofractal*, then $\mathbb{G}$ is *multifractal*. $\square$

For the final topic in this section, we show how to use the $x(s)$ summary vectors to compute $D_\infty \equiv \lim_{q \to \infty} D_q$. Let $x(s) = \sum \mathcal{B}(s)$ be lexico minimal and let $x_1(s)$ be the first element of $x(s)$. Let $m(s)$ be the multiplicity of $x_1(s)$, defined to be the number of occurrences of $x_1(s)$ in $x(s)$. (For example, if $x(s) = (7, 7, 6, 5, 4)$, then $x_1(s) = 7$ and $m(s) = 2$ since there are two occurrences of 7.) Then $x_1(s) > x_j(s)$ for $j > m(s)$. Using (17.4), we have

$$
\begin{aligned}
\lim_{q \to \infty} \frac{\log Z(x(s), q)}{q - 1} &= \lim_{q \to \infty} \left\{ \frac{1}{q-1} \log \left(m(s) \left(\frac{x_1(s)}{N} \right)^q \right. \right. \\
&\qquad\qquad \left. \left. + \sum_{j=m(s)+1}^{B(s)} \left(\frac{x_j(s)}{N} \right)^q \right) \right\} \\
&= \lim_{q \to \infty} \left\{ \frac{1}{q-1} \log \left(m(s) \left(\frac{x_1(s)}{N} \right)^q \right) \right\} \\
&= \lim_{q \to \infty} \frac{\log m(s) + q \log (x_1(s)/N)}{q - 1} \\
&= \log \left(\frac{x_1(s)}{N} \right) . \tag{17.7}
\end{aligned}
$$

Rewriting (17.6) yields

$$
\frac{\log Z(x(s), q)}{q - 1} \approx D_q \log \left(\frac{s}{\Delta} \right) + \frac{c}{q - 1}. \tag{17.8}
$$

Taking the limit of both sides of (17.8) as $q \to \infty$, and using (17.7), we obtain

$$
\log \left(x_1(s)/N \right) \approx D_\infty \log (s/\Delta) . \tag{17.9}
$$

We can use (17.9) to compute D_∞ without having to compute any partition function values. It is well-known [Peitgen 92] that, for geometric multifractals, D_∞ corresponds to the densest part of the fractal. Similarly, (17.9) shows that, for a network, D_∞ is the factor that relates the box size s to $x_1(s)$, the number of nodes in the box with the highest probability.

Using Procedure 17.2, the box counting dimension d_B and D_∞ were compared in [Rosenberg 17b] for three networks. The reason for comparing d_B to D_∞ is that for geometric multifractals we have $d_B = D_0 \geq D_\infty$ (Theorem 16.1).

Procedure 17.2 can be used with any box-counting method; [Rosenberg 17b] used the greedy heuristic of [Song 07], described in Sect. 8.1. Recall that this greedy heuristic first creates the auxiliary graph $\widetilde{\mathbb{G}}_s = (\mathbb{N}, \widetilde{\mathbb{A}}_s)$. Having constructed $\widetilde{\mathbb{G}}_s$, the task is to color the nodes of $\widetilde{\mathbb{G}}_s$, using the minimal number of colors, such that no arc in $\widetilde{\mathbb{A}}_s$ connects nodes assigned the same color. The heuristic used to color $\widetilde{\mathbb{G}}_s$ is the following. Suppose we have ordered the $N = |\mathbb{N}|$ nodes in some order $n_1, n_2, \ldots, n_N$. Define $\mathbb{N} \equiv \{1, 2, \ldots, N\}$. For $i \in \mathbb{N}$, let $\widetilde{\mathbb{N}}(n_i)$ be the set of nodes in $\widetilde{\mathbb{G}}_s$ that are adjacent to node n_i. (This set is empty if n_i is an isolated node.) Let $c(n)$ be the color assigned to node n. Initialize $c(n_1) = 1$ and $c(n) = 0$ for all other nodes. For $i = 2, 3, \ldots, N$, the color $c(n_i)$ assigned to n_i is the smallest color not assigned to any neighbor of n_i:

$$c(n_i) = \min\{k \in \mathbb{N} \,|\, k \neq c(n) \text{ for } n \in \widetilde{\mathbb{N}}(n_i)\}\,.$$

The estimate of the chromatic number $\chi(\widetilde{\mathbb{G}}_s)$, which is also the estimate of $B(s)$, is $\max\{c(n) \,|\, n \in \mathbb{N}\}$, the number of colors used to color the auxiliary graph. All nodes with the same color belong to the same box.

This node coloring heuristic assumes some given ordering of the nodes. Many random orderings of the nodes were generated: for the *dolphins* and *jazz* networks $10N$ random orderings of the nodes were generated, and for the *C. elegans* network N random orderings were generated. The node coloring heuristic was executed for each random ordering. Random orderings were obtained by starting with the permutation array $\pi(n) = n$, picking two random integers $i, j \in \mathbb{N}$, and swapping the contents of π in positions i and j. For the *dolphins* and *jazz* networks $10N$ swaps were executed to generate each random ordering, and for the *C. elegans* network N swaps were executed. In the plots for these three networks, the horizontal axis is $\log(s/\Delta)$, the red line plots $\log(x_1(s)/N)$ versus $\log(s/\Delta)$, and the blue line plots $-\log B(s)$ versus $\log(s/\Delta)$.

Example 17.2 Figure 17.2 (left) provides results for the *jazz* network, with 198 nodes, 2742 arcs, and $\Delta = 6$; see Example 15.5. The red and blue plots are

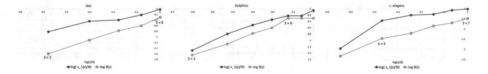

Figure 17.2: D_∞ for three networks

over the range $2 \leq s \leq 6$. Over this range, a linear fit to the red plot yields, from (17.9), $D_\infty = 1.55$, and a linear fit to the blue plot yields $d_B = 1.87$. Since $d_B = D_0$, we have $D_0 > D_\infty$.

Example 17.3 Figure 17.2 (middle) provides results for the *dolphins* network (introduced in Example 7.2), with 62 nodes, 159 arcs, and $\Delta = 8$. The red and blue plots are over the range $2 \leq s \leq 8$. Using (17.9), a linear fit to the red plot over $2 \leq s \leq 6$ (the portion of the plots that appears roughly linear) yields

$D_\infty = 2.29$, and a linear fit to the blue plot over $2 \leq s \leq 6$ yields $d_B = 2.27$, so D_∞ very slightly exceeds d_B.

Example 17.4 Figure 17.2 (right) provides results for the *C. elegans* network, with 453 nodes, 2040 arcs, and $\Delta = 7$. This is the unweighted, undirected version of the neural network of *C. elegans* [Newman]. The red and blue plots are over the range $2 \leq s \leq 7$. Using (17.9), a linear fit to the red plot over $3 \leq s \leq 7$ (the portion of the plots that appears roughly linear) yields $D_\infty = 1.52$, and a linear fit to the blue plot over $3 \leq s \leq 7$ yields $d_B = 1.73$, so $D_0 > D_\infty$.

In each of the three figures, the two lines roughly begin to intersect as s/Δ increases. This occurs since $x_1(s)$ is non-decreasing in s and $x_1(s)/N = 1$ for $s > \Delta$; similarly $B(s)$ is non-increasing in s and $B(s) = 1$ for $s > \Delta$.

Exercise 17.1 An arc-covering based approach, e.g., [Jalan 17, Zhou 07], can be used to define D_q, where $p_j(s)$ is the fraction of all arcs contained in box $B_j \subset \mathcal{B}(s)$. Compare the D_q values obtained using arc-coverings with the D_q values obtained using node-coverings. □

Exercise 17.2 Compute D_q for integer q in $[-10, 10]$ for the *hub and spoke with a tail* network of Example 15.2. □

17.2 Non-monotonicity of the Generalized Dimensions

For a geometric object, we showed (Theorem 16.1) that the generalized dimensions D_q are monotone non-increasing in q. In this section we show that, for a network, the generalized dimensions can fail to be monotone non-increasing in q, even when D_q is computed using the lexico minimal summary vectors $x(s)$. (The fact that D_q can be increasing in q for $q < 0$, or for small positive values of q, was observed in [Wang 11]. For example, for the protein-protein interaction networks studied in [Wang 11], D_q is increasing in q for $q \in [0, 1]$, and D_q peaks at $q \approx 2$.) Moreover, in this section we show that D_q versus q curve can assume different shapes, depending on the range of box sizes used to compute D_q.

Recall that $Z(x(s), q)$ is defined by (17.4), where $x(s) = \sum \mathcal{B}(s)$ is a lexico minimal summary vector for box size s. In Sect. 17.1, we showed that to compute D_q for a given q, we identify a range $[L, U]$ of box sizes, where $L < U$, such that $\log Z(x(s), q)$ is approximately linear in $\log s$ for $s \in [L, U]$. The estimate of D_q is $1/(q - 1)$ times the slope of the linear approximation. More formally, Definition 17.3 required D_q to satisfy

$$\log Z(x(s), q) \approx (q - 1)D_q \log(s/\Delta) + c \tag{17.10}$$

for integer $s \in [L, U]$. For the remainder of this section, assume that (17.10) holds as a strict equality.

Example 17.5 To illustrate how different minimal s-coverings can yield very different D_q versus q curves, consider again the *chair* network of Fig. 17.1, which shows two minimal 3-coverings and a minimal 2-covering. We have $x(2) = \sum \mathcal{B}(2) = (2, 2, 1)$ and $Z(x(2), q) = 2(\frac{2}{5})^q + (\frac{1}{5})^q$. For the 3-covering $\widetilde{\mathcal{B}}(3)$ we have $\widetilde{x}(3) \equiv \sum \widetilde{\mathcal{B}}(3) = (3,2)$ and $Z\big(\widetilde{x}(3), q\big) = (\frac{3}{5})^q + (\frac{2}{5})^q$. With $L = 2$ and $U = 3$, from (17.6) we have

$$\widetilde{D}_q \equiv \left(\frac{1}{q-1}\right) \left(\frac{\log\left(\frac{3^q + 2^q}{5^q}\right) - \log\left(\frac{(2)(2^q)+1}{5^q}\right)}{\log(3/\Delta) - \log(2/\Delta)}\right) = \frac{\log\left(\frac{3^q+2^q}{(2)(2^q)+1}\right)}{\log(3/2)(q-1)}. \quad (17.11)$$

If instead for $s = 3$ we choose the covering $\widehat{\mathcal{B}}(3)$ then $\widehat{x}(3) \equiv \sum \widehat{\mathcal{B}}(3) = (4,1)$ and $Z\big(\widehat{x}(3), q\big) = (\frac{4}{5})^q + (\frac{1}{5})^q$. With $L = 2$ and $U = 3$, but now using $\widehat{x}(3)$ instead of $\widetilde{x}(3)$, we obtain

$$\widehat{D}_q \equiv \left(\frac{1}{q-1}\right) \left(\frac{\log\left(\frac{4^q + 1^q}{5^q}\right) - \log\left(\frac{(2)(2^q)+1}{5^q}\right)}{\log(3/\Delta) - \log(2/\Delta)}\right) = \frac{\log\left(\frac{4^q+1}{(2)(2^q)+1}\right)}{\log(3/2)(q-1)}. \quad (17.12)$$

Figure 17.3 plots $\widetilde{D}_q$ versus q, and $\widehat{D}_q$ versus q over the range $0 \le q \le 15$. Neither curve is monotone non-increasing: the $\widetilde{D}_q$ curve is unimodal, with a

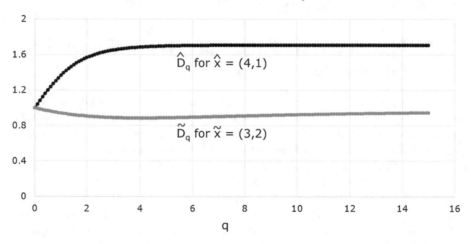

Figure 17.3: Two plots of the generalized dimensions of the *chair* network

local minimum at $q \approx 4.1$, and the $\widehat{D}_q$ curve is monotone increasing. □

In Example 17.5, since $(4,1) \succeq (3,2)$ then $x(3) = (3,2)$ is the lexico minimal summary vector for $s = 3$. Using the $x(s)$ vectors eliminates one source of ambiguity in computing D_q. However, there is yet another source of ambiguity: the selection of the range of box sizes used to compute D_q. What has not been previously recognized is that for a network the choice of L and U can dramatically change the shape of the D_q versus q curve [Rosenberg 17c]; this is the subject of the remainder of this section.

As frequently discussed in previous chapters, the computation of a fractal dimension, whether for a geometric object or for a network, typically involves picking a region of box sizes over which the plot (in double logarithmic coordinates) looks approximately linear. Thus one way to compute D_q for a given

q is to determine a range $[L_q, U_q]$ of s values over which $\log Z\big(x(s), q\big)$ is approximately linear in $\log s$, and then use (17.10) to estimate D_q, e.g., using linear regression. With this approach, to report computational results to other researchers it would be necessary to specify, for each q, the range of box sizes used to estimate D_q. This is certainly not the protocol currently followed in research on generalized dimensions. Instead, the approach taken in [Wang 11] and [Rosenberg 17c] was to pick a single L and U and estimate D_q for all q with this L and U. Moreover, rather than using a technique such as linear regression to estimate D_q over the range $[L, U]$ of box sizes, in [Rosenberg 17c] D_q was estimated using only the two box sizes L and U.[1] With this two-point approach, the estimate of D_q is the slope of the secant line connecting the point $\Big(\log L, \log Z\big(x(L), q\big)\Big)$ to the point $\Big(\log U, \log Z\big(x(U), q\big)\Big)$, where $x(L)$ and $x(U)$ are the lexico minimal summary vectors for box sizes L and U, respectively. Using (17.4) and (17.10), the estimate of D_q is $D_q(L, U)$, defined by

$$
\begin{aligned}
D_q(L, U) &\equiv \frac{\log Z\big(x(U), q\big) - \log Z\big(x(L), q\big)}{(q-1)\big(\log(U/\Delta) - \log(L/\Delta)\big)} \\
&= \frac{1}{\log(U/L)(q-1)} \log\left(\frac{\sum_{B_j \in \mathcal{B}(U)} [x_j(U)]^q}{\sum_{B_j \in \mathcal{B}(L)} [x_j(L)]^q}\right). \quad (17.13)
\end{aligned}
$$

Example 17.6 Figure 17.4 (left) plots box counting results for the *dolphins* network. This figure shows that the $\big(-\log(s/\Delta), \log B(s)\big)$ curve is approximately linear for $2 \le s \le 6$. Figure 17.4 (right) plots $\log Z\big(x(s), q\big)$ versus

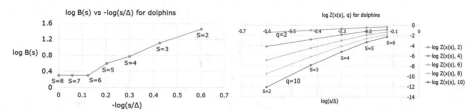

Figure 17.4: Box counting and $\log Z\big(x(s), q\big)$ versus $\log(s/\Delta)$ for the dolphins network

$\log(s/\Delta)$ for $2 \le s \le 6$ and for $q = 2, 4, 6, 8, 10$ ($q = 2$ is the top curve, and $q = 10$ is the bottom curve). This figure shows that, although the $\log Z\big(x(s), q\big)$ versus $\log(s/\Delta)$ curves are less linear as q increases, a linear approximation is quite reasonable. Moreover, we are particularly interested, in the next section, in the behavior of the $\log Z\big(x(s), q\big)$ versus $\log(s/\Delta)$ curve for small positive q, the region where the linear approximation is best. Using (17.13), Fig. 17.5 plots the secant estimate $D_q(L, U)$ versus q for various choices of L and U. Since the

---

[1]Two-point estimates of fractal dimensions for geometric objects were discussed in [Theiler 93], and two-point estimates were used in Sect. 11.4 to compute the correlation dimension of a rectilinear grid network.

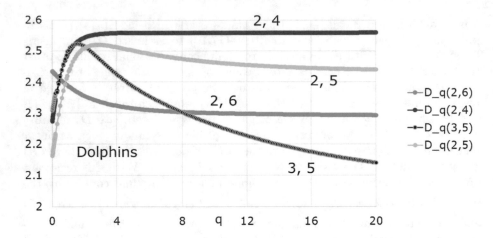

Figure 17.5: Secant estimate of D_q for the dolphins network for different L and U

D_q versus q curve for a geometric multifractal is monotone non-increasing, it is remarkable that different choices of L and U lead to such different shapes for the $D_q(L,U)$ versus q curve.

Recalling that $D_q(L,U)$, defined by (17.13), is the secant estimate of D_q using the box sizes L and U, let $D_0'(L,U)$ denote the first derivative with respect to q of $D_q(L,U)$, evaluated at $q = 0$. A simple closed-form expression for $D_0'(L,U)$ was derived in [Rosenberg 17c]. For box size s, let $x(s) = \sum \mathcal{B}(s)$ summarize the minimal s-covering $\mathcal{B}(s)$. Define

$$
\begin{aligned}
G(s) &\equiv \left(\prod_{j=1}^{B(s)} x_j(s) \right)^{1/B(s)} \\
A(s) &\equiv \frac{1}{B(s)} \sum_{j=1}^{B(s)} x_j(s) \\
R(s) &\equiv \frac{G(s)}{A(s)}
\end{aligned}
\tag{17.14}
$$

so $G(s)$ is the geometric mean of the box masses summarized by $x(s)$, $A(s)$ is the arithmetic mean of the box masses summarized by $x(s)$, and $R(s)$ is the ratio of the geometric mean to the arithmetic mean. By the classical arithmetic-geometric inequality, for each s we have $R(s) \leq 1$. Since $\sum_{j=1}^{B(s)} x_j(s) = N$, then

$$
B(s)\, A(s) = N \, .
\tag{17.15}
$$

Theorem 17.4

$$
D_0'(L,U) = \frac{1}{\log(U/L)} \log \frac{R(L)}{R(U)} \, .
$$

Proof [Rosenberg 17c] Define $J \equiv B(L)$ (the number of boxes in the minimal L-covering $\mathcal{B}(L)$) and define $K \equiv B(U)$. For simplicity of notation, by $\sum_j$ we mean $\sum_{j=1}^{J}$ and by $\sum_k$ we mean $\sum_{k=1}^{K}$. Similarly, by $\prod_j$ we mean $\prod_{j=1}^{J}$ and by $\prod_k$ we mean $\prod_{k=1}^{K}$. Assume that log means the natural log. For $1 \leq j \leq J$, define $a_j \equiv x_j(L)$ and $\alpha_j \equiv \log a_j$. For $1 \leq k \leq K$, define $b_k \equiv x_k(U)$ and $\beta_k \equiv \log b_k$. Define $\omega \equiv 1/\log(U/L)$. For simplicity of notation, define $F(q) \equiv D_q(L, U)$. By $F'(0)$ we mean the first derivative of $F(q)$ evaluated at $q = 0$, so $F'(0) = D_0'(L, U)$. Using the above notation and definitions, from (17.13) we have

$$F(q) = \frac{\omega}{(q-1)} \log \left(\frac{\sum_k e^{\beta_k q}}{\sum_j e^{\alpha_j q}} \right). \tag{17.16}$$

Define $\alpha = (1/J) \sum_j \alpha_j$. Consider the functions $f_1(q) \equiv J e^{\alpha q}$ and $f_2(q) \equiv \sum_j e^{\alpha_j q}$. We have $f_1(0) = f_2(0) = J$. Evaluating the first derivatives at $q = 0$, we have $f_1'(0) = J\alpha = \sum_j \alpha_j$ and $f_2'(0) = \sum_j \alpha_j$. Since $f_1(q)$ and $f_2(q)$ have the same function values and first derivatives at $q = 0$, then $F(0) = M(0)$ and $F'(0) = M'(0)$, where $M(q)$ is defined by

$$M(q) \equiv \frac{\omega}{(q-1)} \log \left(\frac{\sum_k e^{\beta_k q}}{J e^{\alpha q}} \right). \tag{17.17}$$

Simplifying (17.17) yields

$$M(q) = \frac{\omega}{(q-1)} \log \left(\frac{1}{J} \sum_k e^{(\beta_k - \alpha)q} \right).$$

Taking the first derivative with respect to q,

$$M'(q) = \frac{\omega}{(q-1)^2} \left[\frac{(q-1)(1/J) \sum_k [e^{(\beta_k - \alpha)q}(\beta_k - \alpha)]}{(1/J) \sum_k e^{(\beta_k - \alpha)q}} - \log \left(\frac{1}{J} \sum_k e^{(\beta_k - \alpha)q} \right) \right].$$

Plugging in $q = 0$ and using the definition of α yields

$$
\begin{aligned}
M'(0) &= \omega\left(-\frac{1}{K}\sum_k(\beta_k - \alpha) - \log\frac{K}{J}\right) \\
&= \omega\left(-\frac{1}{K}\sum_k \log b_k + \frac{1}{J}\sum_j \log a_j - \log\frac{K}{J}\right) \\
&= \omega\left(-\log\left(\left(\prod_k b_k\right)^{1/K}\right) + \log\left(\left(\prod_j a_j\right)^{1/J}\right) - \log\frac{K}{J}\right) \\
&= \omega\log\left(\frac{J\left(\prod_j a_j\right)^{1/J}}{K\left(\prod_k b_k\right)^{1/K}}\right) \\
&= \frac{1}{\log(U/L)}\log\left(\frac{\frac{J\left(\prod_j a_j\right)^{1/J}}{JA(L)}}{\frac{K\left(\prod_k b_k\right)^{1/K}}{KA(U)}}\right) \quad \text{(by (17.15))} \\
&= \frac{1}{\log(U/L)}\log\frac{R(L)}{R(U)} \quad \text{(by (17.14))}. \quad\square
\end{aligned}
$$

Theorem 17.4 says that the slope of the secant estimate of D_q at $q = 0$ depends on $x(L)$ and $x(U)$ only through the ratio of the geometric mean to the arithmetic mean of the components of $x(L)$, and similarly for $x(U)$. Note that the proof of Theorem 17.4 does not require $x(s)$ to be lexico minimal. Since $U > L$, Theorem 17.4 immediately implies the following corollary.

Corollary 17.1 $D_0'(L,U) > 0$ if and only if $R(L) > R(U)$, and $D_0'(L,U) < 0$ if and only if $R(L) < R(U)$. $\square$

For a given L and U, Theorem 17.5 below provides a sufficient condition for $D_q(L,U)$ to have a local maximum or minimum.

Theorem 17.5 [Rosenberg 17c] Assume (17.10) holds as an equality. If $x(s)$ is a lexico minimal summary vector summarizing the minimal s-covering $\mathcal{B}(s)$, then (i) if $R(L) > R(U)$ and $B(L)/B(U) > x_1(U)/x_1(L)$, then $D_q(L,U)$ has a local maximum at some $q > 0$. (ii) If $R(L) < R(U)$ and $B(L)/B(U) < x_1(U)/x_1(L)$, then $D_q(L,U)$ has a local minimum at some $q > 0$.

Proof We will prove (i); the proof of (ii) proceeds similarly. Since (17.10) is assumed to hold as an equality, then (17.9) holds as an equality, and $D_\infty = \lim_{q\to\infty} D_q$ satisfies $\log\left(x_1(s)/N\right) = D_\infty\log(s/\Delta)$, where $x_1(s)$ is the first coordinate of the lexico minimal summary vector $x(s)$. Defining $\omega \equiv 1/\log(U/L)$, we have

$$
D_\infty = \frac{\log\left(x_1(U)/N\right) - \log\left(x_1(L)/N\right)}{\log(U/\Delta) - \log(L/\Delta)} = \omega\log\frac{x_1(U)}{x_1(L)}. \tag{17.18}
$$

Since $Z\big(x(s),0\big) = B(s)$, from (17.10) with $q = 0$ we obtain

$$D_0 = \frac{\log Z\big(x(L),0\big) - \log Z\big(x(U),0\big)}{\log(U/\Delta) - \log(L/\Delta)} = \omega \log \frac{B(L)}{B(U)}. \qquad (17.19)$$

If $D_0'(L,U) > 0$ and $D_0 > D_\infty$, then $D_q(L,U)$ must have a local maximum at some positive q. By Corollary 17.1, if $R(L) > R(U)$ then $D_0'(L,U) > 0$. By (17.18) and (17.19), if $B(L)/B(U) > x_1(U)/x_1(L)$, then $D_0 > D_\infty$. $\quad\square$

Example 17.7 To illustrate Theorem 17.5, consider again the *dolphins* network of Example 17.2 with $L = 3$ and $U = 5$. We have $B(3) = 13$, $B(5) = 4$, $x_1(3) = 10$, and $x_1(5) = 28$, so $D_0 = \log(13/4)/\log(5/3) = 2.307$ and $D_\infty = \log(28/10)/\log(5/3) = 2.106$. We have $R(3) = 0.773$, $R(5) = 0.660$, and $D_0'(L,U) = 0.311$, so $D_q(3,5)$ has a local maximum, as seen in Fig. 17.5. Moreover, for the *dolphins* network, choosing $L = 2$ and $U = 5$ we have $D_0 = \log(29/4)/\log(5/2) = 2.16$, and $D_\infty = \log(28/3)/\log(5/2) = 2.44$, so $D_0 < D_\infty$, as is evident from Fig. 17.5. Thus the inequality $D_0 \geq D_\infty$, which is valid for geometric multifractals, does not hold for the *dolphins* network with $L = 2$ and $U = 5$. $\quad\square$

If for $s \in \{L,U\}$ we can compute a minimal s-covering with equal box masses, then by Definition 17.3 the network $\mathbb{G}$ is a monofractal but not a multifractal. To see this, suppose all boxes in $\mathcal{B}(L)$ have the same mass, and that all boxes in $\mathcal{B}(U)$ have the same mass. Then for $s \in \{L,U\}$ we have $x_j(s) = N/B(s)$ for $1 \leq j \leq B(s)$, and (17.4) yields

$$Z\big(x(s),q\big) = \sum_{B_j \in \mathcal{B}(s)} \left(\frac{x_j(s)}{N}\right)^q = \sum_{B_j \in \mathcal{B}(s)} \left(\frac{1}{B(s)}\right)^q = [B(s)]^{1-q}.$$

From (17.10), for $q \neq 1$ we have

$$\begin{aligned} D_q &= \frac{\log Z\big((x(U),q\big) - \log Z\big(x(L),q\big)}{(q-1)\big(\log U - \log L\big)} = \frac{\log\big([B(U)]^{1-q}\big) - \log\big([B(L)]^{1-q}\big)}{(q-1)\big(\log U - \log L\big)} \\ &= \frac{\log B(L) - \log B(U)}{\log U - \log L} = D_0 = d_B, \end{aligned} \qquad (17.20)$$

so $\mathbb{G}$ is a monofractal. Thus equal box masses imply $\mathbb{G}$ is a monofractal, the simplest of all fractal structures.

There are several ways to try to obtain equal box masses in a minimal s-covering of $\mathbb{G}$. As discussed in Chap. 15, ambiguity in the choice of minimal coverings used to compute d_I is eliminated by maximizing entropy. Since the entropy of a probability distribution is maximized when all the probabilities are equal, a maximal entropy minimal covering equalizes (to the extent possible) the box masses. Similarly, ambiguity in the choice of minimal s-coverings used to compute D_q is eliminated by minimizing the partition function $Z_q\big(\mathcal{B}(s)\big)$. Since, by Theorem 17.3, for all sufficiently large q the unique lexico minimal vector $x(s)$ summarizes the s-covering that minimizes $Z_q\big(\mathcal{B}(s)\big)$, and since, by

Definition (17.21) differs from definition (16.57) of the sandbox dimension function in that (17.21) employs a limit as $r \to 0$ while (16.57) does not. Interestingly, [Liu 15] characterized the sandbox method as "an extension of the box counting algorithm". Definition (17.21) was then rewritten in the computationally more convenient form

$$\log avg\big(M_{q-1}(r)\big) \propto (q-1)D_q^{sandbox} \log(r/\Delta) + (q-1)\log N . \qquad (17.22)$$

For a given $q \neq 1$, the estimate of $D_q^{sandbox}$ in [Liu 15] is $1/(q-1)$ times the slope of the $\log avg\big(M_{q-1}(r)\big)$ versus $\log(r/\Delta)$ plot over a range of r values, and linear regression was used to obtain this slope. The mass exponent $\tau(q)$ (Sect. 16.4) for $q \neq 1$ was obtained using $\tau(q) = (q-1)D_q^{sandbox}$.

In [Liu 15], the sandbox method was compared to compact box burning (Sect. 8.5) and to a variant of the box-counting method. First the comparison was made for three models of deterministic graphs, including (u, v)-flower networks (Sect. 12.2). For each of these models, the sandbox method CPU time was tiny compared to the other two methods. Plots in [Liu 15] showed that, for $q \in [-10, 10]$, the three methods determined essentially identical estimates of $\tau(q)$ for each of the three network models. For all three network models, the plots of $\tau(q)$ versus q over the range $q \in [-10, 10]$ were not quite linear, but rather were very slightly concave.

The sandbox method was also applied in [Liu 15] to three "real-world" network models. For scale-free networks constructed using the Barabási–Albert preferential attachment model [Barabási 99], $D_q^{sandbox}$ ranged from about $D_{-10} \approx 3.5$ to $D_{10} \approx 1.6$ (although the $D_q^{sandbox}$ versus q curves in [Liu 15] were characterized as convex, they do not appear to be convex for $q \in [-10, 10]$). For small-world networks constructed using the Watts-Strogatz model [Newman 99b], $D_q^{sandbox}$ decreased only slightly as q increased, with $D_{-10} \approx 4.7$ and $D_{10} \approx 4.2$ for $N = 10^4$. The conclusion in [Liu 15] was that multifractality of these small-world networks is not obvious. Finally, Erdős–Rényi random networks were constructed in which pairs of nodes were connected with small probabilities, and the largest connected component was analyzed. For these random networks, the $D_q^{sandbox}$ versus q plot was a straight line with slope zero, so the networks are monofractal.

It was inevitable that the sandbox method would next be applied to multifractal weighted networks, and that step was taken in [Song 15], which also used (17.22) to calculate $D_q^{sandbox}$. The difference between the method of [Song 15] and the method of [Liu 15] described above is that whereas [Liu 15] used integer values of r, [Song 15] used possibly non-integer r values that are determined by the arc lengths. Let c_{ij} be the nonnegative length of the arc connecting nodes i and j, with $c_{ii} = 0$ for each i, and assume symmetric lengths ($c_{ij} = c_{ji}$). To determine the set of radius values to be used, order the arc lengths from smallest to largest. Represent these ordered $A \equiv N(N-1)/2$ arc lengths by $\{c_k\}_{k=1}^A$, where $c_{k_1} \leq c_{k_2}$ for $k_1 < k_2$. Let K be the largest index such that $K \leq A$ and $\sum_{k=1}^K c_k \leq \Delta$, where Δ is the network diameter. Then the radius values to be used are c_1, $c_1 + c_2$, ..., $\sum_{k=1}^{K} c_k$. (A similar method for computing radii values was described in Sect. 8.2 on node coloring for weighted networks.) In [Song 15], using 1000 random sandbox centers and integer $q \in [10, 10]$, linear regression was used to estimate $D_q^{sandbox}$ for three large collaboration networks.

The sandbox method was applied in [Rendon 17] to a network of payments in Estonia. The nodes are companies, and there is an arc between two nodes if there was at least one transaction, in the year 2014, between the two companies. Multifractal analysis was performed on the skeleton *skel*($\mathbb{G}$) (Sect. 2.5) rather than on $\mathbb{G}$, since "by looking at the properties of the skeleton it is easier to appreciate the topological organization of the original network". The skeleton *skel*($\mathbb{G}$) was computed using the arcs with the highest betweenness (Sect. 2.4); the skeleton has diameter 34, compared to $\Delta = 29$ for $\mathbb{G}$. The box counting dimensions of $\mathbb{G}$ and *skel*($\mathbb{G}$) were nearly the same: 2.39 for $\mathbb{G}$ and 2.32 for *skel*($\mathbb{G}$). The range of r used was 2–29, and the range of q was $[-7, 12]$. Since $D_q \approx 7.8$ for $q \in [-7, -4.5]$, at which point D_q decreased rapidly, and since $\min_q D_q = 0.37$, the conclusion in [Rendon 17] was that the network is multifractal.

The sandbox method is not always successful: [Wang 11] found that the sandbox results did not yield a satisfactory fit for linear regression, and instead the results of *Random Sequential Node Burning* (Sect. 17.4) were averaged.

Exercise 17.6 In Sect. 16.6 we presented the barycentric variant of the sandbox method. Is there any way to adapt this method to a network?   □

Exercise 17.7 In Sect. 16.6 we presented the recommendation in [Panico 95] that the calculations of the fractal dimension of an object should use measurements only from the interior of the object. Does this principle have applicability when calculating $D_q^{sandbox}$ for a growing network?   □

Exercise 17.8 Does the fact that $\mathbb{G}$ and *skel*($\mathbb{G}$) have very similar box counting dimensions imply that the multifractal spectra of $\mathbb{G}$ and *skel*($\mathbb{G}$) are very similar? Run some experiments to find out.   □

Chapter 18

Generalized Hausdorff Dimensions of Networks

Scientific research involves going beyond the well-trodden and well-tested ideas and theories that form the core of scientific knowledge. During the time scientists are working things out, some results will be right, and others will be wrong. Over time, the right results will emerge.

Lisa Randall (1962-), American, theoretical physicist, professor at Harvard University, author of "Higgs Discovery: The Power of Empty Space" (2012)

The progression of topics in this book has been to first present a fractal dimension for a geometric object $\Omega \subset \mathbb{R}^E$, and then to show how the same dimension can be defined for a network $\mathbb{G}$. Using this pattern, we defined, for example, the box counting dimension d_B of Ω and then of $\mathbb{G}$, the correlation dimension d_C of Ω and then of $\mathbb{G}$, and the information dimension d_I of Ω and then of $\mathbb{G}$. In this chapter we continue this pattern for two additional dimensions. First, the Hausdorff dimension d_H of Ω was defined in Sect. 5.1, and in this chapter we define d_H of $\mathbb{G}$. Second, the family $\{D_q^H, q \in \mathbb{R}\}$ of generalized dimensions of Ω was defined in Sect. 16.9, and and in this chapter we define the generalized dimensions $\{D_q^H, q \in \mathbb{R}\}$ of $\mathbb{G}$. The definition of d_H of $\mathbb{G}$, and the definition of $\{D_q^H, q \in \mathbb{R}\}$ of $\mathbb{G}$, were introduced in [Rosenberg 18b].

18.1 Defining the Hausdorff Dimension of a Network

Computing the box counting dimension d_B of a network $\mathbb{G}$ requires covering $\mathbb{G}$ by a minimal set $\mathcal{B}(s)$ of boxes of diameter at most s, for a range of values

© Springer Nature Switzerland AG 2020
E. Rosenberg, *Fractal Dimensions of Networks*,
https://doi.org/10.1007/978-3-030-43169-3_18

of s. Then d_B is typically computed from the slope of the $\log B(s)$ versus $\log s$ curve. Although, for a given s, the boxes in $\mathcal{B}(s)$ do not in general have the same diameter or mass, this information is ignored in the computation of d_B. However, the diameters of the boxes in $\mathcal{B}(s)$ can be used to define the Hausdorff dimension d_H of $\mathbb{G}$. The reason for studying d_H of $\mathbb{G}$ is that, for a geometric object, the Hausdorff dimension d_H generalizes the box counting dimension d_B, and d_H is better behaved than d_B. Thus it is of great interest to define and study d_H of $\mathbb{G}$.

To develop a definition of d_H of $\mathbb{G}$, first consider computing d_B of a geometric object Ω. Recall that $B(s)$ is the minimal number of boxes of linear size at most s required to cover Ω and that, as discussed in Sect. 4.2, we can assume that, for each given s, all boxes in the minimal covering of Ω have the same diameter. From definitions (5.1) and (5.2), if each box has diameter s, then

$$v(\Omega, s, d) = B(s)s^d$$

$$v(\Omega, d) = \lim_{s \to 0} B(s)s^d,$$

assuming the limit exists. The box counting dimension d_B of Ω was characterized in ([Schroeder 91], p. 201) as the unique value of d for which

$$\lim_{s \to 0} B(s)s^d = c \qquad (18.1)$$

for some c such that $0 < c < \infty$. More precisely, the existence of d_B does not imply (18.1) but rather, as noted in [Theiler 88] and discussed in Sect. 4.2, only implies $B(s)s^d = \Phi(s)$ for some function $\Phi(\cdot)$ satisfying $\lim_{s \to 0} \log \Phi(s)/\log s = 0$. We thus treat (18.1) as an approximation, rather than as an equality.

Now consider a network $\mathbb{G}$. For integer s, where $2 \leq s \leq \Delta$, consider a minimal s-covering $\mathcal{B}(s)$ of $\mathbb{G}$. By Definition 7.5, the box counting dimension d_B of $\mathbb{G}$ satisfies, over some range of s and for some constant c,

$$\log B(s) \approx -d_B \log(s/\Delta) + c. \qquad (18.2)$$

For reasons that will shortly become clear, it is useful to present a slightly different definition of d_B.

Definition 18.1 $\mathbb{G}$ has box counting dimension d_B if, over some range of s and for some constant c,

$$\log B(s) \approx -d_B \log\big(s/(\Delta + 1)\big) + c. \qquad \square \qquad (18.3)$$

Exercise 18.1 When using linear regression to estimate d_B from a range of $\big(\log s, \log B(s)\big)$ values, do both of the approximations (18.2) and (18.3), when treated as an equality, yield the same value of d_B? $\square$

Corresponding to (18.1) for Ω, and using Definition 18.1, if $\mathbb{G}$ has box counting dimension d_B then for some constant c such that $0 < c < \infty$ we have, over some range of s,

$$B(s)[s/(\Delta+1)]^{d_B} \approx c. \tag{18.4}$$

Hence one way to determine d_B is to compute the value of d which minimizes, over $d > 0$, the standard deviation, over some range of s, of the values $B(s)[s/(\Delta+1)]^d$. We will later illustrate this approach for the *dolphins* network. We are not advocating this approach to computing the box counting dimension d_B, but mention it to show the similarity between this approach to computing d_B of $\mathbb{G}$ and the following approach to defining the Hausdorff dimension d_H of $\mathbb{G}$.

For a given box size s, where $2 \leq s \leq \Delta$, for box $B_j \in \mathcal{B}(s)$ let Δ_j be the diameter of box B_j, and define

$$\delta_j \equiv (\Delta_j + 1)/(\Delta + 1), \tag{18.5}$$

so $0 < \delta_j < 1$. If B_j contains a single node, then $\delta_j = 1/(\Delta + 1)$; if B_j has diameter $\Delta - 1$, then $\delta_j = \Delta/(\Delta + 1)$. Corresponding to definition (5.1) of $v(\Omega, d, s)$ for Ω, define

$$m(\mathbb{G}, s, d) \equiv \sum_{B_j \subset \mathcal{B}(s)} \delta_j^d. \tag{18.6}$$

Note that $m(\mathbb{G}, s, d)$ is independent of the mass of (i.e., number of nodes in) each box. We can regard $m(\mathbb{G}, s, d)$ as the "d-dimensional Hausdorff measure of $\mathbb{G}$ for box size s". Just as d_B is that value of d for which (18.4) holds for some c and some range of s, we can view d_H as the value of d for which

$$m(\mathbb{G}, s, d) \approx c \tag{18.7}$$

holds for some c and some range of s. Expressed differently, d_H is the value of d for which the Hausdorff measures $m(\mathbb{G}, s, d)$ are nearly equal over some range of s. Note that, since $m(\mathbb{G}, s, d)$ is not an explicit function of s, relation (18.7) does not yield a "traditional" fractal scaling law, similar to (1.1), that relates s to some appropriately defined "bulk".

We can now justify why we replaced the approximation (18.2) with (18.3) in the definition of d_B. We want each term in the sum (18.6) to depend on d, so we want $0 < \delta_j < 1$ for each j. Since a box in $\mathcal{B}(s)$ containing only a single node has diameter 0, adding 1 to Δ_j in (18.5) ensures that $\delta_j > 0$ for each j. On the other hand, if a box in $\mathcal{B}(s)$ has diameter $s - 1$, then adding 1 to Δ in (18.5) ensures that $\delta_j < 1$ for each j.

Before we formalize the definition of d_H, we must deal with non-uniqueness of $m(\mathbb{G}, s, d)$. For each s there are, in general, multiple minimal coverings $\mathcal{B}(s)$ of $\mathbb{G}$, all yielding the same minimal value $B(s)$. As discussed in Sect. 17.1, a definition of D_q of $\mathbb{G}$ based on the partition function $Z_q(s) \equiv \sum_{B_j \in \mathcal{B}(s)} [p_j(s)]^q$

can be ambiguous, since different minimal s-coverings can yield different values of $Z_q(s)$, and hence different values of D_q. The ambiguity was resolved by requiring each minimal s-covering used to compute $Z_q(s)$ to be lexicographically minimal. Similarly, as shown by Example 18.1 below, definition (18.6) does not in general unambiguously define $m(\mathbb{G}, s, d)$, since different minimal s-coverings can yield different values of $m(\mathbb{G}, s, d)$.

Example 18.1 Consider the 7 node chain network of Fig. 18.1, for which $\Delta = 6$. From definitions (18.5) and (18.6), for the minimal 3-covering (i) we have

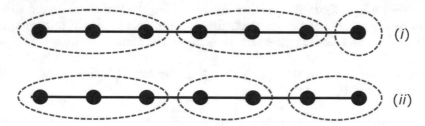

$$(i)$$

$$(ii)$$

Figure 18.1: Two minimal 3-coverings of a 7 node chain

$$\widetilde{m}(\mathbb{G}, 3, d) = 2 \left(\frac{2+1}{6+1} \right)^d + \left(\frac{0+1}{6+1} \right)^d = \frac{3^d + 3^d + 1^d}{7^d}$$

while for the minimal 3-covering (ii) we have

$$\widehat{m}(\mathbb{G}, 3, d) = \left(\frac{2+1}{6+1} \right)^d + 2 \left(\frac{1+1}{6+1} \right)^d = \frac{3^d + 2^d + 2^d}{7^d}.$$

We have $\widetilde{m}(\mathbb{G}, 3, d) > \widehat{m}(\mathbb{G}, 3, d)$ for $d > 1$, with equality when $d = 1$. $\square$

The ambiguity in the definition of $m(\mathbb{G}, s, d)$ can be eliminated by adapting the lexicographic approach of [Rosenberg 17b], described in Sect. 17.1. Consider a minimal s-covering $\mathcal{B}(s)$ and define $J \equiv B(s)$. Recalling that Δ_j is the diameter of box $B_j \in \mathcal{B}(s)$, we associate with $\mathcal{B}(s)$ the point

$$\Delta(J) \equiv (\Delta_1, \Delta_2, \ldots, \Delta_J) \in \mathbb{R}^J,$$

where $\Delta_1 \geq \Delta_2 \geq \cdots \geq \Delta_J$. We call $\Delta(J)$ the "diameter vector" associated with $\mathcal{B}(s)$. An unambiguous choice of a minimal s-covering is to pick the minimal s-covering for which the associated diameter vector is lexicographically smallest. That is, suppose that, for a given s, there are two distinct minimal s-coverings $\widetilde{\mathcal{B}}(s)$ and $\widehat{\mathcal{B}}(s)$. Let $\widetilde{\Delta}(J)$ be the diameter vector associated with $\widetilde{\mathcal{B}}(s)$, and let $\widehat{\Delta}(J)$ be the diameter vector associated with $\widehat{\mathcal{B}}(s)$. Then we choose $\widetilde{\mathcal{B}}(s)$ if $\widehat{\Delta}(J) \succeq \widetilde{\Delta}(J)$, where "$\succeq$" denotes lexicographic ordering. Applying this to the two minimal 3-coverings of Fig. 18.1, the associated diameter vectors are $(2, 2, 0)$ for (i) and $(2, 1, 1)$ for (ii). Since $(2, 2, 0) \succeq (2, 1, 1)$ we choose the covering (ii).

Having unambiguously defined $m(\mathbb{G}, s, d)$, we can now formalize the definition of d_H of $\mathbb{G}$ as that value of d for which $m(\mathbb{G}, s, d)$ is roughly constant over a range of s. Let $\mathcal{S}$ be a given set of integer values of s such that $2 \leq s \leq \Delta$ for $s \in \mathcal{S}$. For example, for integers L, U such that $2 \leq L < U \leq \Delta$ we might set $\mathcal{S} = \{s \mid s \text{ is integer and } L \leq s \leq U\}$. For a given $d > 0$, define $\sigma(\mathbb{G}, \mathcal{S}, d)$ to be the standard deviation (in the usual statistical sense) of the values $m(\mathbb{G}, s, d)$ for $s \in \mathcal{S}$ as follows. Define

$$m_{avg}(\mathbb{G}, \mathcal{S}, d) \equiv \frac{1}{|\mathcal{S}|} \sum_{s \in \mathcal{S}} m(\mathbb{G}, s, d). \qquad (18.8)$$

We can regard $m_{avg}(\mathbb{G}, \mathcal{S}, d)$ as the "average d-dimensional Hausdorff measure of $\mathbb{G}$ for box sizes in $\mathcal{S}$". Next we compute the standard deviation of the Hausdorff measures:

$$\sigma(\mathbb{G}, \mathcal{S}, d) \equiv \sqrt{\frac{1}{|\mathcal{S}|} \sum_{s \in \mathcal{S}} \left[m(\mathbb{G}, s, d) - m_{avg}(\mathbb{G}, \mathcal{S}, d) \right]^2}. \qquad (18.9)$$

We associate with $\mathbb{G}$ the Hausdorff dimension d_H, defined as follows.

Definition 18.2 [Rosenberg 18b] (*Part A.*) If $\sigma(\mathbb{G}, \mathcal{S}, d)$ has at least one local minimum over $\{d \mid d > 0\}$, then d_H is the value of d satisfying (*i*) $\sigma(\mathbb{G}, \mathcal{S}, d)$ has a local minimum at d_H, and (*ii*) if $\sigma(\mathbb{G}, \mathcal{S}, d)$ has another local minimum at $\tilde{d}$, then $\sigma(\mathbb{G}, \mathcal{S}, d_H) < \sigma(\mathbb{G}, \mathcal{S}, \tilde{d})$. (*Part B.*) If $\sigma(\mathbb{G}, \mathcal{S}, d)$ does not have a local minimum over $\{d \mid d > 0\}$, then $d_H \equiv \infty$. $\square$

For all of the networks studied in [Rosenberg 18b], d_H is finite. An interesting open question is whether there exists a network with a finite d_B and infinite d_H.

Definition 18.3 [Rosenberg 18b] If $d_H < \infty$, then the Hausdorff measure of $\mathbb{G}$ is $m_{avg}(\mathbb{G}, \mathcal{S}, d_H)$. If $d_H = \infty$, then the Hausdorff measure of $\mathbb{G}$ is 0. $\square$

These definitions are with respect to a given set $\mathcal{S}$. For a given $\mathbb{G}$, choosing different $\mathcal{S}$ will in general yield different values of d_H; this is not surprising, and a similar dependence was studied in Sect. 17.2. The reason, in *Part A* of Definition 18.2, for requiring d_H to yield a local minimum of $\sigma(\mathbb{G}, \mathcal{S}, d)$ is that by definitions (18.5) and (18.6) for each s we have

$$\lim_{d \to \infty} m(\mathbb{G}, s, d) = 0. \qquad (18.10)$$

Suppose d_H yields a local minimum of $\sigma(\mathbb{G}, \mathcal{S}, d)$. From (18.10), we know that $\sigma(\mathbb{G}, \mathcal{S}, d)$ is not a convex function of d. Thus the possibility exists that $\sigma(\mathbb{G}, \mathcal{S}, d)$ might have multiple local minima. To deal with that possibility, we require that $\sigma(\mathbb{G}, \mathcal{S}, d_H)$ be smaller than the standard deviation at any other local minimum. As for *Part B*, it may be possible that $\sigma(\mathbb{G}, \mathcal{S}, d)$ has no local minimum. In this case, by virtue of (18.10) we define $d_H = \infty$.

By Definition 18.2, the computation of d_H of $\mathbb{G}$ requires a *line search* (i.e., a 1-dimensional minimization) to minimize $\sigma(\mathbb{G}, \mathcal{S}, d)$ over $d > 0$. Since $\sigma(\mathbb{G}, \mathcal{S}, d)$ might have multiple local minima, we cannot use line search methods such as bisection or golden search [Wilde 64] designed for minimizing a unimodal function. Instead, in [Rosenberg 18b] d_H was computed by evaluating $\sigma(\mathbb{G}, \mathcal{S}, d)$

from $d = 0.1$ to $d = 6$, in increments of 10^{-4}, and testing for a local minimum (this computation took at most a few seconds on a laptop). For all of the networks studied, $\sigma(\mathbb{G}, \mathcal{S}, d)$ had exactly one local minimum. While these computational results do not preclude the possibility of multiple local minima for some other network, a convenient working definition is that d_H is the value of d yielding a local minimum of the standard deviation $\sigma(\mathbb{G}, \mathcal{S}, d)$.

In scientific applications of fractal analysis, it is well-established that the ability to calculate a fractal dimension of a geometric object does not convey information about the structure of the object. For example, box counting was used in [Karperien 13] to calculate d_B of microglia (a type of cell occurring in the central nervous system of invertebrates and vertebrates). However, we were warned in [Karperien 13] that the fractal dimension of an object "does not tell us anything about what the underlying structure looks like, how it was constructed, or how it functions". Moreover, the ability to calculate a fractal dimension of an object "does not mean the phenomenon from which it was gleaned is fractal; neither does a phenomenon have to be fractal to be investigated with box counting fractal analysis".

Similarly, the ability to calculate d_H of $\mathbb{G}$ does not provide any information about the structure of $\mathbb{G}$. Indeed, for a network $\mathbb{G}$ the definitions of *fractality* and *self-similarity* are quite different [Gallos 07b]; as discussed in Sect. 7.2, the network $\mathbb{G}$ is *fractal* if $B(s) \sim s^{-d_B}$ for a finite exponent d_B, while $\mathbb{G}$ is *self-similar* if the node degree distribution of $\mathbb{G}$ remains invariant over several generations of renormalization. (Renormalizing $\mathbb{G}$, as discussed in Chap. 21, means computing an s-covering of $\mathbb{G}$, merging all nodes in each box in the s-covering into a supernode, and connecting two supernodes if the corresponding boxes are connected by one or more arcs in the original network.) Moreover, "Surprisingly, this self-similarity under different length scales seems to be a more general feature that also applies in non-fractal networks such as the Internet. Although the traditional fractal theory does not distinguish between fractality and self-similarity, in networks these two properties can be considered to be distinct" [Gallos 07b]. With this caveat on the interpretation of a fractal dimension of $\mathbb{G}$, in the next section we compute d_H of several networks.

18.2   Computing the Hausdorff Dimension of a Network

The following examples from [Rosenberg 18b] illustrate the computation of d_H of some networks, ranging from a small constructed example of 5 nodes to a real-world protein network of 1458 nodes. Since *fractality* and *self-similarity* of $\mathbb{G}$ are distinct characteristics, d_H can be meaningfully computed even for small networks without any regular or self-similar structure.

Example 18.2 Consider the *chair* network of Fig. 18.2 (left), for which $\Delta = 3$. We have $B(3) = 2$ and $B(2) = 3$. Using the two box sizes $s = 2, 3$ we have $d_B = \big(\log B(2) - \log B(3) \big)/(\log 3 - \log 2) = 1$. For $s = 2$ the three boxes in $\mathcal{B}(2)$ have diameter 1, and 1, and 0, so from definitions (18.5) and (18.6) we have $m(\mathbb{G}, 2, d) = (2/4)^d + (2/4)^d + (1/4)^d$. For $s = 3$ the two boxes in $\mathcal{B}(3)$ have diameter 2 and 1, so $m(\mathbb{G}, 3, d) = (3/4)^d + (2/4)^d$. Figure 18.2 (right) plots

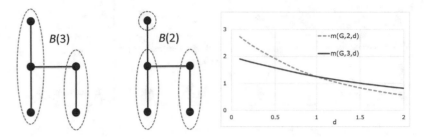

Figure 18.2: *chair* network (left) and d_H calculation (right)

$m(\mathbb{G}, 2, d)$ and $m(\mathbb{G}, 3, d)$ versus d. Set $\mathcal{S} = \{2, 3\}$. Since for $d = 1$ we have $m(\mathbb{G}, 2, d) = m(\mathbb{G}, 3, d) = 5/4$, then $d_H = 1$ and $\sigma(\mathbb{G}, \mathcal{S}, d_H) = 0$. Moreover, $d_H = 1$ is the unique point minimizing $\sigma(\mathbb{G}, \mathcal{S}, d)$. For this network we have $d_H = d_B$. By Definition 18.3, the Hausdorff measure $m(\mathbb{G}, \mathcal{S}, d_H)$ of $\mathbb{G}$ is 5/4. □

Example 18.3 Consider the two networks in Fig. 18.3. The top network $\mathbb{G}_1$ has $\Delta = 4$; the bottom network $\mathbb{G}_2$ has $\Delta = 5$. Figure 18.4 shows box counting

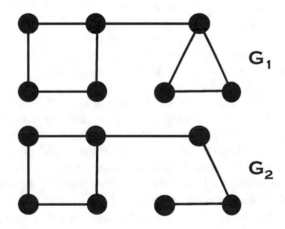

Figure 18.3: Two 7 node networks with the same d_B but different d_H

results, for $s = 2$ and $s = 3$, for the two networks. For $s = 2$ we have, for both networks, $B(2) = 4$, $B(3) = 2$, and $B(4) = 2$. Since $B(3) = B(4)$, to compute d_B we ignore the $s = 4$ results and choose $\mathcal{S} = \{2, 3\}$. This yields, for both networks, $d_B = \big(\log B(2) - \log B(3) \big) / (\log 3 - \log 2) \approx 1.71$.

For $\mathbb{G}_1$, from (18.5) and (18.6) we have $m(\mathbb{G}_1, 2, d) = (3)(2/5)^d + (1/5)^d$ and $m(\mathbb{G}_1, 3, d) = (3/5)^d + (2/5)^d$. For $\mathbb{G}_1$, the plots of $m(\mathbb{G}_1, 2, d)$ versus d and $m(\mathbb{G}_1, 3, d)$ versus d intersect at $d_H = 2$, and $m(\mathbb{G}_1, 2, d_H) = m(\mathbb{G}_1, 3, d_H) = 13/25$. Hence $\sigma(\mathbb{G}_1, \mathcal{S}, d_H) = 0$, and $d_H = 2$ is the unique point minimizing

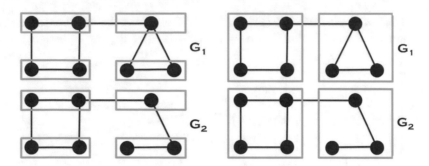

Figure 18.4: Box counting results for $s = 2$ (left) and $s = 3$ (right)

$\sigma(\mathbb{G}_1, \mathcal{S}, d)$. Thus for $\mathbb{G}_1$ and $\mathcal{S} = \{2, 3\}$ we have $d_H = 2 > 1.71 \approx d_B$, so the inequality $d_H \leq d_B$, known to hold for a geometric object [Falconer 03], can fail to hold for a network. By Definition 18.3, the Hausdorff measure $m(\mathbb{G}_1, \mathcal{S}, d_H)$ of $\mathbb{G}_1$ is $13/25$.

Turning now to $\mathbb{G}_2$, from (18.5) and (18.6) we have $m(\mathbb{G}_2, 2, d) = (3)(2/6)^d + (1/6)^d$ and $m(\mathbb{G}_2, 3, d) = (2)(3/6)^d$. The plots of $m(\mathbb{G}_2, 2, d)$ versus d and and $m(\mathbb{G}_2, 3, d)$ versus d intersect at $d_H \approx 1.31$, and $m(\mathbb{G}_2, 2, d_H) = m(\mathbb{G}_2, 3, d_H) \approx 0.81$. Thus for $\mathbb{G}_2$ and $\mathcal{S} = \{2, 3\}$ we have $\sigma(\mathbb{G}_2, \mathcal{S}, d_H) = 0$, and $m(\mathbb{G}_2, \mathcal{S}, d_H) \approx 0.81$. The two networks $\mathbb{G}_1$ and $\mathbb{G}_2$ in Example 18.3 have the same number of nodes and the same d_B, yet have different d_H.   □

Example 18.3 suggests that d_H can be better than d_B for comparing two networks, or for studying the evolution of a network over time. The reason d_H provides a "finer" measure than d_B is that d_B considers only the variation in the number $B(s)$ of boxes needed to cover the network, while d_H recognizes that the different boxes in the covering $\mathcal{B}(s)$ will in general have different diameters. The next two examples, from [Rosenberg 18b], show that the Hausdorff dimension of $\mathbb{G}$ produces expected "classical" results: for a large 1-dimensional chain we have $d_H = 1$, and for a large 2-dimensional rectilinear grid we have $d_H = 2$.

Example 18.4 Let $\mathbb{G}$ be a chain of N nodes, so $\Delta = N - 1$. For $s \ll N$, an s-covering of $\mathbb{G}$ consists of $\lfloor N/s \rfloor$ boxes, each containing s nodes and having diameter $s - 1$, and one box containing $N - s\lfloor N/s \rfloor$ nodes and having diameter $N - s\lfloor N/s \rfloor - 1$. By definitions (18.5) and (18.6), for $s \ll N$ we have

$$m(\mathbb{G}, s, d) = \left\lfloor \frac{N}{s} \right\rfloor \left(\frac{s}{N} \right)^d + \left(\frac{N - s\lfloor N/s \rfloor}{N} \right)^d \approx \left(\frac{s}{N} \right)^{d-1} + \left(\frac{N - s\lfloor N/s \rfloor}{N} \right)^d .$$

Setting $d = 1$ yields $m(\mathbb{G}, s, 1) \approx 1$. Now let $\mathcal{S}$ be a (finite) set of box sizes, such that $s \ll N$ for $s \in \mathcal{S}$. By definitions (18.8) and (18.9), setting $d = 1$ yields $\sigma(\mathbb{G}, \mathcal{S}, 1) \approx 0$, so $d_H = 1$ for a long chain of N nodes.   □

Example 18.5 Consider first an infinite two-dimensional rectilinear lattice such that each node has integer coordinates, and where node (i, j) is adjacent to $(i - 1, j)$, $(i + 1, j)$, $(i, j - 1)$, and $(i, j + 1)$. Let the distance between nodes (x, y) and (u, v) be $|x - u| + |y - v|$ (i.e., the "Manhattan" distance). For $r > 0$, the number of nodes at a distance r from a given node z is $4r$, so there are $n(r) \equiv 1 + 4\sum_{i=1}^{r} = 2r^2 + 2r + 1$ nodes within distance r of node z; the box containing these $n(r)$ nodes has diameter $2r$, and the box size s for this box is given by $s = 2r+1$. Now consider a finite version of the infinite two-dimensional rectilinear lattice, defined as follows. For a positive integer $K \gg 1$, let $\mathbb{G}$ be a $K \times K$ two-dimensional rectilinear lattice such that each node has integer coordinates, where node (i, j) is adjacent to $(i - 1, j)$, $(i + 1, j)$, $(i, j - 1)$, and $(i, j + 1)$, and where each edge of the lattice contains K nodes. The diameter Δ of $\mathbb{G}$ is $2(K - 1)$. Figure 18.5 illustrates a box of radius $r = 1$ in the "interior" of a 6×6 lattice.

For $1 \ll r \ll K$, a $(2r + 1)$-covering of $\mathbb{G}$ using boxes of radius r consists of approximately $K^2/n(r)$ "totally full" boxes with $n(r)$ nodes and diameter $2r$, and $\mathcal{O}(K)$ "partially full" boxes of diameter not exceeding $2r$ and containing fewer than $n(r)$ nodes; these $\mathcal{O}(K)$ partially full boxes are needed around the 4 edges of $\mathbb{G}$. Since $K \gg 1$, we ignore these $\mathcal{O}(K)$ partially full boxes, and

$$m(\mathbb{G}, s, d) \approx \left(\frac{K^2}{n(r)}\right)\left(\frac{2r + 1}{2(K - 1) + 1}\right)^d \approx \left(\frac{K^2}{2r^2}\right)\left(\frac{2r}{2K}\right)^d = \left(\frac{1}{2}\right)\left(\frac{r}{K}\right)^{d-2}.$$

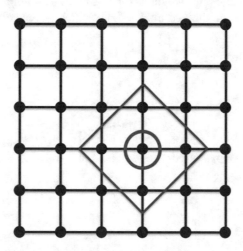

Figure 18.5: Box of radius $r = 1$ in a 6×6 rectilinear grid

Setting $d = 2$ yields $\sigma(\mathbb{G}, \mathcal{S}, 2) \approx 0$, so $d_H = 2$ for a large $K \times K$ rectilinear lattice. $\square$

Example 18.6 Figure 18.6 plots box counting results for the *dolphins* network (introduced in Example 7.2), with 62 nodes, 159 arcs, and $\Delta = 8$. For this network, and for all other networks described in this chapter, the minimal covering for each s was computed using the graph coloring heuristic described in [Rosenberg 17b]. Setting $\mathcal{S} = \{2, 3, 4, 5, 6\}$, Fig. 18.6 shows that the $\log B(s)$ versus $\log s$ curve is approximately linear for $s \in \mathcal{S}$. (In this example, as well as in subsequent examples, the determination of the set $\mathcal{S}$ was done by visual inspection.) Linear regression over this range yields $d_B \approx 2.38$. As discussed in Sect. 18.1, d_B can also be viewed as the value of d minimizing, over $\{d \,|\, d > 0\}$,

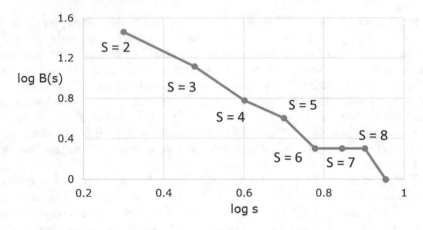

Figure 18.6: $\log B(s)$ versus $\log s$ for the *dolphins* network

the standard deviation of the terms $B(s)[s/(\Delta + 1)]^d$ for $s \in \mathcal{S}$. Figure 18.7 (left) plots $B(s)[s/(\Delta + 1)]^d$ versus d for $s \in \mathcal{S}$ for the *dolphins* network. The standard deviation is minimized at $d \approx 2.41$, which is very close to $d_B \approx 2.38$.

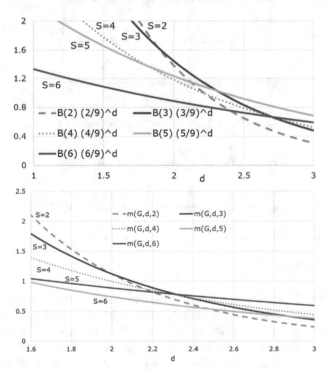

Figure 18.7: $B(s)[s/(\Delta + 1)]^d$ versus d for $s \in \mathcal{S}$ (left) and $m(\mathbb{G}, s, d)$ versus d (right) for *dolphins*

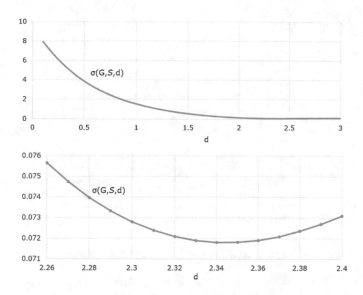

Figure 18.8: $\sigma(\mathbb{G}, \mathcal{S}, d)$ versus d for *dolphins* for $d \in [0,3]$ (left) and $d \in [2.26, 2.40]$ (right)

The same $\mathcal{S}$ was used to calculate d_H of the *dolphins* network. Figure 18.7 (right) plots $m(\mathbb{G}, s, d)$ for $s \in \mathcal{S}$ over the range $d \in [1.6, 3]$. Figure 18.8 (left), which plots $\sigma(\mathbb{G}, \mathcal{S}, d)$ versus d over the range $d \in [0, 3]$, shows that the location of d_H is not immediately evident. Numerical calculation revealed that $\sigma(\mathbb{G}, \mathcal{S}, d)$ has a unique local minimum at $d_H \approx 2.34$. Figure 18.8 (right) shows that $\sigma(\mathbb{G}, \mathcal{S}, d)$ is quite "flat"[1] in the vicinity of d_H, since for $d \in [2.26, 2.40]$ we have $0.071 < \sigma(\mathbb{G}, \mathcal{S}, d) < 0.076$. $\quad\square$

Example 18.7 Figure 18.9 plots box counting results for the *USAir97* network [Batagelj 06], which has 332 nodes, 2126 arcs, and $\Delta = 6$. Setting $\mathcal{S} = \{2, 3, 4, 5, 6\}$, linear regression over this range yields $d_B \approx 4.04$. The shape of the $\sigma(\mathbb{G}, \mathcal{S}, d)$ versus d plot for this network closely resembles the shape in Fig. 18.8 (left). There is a unique local minimum of $\sigma(\mathbb{G}, \mathcal{S}, d)$ at $d_H \approx 4.02$. The standard deviation $\sigma(\mathbb{G}, \mathcal{S}, d)$ is also quite flat in the vicinity of d_H. $\quad\square$

Example 18.8 Figure 18.10 (left) provides box counting results for the *jazz* network (see Example 15.5), with 198 nodes, 2742 arcs, and $\Delta = 6$. Setting $\mathcal{S} = \{2, 3, 4, 5, 6\}$, linear regression over this range yields $d_B \approx 2.66$. Figure 18.10 (right) plots $m(\mathbb{G}, s, d)$ for $s \in \mathcal{S}$ over the range $d \in [1.6, 4]$. There is a unique local minimum of $\sigma(\mathbb{G}, \mathcal{S}, d)$ at $d_H \approx 3.26$. $\quad\square$

Example 18.9 Figure 18.11 (left) plots box counting results for $s \in [2, 7]$ for the *C. elegans* network (see Example 17.4), with 453 nodes, 2040 arcs, and $\Delta = 7$. Excluding the point $\big(\log 2, \log B(2)\big)$, a linear fit is reasonable for the other five points, so we set $\mathcal{S} = \{3, 4, 5, 6, 7\}$. Linear regression over this range yields $d_B \approx 2.68$. Figure 18.11 (right) plots $m(\mathbb{G}, s, d)$ for $s \in \mathcal{S}$ over the range

---

[1] The function f is said to be *flat* in the region X if $|f(x) - f(y)|$ is "small" for $x, y \in X$, e.g., if f is twice differentiable and at a local minimum $x^\star$ of f the second derivative $f''(x^\star)$ is a small positive number, then f is flat in the neighborhood of $x^\star$.

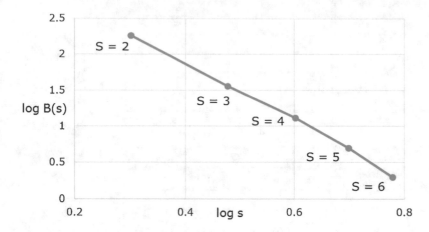

Figure 18.9: $\log B(s)$ versus $\log s$ for the *USAir97* network

$d \in [1.5, 5]$. Figure 18.12 (left) plots $\sigma(\mathbb{G}, \mathcal{S}, d)$ versus d over the range $d \in [0, 5]$. There is a unique local minimum of $\sigma(\mathbb{G}, \mathcal{S}, d)$ at $d_H \approx 3.51$.

Again, the $\sigma(\mathbb{G}, \mathcal{S}, d)$ versus d plot is very flat in the vicinity of d_H, as shown by Fig. 18.12 (right), since for $d \in [3.45, 3.55]$ we have $0.0560 < \sigma(\mathbb{G}, \mathcal{S}, d) < 0.0576$. □

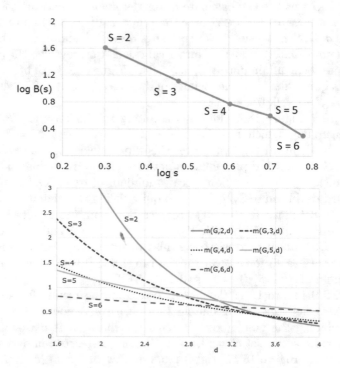

Figure 18.10: $\log B(s)$ versus $\log s$ (left) and $m(\mathbb{G}, s, d)$ versus d (right) for the *jazz* network

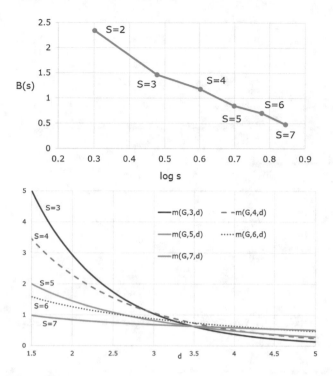

Figure 18.11: $\log B(s)$ versus $\log s$ (left) and $m(\mathbb{G}, s, d)$ versus d (right) for the *C. elegans* network

Example 18.10 Figure 18.13 plots box counting results for the connected giant component of a *protein* network.[2] This component has 1458 nodes, 1948 arcs, and $\Delta = 19$. Choosing the range $\mathcal{S} = \{4, 5, \dots, 14\}$, Fig. 18.13 shows that $\log B(s)$ is close to linear in $\log s$ for $s \in \mathcal{S}$ (the dotted line is the linear least squares fit), and $d_B = 2.44$. The shape of the $\sigma(\mathbb{G}, \mathcal{S}, d)$ versus d plot for *protein* has a shape similar to the plot for *C. elegans*, and for the *protein* network there is a unique local minimum of $\sigma(\mathbb{G}, \mathcal{S}, d)$ at $d_H = 3.08$. □

Figure 18.14, from [Rosenberg 18b], summarizes the above computational results. The proof that for geometric objects we have $d_H \le d_B$ (Theorem 5.1) does not apply to networks with integer box sizes, and indeed Fig. 18.14 shows that for several networks d_H exceeds d_B, sometimes significantly so.

---

[2]the "Pro-Pro" network data set provided in http://moreno.ss.uci.edu/data.html# pro-pro.

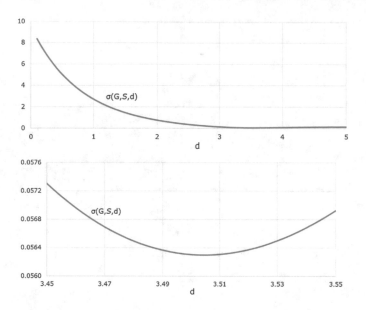

Figure 18.12: $\sigma(\mathbb{G}, \mathcal{S}, d)$ versus d for *C. elegans* for $d \in [0, 5]$ (left) and $d \in [2.26, 2.40]$ (right)

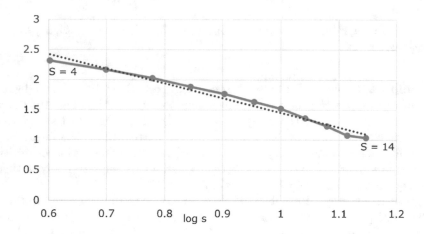

Figure 18.13: $\log B(s)$ versus $\log s$ for $s \in [4, 14]$ for the *protein* network

The above examples show that it is computationally feasible to compute d_H of a network. The numerical value of d_H depends not only on the minimal number $B(s)$ of boxes for each box size s, but also on the diameters of the boxes in $\mathcal{B}(s)$. Any method for computing the box counting dimension d_B of $\mathbb{G}$ suffers from the fact that in general we cannot verify that the computed $B(s)$ is in fact minimal (although a lower bound on $B(s)$ can be computed [Rosenberg 13]). Computing the Hausdorff dimension adds the further complication that there is

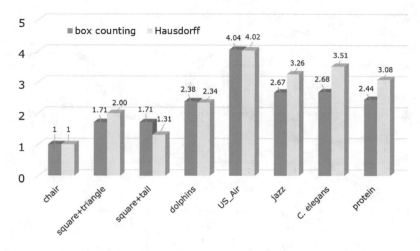

Figure 18.14: Summary of d_B and d_H computations

currently no way to verify that a lexico minimal covering (Sect. 17.1) has been found. Therefore, since for each box size s a randomized box-counting method is typically employed (e.g., the above results used the method described in [Rosenberg 17b]), it is necessary to run sufficiently many randomized iterations until we are reasonably confident that not only has a minimal covering of $\mathbb{G}$ been obtained, but also that a lexicographically minimal covering has been obtained. This extra computational burden is not unique to computing d_H, since computing any fractal dimension of $\mathbb{G}$ other than the box counting dimension requires an unambiguous selection of a minimal covering, e.g., computing the information dimension d_I of $\mathbb{G}$ requires finding a maximal entropy minimal covering (Sect. 15.4), and computing the generalized dimensions D_q of $\mathbb{G}$ requires finding a lexico minimal summary vector.

Example 18.11 [Rosenberg 18b] This example considers the impact on d_B and d_H of perturbing a network $\mathbb{G}$. Let the arc set $\mathbb{A}$ of $\mathbb{G}$ be specified by a list $\mathcal{L}$, so that $\mathbb{G}$ is built, arc by arc, by processing the list $\mathcal{L}$. After each arc is processed, the node degree δ_n of each node n in $\mathbb{G}$ is updated. To perturb $\mathbb{G}$, choose $P \in (0,1)$. After reading arc a from $\mathcal{L}$, pick two random probabilities $p_1 \in (0,1)$ and $p_2 \in (0,1)$. If $p_1 < P$, modify arc a as follows. Let i and j be the endpoints of arc a. If $\delta_i > 1$ and $p_2 < \frac{1}{2}$, replace the endpoint i of arc a with a randomly chosen node distinct from i; otherwise, if $\delta_j > 1$, replace the endpoint j of arc a with a randomly chosen node distinct from j. (Requiring $\delta_i > 1$ ensures i will not become an isolated node.) This perturbation method was applied to the *jazz* network of Example 18.8 for P ranging from 0.01 to 0.08. The results, in Fig. 18.15, show that the d_H versus P plot has roughly the same shape as the d_B versus P plot. Although the range of d_H over the interval $0 \leq P \leq 0.08$ is smaller than the range of d_B, sometimes d_H magnifies the effect of a perturbation, e.g., d_H changed much more than d_B when P increased from 0.02 to 0.03. $\square$

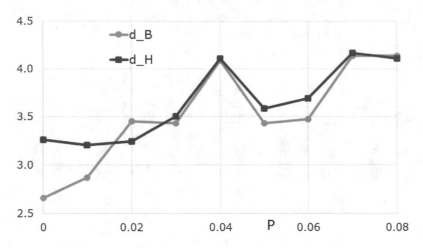

Figure 18.15: Perturbations of the *jazz* network

Example 18.12 [Rosenberg 18b] This example considers a Barábasi–Albert [Albert 02] preferential attachment model of a growing network. Assume that each new node n connects to a single existing node x, and the probability that x is chosen is proportional to the node degree of x. We start with an initial network of 3 nodes connected by 3 arcs, grow the network to 150 nodes, and compute d_B and d_H as the network grows. The diameter of the 150 node network is 11. Figure 18.16 plots the results. The 40 node network is an outlier due to

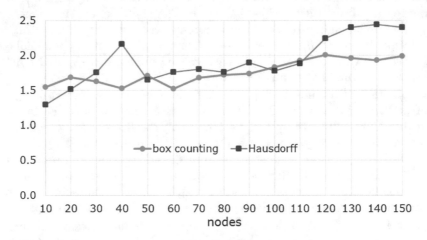

Figure 18.16: d_B and d_H of a network growing with preferential attachment

its high d_H. The range of d_B (as the network grows from 10 to 150 nodes) is 0.48 and the range of d_H is 1.15, so using the Hausdorff dimension, rather than the box counting dimension, magnifies, by a factor of 2.4, differences in the fractal dimensions of the growing network. In particular, the sharp increase in d_H when the network size exceeds 110 nodes indicates that d_H is sensitive to changes in the network topology that are not captured by the simpler measure d_B. $\square$

18.3 Generalized Hausdorff Dimensions of a Network

In Sect. 16.9 we presented the definition in [Grassberger 85] of the generalized Hausdorff dimensions $\{D_q^H\}_{q \in \mathbb{R}}$ of a geometric object. Continuing our approach in this book of defining a fractal dimension for a geometric object and then extending it to a network, and having defined the Hausdorff dimension d_H of a network $\mathbb{G}$ in Sect. 18.1 of this chapter, in this section we study the generalized Hausdorff dimensions $\{D_q^H\}_{q \in \mathbb{R}}$ of $\mathbb{G}$, defined in [Rosenberg 18b]. These dimensions utilize both the diameter and mass of each box in the minimal covering $\mathcal{B}(s)$ of $\mathbb{G}$. Just as the computation of d_H for $\mathbb{G}$ requires a line search, for each q the computation of the generalized Hausdorff dimension D_q^H of $\mathbb{G}$ also requires a line search.

For a given box size s, let $\mathcal{B}(s)$ be a minimal s-covering of $\mathbb{G}$. The probability of box $B_j \in \mathcal{B}(s)$ is, as usual, $p_j(s) \equiv N_j(s)/N$, where $N_j(s)$ is the number of nodes contained in box B_j; for simplicity we write p_j instead of $p_j(s)$. Recalling that δ_j is defined by (18.5), we borrow from [Grassberger 85] and define

$$m(\mathbb{G}, s, q, d) \equiv \begin{cases} \left(\sum_{B_j \in \mathcal{B}(s)} p_j (\delta_j^d / p_j)^{1-q} \right)^{1/(1-q)} & \text{if } q \neq 1 \\ \exp \left(\sum_{B_j \in \mathcal{B}(s)} p_j \log \left(\delta_j^d / p_j \right) \right) & \text{if } q = 1. \end{cases} \quad (18.11)$$

We can regard $m(\mathbb{G}, s, q, d)$ as the "d-dimensional generalized Hausdorff measure of order q of $\mathbb{G}$ for box size s".

Exercise 18.2 Prove that for $q \neq 1$ the derivative with respect to d of $m(\mathbb{G}, s, q, d)$ is negative. $\square$

Extending the approach of Sect. 18.1, we define D_q^H of $\mathbb{G}$ as that value of d for which $m(\mathbb{G}, s, q, d)$ is roughly constant over a given range $\mathcal{S}$ of values of s. For $d > 0$, let $\sigma(\mathbb{G}, \mathcal{S}, q, d)$ be the standard deviation of the values $m(\mathbb{G}, s, q, d)$ for $s \in \mathcal{S}$:

$$m_{avg}(\mathbb{G}, \mathcal{S}, q, d) \equiv \frac{1}{|\mathcal{S}|} \sum_{s \in \mathcal{S}} m(\mathbb{G}, s, q, d)$$

$$\sigma(\mathbb{G}, \mathcal{S}, q, d) \equiv \sqrt{\frac{1}{|\mathcal{S}|} \sum_{s \in \mathcal{S}} \left[m(\mathbb{G}, s, q, d) - m_{avg}(\mathbb{G}, \mathcal{S}, q, d) \right]^2}. \quad (18.12)$$

We associate with $\mathbb{G}$ the generalized Hausdorff dimension of order q, denoted by D_q^H, defined as follows.

Definition 18.4 [Rosenberg 18b] (*Part A.*) If $\sigma(\mathbb{G}, \mathcal{S}, q, d)$ has at least one local minimum over $\{d \mid d > 0\}$, then D_q^H is the value of d satisfying (*i*) $\sigma(\mathbb{G}, \mathcal{S}, q, d)$ has a local minimum at D_q^H, and (*ii*) if $\sigma(\mathbb{G}, \mathcal{S}, q, d)$ has another local minimum at $\widetilde{d}$, then $\sigma(\mathbb{G}, \mathcal{S}, q, D_q^H) < \sigma(\mathbb{G}, \mathcal{S}, q, \widetilde{d})$. (*Part B.*) If $\sigma(\mathbb{G}, \mathcal{S}, q, d)$ does not have a local minimum over $\{d \mid d > 0\}$, then $D_q^H \equiv \infty$. $\square$

Definition 18.5 If $D_q^H < \infty$, then the generalized Hausdorff measure of order q of $\mathbb{G}$ is $m_{avg}(\mathbb{G}, \mathcal{S}, q, D_q^H)$. If $D_q^H = \infty$ then the generalized Hausdorff measure of order q of $\mathbb{G}$ is 0. $\square$

Setting $q = 0$, by (18.6) and (18.11) we have $m(\mathbb{G}, s, d) = m(\mathbb{G}, s, 0, d)$. By (18.9) and (18.12) we have $\sigma(\mathbb{G}, \mathcal{S}, d) = \sigma(\mathbb{G}, \mathcal{S}, 0, d)$. Hence, by Definitions 18.2 and 18.4 we have $D_0^H = d_H$. In words, the generalized Hausdorff dimension of order 0 of $\mathbb{G}$ is equal to the Hausdorff dimension of $\mathbb{G}$.

For a geometric object Ω, it was observed in [Grassberger 85] that if μ is the Hausdorff measure, then $D_q^H = d_H$ for all q, at least for strictly self-similar fractals. The following theorem, the analogue of Grassberger's observation for a network $\mathbb{G}$, says that if the mass of each box is proportional to δ_j, then for $q \neq 1$ we have $D_q^H = 1$.

Theorem 18.1 [Rosenberg 18b] Suppose that for some constant $\alpha > 0$ we have $\delta_j = \alpha p_j$ for $s \in \mathcal{S}$ and for $B_j \in \mathcal{B}(s)$. Then for $q \neq 1$ we have $D_q^H = 1$ and the generalized Hausdorff measure of order q of $\mathbb{G}$ is α.

Proof Suppose $\delta_j = \alpha p_j$ for $s \in \mathcal{S}$ and for $B_j \in \mathcal{B}(s)$. Set $d = 1$ and pick $q \neq 1$. From (18.11), for $s \in \mathcal{S}$ we have

$$m(\mathbb{G}, s, q, 1) = \left[\sum_{B_j \in \mathcal{B}(s)} p_j \left(\frac{\delta_j}{p_j} \right)^{1-q} \right]^{1/(1-q)} = \left[\sum_{B_j \in \mathcal{B}(s)} p_j \left(\frac{\alpha p_j}{p_j} \right)^{1-q} \right]^{1/(1-q)}$$

$$= \left[\alpha^{1-q} \sum_{B_j \in \mathcal{B}(s)} p_j \right]^{1/(1-q)} = \alpha,$$

where the final equality follows as $\sum_{B_j \in \mathcal{B}(s)} p_j = 1$. Since $m(\mathbb{G}, s, q, 1) = \alpha$, by (18.12) we have $\sigma(\mathbb{G}, \mathcal{S}, q, 1) = 0$. By Definition 18.4, we have $D_q^H = 1$. By Definition 18.5, the generalized Hausdorff measure of order q of $\mathbb{G}$ is α. $\square$

Exercise 18.3 Suppose that for some constant $\alpha > 0$ we have $\delta_j = \alpha p_j$ for $s \in \mathcal{S}$ and for $B_j \in \mathcal{B}(s)$. What are the values of d_I and D_1^H? $\square$

Example 18.13 Consider again the *chair* network of Example 18.2. As shown in Fig. 18.2, the box with 1 node has $p_j = 1/5$ and $\delta_j = 1/4$, each box with 2 nodes has $p_j = 2/5$ and $\delta_j = 2/4$, and the box with 3 nodes has $p_j = 3/5$ and $\delta_j = 3/4$. Since the conditions of Theorem 18.1 are satisfied with $\alpha = 5/4$, for $q \neq 1$ we have $D_q^H = 1$, and the generalized Hausdorff measure of order q of the *chair* network is $5/4$. $\square$

Example 18.14 Consider again the network $\mathbb{G}_1$ of Fig. 18.3. Figure 18.17 plots D_q^H versus q for integer $q \in [-10, 10]$. Clearly D_q^H is not monotonic decreasing in q, even over the range $q \geq 0$, since $D_0^H < D_2^H$. Since for a geometric fractal D_q^H is a non-increasing function of q [Grassberger 85], the non-monotonic behavior in Fig. 18.17 is perhaps surprising. However, the generalized dimensions D_q of a network, computed using Definition 17.3, also can fail to be monotone decreasing in q (Sect. 17.2). Thus it is not unexpected that the generalized

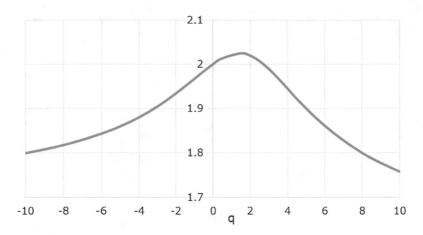

Figure 18.17: D_q^H versus q for the network $\mathbb{G}_1$ of Fig. 18.3

Hausdorff dimensions of a network, computed using Definition 18.4, can also exhibit non-monotonic behavior. □

Example 18.15 Consider again the *dolphins* network of Example 18.6, again with $\mathcal{S} = \{2, 3, 4, 5, 6\}$. Figure 18.18 plots D_q^H versus q. Remarkably, there is a "critical order" $q_c \approx -1.54$ such that D_q^H is finite for $q > q_c$ and D_q^H is infinite for $q < q_c$. □

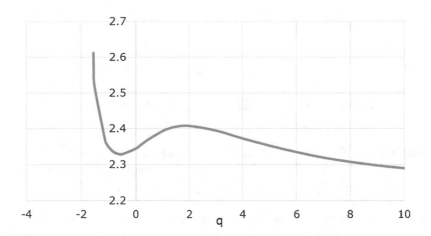

Figure 18.18: D_q^H versus q for the *dolphins* network

Example 18.16 Consider again the *jazz* network of Example 18.8, again with $\mathcal{S} = \{2, 3, 4, 5, 6\}$. Figure 18.19 plots D_q^H versus q. There is a critical interval $Q_c \approx [-3.9, -0.3]$ such that D_q^H is infinite for $q \in Q_c$ and finite otherwise. □

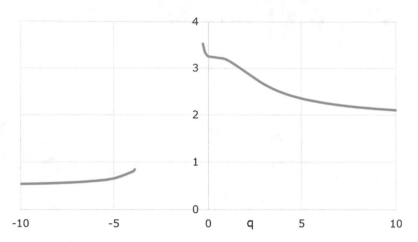

Figure 18.19: D_q^H versus q for the *jazz* network

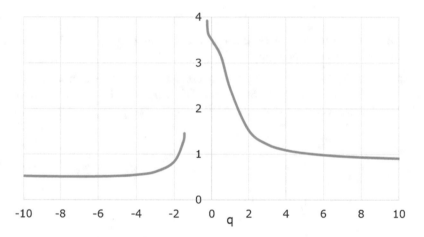

Figure 18.20: D_q^H versus q for the *C. elegans* network

Example 18.17 Consider again the *C. elegans* network of Example 18.9, again with $\mathcal{S} = \{3, 4, 5, 6, 7\}$. Figure 18.20 plots D_q^H versus q. There is a critical interval $Q_c \approx [-1.45, -0.25]$ such that D_q^H is infinite for $q \in Q_c$ and finite otherwise.   □

The above examples show that D_q^H can exhibit complicated behavior for $q < 0$. These findings align with several studies (e.g., [Block 91]) pointing to difficulties in computing and interpreting D_q for a geometric object and the difficulties "imperceptibly hidden" in the negative q domain [Riedi 95]; see Sect. 16.4. Thus, for a network, D_q^H should probably be computed only for $q \geq 0$. The existence of a critical interval in Examples 18.15–18.17 raises many questions, e.g., to characterize the class of networks possessing a critical interval.

To summarize the computational results presented in this chapter, d_H can sometimes be more useful than d_B in quantifying changes in the topology of a network. However, d_H is harder to compute than d_B, and D_q^H is less well behaved than D_q. We conclude that d_H and D_q^H should be added to the set of useful metrics for characterizing a network, but they cannot be expected to replace d_B and D_q.

Chapter 19

Lacunarity

I search for the realness, the real feeling of a subject, all the tex-
ture around it... I always want to see the third dimension of some-
thing... I want to come alive with the object.
Andrew Wyeth (1917–2009), American painter, known for "Christina's
World" (1948)

The *lacunarity* of a pattern is a measure of how uniform it is. The term
was introduced in [Mandelbrot 83b], where it was observed that "texture is an
elusive notion which mathematicians and scientists tend to avoid because they
cannot grasp it". Lacunarity is derived from the Latin *lacuna*, which means lack,
gap, or hole. Patterns with low lacunarity appear relatively homogeneous, while
patterns with high lacunarity appear clumpy, with large gaps. As described in
[Katsaloulis 09]:

> Although two objects might have the same fractal dimension,
> their structure might be statistically very different. ... Instead of
> only defining the percentage of space not occupied by the structure,
> lacunarity is more interested in the holes and their distribution on
> the object under observation. In the case of fractals, even if two
> fractal objects have the d_B, they could have different lacunarity if
> the fractal "pattern" is different.

Lacunarity is illustrated in the following two examples of *Cantor dust*. In
Fig. 19.1, the rectangle (*i*) is divided into four equal size pieces, and the second
and fourth pieces are removed, yielding (*ii*). Each piece in (*ii*) is then divided
into four equal size pieces, and the second and fourth pieces are removed, yielding (*iii*). Continuing this process indefinitely, from (5.8) the resulting fractal
has similarity dimension $d_S = \log(2)/\log\big(1/(1/4)\big) = \log 2/(2\log 2) = 1/2$.

© Springer Nature Switzerland AG 2020
E. Rosenberg, *Fractal Dimensions of Networks*,
https://doi.org/10.1007/978-3-030-43169-3_19

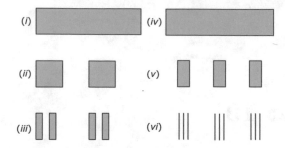

Figure 19.1: Cantor dust lacunarity

For the second example, we start with the rectangle (iv) which is identical to rectangle (i). We divide rectangle (iv) into nine equal size pieces, and keep only the second, fifth, and eight pieces, discarding the other six pieces, yielding (v). Each piece in (v) is then divided into nine equal size pieces, and we keep only the second, fifth, and eight pieces, discarding the other six pieces, yielding (vi). If we continue this process indefinitely, the resulting fractal has similarity dimension $d_S = \log(3)/\log(1/9) = \log 3/(2\log 3) = 1/2$. So both Cantor dust fractals have $d_S = 1/2$ yet they have different textures: the first fractal has a more lumpy texture than the second. The difference in textures is captured using the lacunarity measure. High lacunarity means the object is "heterogeneous", with a wide range of "gap" or "hole" sizes. Low lacunarity means the object is "homogeneous", with a narrow range of gap sizes [Plotnick 93]. The value of lacunarity is that the fractal dimension, by itself, has been deemed insufficient to describe some geometric patterns.

Lacunarity in a Cantor set construction was illustrated in [Pilipski 02]. In generation t, we scale each interval by a factor of 4^{-k} and we take 2^k copies of the scaled interval. For each integer k, the resulting fractal has $d_S = \log(2^k)/\log(4^k) = 1/2$, independent of k. Yet the distribution of "white space" within these figures varies widely. This is illustrated in Fig. 19.2 where the unit

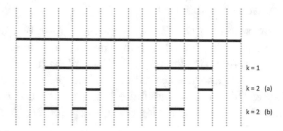

Figure 19.2: Lacunarity for a Cantor set

interval is divided into Sixteen pieces, and three constructions are illustrated, one for $k = 1$ and two for $k = 2$. The two $k = 2$ constructions demonstrate that the construction need not be symmetric.

The fact that both a solid object in $\mathbb{R}^2$ and a Poisson point pattern in $\mathbb{R}^2$ have the same d_B illustrates the need to provide a measure of clumpiness [Halley 04]. Methods for calculating lacunarity were proposed in [Mandelbrot 83b], and subsequently other methods were proposed by other researchers. It was observed in [Theiler 88] that "there is no generally accepted definition of lacunarity, and indeed there has been some controversy over how it should be measured". Sixteen years later, [Halley 04] wrote that "a more precise definition of lacunarity has been problematic", and observed that at least six different algorithms have been proposed for measuring the lacunarity of a geometric object, and a comparison of four of the six methods showed little agreement among them.

In [Theiler 88], lacunarity was defined in relation to the correlation dimension d_C. Assume d_C exists, so by (9.17) we have $d_C = \lim_{r \to 0} \log C(r)/\log r$. For $r > 0$, define $\Phi(r)$ by

$$C(r) = \Phi(r)r^{d_C}$$

(the function $\Phi(r)$ was introduced in Sect. 9.4). For d_C to exist, it is not necessary for $\Phi(r)$ to approach a constant as $r \to 0$; rather, it is only required that $\lim_{r \to 0} \log \Phi(r)/\log r = 0$. The term "lacunarity" was used in [Theiler 88] to refer to the variability in $\Phi(r)$, particularly where $\lim_{r \to 0} \Phi(r)$ is not a positive constant.

19.1 Gliding Box Method

The most widely accepted method for calculating the lacunarity of a geometric object is the *gliding box* method of [Allain 91]. In this method, a box of a given size is moved across the pattern, and the number of points in the box is counted as it moves. Our presentation of this method is based on [Plotnick 93]. Consider the array of 49 numbers (0 or 1) in 7 rows and columns, in Fig. 19.3(i). (The method applies equally well to a rectangular array, but is most easily presented using a square array.) A square box of side length s is superimposed on the matrix, and this box is glided along the top row, and then along each successive row. Since each matrix entry is 0 or 1, the sum of the matrix entries in each box of side length s box ranges from 0 to s^2. Figure 19.3(ii) shows the first box, *box 1* in the upper left corner. The second box is obtained by gliding (i.e., shifting) the first box one column to the right, and so on, until we get to the last box for that row; shown as *box 6* in the upper right corner of Fig. 19.3(ii). Now we glide the box down one row, so the next box is *box 7*, illustrated in Fig. 19.4(i). We continue gliding the box across this row, until we reach the end of the row, illustrated by *box 12* in Fig. 19.4(ii). We glide down to the next row, and continue in this manner, until we reach the beginning of the last row, illustrated by *box 31* in Fig. 19.4(ii). The final box, in the lower right-hand corner, is illustrated by *box 36* in Fig. 19.4(ii).

Suppose the matrix has M rows and M columns. An individual box corresponds to the triplet (i, j, s), where the upper left-hand corner of the box is element (i, j) of the matrix, and s is the box size. The range of s is 1 to M. The range of i and j is 1 to $M - s + 1$, so the total number of boxes of size s is

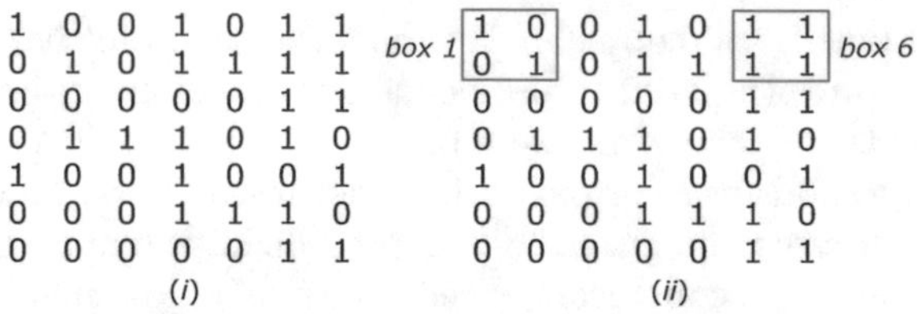

```
1  0  0  1  0  1  1           ┌─────┐                   ┌─────┐
0  1  0  1  1  1  1   box 1   │1  0 │ 0  1  0   box 6   │1  1 │ box 6
0  0  0  0  0  1  1           │0  1 │ 0  1  1           │1  1 │
0  1  1  1  0  1  0           └─────┘ 0  0  0  0  0  1  1
1  0  0  1  0  0  1                   0  1  1  1  0  1  0
0  0  0  1  1  1  0                   1  0  0  1  0  0  1
0  0  0  0  0  1  1                   0  0  0  1  1  1  0
                                      0  0  0  0  0  1  1
         (i)                                   (ii)
```

Figure 19.3: Gliding box method for the top row

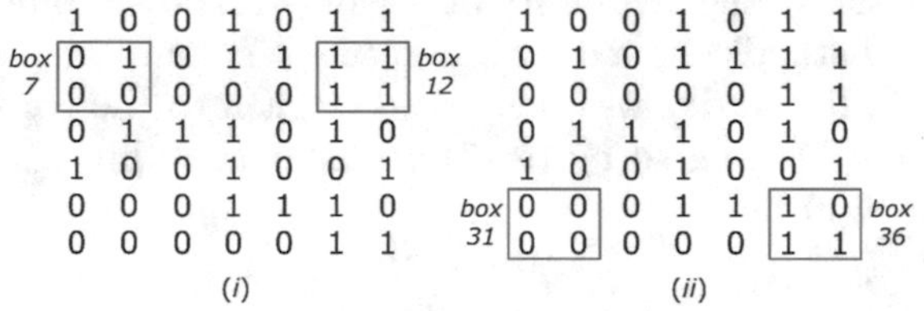

```
        1  0  0  1  0  1  1            1  0  0  1  0  1  1
   box ┌─────┐       ┌─────┐ box       1  0  0  1  0  1  1
    7  │0  1 │0  1  1 │1  1 │  12       0  1  0  1  1  1  1
       │0  0 │0  0  0 │1  1 │           0  0  0  0  0  1  1
       └─────┘        └─────┘          0  1  1  1  0  1  0
       0  1  1  1  0  1  0             1  0  0  1  0  0  1
       1  0  0  1  0  0  1    box ┌─────┐      ┌─────┐ box
       0  0  0  1  1  1  0    31  │0  0 │0  1  1│1  0 │  36
       0  0  0  0  0  1  1        │0  0 │0  0  0│1  1 │
                                 └─────┘       └─────┘
              (i)                           (ii)
```

Figure 19.4: Gliding box example continued

$(M-s+1)^2$. Let $x(i,j,s)$ be the sum of the matrix entries in the box (i,j,s). In our example we have $x(1,1,2) = 2$, $x(1,6,2) = 4$, $x(2,1,2) = 1$, $x(2,6,2) = 4$, $x(6,1,2) = 0$, and $x(6,6,2) = 3$. For a given s and for $k = 0,1,2,\ldots,s^2$, let $N_s(k)$ be the number of boxes of size s for which $x(i,j,s) = k$. Since the total number of boxes of size s is $(M-s+1)^2$ then $\sum_{k=0}^{s^2} N_s(k) = (M-s+1)^2$.

For a given s, we normalize the $N_s(k)$ values to obtain a probability distribution, by defining, for $k = 0,1,\ldots,s^2$,

$$p_s(k) = \frac{N_s(k)}{(M-s+1)^2}. \tag{19.1}$$

The first and second moments of this distribution are

$$\mu_1(s) = \sum_{k=0}^{s^2} k p_s(k) \qquad \mu_2(s) = \sum_{k=0}^{s^2} k^2 p_s(k) \tag{19.2}$$

and the lacunarity $\Lambda(s)$ for box size s was defined in [Plotnick 93] as

$$\Lambda(s) \equiv \frac{\mu_2(s)}{[\mu_1(s)]^2}. \tag{19.3}$$

Let $\sigma^2(s)$ be the variance of the distribution $\{p_s(k)\}_{k=0}^{s^2}$. Using the well-known relationship $\mu_2(s) = [\mu_1(s)]^2 + \sigma^2(s)$ for a probability distribution, we obtain

$$\Lambda(s) = \frac{[\mu_1(s)]^2 + \sigma^2(s)}{[\mu_1(s)]^2} = \frac{\sigma^2(s)}{[\mu_1(s)]^2} + 1. \tag{19.4}$$

For example, the lacunarity of a Poisson process is 2, since the second moment is equal to the square of the mean, i.e., $\langle m^2 \rangle = \langle m \rangle^2$ [Katsaloulis 09]. Figure 19.5, from [Rasband], illustrates three degrees of lacunarity. Some properties of the lacunarity $\Lambda(s)$ are:

low   medium   high

Figure 19.5: Three degrees of lacunarity

1. From (19.4) we see that $\Lambda(s) \geq 1$ for all s. Moreover, for any totally regular array (e.g., an alternating pattern of 0's and 1's), if s is at least as big as the unit size of the repeating pattern, then the $x(i, j, s)$ values are independent of the position (i, j) of the box, which implies $\mu_2(s) = 0$ and hence $\Lambda(s) = 1$ [Plotnick 93].

2. Since there is only one box of size M then there is at most one k value for which $p_M(k)$ is nonzero, namely, $k = x(1, 1, M)$. Thus the variance of the distribution $\{p_M(k)\}_{k=0}^{s^2}$ is 0 and from (19.4) we have $\Lambda(M) = 1$.

3. Suppose each of the M^2 elements in the array is 1 with probability $p = 1/M^2$. For a box size of 1, the expected number of boxes containing a "1" is pM^2, so from (19.1) we have $p_1(1) = pM^2/M^2 = p$. From (19.2) we have $\mu_1(1) = \mu_2(1) = p$. Hence $\Lambda(1) = \mu_2(1)/[\mu_1(1)]^2 = 1/p$ [Plotnick 93].

4. Since $x(i, j, s)$ is the sum of the matrix entries in the box (i, j, s), then $x(i, j, s)$ is a non-decreasing function of s, and hence $\mu_1(s)$ is also non-decreasing in s. Since the relative variance (i.e., the ratio $\sigma^2(s)/\mu_1(s)$) of the $x(i, j, s)$ values is a non-decreasing function of s, then $\Lambda(s)$ is a non-increasing function of s. Or, as expressed in [Krasowska 04], as s increases, the average mass of the box increases, which increases the denominator $\mu_1(s)$ in (19.3); also as s increases the variance $\mu_2(s)$ in (19.3) decreases since all features become averaged into "a kind of noise". Thus $\Lambda(s)$ decreases as s increases.

5. For a given s, as the mean number of "occupied sites" (i.e., entries in the matrix with the value 1) decreases to zero, the lacunarity increases to ∞ [Plotnick 93]. Thus, for a given s, sparse objects have higher lacunarity than dense objects.

6. The numerical value of $\Lambda(s)$ for a single value of s is "of limited value at best, and is probably meaningless as a basis of comparison of different maps". Rather, more useful information can be gleaned by calculating $\Lambda(s)$ over a range of s [Plotnick 93].

7. For a monofractal in $\mathbb{R}^E$, a plot of $\log\left(\Lambda(s)\right)$ versus $\log s$ is a straight line with slope $d_B - E$ [Halley 04].

In [Plotnick 93], lacunarity was studied for a two-dimensional random map created using three levels of hierarchy. The first hierarchy level was created by partitioning a square into a 6×6 array, as illustrated in Fig. 19.6. The second

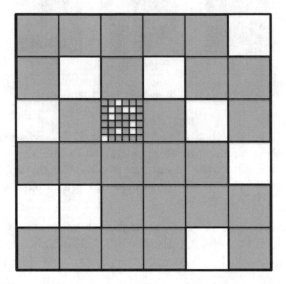

Figure 19.6: Lacunarity with three hierarchy levels

level of hierarchy was created by subdividing each of the 36 first-level regions into a 6×6 array; this is illustrated in Fig. 19.6 for only one of the first-level regions. Finally, a third level of hierarchy was created by subdividing each of the 36^2 second-level regions into a 6×6 array; this is not illustrated. The end result is square $M \times M$ array, where $M = 6^3 = 216$.

Three independently chosen probabilities determine whether a region is occupied. Each first-level region is occupied with probability p_1. If a first-level region R_1 is not occupied, then each of the 36 second-level regions within R_1 is declared "not occupied", and in turn each of the 36 third-level regions within each second-level region in R_1 is also declared "not occupied". On the other hand, if a first-level region R_1 is occupied, then each second-level region within R_1 is occupied with probability p_2. The same process is used to create the third-level regions. If a second-level region R_2 is not occupied, then each of the 36 third-level regions within R_2 is declared "not occupied". If a second-level region R_2 is occupied, then each third-level region within R_2 is occupied with probability p_3. Figure 19.6 shows 9 unoccupied first-level regions (shaded white) and 9 unoccupied second-level regions (shaded white) within one of the occupied first-level regions.

Several different strategies for choosing the probabilities were studied. For example, the "same" strategy sets $p_1 = p_2 = p_3$; this strategy generates statistically self-similar maps. The "bottom" strategy sets $p_1 = p_2 = 1$ and $p_3 = p$ (for some $p \in (0, 1)$) for each third-level region, so all first-level and second-level regions are occupied; with this strategy the contiguous occupied regions are small, as are the gaps between occupied regions. The "top" strategy sets $p_2 = p_3 = 1$, so that if a first-level region R_1 is occupied, all second-level and third-level regions within R_1 are occupied; with this strategy the occupied regions are large, as are the gaps between occupied regions. For each strategy, $\log \Lambda(s)$ versus $\log s$ was plotted (Figures 8, 9, and 10 in [Plotnick 93]) for $r = 2^i$ for i from 0 to 7. The plots for the different strategies form very different curves: e.g., for $p = 1/2$ the "top" curve appears concave, the "bottom" curve appears convex, and the "same" curve appears linear (in accordance with the theoretical result that the slope is $d_B - E$).

The relationship of lacunarity to the generalized dimension D_2 was explored in [Borys 09]. Suppose points are distributed in the unit square $[0,1] \times [0,1]$ and that, for a given box size $s \ll 1$, the gliding box method yields K boxes (i.e., there are K different positions of a box of size s). For $1 \leq k \leq K$, let $x_k(s)$ be the number of points in box k. From (19.4),

$$\Lambda(s) = \frac{\frac{1}{K} \sum_{k=1}^{K} x_k^2(s)}{\frac{1}{K^2} \left(\sum_{k=1}^{K} x_k(s) \right)^2}. \tag{19.5}$$

We approximate the ratio (19.5) using contiguous and non-overlapping boxes of size s, as in the box-counting method. Since the box size is s, there are $J = s^{-2}$ such boxes in the unit square. Let $N_j(s)$ be the number of points in box j. Define $N \equiv \sum_{j=1}^{J} N_j(s)$, so N is the total number of points. We have

$$\Lambda(s) \approx \frac{\frac{1}{J} \sum_{j=1}^{J} N_j^2(s)}{\frac{1}{J^2} \left(\sum_{j=1}^{J} N_j(s) \right)^2} = \frac{J \sum_{j=1}^{J} N_j^2(s)}{\left(\sum_{j=1}^{J} N_j(s) \right)^2}$$

$$= \frac{s^{-2} \sum_{j=1}^{J} N_j^2(s)}{N^2} = s^{-2} \sum_{j=1}^{J} \left(\frac{N_j(s)}{N} \right)^2. \tag{19.6}$$

Recall from (16.7) that the generalized dimension D_2 is given by

$$D_2 = \lim_{s \to 0} \frac{1}{\log s} \log \sum_{j=1}^{J} \left(\frac{N_j(s)}{N} \right)^2. \tag{19.7}$$

Hence for $s \ll 1$ we have $\sum_{j=1}^{J} \left(\frac{N_j(s)}{N} \right)^2 \sim s^{D_2}$. From (19.6) we obtain

$$\Lambda(s) \sim s^{-2} s^{D_2} = s^{D_2 - 2}. \tag{19.8}$$

Since $s \ll 1$ then the lacunarity $\Lambda(s)$ is *decreasing* function of D_2.[1]

The variation of lacunarity with fractal dimension was considered in [Krasowska 04], which mentioned a relation due to Kaye that the local density ρ of a lattice of tiles satisfies $\rho \sim \Lambda d_B$, where Λ is the average of $\Lambda(s)$ over

---

[1] In [Borys 09], D_2 was incorrectly defined as $\lim_{s \to 0} \frac{1}{\log s} \log \sum_{j=1}^{J} N_j^2(s)$.

a range of box sizes. This relation suggests that there is a negative correlation between Λ and d_B. Based on an analysis of polymer surfaces and stains in cancer cells, it was speculated in [Krasowska 04] that the relation $\rho \sim \Lambda d_B$ applies only to surfaces generated by similar mechanisms, which thus show similar types of regularity. When a negative correlation fails to hold, it may be due to the existence of several different surface generation mechanisms.

19.2   Lacunarity Based on an Optimal Covering

A very different approach to lacunarity was proposed in [Tolle 03]. Recall that an ϵ-covering of Ω (Sect. 5.1) is a finite collection $\{X_j\}$ of sets such that $\Omega \subseteq \cup_j X_j$ and $diam(X_j) \leq \epsilon$ for each j. Assume for simplicity of presentation that each set in the covering is a ball, so a covering is a finite set of balls $\{B(c_j, r_j)\}$ with $diam(X_j) = 2r_j$. This is illustrated in Fig. 19.7, where Ω is the set of small solid black circles, the covering for a given ϵ is the set of five balls, and the center of each ball is a blue circle.

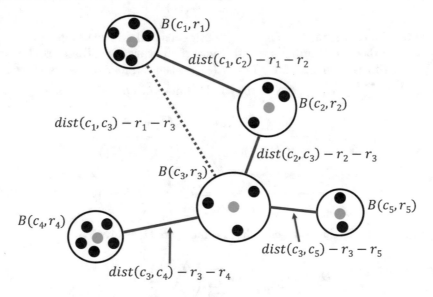

Figure 19.7: A covering of Ω by balls

Suppose that, for a given ϵ, an optimal ϵ-covering $\{B_\epsilon(c_j, r_j)\}_{j=1}^J$ has been computed; we define "optimal" below. The distance between the ball $B(c_i, r_i)$ and the ball $B(c_j, r_j)$ is $c_{ij}(\epsilon) \equiv dist(c_i, c_j) - r_i - r_j$, where $dist(c_i, c_j)$ is the Euclidean distance between c_i and c_j; having defined $c_{ij}(\epsilon)$, for simplicity we now write c_{ij} rather than $c_{ij}(\epsilon)$. Since there are J balls in the covering, there are $J(J-1)/2$ inter-ball distances c_{ij} to compute. Viewing each ball as a "supernode", the J supernodes and the distances between them form a complete graph. Compute a minimal spanning tree (MST) of this graph by any method, e.g., by Prim's method or Kruskal's method. Let $\mathcal{T}$ be the set specifying the $J - 1$ arcs (connecting supernodes) in the MST. This is illustrated in Fig. 19.7. Since there are 5 balls in the cover, then there are 5 supernodes, and 10 arcs interconnecting the supernodes; 5 of these arcs are illustrated (4 using solid red lines, and 1 using a dotted red line). The 4 solid red lines constitute the MST of the complete graph. Thus $\mathcal{T} = \{(1,2),(2,3),(3,4),(3,5)\}$.

Once the MST has been computed for a given ϵ, the lacunarity $\Lambda(\epsilon)$ is defined by

$$\Lambda(\epsilon) \equiv \max\left\{c_{ij} \mid (i,j) \in \mathcal{T}\right\}.$$

In Fig. 19.7 the largest c_{ij} is c_{12}, so $\Lambda(\epsilon) = c_{12}$. An alternative definition [Tolle 03] sets $\Lambda(\epsilon) \equiv \sum_{(i,j)\in\mathcal{T}} c_{ij}$, so for Fig. 19.7 we have $\Lambda(\epsilon) = c_{12} + c_{23} + c_{34} + c_{35}$.

Given a set $\mathbb{N}$ of points in Ω and given $\epsilon > 0$, the major computation burden in the method of [Tolle 03] is computing an optimal ϵ-covering. Recalling the definition of the Hausdorff dimension and (5.1) in Sect. 5.1, for $d > 0$ an ϵ-covering of Ω is optimal if it minimizes $\sum_j \left(diam(X_j)\right)^d$ over all ϵ-coverings of Ω. Since this definition does not in general lead to a useful algorithm, the approach of [Tolle 03] was to define an optimal covering as a solution to the following optimization problem $\mathcal{P}$. Let $\epsilon > 0$ and the number J of balls in the covering be given. Let $b_n \in \mathbb{R}^2$ be the given coordinates of point $n \in \mathbb{N}$, and let $c_j \in \mathbb{R}^2$ be the coordinates (to be determined) of the center of ball j. Define $N \times J$ binary variables x_{nj}, where $x_{nj} = 1$ if point n is assigned to ball j, and $x_{ij} = 0$ otherwise. We require $\sum_{j=1}^{J} x_{nj} = 1$ for $i = 1, 2, \ldots, N$; this says that each point in $\mathbb{N}$ must be assigned to exactly one ball. The objective function, to be minimized, is $\sum_{j=1}^{J} \sum_{n=1}^{N} x_{nj}[dist(b_n, c_j)]^2$, which is the sum of the squared distance from each point to the center of the ball to which the point was assigned. Optimization problem $\mathcal{P}$ can be solved using the *Picard* method [Tolle 03]. After the Picard method has terminated, principal component analysis can be used to compute the inter-cluster distances c_{ij}.

Exercise 19.1 In Sect. 16.8, we discussed the use by Martinez and Jones [Martinez 90] of a MST to compute d_H of the universe. In this section, we discussed the use by Tolle et al. [Tolle 03] of a MST to calculate lacunarity. Contrast these two uses of a MST. □

Exercise 19.2 Although the variables x_{nj} in the optimization problem $\mathcal{P}$ are binary, the Picard method for solving $\mathcal{P}$ treats each x_{nj} as a continuous variable. Is there a way to approximately solve $\mathcal{P}$ which treats each x_{nj} as binary? Does $\mathcal{P}$ become easier to solve if we assume that each ball center c_j is one of the N points? Is that a reasonable assumption? □

19.3 Applications of Lacunarity

Lacunarity has been studied by geologists and environmental scientists (e.g., coexistence through spatial heterogeneity has been shown for both animals and plants [Plotnick 93]), and by chemists and material scientists [Krasowska 04]. We conclude this chapter by describing other applications of lacunarity of a geometric object.

Stone Fences: The aesthetic responses of 68 people to the lacunarity of stone walls, where the lacunarity measures the amount and distribution of gaps between adjacent stones, was studied in [Pilipski 02]. Images were made of stone walls with the same d_B but with different lacunarities, which ranged from 0.1 to 4.0. People were shown pairs of images, and asked which one they preferred. There was a clear trend that the preferred images had higher lacunarity, indicating that people prefer images with more texture, and that the inclusion of some very large stones and more horizontal lines will make a stone wall more attractive.

Scar Tissue: In a study to discriminate between scar tissue versus unwounded tissue in mice, [Khorasani 11] computed d_B and the lacunarity of mice tissue. Microscope photographs were converted to 512×512 pixel images, which were analyzed using the ImageJ package.[2] They found that using d_B and lacunarity was superior to the rival method of Fourier transform analysis.

The Human Brain: The gliding box method was used by Katsaloulis et al. [Katsaloulis 09] to calculate the lacunarity of neural tracts in the human brain. The human brain consists of white matter and gray matter; white matter "consists of neurons which interconnect various areas of brain together, in the form of bundles. These pathways form the brain structure and are responsible for the various neuro-cognitive aspects. There are about 100 billion neurons within a human brain interconnected in complex manners". Four images from MRI scans were mapped to an array of 512×512 pixels. Using box sizes from 1 to 64 pixels, d_B ranged from 1.584 to 1.720. The same range of box sizes was used to compute the lacunarity of each image, where for each box size r the lacunarity $\Lambda(r)$ was defined by (19.5). It was observed in [Katsaloulis 09] that lacunarity analysis of two images provided a finer or secondary measure (in addition to d_B) which may indicate that the neurons originate from different parts of the brain, or may indicate damaged versus healthy tissue. Similarly, [Di Ieva 13] observed that "Lacunarity and d_B have been used to show the physiological decline of neuronal and glial structures in the processes of aging, whereby the d_B decreases and lacunarity increases in neurons, while the contrary occurs in glial cells, giving rise to a holistic representation of the neuroglial network".

Road Networks: In Sect. 6.5, we described the study of [Sun 12], which computed d_B and local dimensions for the road network of Dalian, China. The road network was treated as a geometric object Ω, not as a network, and Ω was superimposed on a 2022×1538 grid of cells. For each cell x in the grid, [Sun 12] created a sub-domain $\Omega(x)$ of size 128×128 cells centered at x; if x is near the boundary, then the sub-domain extended beyond Ω. The sub-domains overlap; the closer the cells x and y are to each other, the greater the overlap. For each x, [Sun 12] restricted attention to $\Omega(x)$ and calculated a local box counting dimension $d_B(x)$ by the usual procedure of subdividing $\Omega(x)$ into boxes of a given

---

[2] Shareware available from the National Institutes of Health website.

size and counting boxes that contain part of the image in $\Omega(x)$. Doing this for each cell x in the grid yielded a 2022×1538 array of $d_B(x)$ values.

The motivation of [Sun 12] for computing these $d_B(x)$ values was that the overall dimension d_B is useful in comparing cities, but not useful in investigating the heterogeneity of different areas within a single urban road network. The $d_B(x)$ were displayed using a color map, with high $d_B(x)$ values getting a more red color, and low values getting a more blue color. They also studied how the $d_B(x)$ values correlate with the total road length within the sub-domain. However, they did not use the term "lacunarity", nor did they compute an overall statistic that combines the $d_B(x)$.

Air Pollution: Lacunarity was used by Arizabalo et al. [Arizabalo 11] to study air pollution in Mexico. Atmospheric sub-micron particle concentrations were measured for particle sizes ranging from 16 to 764 nm using a Scanning Mobility Particle Sizer. The calculated lacunarity values ranged from 1.095 to 2.294. They found that days of low lacunarity indicated a nucleation particle formation process (with particle diameters of 10–20 nm), medium lacunarity indicated an Aitken process (with particle diameters of 20–100 nm), and high lacunarity indicated an accumulation process (with particle diameters of 100–420 nm). The results showed that lacunarity can be a useful tool for "identifying important changes concerning the particle concentrations (roughness) and size distributions (lagoons), respectively, which occur as a result of dynamic factors such as traffic density, industrial activity, and atmospheric conditions".

Pre-sliced Cooked Meat: In [Valous 10], lacunarity was computed for slices of meat that were prepared using different percentages of brine solutions and different processing regimes. It was observed in [Valous 10] that, in theory at least, textural information may be lost in the process of converting grayscale images to binary images, so the lacunarity calculated from grayscale images could provide better accuracy for characterization or classification than the lacunarity calculated using binary images. The main objective of this study was to determine the usefulness of different color scales, and compare them to a binary approach. When using a color scale, each pixel in the 256×256 image is identified by two coordinates, and has an associated integer "intensity" between 0 and 255. The lacunarity analysis was performed by the FracLac software [Rasband], which determined the average intensity of pixels per gliding box. The study confirmed that the intensity-based lacunarity values are a useful quantitative descriptor of visual texture. However, [Valous 10] also cautioned that discerning and measuring texture is challenging, and while lacunarity is useful, it is not a panacea. Finally, [Valous 10] credited Feagin with the observation that while lacunarity analysis may seem similar to multifractal analysis, "multifractals discern a globally consistent value based upon the singularity of local scaling exponents, whereas lacunarity defines the magnitude of local variations not as they scale outwards from those localities, but rather between localities". The extensive reference list in [Valous 10] is an excellent guide to further reading.

Exercise 19.3 The "gliding box" method for computing lacunarity assumed a square matrix. Modify the algorithm to compute the lacunarity for a rectangular matrix.   □

Exercise 19.4 Suppose the square geometric object $\Omega \subset \mathbb{R}^2$ can be partitioned as illustrated in Fig. 19.8, for some square region Ψ of side length L. Let

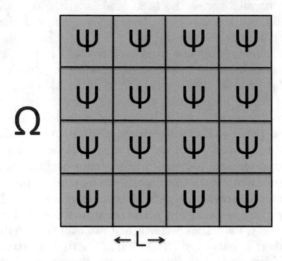

Figure 19.8: Ω is partitioned into 16 identical smaller pieces

$\Lambda(\Omega, s)$ and $\Lambda(\Psi, s)$ be the lacunarity of Ω and Ψ, respectively, for box size s. Can we estimate $\Lambda(\Omega, s)$ using $\Lambda(\Psi, s)$?   □

Exercise 19.5⋆ The only network considered in this chapter is a road network [Sun 12] which was studied as a geometric object. Extend the definition of lacunarity to a spatially embedded network, and apply the definition to an example.   □

Exercise 19.6⋆ Can the definition of lacunarity be extended to a network, not spatially embedded, with arbitrary arc costs?   □

Chapter 20

Other Dimensions

Our ancient experience confirms at every point that everything is linked together, everything is inseparable.
His Holiness the Fourteenth Dalai Lama of Tibet (1935–).

In this chapter we present some other definitions of the dimension of a geometric object. The chapter concludes with a discussion of the uses and abuses of the fractal dimension of a geometric object.

20.1 Spectral Dimension

There are at least four definitions of *spectral dimension* appearing in the literature. First, since d_B does not distinguish between two images which are identical except for the colors of the image, the *spectral fractal dimension* was proposed in [Berke 07] as an alternative to d_B. For example, a comparison of three species of potatoes yielded a spectral fractal dimension of about 2.357 for *agata*, 2.398 for *derby*, and 2.276 for *white lady*.

A second definition of spectral dimension can be applied to a self-affine curve (i.e., a curve exhibiting self-similarity in some but not all dimensions; see Sect. 3.1). After computing the power spectrum, e.g., using the Fast Fourier Transform (FFT), the slope of the log power versus log frequency curve is an estimate of the spectral dimension. Many applications of this method were mentioned in [Klinkenberg 94], e.g., to study geographic features [Power 87]. Contrasting several methods for calculating the fractal dimension of linear features (a *linear feature* is a natural feature, such as a coastline, with $d_T = 1$) in geology, [Klinkenberg 94] offered the following summary of the spectral dimension:

> Although a rigorous method, it is computationally difficult and computer intensive, and requires much more data preprocessing than any of the other methods, all of which work with the data "as is". ... A more significant problem is that several authors have made incorrect assumptions about the appropriate slope of the power spectrum plot, perhaps because there are a variety of ways of expressing

© Springer Nature Switzerland AG 2020
E. Rosenberg, *Fractal Dimensions of Networks*,
https://doi.org/10.1007/978-3-030-43169-3_20

the power or amplitude spectra. ... Nonetheless, the inherent complexity of spectral methods requires that they be carefully implemented [since one study] found several examples where subsequent re-interpretation of previously published spectral analyses have dramatically altered the conclusions.

A third definition of spectral dimension concerns a percolation cluster in $\mathbb{R}^E$ for which the number of occupied sites (i.e., the "mass") at radius r from a random site scales as r^{d_C} and for which the root mean square (rms) displacement L is related to the number S of steps in a random walk by the scaling $S \sim L^{d_{rms}}$. The spectral dimension of the cluster is

$$d_{sp} \equiv \frac{2d_C}{d_{rms}}.$$

Computational results show $d_{sp} \approx 4/3$, and this value appears to be independent of E for $E \geq 2$ [Meakin 83].

A fourth definition of the spectral dimension characterizes the behavior of a random walk or a diffusion. (A *diffusion* is the spreading of objects through a medium over time, e.g., an ink drop in a tank of water or a disease in a population.) Each molecule of water or individual in a population has a certain number of nearest neighbors, which determines the diffusion rate for the molecule or disease. For example, a drop of ink in a 3-dimensional medium grows in size as $t^{3/2}$. But in a twisty shape such as a Sierpiński gasket, for which $d_B = \log 3/\log 2 \approx 1.585$, the ink diffuses more slowly: the size scales as $t^{0.6826}$, and the spectral dimension d_{sp} is $(2)(0.6826) = 1.3652$. The spectral dimension is important in quantum gravity simulations, which drops a "tiny person" into one building block in quantum spacetime. The person walks around at random. The total number of spacetime building blocks the person visits over a given period reveals d_{sp} [Jurkiewicz 08].

Following [Horava 09], who followed the approach of [Ambjorn 05], the spectral dimension d_{sp} of a geometric object Ω is the effective dimension of Ω as seen by an appropriately defined diffusion process (or a random walker) on Ω. Let x and y be points in Ω. The diffusion process is characterized by the probability density $\rho(x, y; t)$ of diffusion from x to y in time t, subject to the initial condition that $\rho(x, y; 0)$, the density at time $t = 0$, depends only on some function of $x - y$. The *average return probability* $P(t)$ is obtained by averaging $p(x, x; t)$ over all x in Ω and defining

$$d_{sp} \equiv -2 \frac{\mathrm{d}\log P(t)}{\mathrm{d}\log t}$$

which implies $P(t) \sim t^{-d_{sp}/2}$ for small t. For example, when $\Omega \subset \mathbb{R}^E$ and the Euclidean metric is used, we have [Horava 09]

$$p(x, y; t) = \frac{e^{-(x-y)^2/4t}}{(4\pi t)^{E/2}}$$

and it can be shown [Modesto 10] that $P(t) = (4\pi t)^{-d_{sp}/2}$ and $d_{sp} = E$ (i.e., the spectral dimension is the dimension of the Euclidean space). Moreover, d_{sp} can be defined in a coordinate-independent way, which makes it applicable to a wide range of geometric objects beyond smooth manifolds, such as fractal objects. For example, for branched polymers we have $d_{sp} = 4/3$ [Horava 09].

20.2 Lyapunov Dimension

The Lyapunov dimension, proposed in [Kaplan 79], is based on the Lyapunov numbers, which quantify the average stability properties of the orbit of a strange attractor.[1] The advantage of the Lyapunov dimension is that it can be calculated directly from the strange attractor of the orbit of a dynamical system, without explicit knowledge of the underlying system of coupled differential equations or the recursive difference equations [Kinsner 07]. Following [Farmer 83], we consider the dynamical system $x_{t+1} = F(x_t)$, where $x \in \mathbb{R}^E$. For $t = 1, 2, \ldots$, define $J_t = J(x_t)J(x_{t-1}) \cdots J(x_1)$, where $J(x_t)$ is the Jacobian[2] of F at the point x_t, so J_t is the product of t square $E \times E$ matrices. Let the magnitudes of the E eigenvalues[3] of J_t be denoted by $\omega_1(t)$, $\omega_2(t)$, $\ldots$, $\omega_E(t)$. For $i = 1, 2, \ldots, E$, define

$$\lambda_i \equiv \lim_{t \to \infty} [\omega_i(t)]^{1/t} \tag{20.1}$$

where $[\lambda_i(t)]^{1/t}$ is the positive real t-th root of $\omega_i(t)$. These numbers, called the *Lyapunov numbers*, depend in general on the choice of the initial condition x_1. By convention, we assume $\lambda_1 \geq \lambda_2 \geq \cdots \geq \lambda_E$. For a dynamical system in two dimensions ($E = 2$), the numbers λ_1 and λ_2 are the average principal stretching factors of an infinitesimal circular arc. This is illustrated in Fig. 20.1, which shows that t iterations of the discrete time system transform the sufficiently small circle of radius r approximately into an ellipse whose major radius is $\lambda_1^t r$ and whose minor radius is $\lambda_2^t r$.

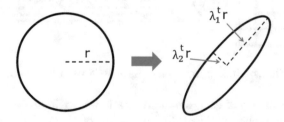

Figure 20.1: Lyapunov numbers

For $i = 1, 2, \ldots, E$, define $\gamma_i \equiv \log \lambda_i$; these are the *Lyapunov exponents* of the system. If $\gamma_i < 0$ (i.e., if $\lambda_i < 1$), there is a contraction in the direction of

---

[1] A.M. Lyapunov (1857–1918) was a Russian mathematician.

[2] Let $F : \mathbb{R}^m \to \mathbb{R}^k$ so $F(x) = (F_1(x), F_2(x), \ldots, F_k(x))$, where each $F_i : \mathbb{R}^m \to \mathbb{R}$. The *Jacobian* matrix J of F is the k by m matrix of partial derivatives defined by $J_{ij} = \partial F_i / \partial x_j$ for $i = 1, 2, \ldots, k$ and $j = 1, 2, \ldots, m$.

[3] Let J be a square matrix. Then the real number λ is an *eigenvalue* of J if for some nonzero vector u (called an *eigenvector* of J) we have $Ju = \lambda u$.

the eigenvector of the Jacobian matrix associated with λ_i, and if $\gamma_i > 0$ (i.e., if $\lambda_i > 1$), there is a repeated expansion and folding in the direction of the associated eigenvector. A dissipative system with at least two variables must have at least one negative Lyapunov exponent, and the sum of all Lyapunov exponents must be negative, indicating an overall contraction. A dissipative physical system with a strange attractor must have at least one positive Lyapunov exponent [Chatterjee 92]. If the equations defining the evolution of the system are known, the Lyapunov exponents can be calculated. If the equations are not known, and only experimental data is available, estimating the exponents is more challenging.

For a chaotic attractor, on average, nearby points initially diverge at an exponential rate, so at least one of the Lyapunov numbers exceeds the value 1. Since λ_1 is by convention the largest of the Lyapunov numbers, we use the condition $\lambda_1 > 1$ as the definition of *chaos*; this quantifies the qualitative phrase "sensitive dependence on initial conditions". Assume that for almost every initial condition (i.e., initial point x_0) in the basin of attraction, the attractor has the same set of Lyapunov numbers.[4] Then the set of Lyapunov numbers may be considered to be a property of an attractor.

The following argument, from [Farmer 83], shows the connection between the Lyapunov numbers and a fractal dimension. Consider a strange attractor $\Omega \subset \mathbb{R}^2$ for which $\lambda_1 > 1 > \lambda_2$. To compute the box counting dimension d_B of Ω, let $\mathcal{B}(s)$ be a covering of Ω by boxes of side length s, so $B(s)$ is the number of such squares needed to cover Ω. Now perform t iterations of the dynamical system. For sufficiently small s, the action of the map $x_{t+1} = F(x_t)$ is roughly linear over each small square in $\mathcal{B}(s)$, and each small square is mapped into (approximately) a long thin parallelogram, as illustrated in Fig. 20.2. After the

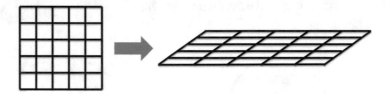

Figure 20.2: F maps a square to a parallelogram

t iterations, the image of each small square is approximately a parallelogram which has been stretched horizontally be a factor of λ_1^t and contracted vertically by a factor of λ_2^t.

Now we cover the $B(s)$ small parallelograms using squares of size $\lambda_2^t s$. Since $\lambda_2 < 1$, this is a finer grid than the square grid of size s. Using this finer grid, the number of boxes needed to cover each small parallelogram is approximately $(\lambda_1/\lambda_2)^t$. Assuming that each small square is transformed the same way, since

---

[4] The *basin of attraction* [Farmer 83] of an attractor Ω is the closure of the set of initial conditions that approach Ω.

$B(s)$ small squares of size s were needed to cover Ω, the number of boxes of size $\lambda_2^t s$ needed to cover the $B(s)$ small parallelograms is $(\lambda_1/\lambda_2)^t B(s)$. Thus

$$B(\lambda_2^t s) \approx (\lambda_1/\lambda_2)^t B(s) . \tag{20.2}$$

For the final step in the heuristic derivation in [Farmer 83] of the Lyapunov dimension, assume that $B(s)$ obeys the scaling law $B(s) \approx c s^{-d_L}$ for some constant c and some exponent d_L. Substituting this into (20.2) yields

$$c \left(\frac{1}{\lambda_2^t s} \right)^{d_L} \approx c \left(\frac{\lambda_1}{\lambda_2} \right)^t \left(\frac{1}{s} \right)^{d_L} . \tag{20.3}$$

Treating the approximation as an equality and solving for d_L yields

$$d_L = 1 + \frac{\log \lambda_1}{\log(1/\lambda_2)} , \tag{20.4}$$

and we call d_L the *Lyapunov dimension*. This discussion assumed a two-dimensional system but can be extended to E dimensions for $E > 2$ by defining

$$d_L \equiv p + \frac{\log(\lambda_1 \lambda_2 \cdots \lambda_p)}{\log(1/\lambda_{p+1})} , \tag{20.5}$$

where p is the largest index for which $\lambda_1 \lambda_2 \cdots \lambda_p \geq 1$. If $\lambda_1 < 1$, then $d_L \equiv 0$, and if $\lambda_1 \lambda_2 \cdots \lambda_E \geq 1$, then $d_L \equiv E$. The Lyapunov dimension was first defined in [Kaplan 79].

The above arguments should not be construed as suggesting that the Lyapunov and box counting dimensions are equal. As discussed in [Farmer 83], the reason we should not expect $d_L = d_B$ is that the Lyapunov numbers are averages, and an average implies that each box should be weighted by a probability. In contrast, with box counting, typically the vast majority of boxes needed to cover the attractor contain very few points. The collection of these infrequent boxes accounts for only an extremely small fraction of the probability on the attractor, yet they account for almost all of $B(r)$. Since the information dimension d_I explicitly incorporates the probability of each box, [Farmer 83] conjectured that for "typical" attractors we have $d_L = d_I$, and presented some evidence supporting this conjecture.

20.3 Three More Dimensions

In this short section we describe three other interesting definitions of the dimension of a geometric object.

Cumulative Intersection Dimension: This dimension, first proposed in 1973 by Stell and Witkovsky in studying the retinal structure of a certain type of fish, considers the branching pattern of an image. Select a random point in Ω

to be the center of a circle. For a set of increasing radii r in a range, count the cumulative number of branching points (i.e., intersections) within the circle with this center and with radius r. Repeat this for many randomly generated circle centers. Let $b(r)$ be the average number of branches at radius r. The estimate of the fractal dimension is the slope of the best fit linear approximation to the $\log b(r)$ versus $\log r$ curve [Jelinek 98, Cornforth 02].

The cumulative intersection method was used in [Jelinek 98] in a comparison of methods to discriminate between three types of cat retinal ganglion cells. The study utilized various methods for calculating a fractal dimension, together with analysis of variance (ANOVA). On binary images, three different box-counting methods analyzed had a success rate for classifying a cell ranging from 47.9 to 59.2%, the dilation method had a 63.5% success rate, and the cumulative intersection method had a 62% success rate; these results agreed with previously published results. The classification rate improved significantly when using binary images rather than skeletonized or border only images. The conclusion in [Jelinek 98] was that fractal dimension is a robust measure allowing differentiation of morphological features, such as the complexity of dendritic branching patterns. Even if the computed fractal dimension of a given method is not very accurate, when the same method is applied to different images the computed dimension is useful and valid for purposes of comparison and classification.

Exercise 20.1 Can the cumulative intersection method described above be considered a variant of the sandbox method? □

α-Box Counting Dimension: For a given α satisfying $0 < \alpha \leq 1$, let $N_\alpha(s)$ be the minimal number of boxes of size s needed to cover a fraction α of the natural measure (Sect. 9.2) of Ω. The α-box counting dimension was defined in [Farmer 83] as

$$d_B(\alpha) \equiv \lim_{s \to 0} \frac{\log N_\alpha(s)}{\log(1/s)} .$$

Thus $d_B(1) = d_B$. Interestingly, for all the examples studied in [Farmer 83], computational results for $0 < \alpha < 1$ showed that $d_B(\alpha)$ is independent of α, but $d_B(\alpha)$ for $0 < \alpha < 1$ may differ from d_B.

Clustering and Transitivity Dimensions: Consider the time series x_t, $t = 1, 2, \ldots N$, where $x_t \in \mathbb{R}^E$, that is generated by sampling the system at a fixed number N of points in time. There is a *recurrence* at time t_j if the system returns to the neighborhood of a previous state, i.e., if x_j is contained in a ball of radius r centered around x_i for some $i < j$. This leads to the definition of the *r-recurrence matrix*

$$R_{ij}(\epsilon) = I_0\Big(\epsilon - dist\big(x_i, x_j\big)\Big), \tag{20.6}$$

where I_0 is defined by (9.7) and where $dist(\cdot, \cdot)$ corresponds to some norm in the metric space containing the attractor of the dynamical system. It was shown in [Donner 11] that $R_{ij}(\epsilon)$ can be used to estimate two new dimensions. The clustering dimension considers the scaling of $\mathcal{C}(x, \epsilon)$, defined for $x \in \Omega$ and $\epsilon > 0$ as follows: for points y and z randomly selected from Ω, let $\mathcal{C}(x, \epsilon)$ be the probability that $dist(y, z) \leq \epsilon$ given that $dist(x, y) \leq \epsilon$ and $dist(x, z) \leq \epsilon$. The transitivity dimension considers the scaling of $\mathcal{T}(\epsilon)$, defined as follows: for points x, y, and z randomly selected from Ω, let $\mathcal{T}(\epsilon)$ be the probability that $dist(y, z) \leq \epsilon$ given that $dist(x, y) \leq \epsilon$ and $dist(x, z) \leq \epsilon$. For example, for $\Omega = [0, 1]^E$ (the E-dimensional unit cube), with points selected from the uniform density, and using the L_∞ norm, we have $\mathcal{C}(x, \epsilon) = \mathcal{T}(\epsilon) = (3/4)^E$. For $x \in \Omega$, the *upper cluster dimension* at x is defined as

$$\limsup_{\epsilon \to 0} \frac{\log \mathcal{C}(x, \epsilon)}{\log(3/4)} \,,$$

with the *lower cluster dimension* at x similarly defined using lim inf. The *upper transitivity dimension* is defined as

$$\limsup_{\epsilon \to 0} \frac{\log \mathcal{T}(\epsilon)}{\log(3/4)} \,,$$

with the *lower transitivity dimension* similarly defined using lim inf. Since $\mathcal{C}(x, \epsilon)$ and $\mathcal{T}(\epsilon)$ are based on geometric three-point interactions, these two dimensions are statistical properties of higher order than the box counting and Rényi dimensions, and even of higher order than the correlation dimension, which is based on two-point correlations in phase space.

20.4   Uses and Abuses of Fractal Dimensions

In this section we discuss the "uses" and "abuses" of the term "fractal dimension".[5] Despite the current enormous interest in fractals, in the early days of fractal analysis not everyone viewed these techniques as applicable. For example, [Shenker 94] argued that, despite the widespread enthusiasm for fractals, the natural world is *not* properly modelled by fractal geometry. The crux of the argument in [Shenker 94] is that fractals are idealized, non-constructible objects, similar to irrational numbers, and hence cannot properly model real-world objects. A fractal view of nature contradicts the physical principle that nature is "atomic", meaning that there are fundamental, indivisible building blocks in nature. Thus the amount of detail in nature may be very large, but it is finite, whereas a fractal has infinite complexity. So to claim that nature is inherently fractal proposes a "radically new, non-atomistic science". Moreover,

> The application of fractal geometry to finitely complex forms, i.e., forms having a finite number of details, is worse than superfluous. Analysis of such forms in terms of non-exact geometries is more exact and necessitates the use of fewer approximations. This is even true for complex forms, as long as their complexity is finite. It is especially significant in the limit of the very small scales used in fractal geometry's definitions.

---

[5] The title of this section is a rephrasing of the title of [Halley 04].

This was written in 1996; the thousands of papers on the theory and successful application of fractals nullify Shenker's objections. One might also wonder
whether Shenker's arguments would similarly discard derivatives and integrals,
which are based upon vanishing infinitesimals, if nature is ultimately discrete.

Indeed, the value of fractal analysis is by now so firmly established that failure to employ fractal analysis invites criticism. For example, [Losa 11] critiqued
a study of the morphology of certain rat brain cells.[6] This study subjectively
classified somata (Fig. 20.3) into triangular, round, and oval shapes.[7] The observation in [Losa 11] was that, for such a study, a fractal perspective is mandatory:

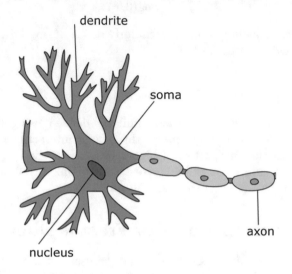

Figure 20.3: Cell structure

> ... during the last two decades, several studies have been performed
> on brain tissue and nervous system cells by adopting fractal concepts
> and methods, which has enabled to quantitatively elucidate most de-
> velopmental, morphological, and spatial pattern avoiding arbitrary
> approximation or smoothing of cellular shapes and structures. .... .
> Biologic structures with irregular shape and complex morphology
> should not be approximated to ideal geometric objects ...

Growth of Ivy: As to the "abuses" of fractals, we first cite [Murray 95], who
discussed the misuse of fractals in neuroscience, starting with a warning: "The
problem with a good name for a new field is that uninformed proselytizing and
inappropriate use can raise unrealistic expectations as to its relevance and applicability". In particular, [Murray 95] cautioned against "a deluge of irrelevant
uses of fractal measurements". For example, consider ivy growing on a wall. In
winter, the ivy sheds its leaves. If the box-counting method is applied to the

---

[6] Morphology is the branch of biology concerned with the form and structure of organisms.
[7] Adapted from http://en.wikipedia.org/wiki/Axon.

winter branching structure, a fractal dimension between 1 and 2 is obtained. However, calculating this dimension says nothing about the actual biological problems of interest. Since in summer the leaves grow back, and the branching structure is space filling, a more interesting question is how the branching is achieved by the plant, and how interbranch space is detected.

Rabbit Retina and DLA: Another example discussed in [Murray 95] concerns the use of a *diffusion limited aggregation* (DLA) model. As described in Sect. 10.5, DLA is a diffusion process in which particles engage in a random walk, and when two particles meet they attach and become stationary. The process is usually started with a seed set of stationary particles onto which a set of moving particle is released. The spatial pattern formed by this dynamic process is fractal. It has been observed that the DLA pattern resembles the pattern of the amacrine cell from the central region of the rabbit retina, which led to the suggestion that the shape of neuronal cells may be determined by a DLA process. However, the visual similarity of the two images does not justify such a conclusion. Any model mechanism, such as DLA, must be compared against other biological spatial pattern generators and by appropriate experiments. Such seduction by hypothetical models (e.g., of ferns and trees) deflects attention from the main purpose of studying biological pattern formations, which is to discover the underlying biological processes [Murray 95].

Goose Down: Application of the sandbox method (Sect. 10.5) determined that the sandbox dimension of goose down is 1.704 [Gao 09]. Since 1.704 is close to the golden mean ratio of $(1 + \sqrt{5})/2 \approx 1.618$, and since the golden mean ratio is known to be pervasive in nature, [Gao 09] concluded that this "assuredly reveals certain structural optimality in goose down", and that the significance of this fractal dimension of goose down "for the survival of geese is quite self-explanatory". While the hypothesis in [Gao 09] about the applicability of the golden ratio to the structure of goose down is interesting, their analysis does not validate this hypothesis, and the significance of this fractal dimension is not "self-explanatory".

Neuroscience: The use of box counting in studying microglia[8] was discussed in [Karperien 13], which warned that knowing d_B of an object does not tell us anything about its underlying structure, its construction, or its function. Moreover, box counting neither finds nor confirms the existence of self-similarity, and being able to calculate d_B for an object does not imply that it arose from a fractal process. Box counting is not always the method of choice: when measuring ramification (i.e., branching), it is useful to compute the skeleton of the image, in which case the dilation method for calculating the Minkowski dimension d_K, rather than the box-counting method for calculating d_B, is the preferred method (as discussed in Sect. 4.5, if d_B exists, then so does d_K and $d_B = d_K$). One suggestion in [Karperien 13] was that the d_B of microglia in spinal fluid could be used to monitor and quantify the pathological status, the

---

[8] Microglia are cells occurring in the central nervous system of invertebrates and vertebrates. These cells mediate immune responses in the central nervous system by clearing cellular debris and dead neurons from nervous tissue through the process of phagocytosis (cell eating). http://www.britannica.com/EBchecked/topic/380412/microglia.

stages of an ongoing response after injury, or the effects of drugs (e.g., for schizophrenia) in patients.

The Music of Bach and Mozart: It has been known since the time of Pythagoras that when the ratio of the length of two strings (of the same thickness) is the ratio of two small whole numbers, plucking both strings simultaneously yields a harmonious (i.e., pleasing and consonant) sound. For example, the ratio of 2:1 produces an octave,[9] the ratio 4:3 produces a fourth,[10] and the ratio 3:2 produces a fifth.[11] In the western diatonic scale, there are 12 semitones between consecutive octaves. For example, starting at the note C, we have (going up in pitch) $C^{\#}$ (otherwise known as $D^{\flat}$), D, $D^{\#}$ (otherwise known as $E^{\flat}$), E, F, $F^{\#}$, G, $G^{\#}$ (otherwise known as $A^{\flat}$), A, $A^{\#}$, and then B, before reaching the next C. Let f be the frequency (meaning "pitch", not meaning probability of occurrence) of the first C and let f^{oct} be the frequency of the C an octave higher, so $f^{oct}/f = 2$.

Let $f(P)$ be the audio frequency of the note with pitch P. If *equal tempered tuning* is used (there are other less common methods), the ratio of the frequencies between any two adjacent pitches (i.e., pitches separated by a semitone) is the same, and that ratio is $2^{1/12}$. That is, $f(C^{\#})/f(C) = 2^{1/12}$, $f(D)/f(C^{\#}) = 2^{1/12}$, $f(D^{\#})/f(D) = 2^{1/12}$, etc. Hence the interval between adjacent notes is equal on a logarithmic scale, so $\log f(C^{\#}) - \log f(C) = (1/12)\log 2$, $\log f(D) - \log f(C^{\#}) = (1/12)\log 2$, etc. The interval from C to G is a perfect fifth, composed of 7 semitone intervals, so $f(G)/f(C) = (2^{1/12})^7 = 2^{7/12} \approx 1.498 \approx 3/2$. The interval from C to the next C (call it C^{oct}) is an octave, composed of 12 semitone intervals, so $f(C^{oct})/f(C) = (2^{1/12})^{12} = 2$. It is certainly possible for the interval between successive notes to exceed one octave, but this is uncommon.

A study by Hsü and Hsü [Hsu 90] considered whether the distribution of intervals in music can be used to categorize music, thus enabling us to mathematically distinguish the music of, for example, the Baroque era composer Bach (1685–1750) from the modern composer Stockhausen (1928–2007). They examined seven compositions: three of Bach, two of Mozart, six Swiss children's songs (considered as one composition), and Stockhausen's *Capricorn*. For each of the seven compositions they determined the exponent d (which they called a fractal dimension) fitting the curve $F_i = c\,i^{-d}$, where c is a constant, i is the interval between successive notes in a composition, and F_i is the frequency (meaning probability of occurrence) of the interval i between successive notes in a composition. The value $i = 0$ signifies that successive notes have the same pitch. For one Bach composition (*Invention no. 1 in C Major, BWV 772*), they calculated $d = 2.4184$ and $c = 2.15$ over the range $2 \leq i \leq 10$. Notable devia-

---

[9] In the song "Over the Rainbow", immortalized by Judy Garland in the movie "The Wizard of Oz", the interval between the first two syllables "some" and "where" is an octave.

[10] In the end of the first line of the "Star Spangled Banner", for the notes "you" and "see", the pitch of "see" is a fourth higher than "you".

[11] In the song "Twinkle Twinkle Little Star", the pitch of the third and fourth notes (the second "twinkle") is a fifth higher than the pitch of the first two notes (the first "twinkle"). That is, the interval is a fifth.

tions from $F_i = (2.15)i^{-2.4184}$ occur for $i = 6$ (which is under-represented in the composition) and $i = 7$ (which is over-represented). The interval $i = 6$, a *diminished fifth*, is difficult to sing or play on a string instrument. The interval $i = 7$ is, as mentioned above, a pleasing fifth. For Bach's *Tocata in $F^\#$ Minor*, they obtained $F_i = (0.376)i^{-1.3403}$ over the range $1 \leq i \leq 7$. For the first movement of Mozart's *Sonata in F Major*, they obtained $F_i = (1.0)i^{-1.7322}$ over the range $2 \leq i \leq 4$. Since $c = 1$ for the Mozart Sonata, and hence $F_i\,i^{1.7322} = 1$, Hsü and Hsü called d a similarity dimension (Sect. 5.2) and suggested that Mozart's music could be considered pictorial. In contrast, since for Bach's *Tocata in $F^\#$* we have $c = 0.376$ and thus the scaling $F_i\,i^d = 1$ does not hold for $d = 1.3403$, they concluded that this d is not a similarity dimension, and therefore that *Tocata in $F^\#$* cannot cannot be represented by a picture, and instead is more mathematical.

Exercise 20.2 The study of [Hsu 90], discussed above, computed $c = 2.15$ for the Bach *Invention no. 1 in C Major* and $c = 0.376$ for the Bach *Tocata in $F^\#$*. Thus it is quite plausible that for other Bach compositions we could have $c \approx 1$. Given how prolific these composers were, is it justified to conclude that Bach is more mathematical than Mozart based on two values of c? (Although any student of classical music would undoubtedly not dispute that assertion.) Also, a similarity dimension is obtained when there is a model of taking K copies of a larger object, each scaled down by some factor α (Sect. 5.2). Since no such model is proposed in [Hsu 90] for the Mozart sonata, does the fact that $c = 1$ for this composition justify the assertion that the corresponding d is a similarity dimension? $\square$

Music Fingerprinting: If one accepts the premise that a power-law relationship automatically confers "fractal" status, then "fractal dimension" becomes just another name for the exponent in the power law. This was the approach taken in [Hughes 12], who studied automated music search and playlist generation. The approach in [Hughes 12] improved on the music fingerprinting application Shazam[12] in that it could also recommend songs, ordered by musical and timbral resemblance.[13] Building upon prior research establishing that, for music, power-law relationships of the form $p(x) \sim ax^{-d}$ (where $p(x)$ is the probability of x) exist for pitch, loudness, and rhythm, [Hughes 12] computed 372 power-law metrics for a wide variety of songs. The variety of songs they studied includes classical (Bach, Beethoven, Mozart) to jazz (Charles Mingus, Miles Davis, John Coltrane) to rock (The Doors, Led Zeppelin) to house-techno (DJ Dado, Eiffel 65, DJ Keoli). Since the 372 metrics tended to be highly correlated, various statistical techniques were used to define a set of approximately 45 metrics that were shown to capture important aspects of music aesthetics and similarity.

---

[12] "Shazam" is a registered trademark of Apple Inc. and its affiliates.

[13] *Timbre* is the quality given to a sound by its overtones: such as the resonance by which the ear recognizes and identifies a voiced speech sound, or the quality of tone distinctive of a particular singing voice or musical instrument. https://www.merriam-webster.com/dictionary/timbre.

Exercise 20.3 Without explicitly defining "fractal dimension", [Hughes 12], discussed above, implicitly used the term to mean the exponent d in the power-law relationship. Is this use of the "fractal dimension" warranted, or is just a fancier name for "power-law exponent"?   □

To conclude this chapter, we quote from [Jelinek 06] who examined "... how methodology in fractal analysis is influenced by diverse definitions of fundamental concepts that lead to difficulties in understanding fundamental issues" and observed that

> There exists disagreement among physicists and mathematicians, [about] what constitutes a dimension, and what measures can be included in this term. Apart from the Hausdorff dimension other so called fractal dimensions such as the Minkowski and the Kolmogorov dimension are not really dimensions in a mathematical sense. However in the literature authors often refer to these two measures as dimensions possibly because of trying to simplify what is really quite a complex area in mathematics. ... Indeed books and articles containing the terms "fractal analysis" in their titles often do not point out that a strict application of the rules of mathematics to the procedures of fractal analysis is not possible. One only needs to observe the intricate definition of the Hausdorff dimension, which is not usable in practice and the shortcomings of the analysis procedures in estimating the Hausdorff dimension using for instance the box-counting method.

As frequently mentioned in this book, different researchers have applied different names to the same fractal dimension, and a cursory reading of a research paper may leave the reader with a false impression of which fractal dimension is actually being investigated. Indeed, it is not unusual to see the box counting dimension called the Hausdorff dimension. Fortunately, this confusion of different names for the same dimension is growing less common over time. Moreover, not all of the above remarks of [Jelinek 06] still hold today, e.g., although computing the Hausdorff dimension of Ω is not in general feasible, computing the Hausdorff dimension (appropriately defined) of a network is quite feasible, as shown in Chap. 18.

Chapter 21

Coarse Graining and Renormalization

> *Any great work of art revives and readapts time and space, and the measure of its success is the extent to which it makes you an inhabitant of that world—the extent to which it invites you in and lets you breathe its strange, special air.*
> Leonard Bernstein (1918–1990), American conductor and composer, best known for the music to "West Side Story".

Murray Gell-Mann, winner of the 1969 Nobel Prize in physics, provided a non-technical introduction [Gell-Mann 88] to coarse graining by discussing a debate among ecologists over "whether complex ecosystems like tropical forests are more robust than comparatively simple ones such as the forest of oaks and conifers near the tops of the San Gabriel Mountains" [near Caltech, where Gell-Mann was a professor]. He wrote

> But what do they mean by simple and complex? To arrive at a definition of complexity for forests, they might count the number of species of trees (less than a dozen near the tops of the San Gabriels compared to several hundred in a tropical forest); they might count the number of species of birds, mammals, or insects. [...] The ecologists might also count the interactions among the organisms: predator–prey, parasite–host, pollinator–pollinated, and so on. Down to what level of detail would they count? Would they look at microorganisms, even viruses? Would they look at very subtle interactions as well as the obvious ones? Clearly, to define complexity you have to specify the level of detail that you are going to talk about and ignore everything else—to do what we in physics call "coarse–graining".

When we apply *coarse graining* to a system, we are "zooming out" and thus losing the *fine-grained* details. The use of coarse graining dates back at least to

© Springer Nature Switzerland AG 2020
E. Rosenberg, *Fractal Dimensions of Networks*,
https://doi.org/10.1007/978-3-030-43169-3_21

J.W. Gibbs (1839–1903) in his proof [Bais 08] of the Second Law of Thermodynamics, which says that the entropy of a closed system increases until it reaches its equilibrium value.[1] Coarse graining is the key to renormalization. In this chapter we first study coarse graining and renormalization for physical systems. Then we show how renormalization can be applied to networks, facilitating the calculation of various network parameters.

21.1   Renormalization Basics

Renormalization is a powerful analytic technique, widely used in statistical physics for investigating critical phenomena such as phase transitions (e.g., the transition from liquid water to ice at 0 °C), and for determining the critical values of parameters governing such phenomena.[2] Kenneth Wilson [Wilson 79], winner of the 1982 Nobel Prize in physics and whose Ph.D. advisor at Caltech was Gell-Mann, cited the work of Bethe, Schwinger, Tomonaga, Feynman, Dyson, and others in the late 1940s as the starting point of the theory. These researchers used renormalization to eliminate divergent quantities arising in quantum electrodynamics, which is the theory of interactions between electrically charged particles and the electromagnetic field.[3] The term *renormalization group* first appeared in a 1953 paper by Stueckelberg and Petermann (they used the term "groupes de normalization", which translates as "renormalization group".) Major contributions were made in the 1960s, largely by Fisher, Kadanoff, and Widom, by Wegner in 1972, and Nobel prize winning work by Wilson in the 1970s. Interestingly, in Wilson's Nobel Prize lecture [Wilson 82], he mentioned a "murky" connection between scaling ideas in critical phenomena and Mandelbrot's fractal theory. A brief non-mathematical introduction to renormalization was provided in [Kadanoff 13]:

> Statistical mechanics describes the equilibrium behavior of thermodynamic systems in terms of a probability distribution, termed an ensemble, that describes the probability of occurrence of all the possible configurations of the system. ... The basic problem faced by workers in statistical mechanics is to compute the observable properties of a system given a specification of its ensemble, usually given in

---

[1] https://www.famousscientists.org/j-willard-gibbs/.

[2] It was suggested in [Goldberger 96] that the abrupt shift around 1140 from a Romanesque style of architecture to a Gothic style, through the creative efforts of Abbot Suger "who, singlehandedly, conceived and directed the rebuilding of the Royal Abbey Church" outside Paris, France, is an example of a phase-like transition. While Romanesque structure are notable for their solidity and filled-in appearance, "the surfaces of Romanesque buildings tend to be smoother and less irregular and spiky than their Gothic counterparts". Fractals capture key features of Gothic architecture, such as its porous "holeyness" and carved-out appearance, its wrinkled, crenelated surfaces, and its overall self-similarity. The Cathédrale Notre-Dame de Paris, badly damaged by fire in April 2019, is the most famous of the Gothic cathedrals of the Middle Ages and is distinguished for its size, antiquity, and architectural interest. https://www.britannica.com/topic/Notre-Dame-de-Paris.

[3] The book *Genius* by James Gleik, a biography of Richard Feynman, frequently mentioned the problem of divergent sums encountered by Feynman and his colleagues.

terms of such basic parameters as temperature, chemical potential, and the form of the interaction among particles. Direct calculation of ensemble properties has proven to be very difficult for systems at or near phase transitions. Here the behavior is dominated by correlated fluctuations extending over regions containing very many particles. As the phase transition is approached, the fluctuations cover larger and larger regions, over distances finally tending to become infinitely large. Because of this infinity, computing the correlations among the different particles presents a calculational problem of very considerable complexity. This problem was unsolved before the development of the renormalization method. This method, brought to its completed form by Wilson, provides an indirect method of approaching the difficulty presented by large fluctuating regions. The renormalization method, a process of averaging the ensemble over small-scale correlations, constructs a new ensemble that describes the same problem, but uses spatially averaged variables to do so. Step by step, the gaze of the investigator focuses upon larger and larger length scales. It is usually true that eventually, after many renormalizations, the system will reach a fixed point, that is, an ensemble that remains unchanged under the successive transformations. Methods have been developed that enable one to calculate the new ensemble, using the larger length scale, from the old one and thereby find the ensembles describing the fixed points.

A slightly more quantitative introduction to renormalization was provided in [Fisher 98] and by Wilson's Nobel Prize lecture. Consider a sealed glass tube of carbon dioxide. At room temperature, there are two phases in the tube: a liquid phase and a gas phase. As the temperature is raised and reaches a critical temperature T_c, the liquid and gas phases become identical and the meniscus boundary between the two phases disappears in "mist" of "critical opalescence" (the words of an 1869 report by Andrews to the Royal Society, as quoted in [Fisher 98]). Above T_c, there is complete "continuity of state", meaning no distinction whatsoever between liquid and gas, and no meniscus. A plot of the liquid density $\rho_L(T)$ and the gas density $\rho_G(T)$ as a function of the temperature T shows that the plot of $\rho_L(T)$ for $T > T_c$ and the plot of $\rho_G(T)$ for $T < T_c$ meet smoothly at the critical point (T_c, ρ_c).

The same phenomena occurs in all elemental and simple molecular fluids and in fluid mixtures, although the critical temperature varies from fluid to fluid. However, for all these liquids, the shapes of the plots of $\rho_L(T)$ and $\rho_G(T)$ versus T become asymptotically *universal* (i.e., having the same functional form) as T_c is reached. In general, the shape of the liquid gas coexistence curve in the neighborhood of the critical region exhibits the power law

$$\rho_L(T) - \rho_G(T) \approx Bt^\beta \text{ as } t \to 0,$$

where $t \equiv (T-T_c)/T_c$, where B is a *non*-universal constant, and where $\beta \approx 0.325$ is a universal constant (i.e., the same for all fluids). Since all fluids exhibit this functional form near their critical temperature, with the same β, this defines a *universality class*. Renormalization theory includes the study of these universality classes.

Such problems are theoretically challenging because they involve very many coupled degrees of freedom, and many length scales (e.g., a wide range of electron wavelengths). The renormalization approach to problems with many length scales is to tackle the problem in steps, one step for each length scale. For example, in critical phenomena like phase transitions, statistical averages are performed beginning with an atomic scale and then moving out to successively larger scales until thermal fluctuations on all scales have been averaged out. As these fluctuations on length scale are averaged out, a new *free energy functional* is generated from the previous functional. We quote from Wilson's Nobel Prize lecture [Wilson 82], making minor changes as indicated to remove mathematical formulas:

> This process is repeated many times. If the [free energy functionals] are expressed in dimensionless form, then one finds that the transformation leading from [one functional] to the [next functional] is repeated in identical form many times. (The transformation group thus generated is called the "renormalization group".) As [the distance scale] becomes large, the [functional] approaches a fixed point of the transformation, and thereby becomes independent of details of the system at the atomic level. This leads to an explanation of the universality of critical behavior for different kinds of systems at the atomic level. Liquid-gas transitions, magnetic transitions, alloy transitions, etc., all show the same critical exponents experimentally; theoretically this can be understood from the hypothesis that the same "fixed point" interaction describes all these systems.

Digging a little deeper into renormalization, we now consider the seminal method of [Kadanoff 66] for mapping a critical or near-critical system onto itself by reduction in the effective number of degrees of freedom [Fisher 98]. Kadanoff studied the two-dimension Ising model, introduced in the 1920s by Wilhelm Lenz and his student Ernest Ising.[4] Consider a two-dimensional rectilinear lattice, where each node has the *spin* value +1 or −1. Each node has an influence on the spin of its neighbors; this influence favors spin alignment (i.e., having the same spin as the neighbor). The alignment is quantified by a *magnetization* $M(T)$, where T is the temperature. At low T, the majority of the spins are +1 or, alternatively, −1, so there is a spontaneous magnetization. As T rises, $M(T)$ decreases, and $M(T)$ reaches 0 at a critical temperature T_c. The goal is to determine T_c.

---

[4] Ising (1900–1998) [Kobe] studied a model of ferromagnetism introduced by Lenz in the 1920s. In his doctoral thesis, Ising considered the special case of a chain of magnetic moments which can assume only an "up" or "down" position, and where interactions occur only between nearest neighbors; this became the famous Ising Model. Dr. Ising and his wife fled Nazi Germany and ultimately arrived in America in 1947. From 1948–1976, he was Professor of Physics at Bradley University in Illinois, where the Ernst Ising Collection is housed. The Ising model and other critical phenomena in networks were examined in [Dorogovtsev 08].

Kadanoff divided the lattice into disjoint blocks, as illustrated in Fig. 21.1. Each block of s^2 nodes (where $s = 4$ in this example) in the left-hand figure

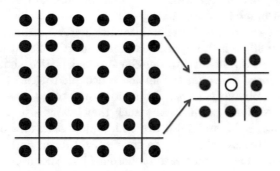

Figure 21.1: Kadanoff's block renormalization

is mapped to a single block node in the right-hand figure. For example, in Fig. 21.1, the central block of 16 nodes is mapped to the block node indicated by the hollow circle (the horizontal and vertical lines have no significance other than to demarcate the blocks). The block node possesses a *block spin*. To relate the block spins of the block nodes to the spins of the original nodes, Kadanoff introduced a *spin rescaling or renormalization* factor $f(s)$; to relate the temperature in the renormalized block node system to the original temperature, he introduced a *thermal rescaling* factor $g(s)$. By choosing $f(s) = s^{-\omega}$ and $g(s) = s^{\lambda}$, where ω and λ characterize the critical point being studied, the statistical properties of the original system can be understood by studying the renormalized system. Since the renormalized system contains fewer degrees of freedom (16 nodes are mapped to one block node), Kadanoff's renormalization opened the door to actual numerical calculation of critical values. It turns out that this approach has some difficulties for 3-dimensional lattices and for other systems; the 1971 work of Wilson resolved these issues. Another perspective on renormalization is that if we view biological processes as repeated applications of duplication and divergence (i.e., branching), then renormalization can be viewed as going backwards in time.

With this very brief introduction to Kadanoff's method, we can now discuss the *renormalization group*. A *group* in mathematics is a set $\mathbb{S}$ of elements, together with a group *operation* denoted by $\cdot$ which combines two elements a and b of $\mathbb{S}$ to form $a \cdot b$, such that four axioms hold. The *closure* axiom requires that if $a, b \in \mathbb{S}$, then $a \cdot b \in \mathbb{S}$. The second axiom requires the existence of an identity element e such that $a \cdot e = e \cdot a = a$ for $a \in \mathbb{S}$. The third axiom requires that each $a \in \mathbb{S}$ must have an inverse $a^{-1} \in \mathbb{S}$ such that $a \cdot a^{-1} = a^{-1} \cdot a = e$. The last axiom is the *associativity* requirement that if $a, b, c \in \mathbb{S}$, then $(a \cdot b) \cdot c = a \cdot (b \cdot c)$. A simple example of a group is obtained by letting $\mathbb{S}$ be the set of all integers, and letting $a \cdot b$ be the sum $a + b$.

If the set of transformations satisfies all the requirements of a group except for the existence of an inverse, the set is called a *semigroup*. Kadanoff's block renormalization transformations form a semigroup, but not a group, since there is no way to recover the set of original spins in a block that has been condensed into a single spin, just as the average of a set of numbers cannot be used to recover the set of numbers being averaged.

The notions of scaling functions, universality, and renormalization were discussed in [Stanley 99], which applied renormalization to find the critical exponent for percolation on an infinite one-dimensional lattice that extends from $-\infty$ to $+\infty$. Suppose a site on the lattice is occupied with probability p and unoccupied with probability $1 - p$. Define a *cluster* to be a set of lattice sites extending from position i to position j such that lattice site k is occupied for $i \leq k \leq j$, but sites $i - 1$ and $j + 1$ are unoccupied. The size of the cluster is $j - i + 1$. As p increases from 0, more clusters will form, and the *characteristic size* $\xi(p)$ of the clusters increases monotonically. When critical value p_c is reached a cluster of infinite size appears. The characteristic size $\xi(p)$ follows the scaling $\xi(p) \sim |p - p_c|^{-\nu}$ and the goal is to determine ν.

Applying Kadanoff's method, a group of s adjacent lattice sites is collapsed into a single super-site with occupation probability p'. Writing $p' = R_s(p)$, the function $R_s(p)$ is termed a "renormalization transformation". For this one-dimensional lattice, a super-site is occupied only if all s lattice sites collapsed into the super-site are occupied (for otherwise, the cluster would be broken). Since the probability of s sites being occupied is p^s then $R_s(p) = p^s$. The fixed points of this transformation are $p = 0$ and $p = 1$. The new correlation length $\xi'(p')$ is smaller than the original correlation length by a factor of s, that is, $\xi'(p') = s^{-1}\xi(p)$. If $p = 1$, then $\xi(p) = \infty$ and the relation $\xi'(p') = s^{-1}\xi(p)$ implies that $\xi'(p') = \infty$. These relations can be used [Stanley 99] to show that $\nu = 1$. In [Kadanoff 90], the renormalization method was used to compute the singularity spectrum $f(\alpha)$ for the multifractal Cantor set studied in Sect. 16.1.

21.2   Renormalizing a Small-World Network

In [Newman 99b], renormalization was applied to a slight modification of the Watts–Strogatz re-wiring model (Sect. 2.2) to determine the critical value of the exponent associated with the probability p of re-wiring an arc in the lattice. Our reason for presenting this model is not a concern with this critical exponent value, but rather to show how renormalization can be applied to networks. In the Newman–Watts model, N nodes are arranged in a ring. Each node is connected to its $2k$ closest neighbors, so each node has degree $2k$, as illustrated for $k = 1$ in Fig. 21.2. (This definition of k is different than the definition of k in the Watts–Strogatz [Watts 98] model presented in Sect. 2.2, where each node is assumed to have degree k.) Since each node has degree $2k$, there are kN arcs. For a given $p \in (0, 1)$, we create pkN shortcut arcs. The endpoints of each shortcut arc are randomly chosen, but distinct, nodes. To create the shortcuts, consider a coin that shows "heads" with probability p and "tails" with probability $1 - p$. We toss this coin kN times. For each toss, if we get "heads" we build a shortcut arc between two distinct randomly selected nodes;

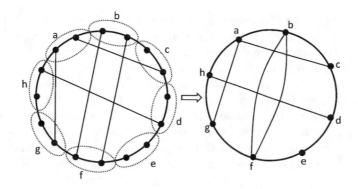

Figure 21.2: Renormalizing the small-world model for $k = 1$

if we get "tails" we do not build a shortcut. The pkN shortcut arcs are added to the network but none of the original arcs are removed; this prevents the network from becoming disconnected.

The quantity of interest is L, the average length of a shortest path in the network, where the average is over all pairs of nodes and over all possible realizations of the coin tosses. There are two regimes of interest. The first regime is for N sufficiently small so that the expected number pkN of shortcuts satisfies $pkN \ll 1$. In this regime, L is governed by the structure of the original lattice (without random connections) and we expect L to be proportional to N. The second regime is for N sufficiently large so that $pkN > 1$. In this regime the results of Watts and Strogatz suggest that L scales as $\log N$ [Newman 99b].

For a given p, assume that the transition point between these two regimes occurs when the number of nodes reaches a critical value N_c for which the expected number pkN of shortcuts is 1. Thus N_c satisfies $N_c \approx (pk)^{-1}$, so $N_c \sim p^{-1}$. However, other researchers (whose worked was cited in [Newman 99b]) have suggested that N_c scales as $p^{-\tau}$ for $\tau \neq 1$. Renormalization analysis can determine if $\tau = 1$ or if the other researchers are right. The renormalization arguments are slightly different for $k = 1$ and for $k > 1$.

Consider first the case $k = 1$. Since, by assumption, L is proportional to N for $N \ll N_c$ and L scales logarithmically with N for $N \gg N_c$, then L should satisfy the scaling law

$$L = N f(N/N_c) \,, \tag{21.1}$$

where $f(\cdot)$ is a "universal scaling function" [Newman 99b] satisfying

$$f(x) \sim \begin{cases} \text{constant} & \text{for } x \ll 1 \\ \log(x)/x & \text{for } x \gg 1 \,. \end{cases} \tag{21.2}$$

Assume that $N_c \sim p^{-\tau}$, where τ is unknown. Substituting this into (21.1) yields, for $N \gg 1$,

$$L = Nf(N/p^{-\tau}) = Nf(p^\tau N) = \frac{N \log(p^\tau N)}{p^\tau N} = \frac{\log(p^\tau N)}{p^\tau} \,. \tag{21.3}$$

Now we tamper with the network. Assume that N is an even number. We create $N/2$ "supernodes", where a supernode is obtained by grouping two nodes adjacent on the original ring lattice. If the nodes are sequentially numbered $1, 2, \ldots, N$ as we circle the ring, then supernode i contains nodes $2i - 1$ and $2i$ for $i = 1, 2, \ldots, N/2$. Having created the $N/2$ supernodes, we create "superarcs" as follows. Supernodes i and j are connected by a superarc if there is a shortcut connecting either of the two nodes comprising i with either of the two nodes comprising j. Supernodes and superarcs are illustrated in Fig. 21.2 (right), where the 16 nodes on the ring on the left are grouped into 9 supernodes on the ring on the right. The two arcs emanating from the a cluster on the left yield the two superarcs (a, c) and (a, g) on the right. The two arcs on the left from the b cluster to the f cluster yield the two superarcs between the supernodes b and f on the right.

The renormalized network has $\widetilde{N} = N/2$ nodes and the same number of shortcuts as the original network. Let $\widetilde{p}$ be the probability of a coin toss resulting in the creation of a shortcut in the renormalized network. Since $k = 1$, we require $\widetilde{p}\widetilde{N} = pkN = pN$. Since $\widetilde{N} = N/2$ then $\widetilde{p} = 2p$. Now consider how the shortest paths (as measured by hop count) change under renormalization. The number of cases for which the geometry of the shortest path changes is negligible for large N and small p [Newman 99b]. The length of a shortest path is, on average, halved along those portions which run around the perimeter of the ring (since the number of arcs on the ring is halved by renormalization). For those portions of a shortest path using shortcuts, the length remains the same, but for large N and small p the portion of the total shortest path length along the shortcuts tends to zero and thus can be neglected. Thus the average path length $\widetilde{L}$ for the renormalized network satisfies, for large N and small p, the relation $\widetilde{L} = L/2$. Thus for $N \gg 1$ the renormalization group equations are

$$\widetilde{N} = N/2 \qquad \widetilde{p} = 2p \qquad \widetilde{L} = L/2 \, .$$

Substituting these in (21.1) yields, for the renormalized network,

$$\widetilde{L} = \widetilde{N} f\!\left(\widetilde{p}^{\,\tau} \widetilde{N}\right) = \frac{\widetilde{N} \log\!\left(\widetilde{p}^{\,\tau} \widetilde{N}\right)}{\widetilde{p}^{\,\tau} \widetilde{N}} = \frac{\log\!\left(\widetilde{p}^{\,\tau} \widetilde{N}\right)}{\widetilde{p}^{\,\tau}} = \frac{\log\!\left((2p)^{\tau} (N/2)\right)}{(2p)^{\tau}} \, . \qquad (21.4)$$

Since $\widetilde{L} = L/2$, from (21.3) and (21.4) we obtain

$$\frac{\log\!\left((2p)^{\tau} (N/2)\right)}{(2p)^{\tau}} = \left(\frac{1}{2}\right) \frac{\log(p^{\tau} N)}{p^{\tau}} \, .$$

This simplifies to

$$\frac{\log\!\left(p^{\tau} N\right)}{\log\!\left(p^{\tau} N 2^{\tau-1}\right)} = 2^{1-\tau} \, ,$$

which has the solution $\tau = 1$.

The above analysis treated the case $k = 1$. For $k > 1$, renormalization groups k adjacent nodes, and the renormalization equations are [Newman 99b]

$$\widetilde{L} = L/k \qquad \widetilde{p} = k^2 p \qquad \widetilde{L} = L.$$

The renormalized network has $k = 1$, so the previous analysis for $k = 1$ applies. Finally, the model can be generalized to an E-dimensional rectilinear lattice. If each node has connections only to its nearest neighbors, then renormalization analysis yields $\tau = 1/E$ for all values of k; for the 1-dimensional ring model studied above we have $\tau = 1$ for all k.

21.3 Renormalizing Unweighted, Undirected Networks

In the previous section, renormalization was used to determine a critical exponent in a highly structured small-world model. Renormalization can also be applied to arbitrary unweighted, undirected networks, to uncover many network characteristics. In [Ichikawa 08], a "two-site scaling method" was used to study how the degree distribution changes under rescaling. The method begins with a pre-processing step: for $n \in \mathbb{N}$, if n has two or more leaf nodes, then remove all leaf nodes of n. After pre-processing, randomly select a pair of adjacent nodes $u, v \in \mathbb{N}$, where neither u nor v is a leaf node, and replace u and v, and the arc connecting them, by a single node w. If either u or v has an arc to a node $n \in \mathbb{N}$, then create an arc from w to n. While [Ichikawa 08] did not specify how many random pairs of nodes to select, it was stated that the number of non-leaf nodes is approximately halved. The last step is to add back to the collapsed network half of the number of leaf nodes.

We illustrate this procedure using Fig. 21.3 (i), which has 16 nodes. In the pre-processing step, we remove nodes 11 and 12 (both adjacent to 10), and nodes 15 and 16 (both adjacent to 14). Node 1 cannot be removed since it is the only leaf node subtending node 2. Now we are ready to combine adjacent non-leaf nodes. *Pair 1*: Combine nodes 2 and 3 and create A, which is connected to 1, 4, 8, and 9. *Pair 2*: Combine nodes 4 and 5 and create B which is connected to A, 9, 6, and 7. *Pair 3*: Combine nodes 6 and 7 (which now are both adjacent to B) and create C which is connected to B. *Pair 4*: Combine 8 and 9 and create node D which is connected to A, B, and 10. *Pair 5*: Combine 13 and 14 and create F which is connected to B, 10, 15, and 16. No more nodes from the original network can be combined. Finally, we add back half the leaf nodes removed, so one leaf node is added to node 10 and one leaf node is added to F. The final renormalized network is shown in Fig. 21.3 (ii).

The reason for not considering leaf nodes when collapsing adjacent nodes into a supernode is that if $\mathbb{G}$ has many leaf nodes, then randomly selected pairs of nodes tend to be pairs consisting of a leaf node and its adjacent node, so contracting these two nodes amounts to removing the leaf node. Wilson's renormalization theory calls for scaling (i.e., coarse graining) homogeneously over the network, and combining a leaf node and its adjacent node (by removing the leaf node) makes the scaling non-homogeneous. Hence only non-leaf

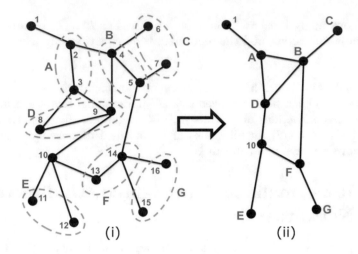

Figure 21.3: Two-site scaling

nodes were coarse grained in [Ichikawa 08]. This method resembles the block spin renormalization (described above) used in condensed matter physics. As observed in [Ichikawa 08], a box-counting method (e.g., [Song 05]) could be utilized to accomplish coarse graining, where all nodes inside a box are collapsed into a single supernode. The advantage of two-site scaling, with the above restriction that only non-leaf nodes are coarse grained, is that the size of each box to be coarse grained is 2, which minimizes the loss of the fine structure of the network [Ichikawa 08].

One goal of [Ichikawa 08] was to compare the node degree distribution before and after renormalization. For example, if $\mathbb{G}$ is scale-free, so that the probability p_k that a node of $\mathbb{G}$ has degree k follows the scaling $p_k \sim k^{-\lambda}$, is the renormalized network also scale-free with the same exponent? Letting $\mathbb{G}_1$ be the renormalized network, we can use the same two-site scaling method to renormalize $\mathbb{G}_1$, obtaining $\mathbb{G}_2$, and study the degree distribution of $\mathbb{G}_2$. If the original network is sufficiently large, we can continue in this manner, renormalizing $\mathbb{G}_2$ to obtain $\mathbb{G}_3$, etc. In [Ichikawa 08], four iterations of renormalization were applied to five networks: a network constructed by the Barabási–Albert preferential attachment scheme (Sect. 2.4), the router level Internet topology, the Autonomous System level Internet topology, an actor's co-starring network, and a protein–protein interaction network. For the Barabási–Albert network, renormalization changed the original power-law distribution to a bimodal distribution, while for the other four "naturally occurring" networks, renormalization preserved the node degree power-law distribution, and the exponent remained relatively constant under renormalization. Although [Ichikawa 08] contrasted their renormalization method to the box-counting method of [Song 07], they did not suggest that their two-site renormalization technique be used for the computation of a fractal dimension of a network. Rather, their renormalization method was intended to extract information, such as critical exponents, for use in classifying networks and uncovering universality classes (Sect. 21.5).

Renormalization was used in [Gallos 13] to study the fractal and modular properties of a subset of the IMDB movie database. To renormalize $\mathbb{G}$, first cover $\mathbb{G}$ with a minimal number of diameter-based boxes of diameter s, or with a minimal number of radius-based boxes of radius r. (As discussed in Sects. 15.4 and 17.1, in general there is not a unique minimal covering, as illustrated by Fig. 21.4, which shows two different diameter-based minimal 3-coverings of the same network.) Define $\mathbb{N}_0 = \mathbb{N}$ and $\mathbb{A}_0 = \mathbb{A}$ and use the minimal covering to create the new graph $\mathbb{G}_1 = (\mathbb{N}_1, \mathbb{A}_1)$ by executing procedure *Network Renormalization*$(\mathbb{N}_0, \mathbb{A}_0)$, specified in Fig. 21.5.

Having created $\mathbb{G}_1$, we can compute statistics for this graph, such as the degree distribution, the maximal and average degree, and various fractal dimensions, e.g., d_B or d_I. We now apply the same renormalization procedure to $\mathbb{G}_1$ by first computing a minimal covering of $\mathbb{G}_1$. If diameter-based boxes of diameter s were used to create $\mathbb{G}_1$, use diameter-based boxes of diameter s to cover $\mathbb{G}_1$. If radius-based boxes of radius r were used to create $\mathbb{G}_1$, use radius-based boxes of radius r to cover $\mathbb{G}_1$. Using this covering, create the new graph

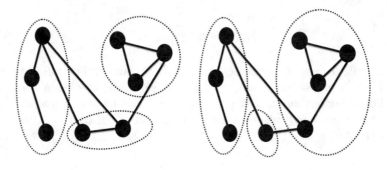

Figure 21.4: Two different minimal 3-coverings of the same network

---

procedure *Network Renormalization*$(\mathbb{N}_t, \mathbb{A}_t)$

1   **initialize:** $\mathbb{N}_{t+1} = \emptyset$ and $\mathbb{A}_{t+1} = \emptyset$;

2   **for** each box in the covering of $(\mathbb{N}_t, \mathbb{A}_t)$:

3   replace the box by the supernode n;

4   add n to $\mathbb{N}_{t+1}$;

5   **for** $a \in \mathbb{A}_t$:

6   let $n_1 \in \mathbb{N}_t$ and $n_2 \in \mathbb{N}_t$ be the endpoints of a;

7   **if** n_1 is in the box replaced by supernode $x \in \mathbb{N}_{t+1}$

            and n_2 is in the box replaced by supernode $y \in \mathbb{N}_{t+1}$:

8   **if** $x \neq y$ and $(x, y) \notin \mathbb{A}_{t+1}$:

9   create arc (x, y) and add (x, y) to $\mathbb{A}_{t+1}$;

10   **return** $(\mathbb{N}_{t+1}, \mathbb{A}_{t+1})$.

Figure 21.5: Procedure *Network Renormalization*

$\mathbb{G}_2 = (\mathbb{N}_2, \mathbb{A}_2)$ by executing procedure *Network Renormalization*$(\mathbb{N}_1, \mathbb{A}_1)$. Compute statistics (e.g., d_B) for $\mathbb{G}_2$, and then apply renormalization to $\mathbb{G}_2$, stopping this iterative process when renormalization has collapsed the original network $\mathbb{G}$ to a single point. This is illustrated by Fig. 21.6, adapted from [Gallos 07b]. Rows 1, 2, and 3 in Fig. 21.6 show renormalization using diameter-based boxes of size s, for $s = 2$, 3, and 5, respectively.

21.4   Renormalization of Scale-Free Networks

This section concerns the application of renormalization to a scale-free network, so assume for the entirety of this section that $\mathbb{G} = (\mathbb{N}, \mathbb{A})$ is scale-free, so the probability p_k that a node of $\mathbb{G}$ has degree k follows the scaling $p_k \sim k^{-\lambda}$. For a given box size s, let $\mathcal{B}(s)$ be a minimal s-covering of $\mathbb{G}$. Let $P_m(s)$ be the probability that a box in $\mathcal{B}(s)$ has mass m (the mass of a box is the number of nodes in the box). Then for some positive number $\eta(s)$ depending on s we have [Kim 07b, Song 05]

$$P_m(s) = m^{-\eta(s)}. \tag{21.5}$$

Suppose that, for this $\mathcal{B}(s)$, we renormalize $\mathbb{G}$ by one application of procedure *Network Renormalization*$(\mathbb{N}, \mathbb{A})$, specified in Fig. 21.5, to obtain the network $\mathbb{G}'(s)$. Following the notation of [Kim 07b], let k denote the degree of a node in $\mathbb{G}$ and let k' denote the degree of a node in $\mathbb{G}'$. Let $\langle N_j(s) \rangle$ denote the average mass of the boxes in $\mathcal{B}(s)$, so $\langle N_j(s) \rangle = (1/B(s)) \sum_{B_j \in \mathcal{B}(s)} N_j$. The average

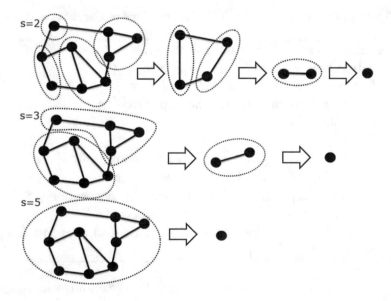

Figure 21.6: Network renormalization with diameter-based boxes

node degree $\langle k'(s) \rangle$ of $\mathbb{G}'(s)$ and the average box mass $\langle N_j(s) \rangle$ were found in [Kim 07b] to be related by the power law

$$\langle k'(s) \rangle \sim \langle N_j(s) \rangle^{\theta(s)} \tag{21.6}$$

for some exponent $\theta(s)$. For example, for the network model in [Goh 06] we have $\theta(2) \approx 1.0$ and $\theta(3) \approx 0.8$ and $\theta(5) \approx 0.6$ [Kim 07b]. Let $\lambda'(s)$ be the node degree exponent of $\mathbb{G}'(s)$, so the probability that a node of $\mathbb{G}'(s)$ has degree k' scales as $(k')^{-\lambda'(s)}$.

From (21.5) and (21.6) we have [Kim 07b]

$$\lambda'(s) = \begin{cases} \lambda & \text{for } \lambda \leq \eta(s) \\ 1 + \big(\eta(s) - 1\big)/\theta(s) & \text{for } \lambda > \eta(s). \end{cases} \tag{21.7}$$

Recalling that $\mathbb{G}$ is assumed in this section to be scale-free, and recalling (21.7), the following definition of self-similarity was given in [Kim 07b].

Definition 21.1 $\mathbb{G}$ is self-similar if $\lambda'(s) = \lambda$ holds for different values of s, or for successive applications of renormalization with a fixed s. $\qquad \square$

Recalling that $\eta(s)$ is defined by (21.5), from Definition 21.1 and (21.7), $\mathbb{G}$ is self-similar if

$$\theta(s) = \big(\eta(s) - 1\big)/(\lambda - 1). \tag{21.8}$$

Equation (21.8) can hold even if $\mathbb{G}$ is not fractal, i.e., even if $\mathbb{G}$ does not have a finite d_B. For example, the Internet at the autonomous system level is known to be both scale-free and non-fractal, yet it exhibits self-similarity [Kim 07b]. Thus, for scale-free networks, fractality (a finite d_B) and self-similarity (invariance of the degree distribution under renormalization) are not equivalent. Emphasizing this point, [Gallos 07b] observed that "Surprisingly, this self-similarity under different length scales seems to be a more general feature that also applies in non-fractal networks such as the Internet. Although the traditional fractal theory does not distinguish between fractality and self-similarity, in complex networks these two properties can be considered to be distinct."[5]

In Sect. 8.6 we discussed why we must allow disconnected boxes when applying box counting to $\mathbb{G}$: we showed that counting each spoke connected to a hub as a separate box creates too many boxes, so *Random Sequential Node Burning* corrects for this phenomenon by allowing nodes in a box to be disconnected within that box. This yields a power-law scaling for the WWW, which is scale-free. More insight into scale-free networks was provided in [Song 05]. Assume now that not only is $\mathbb{G}$ scale-free, but also that $\mathbb{G}$ has a finite d_B (this holds, e.g., for the WWW). Let $\langle N_j(s) \rangle_B$ be the average mass of the boxes in $\mathcal{B}(s)$. Since by assumption $\mathbb{G}$ follows the scaling law $B(s) \sim s^{-d_B}$ over some range of s, then the average box mass $\langle N_j(s) \rangle_B \equiv N/B(s)$ of the boxes in $\mathcal{B}(s)$ follows the scaling law

$$\langle N_j(s) \rangle_B \sim s^{d_B}. \tag{21.9}$$

---

[5] There is recognition that fractality and self-similarity are not equivalent even for geometric objects, as noted in [Bez 11] and discussed in Sect. 6.1.

Recall from (11.6) that the correlation sum $C(r)$ is the average, over all N nodes, of the fraction of nodes, distinct from a randomly chosen node n, whose distance from n is less than r.[6] Define $\langle N_j(r)\rangle_C \equiv NC(r)$, so $\langle N_j(r)\rangle_C$ is the average box mass obtained from the Grassberger–Procaccia method or from the sandbox method. If the correlation dimension d_C of $\mathbb{G}$ were to exist, then over some range of s we would have the power law

$$\langle N_j(r)\rangle_C \sim r^{d_C} . \tag{21.10}$$

However, empirical results for the WWW reveal instead the exponential scaling

$$\langle N_j(r)\rangle_C \sim e^{r/r_0} , \tag{21.11}$$

where r_0 is a characteristic length ($r_0 \approx 0.78$ for the WWW). The reason why cluster growing for the WWW yields the exponential scaling (21.11) and not the power law (21.10) was provided in [Song 05]:

> The topology of scale-free networks is dominated by several highly connected hubs ... implying that most of the nodes are connected to the hubs via one or very few steps. Therefore the average performed in the cluster growing method is biased; the hubs are overrepresented in (21.11) since almost every node is a neighbor of a hub. By choosing the seed of the clusters at random, there is a very large probability of including the hubs in the clusters. On the other hand the box covering method is a global tiling of the system providing a flat average over all the nodes, i.e., each part of the network is covered with an equal probability. Once a hub (or any node) is covered, it cannot be covered again. We conclude that (21.10) and (21.9) are not equivalent for inhomogeneous networks with topologies dominated by hubs with a large degree.

For a *homogeneous* network characterized by a narrow degree distribution, the box-covering method and the cluster growing method are equivalent since each node typically has the same number of neighbors; for such networks, $d_B = d_C$. Scale-free networks, which do not have a narrow degree distribution, enjoy the small-world property [Cohen 03], and for small-world networks (21.11) holds and there is no finite d_C.

The final topic in this section on scale-free networks is a formula derived in [Song 05] that relates d_B and the power-law degree distribution exponent λ of $\mathbb{G}$. Fix the box size s, let $\mathcal{B}(s)$ be a minimal s-covering of $\mathbb{G}$, and let $\mathbb{G}'(s)$ be the renormalized network. Since each box in $\mathcal{B}(s)$ is collapsed into a supernode, the number of supernodes is $B(s)$, so $B(s)$ is also the number of nodes in $\mathbb{G}'(s)$. It was shown in [Song 05] that

$$B(s) = s^{\lambda-1}N . $$

---

[6] If we utilize the sandbox method, then the average is over a random sample of nodes, rather than over all nodes, and we obtain $\widetilde{C}(r)$, as defined by (11.15). In [Song 05], the sandbox method was called the "cluster growing method".

Corresponding to box $B_j \in \mathcal{B}(s)$ is a node $j'(s)$ of $\mathbb{G}'(s)$; let $\delta_{j'(s)}$ be the degree (with respect to the renormalized network $\mathbb{G}'(s)$) of node $j'(s)$. Let $A_j(s)$ be the number of arcs in box B_j. Empirical results showed that for some exponent β, where $\beta > 1$, we have

$$\delta_{j'(s)} \sim s^{-\beta} A_j(s).$$ (21.12)

Values of β were empirically determined in [Song 05] for several networks: 2.5 for the WWW, 5.3 for the actor network, 2.1 for the *E. coli* protein network, and 2.2 for the *H. sapiens* protein network. It was also shown in [Song 05] that

$$\lambda = 1 + \frac{d_B}{\beta}.$$ (21.13)

This equation, which was empirically shown to hold for all the networks analyzed in [Song 05], implies that a scale-free network, rather than being characterized by the degree power-law exponent λ, can be characterized by the "more fundamental" relation (21.13).[7] Equation (21.13) was applied in [Gallos 12a] in a study of functional networks in the human brain.

The analysis and results of [Song 05] were critiqued [Strogatz 05] in a commentary in *Nature*. Strogatz wrote

> These results raise some puzzles of interpretation. First, what do the boxes really mean? Although it's tempting to think of them as something like modules, communities or network motifs, the boxes have no known functional significance, raising the possibility that the observed self-similarity might contain an element of numerology. On the other hand, some real networks ... lack the self-similarity reported by Song et al. indicating that such scaling is not automatic and, consequently, that it can [be] used as a benchmark for testing models of network structure. Second, it's odd that networks should find themselves configured as fractals. In statistical physics, power laws and self-similarity are associated with phase transitions—with systems teetering on the brink between order and chaos. Why do so many of nature's networks live on a razor's edge? Have they self-organized to reach this critical perhaps to optimize some aspect of their performance, or have they merely followed one of the manifold paths to power-law scaling, full of sound and fury, signifying nothing?

---

[7] In [Song 05], the exponent β was denoted by d_k, which is a constant, not a function of k. Also, Eq. (21.12) was written as $k' = s(l_B)k$, where l_B is the box size, $s(l_B) \sim (l_B)^{-d_k}$, k is the number of arcs in box B_j in the covering of $\mathbb{G}$, and k' is the degree of the node in the renormalized network representing B_j. Thus $\delta_{j'(s)}$ in (21.13) corresponds to k' in [Song 05], s in (21.13) corresponds to l_B in [Song 05], β in (21.13) corresponds to d_k in [Song 05], and $A_j(s)$ in (21.13) corresponds to k in [Song 05].

21.5   Universality Classes for Networks

In statistical mechanics, the Hamiltonian function $\mathcal{H}$, which describes the energetics of a system, is used to study the equilibrium properties of the system [Kadanoff 13]. By repeatedly applying renormalization to the system, we end up with a fixed point representing a multi-dimensional continuum of different possible Hamiltonians. These Hamiltonians form a *universality class*, where each Hamiltonian in a given universality class has identical critical properties. There are many different universality classes depending on, e.g., the dimensionality of the problem and the symmetries. For a given universality class, there is a unique fixed point such that the result of repeatedly applying renormalization to any system in that class yields this fixed point. The result of repeatedly applying renormalizing is known as the *renormalization flow*, so we can refer to the renormalization flow of the system. If $\mathcal{R}$ is the renormalization operator, then applying it to the original system $\mathbb{X}$ yields $\mathbb{X}_1 = \mathcal{R}(\mathbb{X})$; applying $\mathcal{R}$ to $\mathbb{X}_1$ yields $\mathbb{X}_2 = \mathcal{R}(\mathbb{X}_1) = \mathcal{R}^2(\mathbb{X})$, and in general $\mathbb{X}_t = \mathcal{R}(\mathbb{X}_{t-1}) = \mathcal{R}^t(\mathbb{X})$.

Since renormalization in statistical mechanics yields universality classes, and since renormalization can be applied to networks, it is natural to ask if successive renormalizations of networks generate fixed points that correspond to a universality class. This question was considered in [Radicchi 09], which determined empirically that universality classes do exist for networks. Self-similarity of the node degree distribution was also studied in [Radicchi 09], using Definition 21.2 below; this definition of network self-similarity differs from Definition 21.1, which does not require the existence of d_B. Recall that a network is *fractal* if it has a finite d_B.

Definition 21.2 [Radicchi 09] $\mathbb{G}$ is self-similar if (i) $\mathbb{G}$ is fractal and (ii) the network $\mathbb{G}'$ obtained by applying Procedure *Network Renormalization* with a minimal s-covering $\mathcal{B}(s)$ is fractal, with the same degree distribution as $\mathbb{G}$, and this continues to hold under successive applications of renormalization.   □

Let $\mathbb{G}_t$ denote the network after t applications of renormalization, where $\mathbb{G}_0 = \mathbb{G}$. We write $\mathbb{G}_t = (\mathbb{N}_t, \mathbb{A}_t)$. Let $N_t \equiv |\mathbb{N}_t|$ be the number of nodes in $\mathbb{G}_t$ and $A_t \equiv |\mathbb{A}_t|$ be the number of arcs in $\mathbb{G}_t$. Let $\delta_t^\star$ be the largest node degree in $\mathbb{G}_t$. The first of the four statistical measures considered in [Radicchi 09] is the cluster coefficient C_t (Sect. 1.2) of $\mathbb{G}_t$. The second and third statistics are

$$\kappa_t \equiv \delta_t^\star/(N_t - 1) \tag{21.14}$$

and

$$\eta_t \equiv A_t/(N_t - 1). \tag{21.15}$$

Since $2A_t$ is the sum of the node degrees of $\mathbb{G}_t$, then η_t is approximately half the average node degree of $\mathbb{G}_t$. The reason for dividing by $N_t - 1$ in (21.15) is that a tree with N nodes has $N - 1$ arcs, so $\eta = 1$. The fourth statistic (called the *susceptibility* in [Radicchi 09]) measures the fluctuations (variation) of κ_t:

$$\chi_t \equiv N_0 \left(\langle \kappa_t^2 \rangle - \langle \kappa_t \rangle^2 \right),$$

where $\langle \kappa_t \rangle$ denotes the average taken over different realizations of box counting (for a given network), and similarly for $\langle \kappa_t^2 \rangle$. These statistics were plotted in [Radicchi 09] as a function of the relative network size $x_t = N_t/N_0$. Let $x^\star = \lim_{t \to \infty} x_t$, so $x^\star = 0$ means the successive renormalizations collapse the network to a single point. Both computer-generated and real-world networks were studied in [Radicchi 09]. Two different renormalization transformations $\mathcal{R}$ were used and compared: *Greedy Coloring* (Sect. 8.1) and *Random Sequential Node Burning* (Sect. 8.3); they found greedy coloring to be "much more effective" than *Random Sequential Node Burning*.

Each computer-generated network studied was found to satisfy

$$\kappa_t = F_\kappa \left((x_t - x^\star) N_0^{1/\nu} \right) \tag{21.16}$$

for some function F_κ which depends on the starting network $\mathbb{G}_0$ and the renormalization $\mathcal{R}$. Similarly,

$$\eta_t = F_\eta \left((x_t - x^\star) N_0^{1/\nu} \right) \text{ and } C_t = F_C \left((x_t - x^\star) N_0^{1/\nu} \right) \tag{21.17}$$

for other functions F_η and F_C which depend on $\mathbb{G}_0$ and $\mathcal{R}$. However, for the susceptibility χ_t a functionally different relation holds:

$$\chi_t = N_0^{\gamma/\nu} F_\chi \left((x_t - x^\star) N_0^{1/\nu} \right) \tag{21.18}$$

for some function F_χ depending on $\mathbb{G}_0$ and $\mathcal{R}$.

For non-self-similar networks, for the final size of the networks they obtained $x^\star = 0$ except for two cases where $x^\star > 0$: when radius-based boxes with $r = 1$ are used (recall from Sect. 7.2 that such boxes consist of a center node and nodes adjacent to this node), or when diameter-based boxes of size $s = 2$ are used (recall from Sect. 7.2 that such boxes consist of a set of nodes that are totally interconnected). For these two cases, the rate of contraction of the network was much slower than if higher values of r or s were used. The fact that $x^\star > 0$ when $r = 1$ or $s = 2$ implies that $\{\mathbb{G}_t\}$ converges to a special stable network $\widetilde{\mathbb{G}}$. The network $\widetilde{\mathbb{G}}$ was obtained after a number t of renormalizations which is logarithmic in N_0, the initial network size. The fixed point $\widetilde{\mathbb{G}}$ has the same topology (a few hubs connected to a large fraction of the nodes) regardless of the topology of $\mathbb{G}_0$.

Summarizing the results of [Radicchi 09], they found that in general the scaling exponent ν does not depend on the particular transformation $\mathcal{R}$ used to renormalize the network, but does depend on the starting network $\mathbb{G}_0$. In particular, for any non-self-similar $\mathbb{G}$ they always obtained $\nu = 2$. For any self-similar $\mathbb{G}$ they always obtained a value $\nu \neq 2$ depending on $\mathbb{G}$ (each self-similar $\mathbb{G}$ is a fixed point of the renormalization flow since the properties of $\mathbb{G}$ are unchanged under repeated renormalization). However, self-similar networks are unstable fixed points of the renormalization transformation, since minimal random perturbations are able to lead the flow out of these fixed points.

The renormalization flow was also studied in [Radicchi 09] for two real-world networks: an actor network from IMDB, and the network of web pages of the domain of the University of Notre Dame. They found that the usual simple measures (e.g., node degree distributions and cluster coefficients) of the network structure, taken after a few iterations of renormalization, reveal that the transformation modifies the topological properties of the network. Furthermore, the repeated renormalization of real networks produced flows converging to the same fixed point structure, when it exists, as the one found in the case of computer-generated non-self-similar networks.

Several studies subsequent to [Radicchi 09] have provided additional insight into the behavior of networks under renormalization. We briefly describe two of these later studies.

The successive renormalization described in Sect. 21.3 requires, in each generation t, computing a minimal s-covering of $\mathbb{G}_t$. A variant of this scheme is *random sequential renormalization* [Son 11]. In each step of random sequential renormalization a "local coarse graining" is performed: a random node n is selected, and all nodes within distance r of n are replaced by a single supernode. Exact combinatorial analysis for one-dimensional networks (whose nodes lie on a line) shows the existence of scaling laws for the cluster mass distribution, and the exponents in the scaling laws depend on r. As observed in [Son 11], "Consequently, and somewhat counter-intuitively, self-similarity observed in random sequential renormalization and similar network renormalization schemes cannot be used to prove that complex networks are themselves self-similar". An analysis in two-dimensions [Christensen 12] used *agglomerative percolation*. In contrast to ordinary percolation in a lattice, in which arcs are added at random, in agglomerative percolation a cluster $\mathcal{C}$ is selected, and arcs from the surface of $\mathcal{C}$ are added to connect $\mathcal{C}$ to all adjacent clusters. One implementation of agglomerative percolation selects, at each time t, a random cluster; another implementation preferentially selects a cluster with a higher mass. Starting with the state where each cluster has size one, the process repeats until the entire lattice (or graph) contains only a single cluster. Agglomerative percolation corresponds to random sequential renormalization of a network. Agglomerative percolation was studied in [Christensen 12] for both square rectilinear lattices and triangular lattices. If the implementation preferentially selects a larger cluster for the local coarse-graining step, the behavior is completely different from ordinary percolation. If each cluster is equally likely to be selected for the local coarse-graining step, then for the rectilinear lattice, only a detailed numerical scaling analysis shows that agglomerative percolation is not in the ordinary percolation universality class; however, for the triangular lattice, agglomerative percolation shares critical exponents with ordinary percolation.

Chapter 22

Other Network Dimensions

Experience is never limited, and it is never complete; it is an immense sensibility, a kind of huge spider-web of the finest silken threads suspended in the chamber of consciousness, and catching every airborne particle in its tissue.
Henry James (1843–1916), American born novelist, author of "The Portrait of a Lady" and "Washington Square".

In this chapter we consider other definitions of the dimension of a network. Some of these are fractal dimensions, some are not. All of them are either more specialized (e.g., applying only to trees) or have received much less attention than the major network fractal dimensions we have considered in previous chapters.

22.1 Dimension of a Tree

A fractal dimension of tree networks was proposed in [Horton 45], well before the seminal book [Mandelbrot 67] on fractals. There are many examples of fractal trees in nature, including river networks, actual plants and trees, root systems, bronchial systems, and cardiovascular systems. Horton recognized scaling relationships in water streams. Consider the stream network in Fig. 22.1, taken from [Pidwirny 06]. Each stream is assigned an *order*. A stream with no upstream tributary is order 1. When two order-1 streams combine, they form

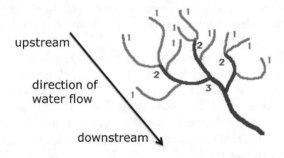

Figure 22.1: Horton's ordering of streams

© Springer Nature Switzerland AG 2020
E. Rosenberg, *Fractal Dimensions of Networks*,
https://doi.org/10.1007/978-3-030-43169-3_22

a stream of order 2. When two order-2 streams combine, they form a stream
of order 3, etc. The figure shows that an order-1 stream can also intersect an
order-2 stream, and an order-2 stream can intersect an order-3 stream. Let S_i
be the number of streams of order i, and let L_i be the average length of streams
of order i. Horton discovered that, for drainage networks, the ratio S_i/S_{i+1},
called the *bifurcation ratio*, is roughly independent of i. Similarly, the ratio
L_i/L_{i+1} is also roughly independent of i. Following [Pelletier 00], the fractal
dimension d of a stream is given by

$$d \equiv \frac{\log\left(S_i/S_{i+1}\right)}{\log\left(L_i/L_{i+1}\right)}. \tag{22.1}$$

Note that d does not relate the number of streams to the stream length; rather,
d relates two dimensionless ratios. We have $d = 1.67$ for the Kentucky River
basin in Kentucky, and $d = 1.77$ for the Powder River basin in Wyoming.

It was shown in 1997 that river networks are optimal networks for transport-
ing run-off with the minimal stream power exerted on the landscape. Because
the structures of leaves (of actual botanical trees, e.g., maple trees) and river
basins are statistically similar, we can infer that the tree structure for transport-
ing nutrients to and from cells in a leaf is also optimal [Pelletier 00]. In human
physiology, "the fractal geometry of the lungs may provide an optimal solution
to the problem of maximizing surface area for diffusion of oxygen and carbon
dioxide" [Goldberger 87]. For example, the fractal structure of the respiratory
system facilitates gas exchange by essentially converting a volume of dimension
three (e.g., blood in large vascular tubes and air in the upper respiratory tract)
into something approaching an alveolar-capillary surface area of dimension two.[1]

It was also observed in [Goldberger 87] that self-similarity and the absence
of a characteristic scale are found in a variety of seemingly unrelated biological
systems possessing a tree topology, e.g., actual trees, the vascular network, the
biliary network, the pancreatic ducts, and the uninary collection system. Accordingly, [Goldberger 87] suggested that, from a developmental viewpoint, the
self-similarity of such biological fractals serves to minimize "construction error"
as such systems are generated by adding the same structure on progressively
smaller scales.

Exercise 22.1 This section has discussed research defining a fractal dimension of river networks. Have these results been used to categorize physiological systems with a tree topology, such as the vascular networks discussed in
[Goldberger 87]?   □

Exercise 22.2 The cumulative intersection method (Sect. 20) was presented
in [Jelinek 98] as a method for the analysis of two-dimensional images. Yet
it can be considered a network method, since it is counting the number of
branches contained in a circle of a given radius. Formalize this method for
networks and use it to define a fractal dimension, which could be called the
branching dimension. Does this branching dimension relate to any other network
dimension studied in this book?   □

---

[1] The adjective "alveolar" means "pertaining to the air sacs where gases are exchanged
during the process of respiration". https://medical-dictionary.thefreedictionary.com/.

22.2 Area-Length Dimension

For a circle of radius r, the ratio of the perimeter P to the square root of the area A is $P/\sqrt{A} = 2\pi r/\sqrt{\pi r^2} = 2\sqrt{\pi}$, so $P \propto A^\alpha$, where $\alpha = 1/2$, and the "fractal dimension" d of the perimeter of a circle is $d = 2\alpha = 1$. For a square of side length s, the ratio of the perimeter P to the square root of the area A is $P/\sqrt{A} = 4s/\sqrt{s^2}$, so again we have $P \propto A^\alpha$, where $\alpha = 1/2$, and the "fractal dimension" of the perimeter of a square is $d = 2\alpha = 1$. For an equilateral triangle with side length s, the ratio of the perimeter P to the square root of the area A is $P/\sqrt{A} = 3s/\sqrt{s^2\sqrt{3}/4}$, so again we have $P \propto A^\alpha$, where $\alpha = 1/2$, and the "fractal dimension" of the perimeter of an equilateral triangle is $d = 2\alpha = 1$. More generally, for any regular polygon we have $P \propto A^\alpha$, and the "fractal dimension" of the perimeter of a regular polygon is $d = 2\alpha = 1$. Similarly, in the analysis of rivers, Hack in 1957 found that the length L of the longest river above a given location and its drainage area A are related by $L \propto A^{d/2}$, where $1 < d < 2.5$ ([Mandelbrot 83b], Chapter 12); we can call d a fractal dimension of the river.

In [Rodin 00], the relationship $L \propto A^{d/2}$ was used to compute a fractal dimension of the streets in each of six zones in Tokyo. Tokyo was studied since that city is constantly referred to as chaotic, and it is said that the original plan of the city was designed as a maze to confuse invading armies nearing the grounds of the Shogun's castle. For each of the six zones, the streets were ordered as follows. Designate a main street as order 0. The order-1 streets proceed from the order-0 street until the first crossing, dead-end, or branching. The order-2 streets proceed from the order-1 streets until the second crossing, dead-end, or branching, and this continues until the highest order streets have no tributaries. Let K be the highest order encountered in the zone. For $k = 1, 2, \ldots, K$, let $L(k)$ be the total length of the order-k streets in the zone, and define $L_k \equiv \sum_{i=1}^{k} L(i)$ so L_1 is the total length of order-1 streets, L_2 is the total length of order-1 and order-2 streets, and so on, until finally L_K is the length of all the streets of order k, for $k = 1, 2, \ldots, K$. (The distances are actual measured distances, and not hop counts.) For $k = 1, 2, \ldots, K$, let $A(k)$ be the area of that portion of the zone containing order-k streets, and define $A_k \equiv \sum_{i=1}^{k} A(i)$ so A_1 is the area containing the order-1 streets, A_2 is the area containing the order-1 and order-2 streets, and so on, until finally A_K is the area containing all the streets of order k, for $k = 1, 2, \ldots, K$, so A_K is the area of the entire zone. If the slope of the best linear fit to the plot of $\log L_k$ versus $\log A_k$ for $k = 1, 2, \ldots, K$ has slope α, then the fractal dimension d of the streets in the zone was defined in [Rodin 00] as $d = 2\alpha$. The fractal dimensions they computed for the six zones are 1.66, 2.4, 2.34, 2.54, 2.8, and 3.24. The zone with dimension 3.24 is the modern city center and the zone with dimension 2.8 has a historically "feudal" street structure, which is of great concern to the authorities due to the difficulty of navigating the streets if an earthquake occurs.

Exercise 22.3 The Tokyo study in [Rodin 00] suggests that an analytic model could be created, where x_1 order-1 streets, each of length c_1, branch off the main road, x_2 order-2 streets, each of length c_2, branch off each order-1 street, and so on. Are there reasonable assumptions that can be made on x_k and c_k, and on the area A_k covered by all the order-k streets, so that an estimate of the fractal dimension of the streets can be obtained?   □

22.3   Metric Dimension

The study of the *metric dimension* of a graph $\mathbb{G}$ was launched by Slater [Slater 75] and independently by Harary and Melter [Harary 76]. The metric dimension of $\mathbb{G}$ is a positive integer. Let $\mathbb{K} \subseteq \mathbb{N}$ be a subset of the nodes of $\mathbb{G}$ and define $K \equiv |\mathbb{K}|$. Denote the nodes in $\mathbb{K}$ by $n_1, n_2, \ldots, n_K$. For $x \in \mathbb{N}$, consider the ordered K-tuple

$$\bigl(dist(x, n_1), dist(x, n_2), \cdots, dist(x, n_K) \bigr),$$

where $dist(x, n_k)$ is the length of the shortest path in $\mathbb{G}$ from x to n_k. Denote this ordered K-tuple by $rep(x|\mathbb{K})$, and call it the *representation*[2] of x with respect to $\mathbb{K}$. The set $\mathbb{K}$ is a *resolving set* for $\mathbb{G}$ if $rep(u|\mathbb{K}) = rep(v|\mathbb{K})$ implies $u = v$ for all $u, v \in \mathbb{N}$; in [Garey 79] a resolving set was called a *metric basis*. Stated differently, $\mathbb{K}$ is a resolving set if all nodes in $\mathbb{N}$ have distinct representations with respect to $\mathbb{K}$. The elements of a resolving set are known as *landmarks*. The *metric dimension* of $\mathbb{G}$, which we denote by $met(\mathbb{G})$, is the minimal cardinality of a resolving set for $\mathbb{G}$.

Example 22.1 Consider the graph of Fig. 22.2[3] and assume that each arc cost is 1. Choosing $\mathbb{K}$ to be the ordered set $\{a, e\}$, we have $K = 2$ and $rep(a|\mathbb{K}) = (0, 1)$, $rep(b|\mathbb{K}) = (1, 1)$, $rep(c|\mathbb{K}) = (2, 1)$, $rep(d|\mathbb{K}) = (1, 1)$, and $rep(e|\mathbb{K}) = (1, 0)$. Since $rep(b|\mathbb{K}) = rep(d|\mathbb{K}) = (1, 1)$, this $\mathbb{K}$ is not a resolving set.

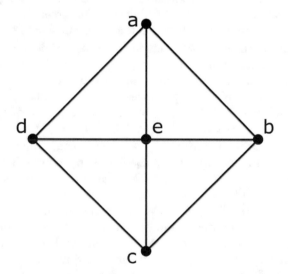

Figure 22.2: Metric dimension example

Now choose $\mathbb{K}$ to be the ordered set $\{a, b, e\}$, so $K = 3$. Then $rep(a|\mathbb{K}) = (0, 1, 1)$, $rep(b|\mathbb{K}) = (0, 1, 0)$, $rep(c|\mathbb{K}) = (2, 1, 1)$, $rep(d|\mathbb{K}) = (1, 2, 1)$, and $rep(e|\mathbb{K}) = (1, 1, 0)$. Since these five 3-tuples are unique, this $\mathbb{K}$ is a resolving set.

Finally, choose $\mathbb{K}$ to be the ordered set $\{a, b\}$, so $K = 2$. Then $rep(a|\mathbb{K}) = (0, 1)$, $rep(b|\mathbb{K}) = (1, 0)$, $rep(c|\mathbb{K}) = (2, 1)$, $rep(d|\mathbb{K}) = (1, 2)$, and $rep(e|\mathbb{K}) = (1, 1)$. Since these five 2-tuples are unique, this $\mathbb{K}$ is a resolving set. It is easy to verify that for this $\mathbb{G}$ a resolving set must have cardinality at least two. Hence $\{a, b\}$ is a minimal resolving set, so $met(\mathbb{G}) = 2$.   □

Determining a minimal resolving set is in general a computationally difficult problem [Diaz 12]. The difficulty arises because the problem is extremely nonlocal, in that a landmark can resolve nodes that are far from the landmark. Moreover, the following problem is NP-complete [Garey 79]: Given the positive integer M, where $M \leq |\mathbb{N}|$, is there a metric basis for $\mathbb{G}$ of cardinality M or less, i.e., is there a subset $\mathbb{K} \subseteq \mathbb{N}$ with $|\mathbb{K}| \leq M$ such that for each pair $u, v \in \mathbb{N}$ there is a $w \in \mathbb{K}$ such that $dist(u, w) \neq dist(v, w)$?

The metric dimension of some graphs is easily obtained. For example, if $\mathbb{G}$ is finite chain of nodes (i.e., each node has degree two except for the two leaf nodes), then $met(\mathbb{G}) = 1$, and a minimal resolving set is either of the nodes of degree 1. If $\mathbb{G}$ has a star topology, with one hub node connected to L spoke nodes of degree 1, then $met(\mathbb{G}) = L - 1$, and a minimal resolving set consists of $L - 1$ spoke nodes.

The metric dimension has applications in pharmaceutical chemistry, robot navigation, combinatorial optimization, and sonar and marine long-range navigation. For example, in chemistry, the structure of a chemical compound can be represented by a graph whose node and arc labels specify the atom and bond types, respectively. A basic problem is to provide mathematical representations for a set of chemical compounds in a way that gives distinct representations to distinct compounds. A graph-theoretic interpretation of this problem is to provide representations for the nodes of a graph in such a way that distinct nodes have distinct representations [Imran 13]. The theoretical study of metric dimension is an active research area.

22.4   Embedding Dimension

The dimension $emb(\mathbb{G})$ of an undirected network $\mathbb{G}$ was defined in [Erdős 65] as the smallest integer E such that G can be embedded in $\mathbb{R}^E$ with each edge of $\mathbb{G}$ having unit arc length.[4] The vertices of $\mathbb{G}$ are mapped onto distinct points of $\mathbb{R}^E$, but there is no restriction on the crossing of arcs.

Example 22.2 The complete graph K_N has N nodes and an arc between each pair of nodes. The graph K_3 can be embedded in $\mathbb{R}^2$ as an equilateral triangle and $emb(K_3) = 2$. Figure 22.3 (left) shows the tetrahedron K_4. For any N we have $emb(K_N) = N - 1$.   □

---

[4] In [Erdős 65] this dimension was denoted by $dim\,\mathbb{G}$; we prefer the more descriptive $emb(\mathbb{G})$, where emb denotes "embedding dimension".

Let $\mathbb{G}_1 = (\mathbb{N}_1, \mathbb{A}_1)$ and $\mathbb{G}_2 = (\mathbb{N}_2, \mathbb{A}_2)$ be graphs. The Cartesian product $G = \mathbb{G}_1 \times \mathbb{G}_2$ is the graph whose set of nodes is $\mathbb{N}_1 \times \mathbb{N}_2$. Two vertices $u = (u_1, u_2)$ and $v = (v_1, v_2)$ are adjacent in $\mathbb{G}$ if and only if $u_1 = v_1$ and (u_2, v_2) is an arc of $\mathbb{G}_2$, or $u_2 = v_2$ and (u_1, v_1) is an arc of $\mathbb{G}_1$.

Example 22.3 The M-cube Q_M is defined as the Cartesian product of M copies of K_2. Since K_2 is the graph of two nodes connected by a single arc, then $Q_1 = K_2$, the graph Q_2 is the unit square, and Q_3 is the unit cube shown in Fig. 22.3 (right). For $M > 1$ we have $emb(Q_M) = 2$. $\square$

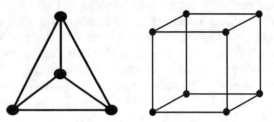

Figure 22.3: Tetrahedron and cube

Let $\mathbb{G}_1 = (\mathbb{N}_1, \mathbb{A}_1)$ and $\mathbb{G}_2 = (\mathbb{N}_2, \mathbb{A}_2)$ be disjoint graphs. The *join* $\mathbb{G}_1 + \mathbb{G}_2$ of $\mathbb{G}_1$ and $\mathbb{G}_2$ is the graph which contains $\mathbb{G}_1$ and $\mathbb{G}_2$ and also has an arc from each node in $\mathbb{N}_1$ to each node in $\mathbb{N}_2$. Let P_M be the polygon with M sides.

Example 22.4 The *wheel* with M spokes is the graph $P_M + K_1$. Figure 22.4 (left) illustrates $P_8 + K_1$, the wheel with 8 spokes. For $M > 6$ we have $emb(P_M + K_1) = 3$. $\square$

Example 22.5 The *complete bicolored graph* $K_{L,M}$ has L nodes of one color, M nodes of another color, and two nodes are adjacent if and only if they have different colors. Figure 22.4 (right) illustrates $K_{3,3}$. We have $emb(K_{1,1}) = 1$, $emb(K_{1,M}) = 2$ for $M > 1$, $emb(K_{2,2}) = 2$, $emb(K_{2,M}) = 3$ for $M \geq 3$, and for all other (L, M) we have $emb(K_{L,M}) = 4$. $\square$

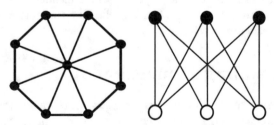

Figure 22.4: Wheel network and complete bicolored graph

Two theorems from [Erdős 65] about $emb(\mathbb{G})$ are stated below. The first theorem is interesting because $\chi(\mathbb{G})$ is the minimal number of boxes needed to color $\mathbb{G}$ for a given box size (Sect. 8.1). The second theorem is interesting since the construction of Sect. 12.1 can generate a tree whose mass dimension is infinite. However, in [Erdős 65] the authors mentioned that they know of no systematic method for determining $emb(\mathbb{G})$ for an arbitrary $\mathbb{G}$, and "Thus the calculation of the dimension of a given graph is at present in the nature of mathematical recreation".

Theorem 22.1 $emb(\mathbb{G}) \leq 2\chi(\mathbb{G})$.

Theorem 22.2 If $\mathbb{G}$ is a tree then $emb(\mathbb{G}) \leq 2$.

22.5 Zeta Dimension

In this section we consider the use of the *zeta function*

$$\zeta(\alpha) \equiv \sum_{i=1}^{\infty} i^{-\alpha} \qquad (22.2)$$

to define the dimension of a network. The zeta function has a rich history [Doty 05, Garrido 09]. It was studied by Euler in 1737 for nonnegative real α and extended in 1859 by Riemann to complex α. We will consider the zeta function only for nonnegative real α. The zeta function converges for $\alpha > 1$ and diverges for $0 \leq \alpha \leq 1$. It is a decreasing function of α and $\zeta(\alpha) \to 1$ as $\alpha \to \infty$ [Shanker 07b].

The use of the zeta function when the domain is $\mathbb{I}^+$, the set of positive integers, was considered in [Doty 05]. An infinite set of positive integers is considered to be "small" if the sum of the reciprocals of its elements is finite. Examples of small sets are the set of perfect squares $\{1^2, 2^2, 3^2, 4^2, \ldots\}$ and the set $\{2^0, 2^1, 2^2, 2^3, \ldots\}$ of nonnegative integer powers of 2. Examples of sets that are not small are the set of prime numbers and the set $\mathbb{I}^+$. The "smallness" of a set $\Omega \subset \mathbb{I}^+$ can be quantified by

$$\zeta(\Omega, \alpha) \equiv \sum_{i \in \Omega} i^{-\alpha}$$

and the *zeta dimension* $d_Z(\Omega)$ of Ω is defined by Doty et al. [Doty 05]

$$d_Z(\Omega) \equiv \inf\{\alpha \,|\, \zeta(\Omega, \alpha) < \infty\}$$

where the subscript "Z" denotes *zeta*. We have $d_Z(\mathbb{I}^+) = 1$ and $d_Z(\Omega) \leq 1$ for any set $\Omega \subset \mathbb{I}^+$. The zeta dimension of any finite subset of $\mathbb{I}^+$ is 0.

A definition of network dimension based on the zeta function was proposed in [Shanker 07b]. Although the zeta dimension of a network has not enjoyed widespread popularity, it has interesting connections to the Hausdorff dimension. Recall from (12.23) that $\partial \mathbb{N}(n, r)$ is the set of nodes whose distance from node n is exactly r, and $|\partial \mathbb{N}(n, r)|$ is the number of such nodes. Define the *graph surface function* by

$$S_r \equiv \frac{1}{N} \sum_{n \in \mathbb{N}} |\partial \mathbb{N}(n, r)| \qquad (22.3)$$

so S_r is the average number of nodes at a distance r from a random node in the network. Define the *graph zeta function* $\zeta(\mathbb{G}, \alpha)$ [Shanker 07b] by

$$\zeta(\mathbb{G}, \alpha) \equiv \frac{1}{N} \sum_{x \in \mathbb{N}} \sum_{\substack{y \in \mathbb{N} \\ y \neq x}} dist(x, y)^{-\alpha} . \qquad (22.4)$$

Since $\mathbb{G}$ is a finite network, then $\zeta(\mathbb{G}, \alpha)$ is finite, and the graph zeta function and the graph surface function are related by

$$
\begin{aligned}
\zeta(\mathbb{G}, \alpha) &= \frac{1}{N} \sum_{x \in \mathbb{N}} \sum_{r \geq 1} |\partial \mathbb{N}(x, r)| r^{-\alpha} \\
&= \sum_{r \geq 1} \left(\frac{1}{N} \sum_{x \in \mathbb{N}} |\partial \mathbb{N}(x, r)| \right) r^{-\alpha} \\
&= \sum_{r \geq 1} S_r r^{-\alpha} . \qquad (22.5)
\end{aligned}
$$

The function $\zeta(\mathbb{G}, \alpha)$ is decreasing in α. For $\alpha = 0$ we have

$$\zeta(\mathbb{G}, 0) = (1/N) \sum_{x \in \mathbb{N}} (N - 1) = N - 1 .$$

For a given $x \in \mathbb{N}$, if $dist(x, y) > 1$, then $dist(x, y)^{-\alpha} \to 0$ as $\alpha \to \infty$, so

$$\lim_{\alpha \to \infty} \sum_{\substack{y \in \mathbb{N} \\ y \neq x}} dist(x, y)^{-\alpha} = \lim_{\alpha \to \infty} \sum_{\substack{y \in \mathbb{N} \\ dist(x,y)=1}} dist(x, y)^{-\alpha} = \delta_x , \qquad (22.6)$$

where δ_x is the node degree of x. Thus $\zeta(\mathbb{G}, \alpha)$ approaches the average node degree as $\alpha \to \infty$.

Since (22.4) defines $\zeta(\mathbb{G}, \alpha)$ only for a finite network, we would like to define $\zeta(\mathbb{G}, \alpha)$ for an infinite network $\mathbb{G}$. If $\mathbb{G}_N$ is a graph with N^E nodes embedded in $\mathbb{R}^E$, and if $\mathbb{G} = \lim_{N \to \infty} \mathbb{G}_N$, then we can define $\zeta(\mathbb{G}, \alpha) \equiv \lim_{N \to \infty} \zeta(\mathbb{G}_N, \alpha)$, as was implicitly done in [Shanker 07b]. For example, $\mathbb{G} = \lim_{N \to \infty} \mathbb{G}_N$ holds when $\mathbb{G}_N$ is a finite E-dimensional rectilinear lattice for which each edge has N nodes, and $\mathbb{G}$ is an infinite E-dimensional rectilinear lattice.

For an infinite one-dimensional rectilinear lattice, each node has two neighbors, so from (22.5) we have $\zeta(\mathbb{G}, \alpha) = 2\zeta(\alpha)$, where $\zeta(\alpha)$ is defined by (22.2). For an infinite E-dimensional rectilinear lattice, and using the L^1 norm $dist(x, y) = \sum_{i=1}^{E} |x_i - y_i|$, we have the recursion [Shanker 07b]

$$S_r^{E+1} = 2 + S_r^E + 2 \sum_{i=1}^{r-1} S_i^E .$$

This recursion is visually interpreted by choosing a given lattice axis and summing over cross sections for the allowed range of distances along the chosen axis.

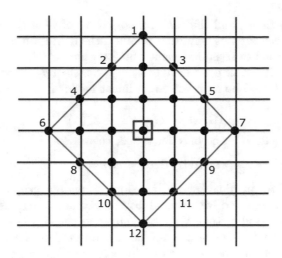

Figure 22.5: Computing S_r^E by summing over cross sections

Table 22.1: $\zeta(\mathbb{G}, \alpha)$ for an infinite rectilinear lattice and the L_1 norm

E	S_r	$\zeta(\mathbb{G}, \alpha)$
1	2	$2\zeta(\alpha)$
2	$4r$	$4\zeta(\alpha - 1)$
3	$4r^2 + 2$	$4\zeta(\alpha - 2) + 2\zeta(\alpha)$
4	$(8/3)r^3 + (16/3)r$	$(8/3)\zeta(\alpha - 3) + (16/3)\zeta(\alpha - 1)$
$r \to \infty$	$O\big(2^E r^{E-1}/\Gamma(E)\big)$	$O\big(2^E \zeta(\alpha - E + 1)/\Gamma(E)\big)$

The recursion is illustrated in Fig. 22.5 for $E = 1$ and $r = 3$, where the random node is the node in the red box. We choose the horizontal axis and take cross sections. For $i = 1$ the term $2S_i^E$ "counts" the four nodes 2, 3, 10, and 11 whose horizontal distance from the random node is 1. For $i = 2$ the term $2S_i^E$ counts the four nodes 4, 5, 8, and 9 whose horizontal distance from the random node is 2. For $i = 3$ the term S_i^E counts the two nodes 6 and 7 whose horizontal distance from the random node is 3. The constant 2 counts the two nodes 1 and 12 whose horizontal distance from the random node is 0. As $r \to \infty$ we have $S_r^E \sim 2^E r^{E-1}/\Gamma(E)$, where $\Gamma(\cdot)$ is the gamma function, so $\Gamma(E) = (E - 1)!$. Table 22.1, from [Shanker 07b], provides $\zeta(\mathbb{G}, \alpha)$ for an infinite rectilinear lattice and the L_1 norm.

We are interested in infinite networks $\mathbb{G}$ for which $\zeta(\mathbb{G}, \alpha)$ can be infinite. Since $\zeta(\mathbb{G}, \alpha)$ is a decreasing function of α, if $\zeta(\mathbb{G}, \alpha)$ is finite for some value α, it is finite for $\alpha' > \alpha$. If $\zeta(\mathbb{G}, \alpha)$ is infinite for some value α, it is infinite for $\alpha' < \alpha$. Thus there is at most one value of α for which $\zeta(\mathbb{G}, \alpha)$ transitions from being infinite to finite, and the *zeta dimension* d_Z of $\mathbb{G}$ is the value at which this transition occurs. This definition parallels the definition in Sect. 5.1 of the Hausdorff dimension of Ω as that value of d for which $v^\star(\Omega, d)$ transitions from infinite to finite. If for all α we have $\zeta(\mathbb{G}, \alpha) = \infty$, then d_Z is defined to be ∞.

Example 22.6 Let $\mathbb{G}$ be an infinite rectilinear lattice in $\mathbb{R}^3$. From Table 22.1 we have

$$\zeta(\mathbb{G}, \alpha) = 4\zeta(\alpha - 2) + 2\zeta(\alpha).$$

Since $4\zeta(\alpha - 2) + 2\zeta(\alpha) < \infty$ only for $\alpha > 3$, then

$$d_Z = \inf\{\alpha \,|\, \zeta(\mathbb{G}, \alpha) < \infty\} = 3. \quad \square$$

Example 22.7 As in [Shanker 13], consider a random graph in which each pair of nodes is connected with probability p. For any three distinct nodes x, y, and z, the probability that z is not connected to both x and y is $1 - p^2$. The probability that x and y are not both connected to some other node is $(1 - p^2)^{N-2}$, which approaches 0 as $N \to \infty$. Thus for large N each pair of nodes is almost surely connected by a path of length at most 2. For large N, each node has $p(N - 1)$ neighbors, so from (22.3) we have $S_1 \approx p(N - 1)$. For large N, the number S_2 of nodes at distance 2 from a random node is given by $S_2 \approx (N - 1) - S_1 = (1 - p)(N - 1)$. By (22.5),

$$\zeta(\mathbb{G}_N, \alpha) \approx p(N - 1) + (1 - p)(N - 1)2^{-\alpha}.$$

Since $\lim_{N \to \infty} \zeta(\mathbb{G}_N, \alpha) = \infty$ for all α then $d_Z = \infty$. $\quad \square$

Exercise 22.4 The following is an informal argument that $d_Z = d_C$. By (11.8) we have $C(r) \sim r^{d_C}$, where $C(r)$ is the fraction of nodes, distinct from a randomly chosen node n, whose distance from n is less than r. Suppose for some constant k the stronger condition $C(r) = kr^{d_C}$ holds. Then $NC(r) = Nkr^{d_C}$ is number of nodes, distinct from a randomly chosen node n, whose distance from n is at most $r - 1$. By (22.3) we have $S_r = N[C(r+1) - C(r)]$. Treating r as a continuous variable, and letting $C'(r)$ denote the derivative of $C(r)$,

$$S_r = N[C(r+1) - C(r)] \approx NC'(r)[(r+1) - r] = NC'(r) = Nkd_C r^{d_C - 1}.$$

Using (22.5),

$$\zeta(\mathbb{G}, \alpha) = \sum_{r \geq 1} S_r r^{-\alpha} \approx Nkd_C \sum_{r \geq 1} r^{d_C - 1} r^{-\alpha} = Nkd_C \sum_{r \geq 1} r^{-\left(1 + \alpha - d_C\right)}.$$

The sum $\sum_{r \geq 1} r^{-(1 + \alpha - d_C)}$ converges for $1 + \alpha - d_C > 1$ or, equivalently, for $\alpha > d_C$. Since d_Z is by definition the smallest value of α for which this sum is finite, then $d_Z = d_C$. Is this informal argument a valid proof that $d_Z = d_C$? If not, can it be reworked to create a valid proof? $\quad \square$

An alternative definition of the zeta dimension of an infinite network, that does not require averages over all the nodes of the network, was given in [Shanker 13]. For $n \in \mathbb{N}$, define

$$\zeta(\mathbb{G}, n, \alpha) = \sum_{\substack{x \in \mathbb{N} \\ x \neq n}} dist(n, x)^{-\alpha}.$$

There is exactly one value of α at which $\zeta(\mathbb{G}, n, \alpha)$ transitions from being infinite to finite; denote this value by $d_Z(n)$. The alternative definition of the zeta dimension of $\mathbb{G}$ is

$$d_Z \equiv \limsup_{n \in \mathbb{N}} d_Z(n).$$

This definition is not always identical to the above definition of d_Z as the value at which $\zeta(\mathbb{G}, \alpha)$ transitions from infinite to finite [Shanker 13].

22.6   Dimensions Relating to Graph Curvature

The dimension of a class of graphs embedded in the infinite two-dimensional triangular lattice $\mathcal{L}$, illustrated by Fig. 22.6, was defined in [Knill 12]. The nodes

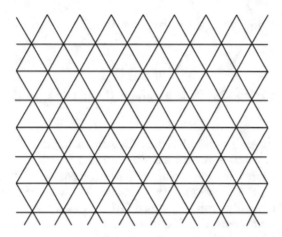

Figure 22.6: Two-dimensional triangular lattice $\mathcal{L}$

of $\mathcal{L}$ are the points $i(1,0) + j(1, \sqrt{3})/2$, where i and j are integers. There is an arc between each pair of nodes separated by a Euclidean distance of 1, and each node has 6 neighbors. Consider a graph $\mathbb{G} = (\mathbb{N}, \mathbb{A})$ such that $\mathbb{N}$ is a finite subset of the nodes of $\mathcal{L}$, and $\mathbb{A}$ is a finite subset of the arcs of the lattice $\mathcal{L}$ connecting nodes in $\mathbb{N}$, as illustrated by the graph of Fig. 22.7, where the arcs of $\mathbb{G}$ are indicated by the bold lines.

Let r be a positive integer and let $n \in \mathbb{N}$. Define $\mathbb{G}(n, r)$ to be the subgraph of $\mathbb{G}$ such that x is a node of $\mathbb{G}(n, r)$ if $x \in \mathbb{G}$ and $dist(n, x) = r$ (where by distance we mean the shortest distance in $\mathbb{G}$), and a is an arc of $\mathbb{G}(n, r)$ if $a \in \mathbb{A}$ and a connects two nodes of $\mathbb{G}(n, r)$ separated by a distance of 1. For example, in Fig. 22.7, choosing n to be the central blue node and choosing $r = 1$, the nodes of $\mathbb{G}(n, 1)$ are the four red nodes, and the arcs of $\mathbb{G}(n, 1)$ are the arcs a, b, and c. With the same n and choosing $r = 2$, the nodes of $\mathbb{G}(n, 2)$ are the five green nodes and the arcs of $\mathbb{G}(n, 2)$ are d, e, and f.

Various dimensions were defined in [Knill 12] as follows. A node $n \in \mathbb{G}$ has dimension 0 if n is not connected to any other node in $\mathbb{G}$. A graph $\mathbb{G}$

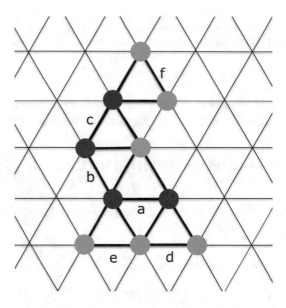

Figure 22.7: A finite graph $\mathbb{G}$ embedded in $\mathcal{L}$

has dimension 0 if each node of $\mathbb{G}$ has dimension 0 (i.e., if $\mathbb{G}$ is a collection of isolated nodes). A node $n \in \mathbb{G}$ has dimension 1 if $\mathbb{G}(n,1)$ has dimension 0, and $\mathbb{G}$ has dimension 1 if *some* node in $\mathbb{G}$ has dimension 1. In Fig. 22.8 (left), considering node 5, the subgraph $\mathbb{G}(5,1)$ contains only node 1, and contains no arcs; thus $\mathbb{G}(5,1)$ has dimension 0. Since $\mathbb{G}(5,1)$ has dimension 0, then node 5 has dimension 1, and hence $\mathbb{G}$ has dimension 1.

A node $n \in \mathbb{G}$ has dimension 2 if $\mathbb{G}(n,1)$ has dimension 1, and $\mathbb{G}$ has dimension 2 if *each* node in $\mathbb{G}$ has dimension 2. For example, in Fig. 22.8 (right), the arcs of $\mathbb{G}(n_1,1)$ are a_1 and a_2, and $\mathbb{G}(n_1,1)$ has dimension 1; hence n_1 has dimension 2. Since $\mathbb{G}(n,1)$ has dimension 1 for each $n \in \mathbb{G}$, then $\mathbb{G}$ has dimension 2.

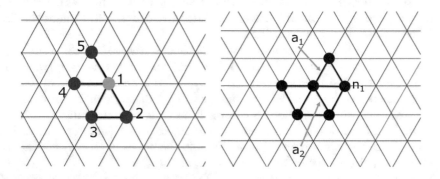

Figure 22.8: 1-dimensional (left) and 2-dimensional (right) graphs

As another example, consider the platonic solids viewed as graphs. Then the cube (which has 6 faces) and the dodecahedron (12 faces) have dimension 1. The isocahedron (20 faces) and the octahedron (8 faces) have dimension 2. Each face of the tetrahedron (4 triangular faces) has dimension 2.

These definitions were used in [Knill 12] to study the curvature of graphs. For $n \in \mathbb{G}$, the curvature $K(n)$ of n is $K(n) \equiv 2|\mathbb{A}(n,1)| - |\mathbb{A}(n,2)|$, where $|\mathbb{A}(n,1)|$ and $|\mathbb{A}(n,2)|$ are the number of arcs in $\mathbb{G}(n,1)$ and $\mathbb{G}(n,2)$, respectively. Thus for a node n in the infinite lattice $\mathcal{L}$ we have $K(n) = 2 \cdot 6 - 12 = 0$. The study of $\sum_{n \in \partial \mathbb{G}} K(n)$, where $\partial \mathbb{G}$ is the appropriately defined boundary of $\mathbb{G}$, has connections to differential geometry.

22.7 Spectral Dimension of Networks

In Sect. 20.1 we briefly considered the spectral dimension of a geometric object. A spectral dimension can also be defined for a network. The spectral dimension of a network can be viewed as the dimension of the network as experienced by a random walker on the network. The "return to origin" probability $P(t)$ is the probability that the walker, starting at some given node at time $t = 0$, returns to the given node at time t. For small t we have $P(t) \sim t^{-d_{sp}/2}$ [Burda 01]. For E-dimensional regular lattices, $P(t) \sim t^{-E/2}$ when t is even and $P(t) = 0$ when t is odd [Hwang 10]. Using a renormalization group approach (Chap. 21), [Hwang 10] calculated d_{sp} for a weighted (u, v)-flower network (Sect. 12.1).

Another definition of the spectral dimension of a network concerns the eigenvalues of the Laplacian $\mathcal{L}$ of the network $\mathbb{G}$. Recalling that δ_i is the node degree of node i, the Laplacian $\mathcal{L}$ is the $N \times N$ matrix defined by

$$\mathcal{L}_{ij} = \begin{cases} \delta_i & \text{if } i = j, \\ -1 & \text{if } i \text{ and } j \text{ are adjacent}, \\ 0 & \text{otherwise}. \end{cases}$$

(We will again encounter the Laplacian in Sect. 23.6.) If the density $g(\lambda)$ of eigenvalues of $\mathcal{L}$ scales as

$$g(\lambda) \sim \lambda^{(d_{sp}/2)-1},$$

then d_{sp} is the spectral dimension of $\mathbb{G}$. For regular E-dimensional lattices we have $d_{sp} = E$ [Wu 15].

The spectral dimension was used in [Reuveni 10] to study proteins. There has been an accumulation of evidence that proteins are fractal-like objects, due to the way they vibrate and the way they fill space. The spatial structure of proteins was characterized in [Reuveni 10] by two parameters, called the "fractal dimension" and the spectral dimension d_{sp}. The approach of [Reuveni 10] was based upon deducing the innate properties of a network by studying the behavior of a random walker on the network.[5] These properties include the "return to origin" probability $P(t)$ mentioned above, the mean square displacement from

---

[5] Using a random walk on a network to compute d_C was discussed in Sect. 11.3.

the origin, and the first passage time between two tagged points [Nakayama 94]. Proteins are characterized by $d_{sp} < 2$, which leads to a relatively high ability for the protein to flex. The mean square displacement $msd(t)$ of a random walker at time t follows the scaling $msd(t) \sim t^{2/d_w}$, where d_w is the *walk dimension*. Define the mean first passage time $mfpt(s)$ to be the average time for a random walker to travel between two tagged nodes (amino acids) when the distance between them is s; the average is over all nodes separated by a distance s. Defining $\alpha \equiv d_w[1 - (d_{sp}/2)]$, for a finite fractal structure of N nodes with $d_{sp} < 2$ we have $mfpt(s) = N(A + Bs^\alpha)$, where A and B are structure specific constants.

22.8   Other Avenues

We conclude this chapter by briefly describing some other ways to think about network fractal dimensions. Since we present no formal definitions, these topics are potential avenues for future investigation.

Avenue 1 In Sect. 20.1, we mentioned the use of the spectral dimension in studying quantum gravity. Another definition of the dimension of a graph arises in the study of quantum gravity [Ambjorn 92]. Suppose we triangulate $\mathbb{R}^2$ with a random, nonuniform triangulation. Define the distance between two vertices of the triangulation as the length of a shortest path within the triangulation which connects the two vertices. Define the distance between two triangles to be the smallest number of triangles which constitute a connected path of triangles between the two. Let $V(r)$ be the number of triangles within radius r of a given vertex or triangle. If the relation $\langle V(r) \rangle \sim r^d$ holds, where the average $\langle V(r) \rangle \sim r^d$ is over the vertices or triangles, then, in the terminology of [Ambjorn 92], the ensemble of triangulations is said to have Hausdorff dimension d. In this definition of dimension, the "bulk", as in (1.1), is a number of triangles. It was observed in [Ambjorn 92] that the Hausdorff dimension is a very rough measure of the geometrical properties of the systems considered in [Ambjorn 92], and the Hausdorff dimension might not tell much about the geometry.

Avenue 2 In [Burda 01], an ensemble of graphs was considered, where each graph was given a weight which is proportional to the product of the weights attached to individual nodes, and which depends on the node degrees. An ensemble was characterized by its spectral dimension d_{sp} and by a fractal dimension which in [Burda 01] was called the Hausdorff dimension. Feynman diagrams, used in field theory methods of physics, are the graphs in an ensemble. (The analysis and results in [Burda 01] are not easily accessible to the non-physicist.) One topic in [Burda 01] is the study of trees such that each node in the tree has degree 1 or 3; such a tree is illustrated in Fig. 22.9. For such trees, the Hausdorff dimension is 2 and typically the diameter $\sim \sqrt{N}$.

Avenue 3 In [Tang 06] a 2-dimensional image was represented by a function $f(x, y)$, where x and y are the pixel coordinates. The 2-dimensional discrete Fourier transform can be applied to $f(x, y)$ to obtain the *Fourier transform* fractal dimension.

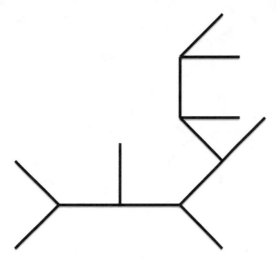

Figure 22.9: A tree whose nodes have degree 1 or 3

Avenue 4 In [Luke 07] it was suggested that there are three broad categories of network analysis in public health: analysis of networks that transmit disease or information, analysis of how social structure and relationships act to promote or influence health and health behavior, and analysis of agencies and organizations rather than individuals. Not surprisingly, considerable attention has been paid to disease transmission networks. The identification of network characteristics can spur the development and implementation of more efficient and effective strategies. The structure of disease transmission networks can change during the progression of an epidemic. For example, in a study of sexually transmitted disease (STD) networks, early stages of an epidemic manifest as cyclic (closed loop) structures, while dendritic (tree-like) structures indicate later phases of the epidemic in which the rate of new cases is decreasing. Also, different STDs show different network characteristics. The computation of network fractal dimensions, in addition to the network metrics cited in [Luke 07], may provide additional useful ways to characterize such networks.

Chapter 23

Supplemental Material

Humankind has not woven the web of life. We are but one thread within it. Whatever we do to the web, we do to ourselves. All things are bound together. All things connect.
Chief Seattle (1786–1866), a leader of the Suquamish and Duwamish Native American tribes in what is now Washington state.

This chapter provides supplèmental information on six topics: (i) rectifiable arcs and arc length, to supplement the material on jagged coastlines in the beginning of Chap. 3 and also used in Sect. 6.4 on the box counting dimension of a spiral, (ii) the Gamma function and hypersphere volumes, used in Sect. 5.1 on the Hausdorff dimension, (iii) Laplace's method of steepest descent, used in Sect. 16.3 on the spectrum of the multifractal Cantor set, (iv) Stirling's approximation, used in Sect. 16.3, (v) convex functions and the Legendre transform, used in Sect. 16.4 on multifractals, and (vi) graph spectra (the study of the eigenvectors and eigenvalues of the adjacency matrix of $\mathbb{G}$), used in Sect. 20.1.

23.1 Rectifiable Arcs and Arc Length

In this section we compute the length of an arc, where "arc" refers not to a connection between two nodes in a graph, but rather to a curvy line segment. Consider the curve shown in Fig. 23.1 and the points A and B on the curve. As shown in this figure, divide the arc between A and B into K pieces by choosing $K - 1$ points $P_1, P_2, \ldots, P_{K-1}$ on the curve. Draw the K chords $AP_1, P_1P_2, P_2P_3, \ldots, P_{K-1}B$. Let $L(AP_1)$ be the length of chord AP_1. For $k = 1, 2, \ldots, K-2$, let $L(P_kP_{k+1})$ be the length of chord P_kP_{k+1}. Let $L(P_{K-1}B)$ be the length of chord $P_{K-1}B$. Define $L_K \equiv L(AP_1) + L(P_1P_2) + \cdots + L(P_{K-1}B)$, so L_K is the sum of the K chord lengths. Let δ be the length of the longest of these K chords, and let the number of points dividing the arc AB be increased such that $\delta \to 0$ as $K \to \infty$. If $\lim_{\delta \to 0} L_K$ exists and is independent of the choice of the points $P_1, P_2, \cdots, P_{K-1}$, then arc AB is said to be *rectifiable*, and this limit is defined to be the length of the arc [Smail 53].

© Springer Nature Switzerland AG 2020
E. Rosenberg, *Fractal Dimensions of Networks*,
https://doi.org/10.1007/978-3-030-43169-3_23

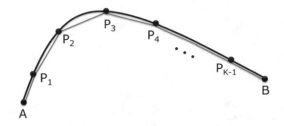

Figure 23.1: Rectifiable Arc

In Sect. 6.4 we need to calculate the length of an arc. Let the function $f(x)$ and its derivative $f'(x)$ be continuously differentiable on the interval (S, T). Then the length L of the arc of $f(x)$ from S to T is

$$L = \int_S^T \sqrt{1 + [f'(x)]^2}\ dx\,. \tag{23.1}$$

For an informal derivation of this formula, consider Fig. 23.2 and the two points

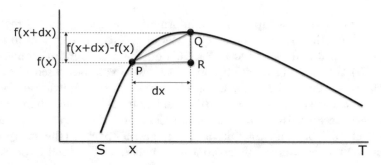

Figure 23.2: Computing the length of a rectifiable arc

$x \in (S, T)$ and $x + dx$, where $dx \ll 1$. The length of the arc from x to $x + dx$ is approximated by the length $L(x)$ of the chord connecting the points $\big(x, f(x)\big)$ and $\big(x + dx, f(x + dx)\big)$. We have

$$
\begin{aligned}
L(x) &= \sqrt{(dx)^2 + [f(x + dx) - f(x)]^2} \\
&= \sqrt{(dx)^2[1 + [\big(f(x + dx) - f(x)\big)/dx]^2} \\
&\approx \sqrt{1 + [f'(x)]^2}\ dx\,.
\end{aligned}
$$

Now divide the interval $[S, T]$ into K pieces, each of size $(T - S)/K$, by inserting points x_k, $k = 1, 2, \ldots, K - 1$ such that $S < x_1 < x_2 < \cdots < x_{K-1} < T$. Define $x_0 = S$ and $x_K = T$. Define $dx = (T - S)/K$. Then the arc length of $f(x)$ from S to T is approximated by the sum of the K chord lengths:

$$L \approx \sum_{k=0}^{K-1} \sqrt{1 + [f'(x_k)]^2}\ dx\,.$$

Letting $K \to \infty$ yields the integral (23.1). A rigorous proof would invoke the mean value theorem for derivatives [Smail 53].

To derive the arc length for a curve defined using polar coordinates, let $y = f(x)$, so $f'(x) = dy/dx$. Rewrite (23.1) as

$$L = \int_S^T \sqrt{1 + (dy/dx)^2}\, dx\,.$$

Since $x = r\cos(\theta)$ and $y = r\sin(\theta)$, then $dx/d\theta = (dr/d\theta)\cos(\theta) - r\sin(\theta)$ and $dy/d\theta = (dr/d\theta)\sin(\theta) + r\cos(\theta)$. From this we obtain, after some algebra,

$$(dx/d\theta)^2 + (dy/d\theta)^2 = r^2 + (dr/d\theta)^2\,. \tag{23.2}$$

$$L = \int_S^T \sqrt{1 + (dy/dx)^2}\, dx = \int_S^T \sqrt{1 + \left(\frac{dy/d\theta}{dx/d\theta}\right)^2}\, dx$$

$$= \int_S^T \sqrt{\frac{(dx/d\theta)^2 + (dy/d\theta)^2}{(dx/d\theta)^2}}\, dx$$

$$= \int_S^T \sqrt{\left(\frac{dx}{d\theta}\right)^2 + \left(\frac{dy}{d\theta}\right)^2}\frac{d\theta}{dx}\, dx = \int_{\theta(S)}^{\theta(T)} \sqrt{r^2 + (dr/d\theta)^2}\, d\theta\,, \tag{23.3}$$

where the final inequality uses (23.2), and $\theta(S)$ and $\theta(T)$ are the limits of integration for the integral over θ.

23.2 The Gamma Function and Hypersphere Volumes

The *gamma* function Γ is defined for positive real x by

$$\Gamma(x) \equiv \int_0^\infty t^{x-1} e^{-t} dt\,. \tag{23.4}$$

Integration by parts yields

$$\Gamma(x+1) = \int_0^\infty t^x e^{-t} dt = -t^x e^{-t}\Big|_{t=0}^\infty + x\int_0^\infty t^{x-1} e^{-t} dt = x\Gamma(x)\,.$$

Since $\Gamma(1) = 1$, it is easily verified that $\Gamma(n+1) = n!$ if n is a positive integer.

The volume $V_E(r)$ of a hypersphere in $\mathbb{R}^E$ with radius r is [Manin 06]

$$V_E(r) = \frac{\Gamma(1/2)^E}{\Gamma\big(1 + (E/2)\big)}\, r^E\,.$$

To evaluate $V_E(r)$, we first evaluate $\Gamma(1/2)$. We have

$$\Gamma(1/2) = \int_0^\infty t^{-1/2} e^{-t}\, dt = 2\int_0^\infty e^{-t}\frac{1}{2\sqrt{t}}\, dt.$$

With the change of variables $u = \sqrt{t}$, we have $t = u^2$ and

$$du = \frac{1}{2\sqrt{t}}\, dt$$

yielding

$$\Gamma(1/2) = 2\int_0^\infty e^{-u^2}\, du = \int_{-\infty}^\infty e^{-u^2}\, du = \sqrt{\pi}\,,$$

where the last equality follows as the area under the normal curve is $\sqrt{\pi}$. Using the identity $\Gamma(x+1) = x\Gamma(x)$,

$$\begin{aligned}
\Gamma(3/2) &= \Gamma((1/2)+1) = (1/2)\sqrt{\pi} \\
\Gamma(5/2) &= \Gamma((3/2)+1) = (3/2)(1/2)\sqrt{\pi} = (3/4)\sqrt{\pi}
\end{aligned}$$

and the volume of a sphere in $\mathbb{R}^3$ is

$$V_3(r) = \frac{\Gamma(1/2)^3}{\Gamma(5/2)}\, r^3 = (4/3)\pi r^3\,. \tag{23.5}$$

23.3   Laplace's Method of Steepest Descent

Laplace's method, also called the method of *steepest descent*, is a way to determine the leading order behavior of the integral

$$F(\lambda) = \int_a^b f(x)\, e^{-\lambda g(x)}\, dx$$

as $\lambda \to \infty$. The essence of the method is that, for $\lambda \gg 1$, the integral $F(\lambda)$ can be closely approximated by a simpler integral whose limits of integration are a neighborhood $[y-\epsilon, y+\epsilon]$ around a point y which minimizes g over $[a,b]$, and the simpler integral itself can be easily approximated.[1] Even if g is relatively flat over $[a,b]$, when $\lambda \gg 1$ the term $e^{-\lambda g(x)}$ is much larger in the neighborhood of y than elsewhere on $[a,b]$.

We require three assumptions. Assume that (i) $F(\lambda)$ exists for all sufficiently large λ; (ii) the functions f and g are twice differentiable on (a,b); and (iii) the minimal value of g over $[a,b]$ is attained at some $y \in (a,b)$ satisfying $g''(y) > 0$ and $f(y) \neq 0$. Assumption (ii) allows us to replace f and g by their second-order Taylor series expansions. Assumption (iii) implies that $g'(y) = 0$ and that g is strictly convex at y, so y is an isolated local minimum of g.

---

[1] Our presentation is based on www.math.unl.edu/~scohn1/8423/intasym1.pdf.

Rewriting $F(\lambda)$ and then applying several approximations, as $\lambda \to \infty$ we have

$$
\begin{aligned}
F(\lambda) &= e^{-\lambda g(y)} \int_a^b f(x)\, e^{-\lambda[g(x)-g(y)]} dx \\[2mm]
&\approx e^{-\lambda g(y)} \int_{y-\epsilon}^{y+\epsilon} f(x)\, e^{-\lambda[g(x)-g(y)]} dx \\[2mm]
&\approx e^{-\lambda g(y)} f(y) \int_{y-\epsilon}^{y+\epsilon} e^{-\lambda[g(x)-g(y)]} dx \\[2mm]
&\approx e^{-\lambda g(y)} f(y) \int_{y-\epsilon}^{y+\epsilon} e^{-\lambda[g'(y)(x-y)+(1/2)g''(y)(x-y)^2]} dx \\[2mm]
&= e^{-\lambda g(y)} f(y) \int_{y-\epsilon}^{y+\epsilon} e^{-(\lambda/2)g''(y)(x-y)^2} dx \\[2mm]
&\approx e^{-\lambda g(y)} f(y) \int_{-\infty}^{\infty} e^{-(\lambda/2)g''(y)(x-y)^2} dx \\[2mm]
&= e^{-\lambda g(y)} f(y) \int_{-\infty}^{\infty} e^{-(\lambda/2)g''(y)z^2]} dz \\[2mm]
&= e^{-\lambda g(y)} f(y) \sqrt{\frac{2\pi}{\lambda g''(y)}}
\end{aligned}
$$

where the final equality follows from the area under a normal curve:

$$
\int_{-\infty}^{\infty} e^{-az^2} dz = \sqrt{\frac{\pi}{a}} .
$$

We thus have the final result that, as $\lambda \to \infty$,

$$
\int_a^b f(x)\, e^{-\lambda g(x)} dx \approx e^{-\lambda g(y)} f(y) \sqrt{\frac{2\pi}{\lambda g''(y)}} . \tag{23.6}
$$

23.4 Stirling's Approximation

In Chap. 16 on multifractals, we made frequent use of the approximation $\log n! \approx n \log n$ for large n, and often this is referred to as *Stirling's approximation*. The actual approximation credited to Stirling is

$$
n! \approx \sqrt{2\pi}\, n^{n+(1/2)}\, e^{-n} \text{ as } n \to \infty . \tag{23.7}
$$

We now derive (23.7). From (23.4), for $x > 0$ we have

$$
\begin{aligned}
\Gamma(x+1) &= \int_{t=0}^{\infty} t^x\, e^{-t} dt \\
&= \int_{t=0}^{\infty} e^{x\ln t}\, e^{-t} dt \\
&= \int_{t=0}^{\infty} e^{-x[(t/x)-\ln t]} dt \\
&= x\int_{z=0}^{\infty} e^{-x[z-\ln(xz)]} dz \quad (\text{using } t = xz) \\
&= x e^{x\ln x} \int_{z=0}^{\infty} e^{-x(z-\ln z)} dz \\
&= x^{x+1} \int_{z=0}^{\infty} e^{-x(z-\ln z)} dz
\end{aligned}
$$

so

$$
\Gamma(x+1) = x^{x+1} \int_{z=0}^{\infty} e^{-x(z-\ln z)} dz\,. \tag{23.8}
$$

Now apply Laplace's approximation (23.6) to the integral in (23.8), with z taking the place of x, with x taking the place of λ, with $f(z) = 1$, and with $g(z) = z - \ln z$. The point $z = 1$ is the unique global minimum of $g(z)$ over $\{z \mid z > 0\}$, and $g(1) = 1$, $g'(1) = 0$, and $g''(1) = 1$. Since all three assumptions of Sect. 23.3 are satisfied, by (23.6) we have, as $x \to \infty$,

$$
\int_{z=0}^{\infty} e^{-x(z-\ln z)} dz \approx e^{-x} \sqrt{\frac{2\pi}{x}}\,. \tag{23.9}
$$

From (23.8) and (23.9) we have, as $x \to \infty$,

$$
\Gamma(x+1) \approx x^{x+1}\, e^{-x} \sqrt{\frac{2\pi}{x}} = \sqrt{2\pi}\, x^{x+(1/2)}\, e^{-x}\,.
$$

If x is the positive integer n, then $\Gamma(n+1) = n!$, which yields (23.7).

Exercise 23.1 Plot the error incurred by using (23.7) to approximate $n!$ as a function of the positive integer n. □

23.5 Convex Functions and the Legendre Transform

In this section we present the Legendre transform of a convex function of a single variable, drawing on [Kennerly 11, Mangasarian 69, Rockafellar 74, Zia 09]. Let $F : \mathbb{R} \to \mathbb{R}$ be convex. For simplicity, assume for the remainder of this section that (i) F is twice differentiable and (ii) $F''(x) > 0$ for all

x. These two assumptions imply that F is strictly convex and that $F'(x)$ is monotone increasing with x.[2] For this discussion it is important to distinguish between the function F and the value $F(x)$ of the function at the point x. Since F is differentiable and convex, then

$$F(y) \geq F(x) + F'(x)(y - x)$$

for all x and y. As illustrated by Fig. 23.3, this inequality means that F lies above the line with slope $F'(x)$ and vertical intercept $F(x) - xF'(x)$ which is tangent to F at x. Since F lies above this line, and since F intersects the line,

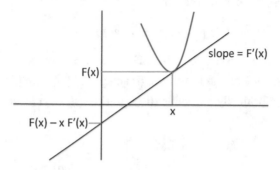

Figure 23.3: A differentiable convex function lies above each tangent line

the line is called a *supporting hyperplane* for F. Figure 23.4 illustrates two supporting hyperplanes of a convex function. Each convex function (whether or not

Figure 23.4: Two supporting hyperplanes of a convex function

it is differentiable) can be characterized in terms of its supporting hyperplanes [Rockafellar 74].

---

[2] Basic definitions relating to convexity were provided in Sect. 1.2.

Since $F'(x)$ is monotone increasing with x, for each p there is a unique x such that $F'(x) = p$. For example, if $F(x) = (x-1)^2$, setting $F'(x) = 2(x-1) = p$ yields $x = 1 + (p/2)$. Let $\tilde{x} : \mathbb{R} \to \mathbb{R}$ be the function defined by

$$\tilde{x}(p) = \{x \mid F'(x) = p\}.$$

That is, for each p the value $\tilde{x}(p)$ is the x value at which the derivative of F is p. For example, if $F(x) = (x-1)^2$, then $\tilde{x}(p) = 1 + (p/2)$. Note that x is a point and $\tilde{x}$ is a function. Having defined the function $\tilde{x}$, we can write $F\big(\tilde{x}(p)\big)$, which is the value of F at the point $\tilde{x}(p)$. Finally, the *Legendre transform* of F is the function $G : \mathbb{R} \to \mathbb{R}$ defined by

$$G(p) \equiv p\,\tilde{x}(p) - F\big(\tilde{x}(p)\big). \tag{23.10}$$

These definitions are illustrated in Fig. 23.5, which shows that $G(p)$ is the negative of the y intercept of the line through the point $\Big(x(p), F\big(\tilde{x}(p)\big)\Big)$ with slope p. To see this algebraically, consider the equation $(y - y_0) = m(x - x_0)$ describ-

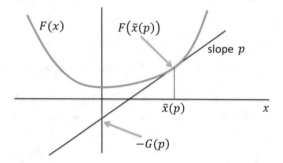

Figure 23.5: The Legendre transform G of F

ing a line with slope m through the point (x_0, y_0). The y-intercept of this line is $y_0 - mx_0$. Choosing $m = p$ and $(x_0, y_0) = \Big(\tilde{x}(p), F\big(\tilde{x}(p)\big)\Big)$, the y-intercept is $F\big(\tilde{x}(p)\big) - p\tilde{x}(p)$, which by (23.10) is $-G(p)$.

Example 23.1 Consider the function F defined by $F(x) = x^6$. We have $F'(x) = 6x^5$. Setting $6x^5 = p$ yields $\tilde{x}(p) = (p/6)^{1/5}$. From (23.10) we have $G(p) = p(p/6)^{1/5} - [(p/6)^{1/5}]^6 = [6^{-1/5} - 6^{-6/5}]p^{6/5}$. □

Exercise 23.2 Derive an expression for the Legendre transform of $1 + x^4$, and plot $G(p)$ versus p. □

Exercise 23.3 Let G be the Legendre transform of F. Prove that the Legendre transform of G is F itself. □

Suppose F satisfies, in addition to assumptions (i) and (ii) above, the additional assumptions that $F(x) \to \infty$ as $|x| \to \infty$, and that $F'(x) = 0$ at some point x. Then the following informal argument suggests that G is convex. Under these additional assumptions, if $x \ll 0$, then $p = F'(x) \ll 0$, so the intercept $-G(p) \ll 0$, which means $G(p) \gg 0$. As x increases then $p = F'(x)$

increases, so the intercept $-G(p)$ increases, which means $G(p)$ decreases. As x increases beyond the point where $F'(x) = 0$, the derivative $p = F'(x)$ continues to increase, so the intercept $-G(p)$ decreases, which means $G(p)$ increases. The assumptions in the above informal argument are stronger than required: it was shown in [Rockafellar 74] that if $F : \mathbb{R}^E \to \mathbb{R}$ is a finite and differentiable convex function such that for each $p \in \mathbb{R}^E$ the equation $\nabla F(x) = p$ has a unique solution $\widetilde{x}(p)$, then G is a finite and differentiable convex function.

The Legendre transform of F is a special case of the *conjugate* $F^\star : \mathbb{R} \to \mathbb{R}$ of F, defined in [Rockafellar 74]:

$$F^\star(p) = \sup_{x \in \mathbb{R}} \{px - F(x)\} . \tag{23.11}$$

Since the conjugate $F^\star$ is the supremum of an (infinite) collection of linear functions, where each linear function maps p to $px - F(x)$, then $F^\star$ is convex. For a given p, to evaluate $F^\star(p)$ we maximize $px - F(x)$ over $x \in \mathbb{R}$. The definition of $F^\star$ does not require F to be differentiable. However, since in this section we do assume that F is twice differentiable, to maximize $px - F(x)$ we take the derivative with respect to x and equate the derivative to 0. The maximum is achieved at the point x at which $F'(x) = p$; this point is, by definition, $\widetilde{x}(p)$. Since F is strictly convex, then $\widetilde{x}(p)$ is the unique point yielding a global maximum, over $\mathbb{R}$, of the concave function $px - F(x)$. This is illustrated in Fig. 23.6, where $p\,\widetilde{x}(p) - F\big(\widetilde{x}(p)\big)$ (the length of the green vertical bar) exceeds $p\,\hat{x} - F\big(\hat{x}\big)$ (the length of the red vertical bar), where $\hat{x}$ is an arbitrary point not equal to $\widetilde{x}(p)$. From (23.11) we have $F^\star(p) = p\,\widetilde{x}(p) - F\big(\widetilde{x}(p)\big)$, which agrees

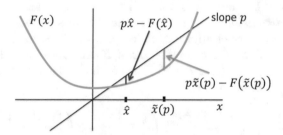

Figure 23.6: The conjugate $F^\star$ of F

with (23.10), since G is a special case of $F^\star$. Thus $G(p) = F^\star(p)$.

We have shown that, under assumptions (i) and (ii) on the convex function F, we can describe F in two different ways. The usual way is to specify the functional form of F. The Legendre transform provides a second way by specifying, for each possible value of $F'(x)$, the vertical intercept of the line tangent to F with slope $F'(x)$. While the above discussion considered only a function of a single variable, the theory of Legendre transforms and conjugate functions holds for multivariate functions [Rockafellar 74, Zia 09].

23.6   Graph Spectra

The *spectrum* of $\mathbb{G}$ is the set of eigenvectors and eigenvalues of the node-node adjacency matrix M of $\mathbb{G}$ (Sect. 2.1). Since $\mathbb{G}$ is undirected and has no self-loops, then the diagonal elements of M are zero, and M is symmetric $(M_{ij} = M_{ji})$.[3] There is a large body of work, including several books, on graph spectra. To motivate our brief discussion of this subject, we consider the *1-dimensional quadratic assignment* problem studied in [Hall 70]. This is the problem of placing the nodes of $\mathbb{G}$ on a line to minimize the sum of the squared distances between adjacent nodes. (Here "adjacent" refers to adjacency in $\mathbb{G}$.) A single constraint is imposed to render infeasible the trivial solution in which all nodes are placed at the same point. Hall's algorithm uses the method of Lagrange multipliers for constrained optimization problems, and Hall's algorithm can be extended to the placement of the nodes of $\mathbb{G}$ in two dimensions. Thus spectral graph analysis is a key technique in various methods for drawing graphs. Spectral methods are also used for community detection in networks, particularly in social networks.

Let x denote an N-dimensional column vector (i.e., a matrix with 1 column and N rows). The transpose x' of x is an N-dimensional row vector (i.e., a matrix with 1 row and N columns). Consider the optimization problem of minimizing the objective function

$$\frac{1}{2} \sum_{i=1}^{N} \sum_{j=1}^{N} \left(x_i - x_j \right)^2 M_{ij} \tag{23.12}$$

subject to the single constraint $\sum_{i=1}^{N} x_i^2 = 1$. This constraint is compactly expressed as $x'x = 1$. Thus the problem is to place N nodes on the line (hence the adjective *1-dimensional*) to minimize the sum of squared distances, where the sum is over all pairs of adjacent nodes (i.e., adjacent in $\mathbb{G}$). For example, let $N = 3$ and let $\mathbb{G}$ be the complete graph K_3. Suppose the three nodes are placed at positions $1/6$, $1/3$, and $1/2$. Since each off-diagonal element of M_{ij} is 1, the objective function value is

$$\left(\frac{1}{2} \right) \left[2 \left(\frac{1}{6} - \frac{1}{3} \right)^2 + 2 \left(\frac{1}{6} - \frac{1}{2} \right)^2 + 2 \left(\frac{1}{3} - \frac{1}{2} \right)^2 \right] .$$

Clearly, $x = (\frac{1}{\sqrt{N}}, \cdots , \frac{1}{\sqrt{N}})'$ solves the optimization problem of minimizing (23.12) subject to the constraint $x'x = 1$, since this x yields an objective function value of 0. Since this x places each node at the same position, we exclude this x and seek a non-trivial solution.

Let D be the $N \times N$ diagonal matrix defined by $D_{ii} = \sum_{j=1}^{N} M_{ij}$ for $i = 1, 2, \ldots, N$ and $D_{ij} = 0$ for $i \neq j$. Thus D_{ii} is the sum of the elements in row i

---

[3] Let M be a symmetric matrix. Then λ is an eigenvalue of M, with associated eigenvector x, if $Mx = \lambda x$. Each eigenvalue of a symmetric matrix is a real number. If M is positive semi-definite, i.e., if $x'Mx \geq 0$ for all x, then each eigenvalue of M is nonnegative.

of M. Since M is symmetric, D_{ii} is also the sum of the elements in column i of M. Compute the $N \times N$ matrix $D - M$, where $(D - M)_{ij} = D_{ij} - M_{ij}$.

Example 23.2 [Hall 70] Consider the $\mathbb{G}$ illustrated in Fig. 23.7. Suppose columns 1, 2, 3, and 4 of M correspond to nodes a, b, c, and d, respectively. Then

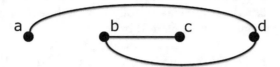

Figure 23.7: Small network to illustrate Hall's method

$$M = \begin{pmatrix} 0 & 0 & 0 & 1 \\ 0 & 0 & 1 & 1 \\ 0 & 1 & 0 & 0 \\ 1 & 1 & 0 & 0 \end{pmatrix} \qquad D = \begin{pmatrix} 1 & 0 & 0 & 0 \\ 0 & 2 & 0 & 0 \\ 0 & 0 & 1 & 0 \\ 0 & 0 & 0 & 2 \end{pmatrix}$$

$$D - M = \begin{pmatrix} 1 & 0 & 0 & -1 \\ 0 & 2 & -1 & -1 \\ 0 & -1 & 1 & 0 \\ -1 & -1 & 0 & 2 \end{pmatrix} \qquad \square$$

Consider the objective function (23.12). For simplicity, we write $\sum_i$ to denote $\sum_{i=1}^{N}$ and $\sum_j$ to denote $\sum_{j=1}^{N}$. We have

$$
\begin{aligned}
\frac{1}{2}\sum_i\sum_j \left(x_i - x_j\right)^2 M_{ij} &= \frac{1}{2}\sum_i\sum_j \left(x_i^2 - 2x_i x_j + x_j^2\right)M_{ij} \\
&= \frac{1}{2}\left(\sum_i x_i^2 D_{ii} - 2\sum_i\sum_j x_i x_j M_{ij} + \sum_j x_j^2 D_{jj}\right) \\
&= \sum_i x_i^2 D_{ii} - \sum_i\sum_j x_i x_j M_{ij} \\
&= x'Dx - x'Mx \\
&= x'(D - M)x .
\end{aligned}
\tag{23.13}
$$

Since $\sum_i\sum_j (x_i - x_j)^2 M_{ij} \geq 0$ for all x then from (23.13) we have $x'(D-M)x \geq 0$ for all x; that is, the matrix $D - M$ is positive semi-definite. Thus the N eigenvalues of $D - M$ are nonnegative.

Let ϕ_0 be the N-dimensional column vector each of whose components is 1. The transpose ϕ_0' of ϕ_0 is the N-dimensional row vector $(1, 1, \ldots, 1)$. By definition of D we have $\phi_0'(D - M) = 0$, where the "0" on the right-hand side of this equality is the N-dimension row vector each of whose components is the number zero. Hence ϕ_0 is an eigenvector of $D - M$ with associated eigenvalue 0. It was proved in [Hall 70] that for a connected graph $\mathbb{G}$ the matrix $D - M$

has rank $N-1$, which implies that the other $N-1$ eigenvalues of $D-M$ are positive.

From (23.13), minimizing (23.12) subject to $x'x = 1$ is equivalent to minimizing $x'(D-M)x$ subject to $x'x = 1$. We write $x'x = 1$ as $x'Ix-1 = 0$, where I is the $N \times N$ identity matrix. Invoking the classical theory of constrained optimization, we associate the Lagrange multiplier λ with the constraint $x'Ix-1 = 0$ and form the Lagrangian $L(x,\lambda)$ defined by

$$L(x,\lambda) \equiv x'(D-M)x - \lambda(x'Ix - 1) = x'(D - M - \lambda I)x - \lambda .$$

Each solution of the optimization problem must satisfy, for $i = 1, 2, \ldots, N$,

$$\frac{\partial L(x,\lambda)}{\partial x_i} = 0$$

which implies $(D - M - \lambda I)x = 0$, that is, $(D-M)x = \lambda x$. Thus (x, λ) solves the optimization problem only if (i) $x'x = 1$ and (ii) x is an eigenvector of $D - M$ and λ is the associated eigenvalue. If x does satisfy (i) and (ii), then

$$x'(D-M)x = \lambda x'Ix = \lambda$$

so the objective function value at x is equal to the associated eigenvalue. Since we want to minimize the objective function, we want the smallest eigenvalue. But, as we have shown, the smallest eigenvalue is 0 and corresponds to the trivial solution with all x_i equal. Hence the desired solution is the eigenvector corresponding to the second smallest eigenvalue, and the minimal objective function value is the second smallest eigenvalue.

Example 23.2 (Continued) For the G illustrated in Fig. 23.7, the four eigenvalues and eigenvectors are given in Table 23.1. The solution is $x = \phi_2$, which yields an objective function value of 0.586; this solution "unravels" $\mathbb{G}$, placing node a to the left of node d, node d to the left of node b, and node b to the left of node c.   □

Table 23.1: Eigenvalues and eigenvectors for the example

Eigenvalue	Eigenvector
$\lambda_1 = 0.0$	$\phi_1 = (0.5,\ 0.5,\ 0.5,\ 0.5)$
$\lambda_2 = 0.586$	$\phi_2 = (-0.653,\ 0.270,\ 0.653,\ -0.270)$
$\lambda_3 = 2.0$	$\phi_3 = (-0.5,\ 0.5,\ -0.5,\ 0.5)$
$\lambda_4 = 3.414$	$\phi_4 = (0.270,\ 0.653,\ -0.270,\ -0.653)$

Exercise 23.4 Apply Hall's method to find the optimal 1-dimensional layout of $\mathbb{G}$ when $\mathbb{G}$ is the network of three nodes connected in a ring.   □

Exercise 23.5 Apply Hall's method to find the optimal 1-dimensional placement of nodes for the graph of Fig. 23.8.   □

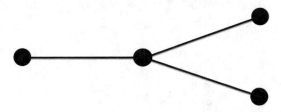

Figure 23.8: Small hub and spoke network

Bibliography

[Aaronson 02] S. Aaronson, "Book Review on 'A New Kind of Science' by Stephen Wolfram", *Quantum Information and Computation* **2** (2002) pp. 410–423

[Abraham 86] N.B. Abraham, A.M. Albano, B. Das, R.S. Gioggia, G. De Guzman, G.P. Puccioni, J.R. Tredicce, and S. Yong, "Calculating the Dimension of Attractors from Small Data Sets", *Physics Letters* **114A** (1986) pp. 217–221

[Adami 02] C. Adami and J. Chu, "Critical and Near-Critical Branching Processes", *Physical Review E* **66** (2002) 011907

[Adrangi 04] B. Adrangi, A. Chatrath, R. Kamath, and K. Raffiee, "Emerging Market Equity Prices and Chaos: Evidence from Thailand Exchange", *International Journal of Business* **9** (2004) pp. 159–178

[Aho 93] R.K. Ahuja, T.L. Magnanti, and J.B. Orlin, *Network Flows* (Prentice-Hall, New Jersey, 1993)

[Albert 02] R. Albert and A.-L. Barabási, "Statistical mechanics of complex networks", *Reviews of Modern Physics* **74** (2002) pp. 47–97

[Albert 99] R. Albert, H. Jeong, and A.-L. Barábasi, "Diameter of the World-Wide Web", *Nature* **401** (1999) pp. 130–131

[Albert 00] R. Albert, H. Jeong, and A.-L. Barábasi, "Error and Attack Tolerance of Complex Networks", *Nature* **406** (2000) pp. 378–382

[Allain 91] C. Allain and M. Cloitre, "Characterizing the Lacunarity of Random and Deterministic Fractal Sets", *Phys. Rev. A* **44** (1991) pp. 3552–3558

[Alon 66] N. Alon, Y. Matias, and M. Szegedy, "The Space Complexity of Approximating the Frequency Moments", in *Proc. 28th ACM Symp. on the Theory of Computing*, May 1966, pp. 20–29

[Alsaif 09] K.I. Alsaif and K.H. AL-Bajary, "Arabic Character Recognition Using Fractal Dimension", *Raf. J. of Compt. & Math's* **6** (2009) pp. 169–178

[Amarasinghe 15] A. Amarasinghe, U. Sonnadara, M. Berg and V. Cooray, "Fractal Dimension of Long Electrical Discharges", *Journal of Electrostatics* **73** (2015) pp. 33–37

© Springer Nature Switzerland AG 2020
E. Rosenberg, *Fractal Dimensions of Networks*,
https://doi.org/10.1007/978-3-030-43169-3

[Ambjorn 92] J. Ambjørn, B. Durhuus, T. Jónsson, and G. Thorleifsson, "Matter Fields with $c > 1$ Coupled to $2d$ Gravity", NBI-HE-92-35, August 1992, http://arxiv.org/pdf/hep-th/9208030.pdf

[Ambjorn 05] J. Ambjørn, J. Jurkiewicz, and R. Loll, "Spectral Dimension of the Universe is Scale Dependent", *Physical Review Letters* **95** 171301

[Anand 09] K. Anand and G. Bianconi, "Entropies of complex networks: toward an information theory of complex topologies", *Physical Review E (Rapid Communication)* **80** (2009) 045102

[Argyrakis 87] P. Argyrakis, "Information Dimension in Random-Walk Processes", *Physical Review Letters* **59** (2009) pp. 1729–1732

[Ariza 13] A.B. Ariza-Villaverde, F.J. Jiménez-Hornero, and E.G. De Ravé, "Multifractal Analysis of Axial Maps Applied to the Study of Urban Morphology", *Computers, Environment and Urban Systems* **38** (2013) pp. 1–10

[Arizabalo 11] R.D Arizabalo, E. González-Avalos, and G.S. Iglesias, "The Hurst Coefficient and Lacunarity Applied to Atmospheric Sub-micron Particle Data", *Latin-American Journal of Physics Education* **5** (2011) pp. 450–456

[Arlt 03] C. Arlt, H. Schmid-Schönbein, and M. Baumann, "Measuring the Fractal Dimension of the Microvascular Network of the Chorioallantoic Membrane", *Fractals* **11** (2003) pp. 205–212

[Austin 05] D. Austin, "Up and Down the Tiles", *Notices of the AMS* **52** (2005) pp. 600–601

[Ayala-Landeros 12] G. Ayala-Landeros, F. Carrion, J. Morales, V. Torres-Argüelles, D. Alaniz-Lumbreras, and V.M. Castaño, "Optical assessment of vibrating systems: A fractal geometry approach", *Optik - Int. J. Light Electron Opt.* **124** (2013) pp. 2260–2268

[Badii 85] R. Badii and A. Politi, "Statistical Description of Chaotic Attractors: The Dimension Function", *Journal of Statistical Physics* **40** (1985) pp. 725–750

[Baez 11] J.C. Baez, "Rényi Entropy and Free Energy", arXiv:1102.2098v3 [quant-ph], June 6, 2011, http://arxiv.org/abs/1102.2098

[Bais 08] F.A. Bais and J.D. Farmer, "The Physics of Information", chapter in *Philosophy of Information*, edited by P. Adriaans and J. van Benthem (Elsevier, Oxford, United Kingdom, 2008), available as arXiv:0708.2837v2

[Baish 00] J.W. Baish and R.K. Jain, "Fractals and Cancer", *Cancer Research* **60** (2000) pp. 3683–3688

[Baker 90] G.L. Baker and J.P. Gollub, *Chaotic Dynamics: an introduction* (Cambridge University Press, Cambridge, 1990)

[Barabási 02] A.-L. Barabási, *Linked* (Plume, NY, 2002)

[Barabási 09] A.-L. Barabási, "Scale-Free Networks: A Decade and Beyond", *Science* **325** (2009) pp. 412–413

[Barabási 18] A.-L. Barabási, "Love is All You Need: Clauset's fruitless search for scale-free networks", March 6, 2018 https://www.barabasilab.com/post/love-is-all-you-need

[Barabási 99] A.-L. Barabási and R. Albert, "Emergence of Scaling in Random Networks", *Science* **286** (1999) pp. 509–512

[Barabási 03] A.-L. Barabási and E. Banabeau, "Scale-Free Networks", *Scientific American* **288** (2003) pp. 50–59

[Barbaroux 01] J. Barbaroux, F. Germinet, and S. Tcheremchantsev, "Generalized Fractal Dimensions: Equivalences and Basic Properties", *Journal de Mathématiques Pures et Appliquées* **80** (2001) pp. 977–2012

[Barkoulas 98] J. Barkoulas and N. Travlos, "Chaos in an emerging capital market? The case of the Athens Stock Exchange", *Applied Financial Economics* **8** (1998) pp. 231–243

[Barnsley 12] M.F. Barnsley, *Fractals Everywhere: New Edition* (Dover Publications, Mineola, New York, 2012)

[Barrow 85] J.D. Barrow, S.P. Bhavsar, and D.H. Sonada, "Minimal Spanning Trees, Filaments and Galaxy Clustering", *Monthly Notices of the Royal Astronomical Society* **216** (1985) pp. 17–35

[Bass 99] T.A. Bass, "Black Box", first published in *The New Yorker*, April 26 & May 3, 1999, http://www.thomasbass.com/

[Bassett 06] D.S. Bassett and E. Bullmore, "Small-World Brain Networks", *The Neuroscientist*, Vol 12 (2006) pp. 512–523

[Bassett 06b] D.S. Bassett, A. Meyer-Lindenberg, S. Achard, T. Duke, and E. Bullmore, "Adaptive Reconfiguration of Fractal Small-World Human Brain Functional Networks", *PNAS* **103** (2006) 19518–19523

[Batagelj 06] V. Batagelj and A. Mrvar "Pajek datasets" (2006) http://vlado.fmf.uni-lj.si/pub/networks/data/

[Beardwood 59] J. Beardwood, J.H. Halton, and J.M. Hammersley, "The Shortest Path Through Many Points", *Proc. Cambridge Philosophical Society* **55** (1959) pp. 299–327

[Beck 09] C. Beck, "Generalized Information and Entropy Measures in Physics", *Contemporary Physics* **50** (2009) pp. 495–510

[Benguigui 92] L. Benguigui, "The Fractal Dimension of Some Railway Networks", *Journal de Physique I France* **2** (1992) pp. 385–388

[Berger 05] A. Berger, L.A. Bunimovich, and T.P. Hill, "One-Dimensional Dynamical Systems and Benford's Law", *Trans. Amer. Math. Soc.* **357** (2005) pp. 197–219

[Berger 11] A. Berger and T.P. Hill, "A basic theory of Benford's Law", *Probability Surveys* **8** (2011) pp. 1–126

[Berke 07] J. Berke, "Measuring of Spectral Fractal Dimension", *New Mathematics and Natural Computation* **3** (2007) pp. 409–418

[Berntson 97] G.M. Berntson and P. Stoll, "Correcting for Finite Spatial Scales of Self-Similarity when Calculating the Fractal Dimensions of Real-World Structures", *Proc. Royal Society London B* **264** (1997) pp. 1531–1537

[Berthelsen 92] C. Berthelsen, J.A. Glazier, and M.H. Skolnick, "Global Fractal Dimension of Human DNA Sequences Treated as Pseudorandom Walks", *Physical Review A* **45** (1992) pp. 8902–8913

[Bez 11] N. Bez and S. Bertrand, "The Duality of Fractals: Roughness and Self-Similarity", *Theoretical Ecology* **4** (2011) pp. 371–383

[Bian 18] T. Bian and Y. Deng, "Identifying Influential Nodes in Complex Networks: A Node Information Dimension Approach", *Chaos* **28** (2018) 043109

[Bianconi 07] G. Bianconi, "The Entropy of Randomized Network Ensembles", *Europhysics Letters* **81** (2007) 28005

[Bianconi 09] G. Bianconi, "Entropy of network ensembles", *Physical Review E* **79** (2009) 036114

[BioGRID] BioGRID: http://thebiogrid.org/download.php

[Birbaumer 96] N. Birbaumer, W. Lutzenberger, H. Rau, G. Mayer-Kress, I. Choi, and C. Braun, "Perception of Music and Dimensional Complexity of Brain Activity", *International Journal of Bifurcation and Chaos* **6** (1996) pp. 267–278

[Block 90] A. Block, W. von Bloh, and H.J. Schellnhuber, "Efficient Box-Counting Determination of Generalized Fractal Dimension", *Physical Review A* **42** (1990) pp. 1869–1874

[Block 91] A. Block, W. von Bloh, and H.J. Schellnhuber, "Aggregation by Attractive Particle-Cluster Interaction", *J. Phys. A: Math. Gen.* **24** (1991) pp. L1037–L1044

[Boccaletti 06] S. Boccaletti, V. Latora, Y. Moreno, M. Chavez, and D.U. Hwang, "Complex Networks: Structures and Dynamics", *Physics Reports* **424** (2006) pp. 175–308

[Bogomolny 98] A. Bogomolny, "Collage Theorem", March, 1998, in *Interactive Mathematics Miscellany and Puzzles*, http://www.cut-the-knot.org/ctk/ifs.shtml

[Bollobas 85] B. Bollobás, *Random Graphs* (Academic, London, 1985)

[Bollobas 82] B. Bollobás and W. F. de la Vega, "The Diameter of Random Regular Graphs", *Combinatorica* **2** (1982) pp. 125–134

[Bonchev 05] D. Bonchev and G.A. Buck, "Quantitative Measures of Network Complex", Chapter 5 in *Complexity in Chemistry, Biology, and Ecology*, D. Bonchev and D.H. Rouvray, eds. (Springer, New York, 2005)

[Boon 08] M.Y. Boon, B.I. Henry, C.M. Suttle, and S.J. Dain, "The Correlation Dimension: A Useful Objective Measure of the Transient Visual Evoked Potential?", *Journal of Vision* **8** (2008) pp. 1–21

[Borys 09] P. Borys, "On the Relation between Lacunarity and Fractal Dimension", *Acta Physica Polonica B* **40** (2009) pp. 1485–1489, Erratum *Acta Physica Polonica B* **42** (2011) 2045

[Boser 05] S.R. Boser, H. Park, S.F. Perry, M.C. Menache, and F.H.Y. Green, "Fractal Geometry of Airway Remodelling in Human Asthma", *American Journal of Respiratory and Critical Care Medicine* **172** (2005) pp. 817–823

[Briggs 92] J. Briggs, *Fractals, the Patterns of Chaos: Discovering a New Aesthetic of Art, Science, and Nature* (Simon & Schuster, New York, 1992)

[Broido 18] A.D. Broido, and A. Clauset, "Scale-Free Networks are Rare", *arXiv:1801.03400 [physics.soc-ph]*, January 9, 2018

[Brown 05] C. T. Brown, W. R. T. Witschey, and L. S. Liebovitch "The Broken Past: Fractals in Archaeology", *Journal of Archaeological Method and Theory* **12** (2005) pp. 37–78

[Brown 02] J.H. Brown, V.K. Gupta, B.-L. Li, B.T. Milne, C. Restrepo, and G.B. West, "The Fractal Nature of Nature: Power Laws, Ecological Complexity, and Biodiversity", *Phil. Trans. R. Soc. Lond. B* **357** (2002) pp. 619–626

[Buhl 09] J. Buhl, K. Hicks, E.R. Miller, S. Persey, O. Alinvi, and D.J.T. Sumpter, "Shape and Efficiency of Wood Ant Foraging Networks," *Behavioral Ecology and Sociobiology* **63** (2009) pp. 451–460

[Burda 01] Z. Burda, J.D. Correia, and A. Krzywicki, "Statistical Ensemble of Scale-Free Random Graphs", *Physical Review E* **64** (2001) 046118

[Camastra 97] F. Camastra and A. M. Colla, "Short-Term Load Forecasting based on Correlation Dimension Estimation and Neural Nets", *Artificial Neural Networks - ICANN '97, Lecture Notes in Computer Science* **1327** (1997) pp. 1035–1040

[Camastra 02] F. Camastra and A. Vinciarelli, "Estimating the Intrinsic Dimension of Data with a Fractal-Based Method", *IEEE Trans. on Pattern Analysis and Machine Intelligence* **24** (2002) pp. 1404–1407

[Caniego 05] F.J. Caniego, R. Espejo, M.A. Martín, and F. San José, "Multifractal Scaling of Soil Spatial Variability", *Ecological Modelling* **182** (2005) pp. 291–303

[Carter 14] T. Carter, "An Introduction to Information Theory and Entropy", *Complex Systems Summer School*, Santa Fe Institute, March 14, 2014 https://www.csustan.edu/sites/default/files/CS/Carter/documents/info-lec.pdf

[Cawley 92] R. Cawley and R.D. Mauldin, "Multifractal Decompositions of Moran Fractals", *Advances in Mathematics* **92** (1992) pp. 196–236

[Chang 08] K. Chang, "Edward N. Lorenz, a Meteorologist and a Father of Chaos Theory, Dies at 90", *The New York Times*, 17 April 2008

[Chang 06] K.H. Chang, B.C. Choi, S.-M. Yoon, and K. Kim, "Multifractal Measures on Small-World Networks", *Fractals* **14** (2006) pp. 119–123

[Chapman 06] R. W. Chapman, J. Robalino†, and H. F. Trent III, "EcoGenomics: Analysis of Complex Systems via Fractal Geometry", *Integrative and Comparative Biology* **46** (2006) pp. 902–911

[Chatterjee 92] S. Chatterjee and M.R. Yilmaz, "Chaos, Fractals, and Statistics", *Statistical Science* **7** (1992) pp. 49–68.

[Chen 11] Y. Chen, "Modeling Fractal Structure of City-Size Distributions using Correlation Functions", *PLoS ONE* **6** (2011) pp. 1–9

[Chen 18] Y.-G. Chen, "Logistic Models of Fractal Dimension Growth of Urban Morphology", *Fractals* **26** (2018) 1850033

[Chhabra 89] A.B. Chhabra, C. Meneveau, R.V. Jensen, and K.R. Sreenivasan, "Direct Determination of the $f(\alpha)$ Singularity Spectrum and its Application to Fully Developed Turbulence", *Physical Review A* **40** (1989) pp. 5284–5294

[Christensen 12] C. Christensen, G. Bizhani, S.-W. Son, M. Paczuski, and P. Grassberger, "Agglomerative Percolation in Two Dimensions", *Europhysics Letters* **97** (2012) 16004

[Chvatal 79] V. Chvátal, "A Greedy Heuristic for the Set Covering Problem", *Mathematics of Operations Research* **4** (1979) pp. 233–235

[Chvatal 83] V. Chvátal, *Linear Programming* (W.H. Freeman, 1983)

[Claps 96] P. Claps and G. Oliveto, "Reexamining the Determination of the Fractal Dimension of River Networks", *Water Resources Research* **32** (1996) pp. 3123–3135

[Clauset 09] A. Clauset, C.R. Shalizi, and M.E.J. Newman, "Power-Law Distributions in Empirical Data", *SIAM Review* **51** (2009) pp. 661–703

[Cohen 10] E.R. Cohen and P. Giacomo, "Symbols, Units, Nomenclature, and Fundamental Constants in Physics, 1987 Revision(2010) Preprint", International Union of Pure and Applied Physics, Commission C2 - SUNAMCO, http://iupap.org/wp-content/uploads/2014/05/A4.pdf

[Cohen 03] R. Cohen and S. Havlin, "Scale-Free Networks are Ultrasmall", *Physical Review Letters* **90** (2003) 058701

[Cohen 04] R. Cohen and S. Havlin, "Fractal Dimensions of Percolating Networks", *Physica A* **336** (2004) pp. 6–13

[Cohen 08] R. Cohen and S. Havlin, "Scaling Properties of Complex Networks and Spanning Trees", Chapter 3 in *Handbook of Large-Scale Random Networks*, B. Bollobás, P. Kozma, and D. Miklós, eds. (János Bolyai Mathematical Society and Springer-Verlag, 2008) pp. 153–169

[Comellas 02] F. Comellas and M. Sampels, "Deterministic Small-World Networks", *Physica A: Statistical Mechanics and its Applications* **309** (2002) pp. 231–235

[Concas 06] G. Concas, M.F. Locci, M. Marchesi, S. Pinna, and I. Turnu, "Fractal Dimension in Software Networks", *Europhysics Letters* **76** (2006) pp. 1221–1227

[Cornforth 02] D. Cornforth, H. Jelinek, and L. Peichl, "Fractop: a tool for automated biological image classification", Presented at the 6th Australasia-Japan Joint Workshop ANU, Canberra, Australia 30 November - 1 December 2002, https://researchoutput.csu.edu.au/

[Corr 01] "Correspondence re: J.W. Baish and R.K. Jain, Fractals and Cancer. Cancer Res., 60: 3683–3688, 2000.", Letters to the Editor, *Cancer Research* **61** (2001) pp. 8347–8351

[Costa 04] L. da F. Costa, "Voronoi and Fractal Complex Networks and Their Characterization", *International Journal of Modern Physics C* **15** (2004) 175

[Costa 11] L. da F. Costa, O.N. Oliveira Jr., G. Travieso, F. A. Rodrigues, P.R.V. Boas, L. Antiqueira, M.P. Viana, and L.E.C. Rocha, "Analyzing and Modeling Real-World Phenomena with Complex Networks: a Survey of Applications", *Advances in Physics* **60** (2011) pp. 329–412

[Costa 07] L. da F. Costa, F. A. Rodrigues, G. Travieso and P.R.V. Boas, "Characterization of Complex Networks: A Survey of Measurements", *Advances in Physics* **56** (2007) pp. 167–242

[Cronin 69] J. Cronin-Scanlon, *Advanced Calculus* (D.C. Heath and Company, Lexington, MA, 1969)

[Curl 86] R.L. Curl, "Fractal Dimensions and Geometries of Caves", *Mathematical Geology* **18** (1986) pp. 765–783

[Cytoscape] Cytoscape software: https://js.cytoscape.org/

[Daccord 86] G. Daccord, J. Nittmann, and H.E. Stanley, "Radial Viscous Fingers and Diffusion-Limited Aggregation: Fractal Dimension and Growth Sites", *Physical Review Letter* **56** (1986) pp. 336–339

[Dam 03] E.R. van Dam and W.H. Haemers, "Which Graphs are Determined by Their Spectrum?", *Linear Algebra and its Applications* **373** (2003) pp. 241–272

[Daqing 11] L. Daqing, K. Kosmidis, A. Bunde, and S. Havlin, "Dimension of Spatially Embedded Networks", *Nature Physics*, 27 February 2011

[Das 05] A. Das and P. Das, "Classification of Different Indian Songs Based on Fractal Analysis", *Complex Systems* **15** (2005) pp. 253–259

[Da Silva 06] D. Da Silva, F. Boudon, C. Godin, P. Puech, C. Smith, and H. Sinoquet, "A Critical Appraisal of the Box Counting Method to Assess the Fractal Dimension of Tree Crowns", in *Advances in Visual Computing*, Lecture Notes in Computer Science **4291** (2006) pp. 751–760

[Daxer 93] A. Daxer, "Characterisation of the Neovascularisation Process in Diabetic Retinopathy by Means of Fractal Geometry: Diagnostic Dmplications", *Graefe's Archive for Clinical and Experimental Ophthalmology* **231** (1993) pp. 681–686

[DeBartolo 04] S.G. De Bartolo, R. Gaudio and S. Gabriele, "Multifractal Analysis of River Networks: Sandbox Approach", *Water Resources Research* **40** (2004) W02201

[Luke 07] D.A. Luke and J.K. Harris, "Network Analysis in Public Health: History, Methods, and Applications", *Annual Review of Public Health* **28** (2007) pp. 69–93

[Lusseau 03] D. Lusseau, K. Schneider, O. J. Boisseau, P. Haase, E. Slooten, and S. M. Dawson, "The Bottlenose Dolphin Community of Doubtful Sound Features a Large Proportion of Long-Lasting Associations", *Behavioral Ecology and Sociobiology* **54** (2003) pp. 396–405

[Mandelbrot 67] B. Mandelbrot, "How Long is the Coast of Britain? Statistical Self-Similarity and Fractional Dimension", *Science* **156** (1967) pp. 636–638

[Mandelbrot 75] B. Mandelbrot, *Les Objets Fractals: forme, hazard et dimension* (Flammarion, Paris, 1975)

[Mandelbrot 77] B. Mandelbrot, *Fractals: forms, chance and dimension* (Freeman, San Francisco, 1977)

[Mandelbrot 83a] B. Mandelbrot, "On Fractal Geometry, and a Few of the Mathematical Questions it has Raised", *Proceedings of the International Congress of Mathematicians*, August 16–24, 1983, Warszawa

[Mandelbrot 83b] B. Mandelbrot, *The Fractal Geometry of Nature* (W.H. Freeman, New York, 1983)

[Mandelbrot 85] B. Mandelbrot, "Self-Affine Fractals and Fractal Dimension", *Physica Scripta* **32** (1985) pp. 257–260

[Mandelbrot 86] B. Mandelbrot, "Self-Affine Fractal Sets", in *Fractals in Physics*, L. Pietronero and E. Tosatti, eds. (North Holland, Amsterdam, 1986) pp. 3–28

[Mandelbrot 89] B. Mandelbrot, "Fractal Geometry: what is it, and what does it do?", *Proc. R. Soc. Lond. A* **423** (1989) pp. 3–16

[Mandelbrot 90] B. Mandelbrot, "Negative Fractal Dimensions and Multifractals", *Physica A* **163** (1990) pp. 306–315

[Mandelbrot 97] B. Mandelbrot, *Fractals and Scaling in Finance* (Springer-Verlag, New York, 1997)

[Mandelbrot 98] B. Mandelbrot, P. Pfeifer, O. Biham, O. Malaci, D. Lidar, and D. Avnir, "Is Nature Fractal?", *Science*, 6 February 1998: 783.

[Mangasarian 69] O.L. Mangasarian, *Nonlinear Programming* (McGraw-Hill, New York, 1969)

[Manin 06] Y.I. Manin, "The Notion of Dimension in Geometry and Algebra", *Bulletin (New Series) of the American Mathematical Society* **43** (2006) pp. 139–161

[Marana 99] A.N. Marana, L. da F. Costa, R.A. Lotufo, and S.A. Velastin, "Estimating Crowd Density with Minkowski Fractal Dimension", *Proc. 1999 IEEE International Conference on Acoustics, Speech, and Signal Processing* (ICASSP99) 15–19 March 1999

[Marliere 01] C. Marlière, T. Woignier, P. Dieudonné, J. Primera, M. Lamy, and J. Phalippou, "Two Fractal Structures in Aerogel", *Journal of Non-Crystalline Solids* **285** (2001) pp. 175–180

[Martin 08] M.A. Martin and M. Reyes, "A Fractal Interaction Model for Winding Paths through Complex Distributions: Applications to Soil Drainage Networks", *Pure and Applied Geophysics* **165** (2008) pp. 1153–1165

[Martinerie 92] J.M. Martinerie, A.M. Albano, A.I. Mees, and P.E. Rapp, "Mutual Information, Strange Attractors, and the Optimal Estimation of Dimension", *Physical Review A* **45** (1992) 7058

[Martinez 90] V.J. Martinez and B.J.T. Jones, "Why the Universe is not a Fractal", *Monthly Notices of the Royal Astronomical Society* **242** (1990) pp. 517–521

[Martin-Garin 07] B. Martin-Garin, B. Lathuilière, E.P. Verrecchia, and J. Giester, "Use of Fractal Dimensions to Quantify Coral Shape", *Coral Reefs* **26** (2007) pp. 541–550

[Mayfield 92] E.S. Mayfield and B. Mizrach, "On Determining the Dimension of Real-Time Stock-Price Data", *Journal of Business and Economic Studies* **10** (1992) pp. 367–374

[Meakin 83] P. Meakin and H.E. Stanley, "Spectral Dimension for the Diffusion-Limited Aggregation Model of Colloid Growth", *Physical Review Letters* **51** (1983) pp. 1457–1460

[Meyer-Ortmanns 04] H. Meyer-Ortmanns, "Functional Complexity Measure for Networks", *Physica A* **337** (2004) pp. 679–690

[Milgram 67] S. Milgram, "The Small-World Problem", *Psychology Today* **2** (1967) pp. 61–69

[Mizrach 92] B. Mizrach, "The State of Economic Dynamics", *Journal of Economic Dynamics and Control* **16** (1992) pp. 175–190

[Modesto 10] L. Modesto and P. Nicolini, "Spectral Dimension of a Quantum Universe", *Physical Review D* **81** (2010) 104040

[Mohammadi 11] S. Mohammadi, "Effect of Outliers on the Fractal Dimension Estimation", *Fractals* **19** (2011) pp. 233–241

[Murdzek 07] R. Murdzek, "The Box-Counting Method in the Large Scale Structure of the Universe", *Rom. Journ. Phys.* **52** (2007) pp. 149–154

[Murr 05] M.M. Murr and D.E. Morse, "Fractal Intermediates in the Self-Assembly of Silicatein Filaments", *PNAS* **102** (2005) pp. 11657–11662

[Murray 95] J.D. Murray, "Use and Abuse of Fractal Theory in Neuroscience", *The Journal of Comparative Neurology* **361** (1995) pp. 369–371

[Nakayama 94] T. Nakayama, K. Yakubo, and R.L. Orbach, "Dynamical Properties of Fractal Networks: Scaling, Numerical Simulations, and Physical Realizations", *Reviews of Modern Physics* **66** (1994) pp. 381–443

[Nasr 02] F.B. Nasr, I. Bhouri, and Y. Heurteaux, "The Validity of the Multifractal Formalism: Results and Examples", *Advances in Mathematics* **165** (2002) pp. 264–284

[Nematode] "What are nematodes?", University of Nebraska-Lincoln, http://nematode.unl.edu/wormgen.htm

[Newman] M.E.J. Newman, "Network Data", http://www-personal.umich.edu/~mejn/netdata/

[Newman 00] M.E.J. Newman, "Models of the Small World", *Journal of Statistical Physics* **101** (2000) pp. 819–841

[Newman 03] M.E.J. Newman, "The Structure and Function of Complex Networks", *SIAM Review* **45** (2003) pp. 167–256

[Newman 99a] M.E.J. Newman and D.J. Watts, "Scaling and Percolation in the Small-World Network Model", *Physical Review E* **60** (1999) 7332

[Newman 99b] M.E.J. Newman and D.J. Watts, "Renormalization Group Analysis of the Small-World Network Model", *Physics Letters A* **263** (1999) pp. 341–346

[Nowotny 88] T. Nowotny and M. Requardt, "Dimension Theory of Graphs and Networks", *J. Phys. A: Math. Gen.* **31** (1988) pp. 2447–2463

[Nowotny 07] T. Nowotny and M. Requardt, "Emergent Properties in Structurally Dynamic Disordered Cellular Networks", *Journal of Cellular Automata* **2** (2007) pp. 273 289

[O'Connor 04] http://www-history.mcs.st-andrews.ac.uk/Biographies/Hausdorff.html

[Oiwa 02] N.N. Oiwa and J.A. Glazier, "The Fractal Structure of Mitochondrial Genomes", *Physica A* **311** (2002) pp. 221–230

[Olsen 95] L. Olsen, "A Multifractal Formalism", *Advances in Mathematics* **116** (1995) pp. 82–196

[Oppelt 00] A.L. Oppelt, W. Kurth, H. Dzierzon, G. Jentschke, and D.L. Godbold, "Structure and Fractal Dimensions of Root Systems of Four Co-Occurring Fruit Tree Species from Botswana", *Ann. For. Sci.* **57** (2000) pp. 463–475

[Orozco 15] C.D.V. Orozco, J. Golay,, and M. Kanevski, "Multifractal Portrayal of the Swiss Population", *Cybergeo: European Journal of Geography*, **714** (2015) http://cybergeo.revues.org/26829

[Ostwald 09] M.J. Ostwald and J. Vaughan, "Calculating Visual Complexity in Peter Eisenman's Architecture", *Proc. 14-th International Conference on Computer Aided Architectural Design Research in Asia* (Yunlin, Taiwan, 22–25 April 2009) pp. 75–84

[Pal 08] P.K. Pal, "Geomorphological, Fractal Dimension and *b*-Value Mapping in Northeast India", *J. Ind. Geophys. Union* **12** (2008) pp. 41–54

[Palla 10] G. Palla, L. Lovász, and T. Vicsek, "Multifractal Network Generator", *PNAS* **107** (2010) pp. 7640–7645

[Palla 11] G. Palla, P. Pollner, and T. Vicsek, "Rotated Multifractal Network Generator", *Journal of Statistical Mechanics* (2011) P02003

[Panas 01] E. Panas, "Long Memory and Chaotic Models of Prices on the London Metal Exchange", *Resources Policy* **27** (2001) pp. 235–246

[Panico 95] J. Panico and P. Sterling, "Retinal Neurons and Vessels Are Not Fractal But Space-Filling", *The Journal of Comparative Neurology* **361** (1995) pp. 479–490

[Pansiot 98] J.J. Pansiot and D. Grad, "On Routes and Multicast Trees in the Internet", *ACM Computer Communication Review* **28** (1998) pp. 41–50

[Pearse] E. Pearse, "An Introduction to Dimension Theory and Fractal Geometry: Fractal Dimensions and Measures", www.math.cornell.edu/~erin/docs/dimension.pdf

[Peitgen 92] H.O. Peitgen, H. Jürgens, and D. Saupe, *Chaos and Fractals* (Springer-Verlay, New York, 1992)

[Pelletier 00] J.D. Pelletier and D.L. Turcotte, "Shapes of River Networks and Leaves: Are They Statistically Similar?", *Phil. Trans. R. Soc. Lond. B* **355** (2000) pp. 307–311

[Pidwirny 06] M. Pidwirny, "Stream Morphometry" in *Fundamentals of Physical Geography, 2nd Ed.* (2006) http://www.physicalgeography.net/fundamentals/10ab.html

[Pilipski 02] M. Pilipski, "The Fractal Lacunarity of Dry Walls and Their Matrix Stones: Why Some Walls Look Good and Others Do Not", Proc. 8th International Drystone-Walling Congress in Visp – Valais (CH) August/September 2002

[Pita 02] J.R. Castrejon Pita, A. Sarmiento Galan, and R. Castrejon Garcia, "Fractal Dimension and Self-Similarity in Asparagus Plumosus", *Fractals* **10** (2002) pp. 429–434

[Plotnick 93] R.E. Plotnick, R.H. Gardner, and R.V. O'Neill, "Lacunarity Indices as Measures of Landscape Texture", *Landscape Ecology* **8** (1993) pp. 201–211

[Power 87] W.L. Power, T.E. Tullis, S.R. Brown, G.N. Boitnott, and C.H. Scholz, "Roughness of Natural Fault Surfaces", *Geophysical Research Letters* **14** (1987) pp. 29–32

[Prigarin 13] S.M. Prigarin, K. Sandau, M. Kazmierczak, and K. Hahn, "Estimation of Fractal Dimension: A Survey with Numerical Experiments and Software Description", *International Journal of Biomathematics and Biostatistics* **2** (2013) pp. 167–180

[Proulx 05] S.R. Proulx, D.E.L. Promislow, and P.C. Phillips "Network Thinking in Ecology and Evolution", *Trends in Ecology and Evolution* **20** (2005) pp. 345–353

[Radicchi 09] F. Radicchi, A. Barrat, S. Fortunato, and J.J. Ramasco, "Renormalization Flows in Complex Networks", *Physical Review E* **79** (2009) 026104

[Ramachandrarao 00] P. Ramachandrarao, A. Sinha, and D. Sanyal, "On the Fractal Nature of Penrose Tiling", *Current Science* **79** (2000) pp. 365–366

[Ramirez 19] A. Ramirez-Arellano, S. Bermúdez-Gómez, L. Manuel Hernández-Simón, and J. Bory-Reyes, "D-Summable Fractal Dimensions of Complex Networks", *Chaos, Solitons and Fractals* **119** (2019) pp. 210–214

[Ramsey 90] J.B. Ramsey, C.L. Sayers, and P. Rothman, "The Statistical Properties of Dimension Calculations Using Small Data Sets: Some Economic Applications", *International Economic Review* **31** (1990) pp. 991–1020

[Ramsey 89] J.B. Ramsey and H.-J. Yuan, "Bias and Error Bars in Dimension Calculations and their Evaluation in some Simple Models", *Physics Letters A* **134** (1989) pp. 287–297

[Rasband] W.S. Rasband, "ImageJ" U. S. National Institutes of Health, Bethesda, Maryland, USA (1997–2016) http://imagej.nih.gov/ij/

[Reijneveld 07] J.C. Reijneveld, S.C. Ponten, H.W. Berendse, and C.J. Stam, "The Application of Graph Theoretical Analysis to Complex Networks in the Brain", *Clinical Neurophysiology* **118** (2007) pp. 2317–2331

[Rendon 17] S. Rendón de la Torre, J. Kalda, R. Kitt, and J. Engelbrecht, "Fractal and Multifractal Analysis of Complex Networks: Estonian Network of Payments", *European Physics Journal B* **90** (2017) 234

[Rényi 70] A. Rényi, *Probability Theory* (North Holland, Amsterdam, 1970)

[Requardt 06] M. Requardt, "The Continuum Limit of Discrete Geometries", *Int. J. Geom. Methods Mod. Phys.* **3** (2006) pp. 285–313

[Reuveni 10] S. Reuveni, R. Granek, and J. Klafter, "Anomalies in the Vibrational Dynamics of Proteins are a Consequence of Fractal-Like Structure", *PNAS* **107** (2010) pp. 13696–13700

[Riddle 19] L. Riddle, "Classic Iterated Function Systems", http://ecademy.agnesscott.edu/~lriddle/ifs/siertri/siertri.htm

[Riedi 95] R. Riedi, "An Improved Multifractal Formalism and Self-Similar Measures", *Journal of Mathematical Analysis and Applications* **189** (1995) pp. 462–490

[Rigaut 90] J.P. Rigaut, "Fractal Models in Biomedical Image Analysis and Vision", *Acta Stereol* **9** (1990) pp. 37–52

[Rioul 14] O. Rioul and J.C. Magossi, "On Shannon's Formula and Hartley's Rule: Beyond the Mathematical Coincidence", *Entropy* **16** (2014) pp. 4892–4910

[Roberts 05] A.J. Roberts, "Use the Information Dimension, not the Hausdorff", arXiv:nlin/0512014v1 [nlin.PS], revised 13 June 2018

[Rockafellar 74] R.T. Rockafellar, *Conjugate Duality and Optimization* (Society for Industrial and Applied Mathematics, Philadelphia, Pennsylvania, 1974)

[Rodin 00] V. Rodin and E. Rodina, "The Fractal Dimension of Tokyo's Streets", *Fractals* **8** (2000) pp. 413–418

[Rosenberg 01] E. Rosenberg, "Dual Ascent for Uncapacitated Network Design with Access, Backbone, and Switch Costs", *Telecommunications Systems* **16** (2001) pp. 423–435

[Rosenberg 02] E. Rosenberg, "Capacity Requirements for Node and Arc Survivable Networks", *Telecommunications Systems* **20** (2002) pp. 107–131

[Rosenberg 05] E. Rosenberg, "Hierarchical Topological Network Design", *IEEE/ACM Trans. on Networking* **13** (2005) pp. 1402–1409

[Rosenberg 12] E. Rosenberg, *A Primer of Multicast Routing* (Springer, New York, 2012)

[Rosenberg 13] E. Rosenberg, "Lower Bounds on Box Counting for Complex Networks", *Journal of Interconnection Networks* **14** (2013) 1350019

[Rosenberg 16a] E. Rosenberg, "Minimal Box Size for Fractal Dimension Estimation", *Community Ecology* **17** (2016) pp. 24–27

[Rosenberg 16b] E. Rosenberg, "The Correlation Dimension of a Rectilinear Grid", *Journal of Interconnection Networks* **16** (2016) 1550010

[Rosenberg 17a] E. Rosenberg, "Maximal Entropy Coverings and the Information Dimension of a Complex Network", *Physics Letters A* **381** (2017) pp. 574–580

[Rosenberg 17b] E. Rosenberg, "Minimal Partition Coverings and Generalized Dimensions of a Complex Network", *Physics Letters A* **381** (2017) pp. 1659–1664

[Rosenberg 17c] E. Rosenberg, "Non-monotonicity of the Generalized Dimensions of a Complex Network", *Physics Letters A* **381** (2017) pp. 2222–2229

[Rosenberg 17d] E. Rosenberg, "Erroneous Definition of the Information Dimension in Two Medical Applications", *International Journal of Clinical Medicine Research* **4** (2017) pp. 72–75

[Rosenberg 18a] E. Rosenberg, *A Survey of Fractal Dimensions of Networks* (Springer International Publishing AG, 2018)

[Rosenberg 18b] E. Rosenberg, "Generalized Hausdorff Dimensions of a Complex Network", *Physica A* **511** (2018) pp. 1–17

[Rosenberg 13a] E. Rosenberg and J. Uttaro, "A Fast Re-Route Method", *IEEE Communications Letters* **17** (2013) pp. 1656–1659

[Rovelli 16] C. Rovelli, *Seven Brief Lectures on Physics* (Riverhead, New York, 2016)

[Royden 68] H.L. Royden, *Real Analysis, 2nd ed.* (Macmillan, New York, 1968)

[Rozenfeld 09] H. D. Rozenfeld, L. K. Gallos, C. Song, and H. A. Makse, "Fractal and Transfractal Scale-Free Networks", in *Encyclopedia of Complexity and Systems Science*, R. A. Meyers, ed. (Springer, 2009) pp. 3924–3943

[Ruelle 80] D. Ruelle, "Strange Attractors", *The Mathematical Intelligencer* **2** (1980) pp. 126–137

[Ruelle 90] D. Ruelle, "Deterministic Chaos: The Science and the Fiction (The 1989 Claude Bernard Lecture)", *Proc. R. Soc. Lond. A* **427** (1990) pp. 241–248

[Salingaros 12] N. Salingaros, "Fractal Art and Architecture Reduce Psychological Stress", *Journal of Biourbanism* **2** (2012) pp. 11–28

[Schlamm 10] A. Schlamm, R.G. Resmini, D. Messinger, and W. Basener, "A Comparison Study of Dimension Estimation Algorithms", *Proc. SPIE 7695, Algorithms and Technologies for Multispectral, Hyperspectral, and Ultraspectral Imagery XVI* 76952D (12 May 2010)

[Schleicher 07] D. Schleicher, "Hausdorff Dimension, Its Properties, and Its Surprises", *The American Mathematical Monthly* **114** (2007) pp. 509–528

[Schneider 03] T.D. Schneider, "Information Theory Primer", version=2.54 of primer.tex, 6 January 2003

[Schroeder 91] M. Schroeder, *Fractals, Chaos, Power Laws: Minutes from an Infinite Paradise* (W.H. Freeman, New York, 1991)

[Seary] A.J. Seary and W.D. Richards, "Spectral Methods for Analyzing and Visualizing Networks: an Introduction", http://www.sfu.ca/personal/archives/richards/Pages/NAS.AJS-WDR.pdf

[Serrano 13] S. Serrano, F. Perán, F.J. Jiménez-Hornero, and E. Gutiérrez de Ravé, "Multifractal Analysis Application to the Characterization of Fatty Infiltration in Iberian and White Pork Sirloins", *Meat Science* **93** (2013) pp. 723–732

[Seuront 14] L. Seuront and H.E. Stanley, "Anomalous Diffusion and Multifractality Enhance Mating Encounters in the Ocean", *PNAS* **111** (2014) pp. 2206–2211

[Shanker 07a] O. Shanker, "Defining Dimension of a Complex Network", *Modern Physics Letters B* **21** (2007) pp. 321–326

[Shanker 07b] O. Shanker, "Graph Zeta Function and Dimension of Complex Network", *Modern Physics Letters B* **21** (2007) pp. 639–644

[Shanker 13] O. Shanker, "Dimension Measure for Complex Networks", in *Advances in Network Complexity*, M. Dehmer, A. Mowshowitz, and F. Emmert-Streib, eds. (Wiley-VCH Verlag GmbH & Co. KGaA, 2013)

[Shannon 01a] "Claude Shannon, Father of Information Theory, Dies at 84", www.bell-labs.com/news/2001/february/26/1.html

[Shannon 01b] "MIT Professor Claude Shannon Dies: Was Founder of Digital Communications", web.mit.edu/newsoffice/2001/shannon.html, 27 February 2001

[Shen 02] Q. Shen, "Fractal Dimension and Fractal Growth of Urbanized Areas", *Int. J. Geographical Information Science* **16** (2002) pp. 419–437

[Shenker 94] O.R. Shenker, "Fractal Geometry is Not the Geometry of Nature", *Stud. Hist. Phil. Sci.* **25** (1994) pp. 967–981

[Shiozawa 15] Y. Shiozawa, B.N. Miller, and J.L. Rouet, "Fractal Dimension Computation From Equal Mass Partitions", *Chaos* **24** (2014) 033106

[Shirer 97] H.N. Shirer, C.J. Fosmire, R. Wells, and L. Suciu, "Estimating the Correlation Dimension of Atmospheric Time Series", *Journal of the Atmospheric Sciences* **54** (1997) pp. 211–230

[Silva 13] F.N. Silva and L. da F. Costa, "Local Dimension of Complex Networks", https://arxiv.org/pdf/1209.2476.pdf, 11 August 2013

[Simon 55] H.A. Simon, "On a Class of Skew Distribution Functions", *Biometrika* **42** (1955) pp. 425–440

[Simpelaere 98] D. Simpelaere, "Correlation Dimension", *Journal of Statistical Physics* **90** (1998) pp. 491–509

[Slater 75] P. J. Slater, "Leaves of Trees", *Congr. Numerantium* **14** (1975) pp. 549–559

[Smail 53] L.L. Smail, *Analytic Geometry and Calculus* (Appleton-Century-Crofts, New York, 1953)

[Smith 88] L.A. Smith, "Intrinsic Limits on Dimension Calculations", *Physics Letters* **A133** (1988) pp. 283–288

[Smith 92] R.L. Smith, "Comment: Relation Between Statistics and Chaos", *Statistical Science* **7** (1992) pp. 109–113

[Smits 16] F.M. Smits, C. Porcaro, C. Cottone, A. Cancelli, P.M. Rossini, and F. Tecchio, "Electroencephalographic Fractal Dimension in Healthy Ageing and Alzheimer's Disease", *PLOS ONE* **11** (2016) e0149587

[Snyder 95] T.L. Snyder and J.M. Steele, "Probabilistic Networks and Network Algorithms", in M.O. Ball et al., eds., *Handbooks in Operations Research and Management Science, Vol. 7* (Elsevier, New York, 1995)

[Sola 78] I. de Sola Pool and M. Kochen, "Contacts and Influence", *Social Networks* **1** (1978/1979) pp. 5–51

[Sommer 94] J.-U. Sommer, T.A. Vilgis, and G. Heinrich, "Fractal Properties and Swelling Behavior of Polymer Networks", *The Journal of Chemical Physics* **100** (1994) pp. 9181–9191

[Son 11] S.-W. Son, G. Bizhani, C. Christensen, P. Grassberger, and M. Paczuski, "Irreversible Aggregation and Network Renormalization", *Europhysics Letters* **95** (2011) 58007

[Song 07] C. Song, L.K. Gallos, S. Havlin, and H.A. Makse, "How to Calculate the Fractal Dimension of a Complex Network: the Box Covering Algorithm", *Journal of Statistical Mechanics* (2007) P03006

[Song 05] C. Song, S. Havlin, and H.A. Makse, "Self-similarity of Complex Networks", *Nature* **433** (2005) pp. 392–395

[Song 06] C. Song, S. Havlin, and H.A. Makse, "Origins of Fractality in the Growth of Complex Networks", *Nat. Phys.* **2** (2006) pp. 275–281

[Song 15] Y.-Q. Song, J.-L. Liu, Z.-G. Yu, and B.-G. Li, "Multifractal Analysis of Weighted Networks by a Modified Sandbox Algorithm", *Scientific Reports* **5** (2015) 17628

[Sorensen 97] C.M. Sorensen and G.C. Roberts, "The Prefactor of Fractal Aggregates", *Journal of Colloid and Interface Science* **186** (1997) pp. 447–452

[Stanley 92] H.E. Stanley, "Fractal Landscapes in Physics and Biology", Physica A **186** (1992) pp. 1–32

[Stanley 99] H.E. Stanley, "Scaling, Universality, and Renormalization: Three Pillars of Modern Critical Phenomena", *Reviews of Modern Physics* **71** (1999) pp. S358–S366

[Stanley 88] H.E. Stanley and P. Meakin, "Multifractal Phenomena in Physics and Chemistry", *Nature* **335** (29 September 1988)

[Strogatz 94] S.H. Strogatz, *Nonlinear Dynamics and Chaos* (Perseus Books Publishing, Cambridge, MA, 1994)

[Strogatz 05] S.H. Strogatz, "Romanesque Networks", *Nature* **433** 27 January 2005, pp. 365–366

[Strunk 09] G. Strunk, "Operationalizing Career Complexity", *Management Revue* **20** (2009) pp. 294–311

[Sun 14] Y. Sun and Y. Zhao, "Overlapping-box-covering Method for the Fractal Dimension of Complex Networks", *Physical Review E* **89** (2014) 042809

[Sun 12] Z. Sun, J. Zheng, and H. Hu, "Fractal Patterns in Spatial Structure of Urban Road Networks", *International Journal of Modern Physics B* **26** (2012) 1250172

[Takens 85] F. Takens, *Dynamical Systems and Bifurications*, Lecture Notes in Mathematics **1125** (Springer, Berlin, 1985)

[Tang 06] D. Tang and A.G. Marangoni, "Computer Simulation of Fractal Dimensions of Fat Crystal Networks", *JAOCS* **83** (2006) pp. 309–314

[Taylor 91] C.C. Taylor and S.J. Taylor, "Estimating the Dimension of a Fractal", *Journal of the Royal Statistical Society, Series B* **53** (1991) pp. 353–364

[Tél 88] T. Tél, "Fractals, Multifractals, and Thermodynamics: An Introductory Review", *Z. Naturforsch* **43a** (1988) pp. 1154–1174

[Tél 89] T. Tél, Á. Fülöp, and T. Vicsek, "Determination of Fractal Dimensions for Geometrical Multifractals", *Physica A* **159** (1989) pp. 155–166

[Tél 87] T. Tél and T. Vicsek, "Geometrical Multifractality of Growing Structures", *J. Phys. A: Math. Gen.* **20** (1987) pp. L835–L840

[Theiler 88] J. Theiler, "Lacunarity in a Best Estimator of Fractal Dimension", *Physics Letters A* **133** (1988) pp. 195–200

[Theiler 90] J. Theiler, "Estimating Fractal Dimension", *J. Optical Society of America A* **7** (1990) pp. 1055–1073

[Theiler 93] J. Theiler and T. Lookman, "Statistical Error in a Chord Estimator of Correlation Dimension: the 'Rule of Five'", *Int. J. of Bifurcation and Chaos* **3** (1993) pp. 765–771

[Tolle 03] C.R. Tolle, T.R. McJunkin, D.T. Rohrbaugh, and R.A. LaViolette, "Lacunarity Definition for Ramified Data Sets Based on Optimal Cover", *Physica D* **179** (2003) pp. 129–152

[Travers 69] J. Travers and S. Milgram, "An Experimental Study of the Small World Problem", *Sociometry* **32** (1969) pp. 425–443

[Uzzi 07] B. Uzzi, L.A.N. Amaral, and F. Reed-Tsochas, "Small-World Networks and Management Science Research: A Review", *European Management Review* **4** (2007) pp. 77–91

[Valous 10] N. A. Valous, D. Sun, P. Allen, and F. Mendoza, "The Use of Lacunarity for Visual Texture Characterization of Pre-sliced Cooked Pork Ham Surface Intensities", *Food Research International* **43** (2010) pp. 387–395

[Vassilicos 91] J.C. Vassilicos and J.C.R. Hunt, "Fractal Dimensions and Spectra of Interfaces with Applications to Turbulence", *Proc. R. Soc. Lond. A* **435** (1991) pp. 505–534

[Vespignani 18] A. Vespignani, "Twenty Years of Network Science", *Nature* **558** (2018) pp. 528–529

[Vicsek 89] T. Vicsek, *Fractal Growth Phenomena, 2nd ed.* (World Scientific, Singapore, 1989)

[Vicsek 90] T. Vicsek, F. Family, and P. Meakin, "Multifractal Geometry of Diffusion Limited Aggregates", *Europhysics Letters* **12** (1990) pp. 217–222

[Wang 13] X. Wang, Z. Liu, and M. Wang, "The Correlation Fractal Dimension of Complex Networks", *International Journal of Modern Physics C* **24** (2013) 1350033

[Wang 11] D.-L. Wang, Z.-G. Yu, and V. Anh, "Multifractality in Complex Networks", *Chinese Physics B* **21** (2011) 080504

[Watts 99a] D.J. Watts, *Small Worlds: The Dynamics of Networks between Order and Randomness* (Princeton University Press, Princeton, NJ, 1999)

[Watts 99b] D.J. Watts, "Networks, Dynamics, and the Small-World Phenomenon", *American Journal of Sociology* **105** (1999) pp. 493–527

[Watts 03] D.J. Watts, *Six Degrees: The Science of a Connected Age* (W.W. Norton & Company, New York, 2003)

[Watts 98] D.J. Watts and S.H. Strogatz, "Collective Dynamics of 'Small-World' Networks", *Nature* **393** (1998) pp. 440–442

[Wei 13] D.-J. Wei, Q. Liu, H.-X. Zhang, Y. Hu, Y. Deng, and S. Mahadevan, "Box-Covering Algorithm for Fractal Dimension of Weighted Networks", *Scientific Reports* **3** (2013) 3049

[Wei 14a] D. Wei, B. Wei, Y. Hu, H. Zhang, and Y. Deng, "A New Information Dimension of Complex Networks", *Physics Letters A* **378** (2014) pp. 1091–1094

[Wei 14b] D. Wei, B. Wei, H. Zhang, C. Gao, and Y. Deng, "A Generalized Volume Dimension of Complex Networks", *Journal of Statistical Mechanics* (2014) P10039

[Wen 18] T. Wen and W. Jiang, "An Information Dimension of Weighted Complex Networks", *Physica A* **501** (2018) pp. 388–399

[West 12] J. West, G. Bianconi, S. Severini, and A.E. Teschendorff, "Differential Network Entropy Reveals Cancer System Hallmarks", *Scientific Reports* **2** (2012) 802

[Wilde 64] D.J. Wilde, *Optimum Seeking Methods* (Prentice Hall, New York, 1964)

[Willinger 09] W. Willinger, D. Alderson, and J.C. Doyle, "Mathematics and the Internet: A Source of Enormous Confusion and Great Potential", *Notices of the AMS* **56** (2009) pp. 586–599

[Wilson 79] K.G. Wilson, "Problems in Physics with Many Scales of Length", *Scientific American*, August 1979, pp. 158–179

[Wilson 82] K.G. Wilson, "The Renormalization Group and Critical Phenomena", *Reviews of Modern Physics* **55** (1983) 583–600

[Witten 81] T.A. Witten and L.M. Sander, "Diffusion-Limited Aggregation, a Kinetic Critical Phenomenon", *Physical Review Letters* **47** (1981) pp. 1400–1403

[Wong 05] A. Wong, L. Wu, P.B. Gibbons, and C. Faloutsos, "Fast Estimation of Fractal Dimension and Correlation Integral on Stream Data", *Information Processing Letters* **93** (2005) pp. 91–97

[Wu 15] Z. Wu, G. Menichetti, C. Rahmede, and G. Bianconi, "Emergent Complex Network Geometry", *Scientific Reports* **5** (2015) 10073

[Wyss 04] M. Wyss, C.G. Sammis, R.M. Nadeau, and S. Wiemer, "Fractal Dimension and *b*-Value on Creeping and Locked Patches of the San Adreas Fault near Parkfield, California", *Bulletin of the Seismological Society of America* **94** (2004) pp. 410–421

[Yadav 10] J.K. Yadav, J.S. Bagla, and N. Khandai, "Fractal Dimension as a Measure of the Scale of Homogeneity", *Monthly Notices of the Royal Astronomical Society* **405** (2010) pp. 2009–2015

[Yu 01] Z.G. Yu, V. Anh, and K.-S. Lau, "Measure Representation and Multifractal Analysis of Complete Genomes", *Phys. Rev. E* **64** (2001) 031903

[Yu 09] Y.Y Yu, H. Chen, C.H. Lin, C.M. Chen, T. Oviir, S.K. Chen, and L. Hollender, "Fractal Dimension Analysis of Periapical Reactive Bone in Response to Root Canal Treatment", *Oral Surgery, Oral Medicine, Oral Pathology, Oral Radiology, and Endodontology* **107** (2009) pp. 283–288

[Yule 24] G.U. Yule, "A Mathematical Theory of Evolution, Based on the Conclusions of Dr. J.C. Willis, F.R.S." *Philosophical Transactions of the Royal Society of London. Series B*, **213** (1924) pp. 21–87

[Zhang 12] H. Zhang and Z. Li, "Fractality and Self-Similarity in the Structure of Road Networks", *Annals of the Association of American Geographers* **102** (2012) pp. 350–365

[ZhangLi 16] Q. Zhang and M. Li, "A New Structure Entropy of Complex Networks Based on Nonextensive Statistical Mechanics", *International Journal of Modern Physics C* **27** (2016) 1650118

[Zhang 16] H. Zhang, D. Wei, Y. Hu, X. Lan, and Y. Deng, "Modeling the Self-Similarity in Complex Networks Based on Coulomb's Law", *Communications in Nonlinear Science and Numerical Simulation* **35** (2016) pp. 97–104

[Zhang 08] Z. Zhang, S. Zhou, L. Chen, and J. Guan, "Transition from Fractal to Non-Fractal Scalings in Growing Scale-Free Networks", *The European Physics Journal B* **64** (2008) pp. 277–283

[Zhao 16] G. Zhao, K. Denisova, P. Sehatpour, J. Long, W. Gui, J. Qiao, D.C. Javitt, and Z. Wang, "Fractal Dimension Analysis of Subcortical Gray Matter Structures in Schizophrenia", PLOS ONE **11** (2016) e0164910

[Zhao 14] X.-Y. Zhao, B. Huang, M. Tang, H.-F. Zhang, and D.-B. Chen, "Identifying Effective Multiple Spreaders by Coloring Complex Networks", *Europhysics Letters* **108** (2014) 68005

[Zhao 11] K. Zhao, M. Karsai, and G. Bianconi, "Entropy of Dynamical Social Networks", *PLoS ONE* **6** (2011) e28116

[Zhou 07] W.-X. Zhou, Z.-Q. Jiang, and D. Sornette, "Exploring Self-Similarity of Complex Cellular Networks: The Edge-Covering Method with Simulated Annealing and Log-Periodic Sampling", *Physica A: Statistical Mechanics and its Applications* **375** (2007) pp. 741–752

[Zhuang 04] X. Zhuang and Q. Meng, "Local Fuzzy Fractal Dimension and its Application in Medical Image Processing", *Artificial Intelligence in Medicine* **32** (2004) pp. 29–36

[Zia 09] R.K.P. Zia, E.F. Redish, and S.R. McKay, "Making Sense of the Legendre Transform", *American Journal of Physics* **77** (2009) 614

[Zlatintsi 11] A. Zlatintsi and P. Maragos, "Musical Instruments Signal Analysis and Recognition Using Fractal Features", *Proc. 19-th European Signal Processing Conference (EUSIPCO 2011)*, Barcelona, Spain (2011) pp. 684–688